PROTEIN METABOLISM

INFLUENCE OF GROWTH HORMONE, ANABOLIC STEROIDS, AND NUTRITION IN HEALTH AND DISEASE

AN INTERNATIONAL SYMPOSIUM

LEYDEN, 25th-29th JUNE, 1962

SPONSORED BY

CIBA

CHAIRMAN

A. QUERIDO

LEYDEN

EDITED BY

F. GROSS

BASLE

WITH 159 FIGURES

SPRINGER-VERLAG
BERLIN HEIDELBERG GMBH

1962

ISBN 978-3-642-53149-1   ISBN 978-3-642-53147-7 (eBook)
DOI 10.1007/978-3-642-53147-7

© by Springer-Verlag Berlin Heidelberg 1962

Originally published by Springer-Verlag OHG Berlin · Göttingen · Heidelberg in 1962

Softcover reprint of the hardcover 1st edition 1962

Library of Congress Catalog Card Number 62-120 57

Preface

The Symposium on "Protein Metabolism: Influence of Growth Hormone, Anabolic Steroids, and Nutrition in Health and Disease" is the fourth in the series of International Symposia sponsored by CIBA Limited, Basle. As in the case of the previous conferences, it was planned and organised with the help of experts in the field concerned. Special thanks are due to Prof. A. QUERIDO and Dr. A. A. H. KASSENAAR who, once the idea of the Symposium had been conceived in the course of joint discussions, embarked upon the project with enthusiasm and inspiration, although they must have known full well what a great deal of time and trouble the organisation of such a meeting would inevitably cost them. For their untiring efforts, for the judicious manner in which they contrived to select precisely those subjects on which interest is chiefly centred today, and — last but not least — for their success in finding competent specialists to participate in the proceedings, we wish to assure them of our sincere gratitude. To all the members of the Department of Clinical Endocrinology and Diseases of Metabolism, at the University Hospital in Leyden, who helped in preparing the meeting, we would likewise extend a warm vote of thanks.

The fact that the present volume, featuring the papers and discussions of the Symposium, has been published only a few months after the event, was made possible thanks to the co-operative help of all who participated. In this connection, we are particularly indebted to Mr. H. D. PHILPS, M. A., and Miss S. R. NAEGELI, who devoted all their available time to the task of meeting the dead-line which we had set ourselves, and to Dr. WIL-TRUD HATZINGER, who kindly prepared the subject index for us. We should also like to express our appreciation to Dr. H. GÖTZE for the co-operation and understanding with which he and his associates of Springer-Verlag met all our demands.

One difficulty with which we were faced was that of the difference between English and American spelling. Lest we should be accused by our American readers of ignoring the welcome simplifications which in their country have been introduced into the complexities of English orthography, we have adhered to the American usage in the case of the papers read by American authors.

Basle, December, 1962   F. G.

Contents

Protein metabolism in human pathological states

The effects of anabolic agents in man

VIII   Contents

Participants in the Symposium

ARIAS, I. M. . . . . . . Department of Medicine, Albert Einstein College of Medicine, New York, N. Y. (U.S.A.)

ASCHKENASY, A. . . . . Hôpital de la Pitié, Paris (France)

BAUER, G. C. H. . . . . Malmö Allmänna Sjukhus, Malmö (Sweden)

BLOM, P. S. . . . . . . Afdeling Interne Geneeskunde, Zuidwal Ziekenhuis, 's-Gravenhage (The Netherlands)

CAVALLERO, C. . . . . . Istituto di Anatomia e Istologia Patologica, Università di Pavia, Pavia (Italy)

DESAULLES, P. A.. . . . Forschungslaboratorien der CIBA Aktiengesellschaft, Basel (Switzerland)

DIRSCHERL, W. . . . . . Physiologisch-chemisches Institut der Universität Bonn, Bonn (Germany)

DREYFUS, J.-C. . . . . . Laboratoire de Recherches de Biochimie Médicale, Hôpital des Enfants-Malades, Paris (France)

DYMLING, J.-F. . . . . . Malmö Allmänna Sjukhus, Malmö (Sweden)

FRASER, T. R. C. . . . . Postgraduate Medical School, London (Great Britain)

GEMZELL, C. A. . . . . . Akademiska Sjukhuset, Uppsala (Sweden)

GROSS, F. . . . . Forschungslaboratorien der CIBA Aktiengesellschaft, Basel (Switzerland)

HAAK, A. . . . . Interne Kliniek, Afdeling Stofwisselingsziekten en Endocrinologie, Academisch Ziekenhuis, Leiden (The Netherlands)

HOFFENBERG, R. . . . . Department of Medicine, Groote Schuur Hospital, Cape Town (South Africa)

IKKOS, D. . . . . Department of Endocrinology, Evangelismos Hospital, Athens (Greece)

JONGH, S. E. DE . . . . Afdeling Pharmacologie, Academisch Ziekenhuis, Leiden (The Netherlands)

KASSENAAR, A. A. H. . . Interne Kliniek, Afdeling Stofwisselingsziekten en Endocrinologie, Academisch Ziekenhuis, Leiden (The Netherlands)

KINNEY, J. M. . . . . . Peter Bent Brigham Hospital, Boston, Mass. (U.S.A.)

KORNER. A. . . . . . . Department of Biochemistry, University of Cambridge, Cambridge (Great Britain)

KOWALEWSKI, K. . . . . Surgical-Medical Research Institute, University of Alberta, Edmonton (Canada)

LABHART, A. . . . . . . Stoffwechselabteilung der Medizinischen Universitätsklinik, Zürich (Switzerland)

LARON, Z. . . . . Pediatric Metabolic and Endocrine Clinic, Beilinson Medical Center, Petah Tiqva (Israel)

LEATHEM, J. H.. . . . . Bureau of Biological Research, Rutgers — The State University, New Brunswick, N. J. (U.S.A.)

MANDEMA, E.. . . . . . Interne Kliniek, Algemeen Provinciaal-, Stadsen Academisch Ziekenhuis, Groningen (The Netherlands)

McCANCE, R. A. . . . . Department of Experimental Medicine, University of Cambridge, Cambridge (Great Britain)

OVERBEEK, G. A. . . . . Pharmacologisch Laboratorium, N.V. Organon, Oss (The Netherlands)

PRADER, A.. . . . . . . Universitäts-Kinderklinik, Zürich (Switzerland)

QUERIDO, A. . . . . . . Interne Kliniek, Afdeling Stofwisselingsziekten en Endocrinologie, Academisch Ziekenhuis, Leiden (The Netherlands)

RABEN, M. S.. . . . . . Pratt Clinic, New England Center Hospital, Boston, Mass. (U.S.A.)

RUNDLES, R. W. . . . . Department of Medicine, Duke University Medical Center, Durham, N.C. (U.S.A.)

SALA, G. . . . . Laboratori Ricerche Farmacologiche, Farmitalia, Milano (Italy)

SCHREIER, K.. . . . . . Universitäts-Kinderklinik, Heidelberg (Germany)

STAEHELIN, M. . . . . . Forschungslaboratorien der CIBA Aktiengesellschaft, Basel (Switzerland)

STEPHEN, J. M. L.. . . . Department of Chemical Pathology, St. Mary's Hospital Medical School, London (Great Britain)

SUBRAMANIAM, R.. . . . Department of Medicine, Madras Medical College, Madras (South India)

SZIRMAI, J. A. . . . . . Wetenschappelijk Laboratorium, Afdeling Rheumatologie, Academisch Ziekenhuis, Leiden (The Netherlands)

TANNER, J. M. . . . . . Institute of Child Health, The Hospital for Sick Children, London (Great Britain)

THAYSEN, J. HESS . . . Medicinsk afdeling A., Rigshospitalet, København (Denmark)

TREMOLIÈRES, J. . . . . L'Unité de Recherches Diététique, Hôpital Bichat, Paris (France)

TUCHMANN-DUPLESSIS, H. Laboratoire d'Embryologie, Faculté de Médecine de Paris, Paris (France)

VERMEULEN, A.. . . . . . Interne Kliniek, Rijksuniversiteit, Gent (Belgium)

WATERLOW, J. C. . . . . Department of Chemical Pathology, St. Mary's Hospital Medical School, London (Great Britain)

WILSON, J. D. . . . . . Department of Internal Medicine, The University of Texas Southwestern Medical School, Dallas, Texas (U.S.A.)

YOUNG, F. G.. . . . . . Department of Biochemistry, University of Cambridge, Cambridge (Great Britain)

Guests attending part of the Symposium:

RITTENBERG, D. . . . . Department of Biochemistry, College of Physicians and Surgeons of Columbia University, New York, N. Y. (U.S.A.)

WERFF TEN BOSCH, J. J.
VAN DER . . . . Interne Kliniek, Afdeling Stofwisselingsziekten en Endocrinologie, Academisch Ziekenhuis, Leiden (The Netherlands)

Protein metabolism
Some problems brought into focus

Opening remarks

By

A. QUERIDO

"However elegant and memorable, brevity
can never in the nature of things do justice
to all the facts of a complex situation".
ALDOUS HUXLEY*

Many of the participants, when invited to attend this Symposium, were kind enough to praise the programme. I can assure you that it was not so difficult, with the efficient aid of Dr. GROSS and Dr. KASSENAAR, to arrive at the final scheme. It was, though, more difficult to think of a concise title. We finally chose "Protein Metabolism", adding, however, the subtitle "Influence of growth hormone, anabolic steroids, and nutrition in health and disease".

You might consider these remarks futile for this distinguished gathering, but they are made to underline a rather important point. The title of this Symposium does not clearly indicate what the organisers were driving at when they invited you to participate.

What, then, was the idea behind this programme? To say it in a risky way, it was to assemble together a number of experienced workers in the field of protein metabolism, in order to *discuss* biological mechanisms and conditions related to protein metabolism at the cellular level and in the organism in a state of health, in a state of disease, and in a state of repair. It is risky for me to put it that way, because I have only mentioned the workers and not their papers — even though the latter have undoubtedly been composed with great care, if perhaps grudgingly. I hope that, having examined these papers thoroughly, we shall consider them only as a framework for the development of our thought. Unfortunately, I must dispel any impression you may have that you had finished your task once you had prepared and handed in your

---

* "Brave New World Revisited" (Harper, New York).

manuscripts, because I sincerely hope that a major and essential part of the work will be done during the actual meeting.

However, it is quite clear that we cannot discuss all aspects of protein metabolism, and the action of physiological and pharmacological agents upon it. I shall therefore try to put forward some thoughts and problems that may be relevant to this Symposium.

It is not necessary before this audience to stress the significance of proteins for life. A short quotation from PAULING (6) may serve our purpose here: "The molecules that compose the body of a human being may be conveniently divided into two classes: small molecules and large molecules. Small molecules are molecules containing 10 or 20 or perhaps 100 atoms" . . . " large molecules are molecules containing hundreds or thousands or tens of thousands of atoms; examples are the proteins and the nucleic acids." . . . "The large molecules are especially important, because it is they that carry biological specificity." To list a few functions which proteins fulfil: they are enzymes or essential parts of enzymes; they take part in many different forms in the intracellular and extracellular structure of the body. Antibodies are proteins, as are a number of pituitary hormones. In combination with nucleic acids they harbour inheritance factors. Another important aspect of proteins is that small alterations in their structure may lead to disease.

Despite all these different functions, proteins have some characteristics in common. All are constructed with a limited assortment of building blocks, the amino acids. They are all manufactured by the cells in the same way, and they are in a dynamic state of break-down and renewal. This dynamic state, so clearly expounded by BORSOOK (2) and SCHOENHEIMER (8) more than 20 years ago, which shows highly different rates of renewal and break-down for the different proteins and under different conditions, makes the study of the organism as a whole extremely difficult.

Table 1. *Large protein stores in the body*
♂, 168.5 cm.; 53.8 kg.
From FORBES et al.: J. Biol. Chem. (U.S.A.) **203**, 359 (1953)

	g.
Total protein (N × 6.25) . . .	10,006
Striated muscle	4,680
Skeleton	1,864
Skin	924
Adipose tissue	361
Estimate of blood:	
Haemoglobin	750
Albumin	250

Table 1, compiled from one of the few available analyses which have been made of the human body, lists the body's major protein stores.

The total amount of body protein is approximately 19% of the fresh weight. 45% of this protein is present in the muscle and 18% in the skeleton, while skin and adipose tissue account for another 10% and 4% respectively. These figures are quite interesting, but, upon closer examination, provide very little information. They would gain in importance if it were known how much is renewed daily, and what factors regulate the renewal rate. Then it would be possible to obtain an estimate of the extent to which these stores participate in daily protein metabolism, since their amino acids have to enter and mix with the amino-acid pool. Surprisingly little exact information on this question is available.

When talking about the renewal of proteins, we have to distinguish between the dynamic state of protein intracellularly or extracellularly and the renewal of whole cells, i.e. replacement of dying cells. The red blood cell may serve as a nice example of this last form of renewal. Haemoglobin seems to be stable during the life-span of an erythrocyte, but $1/_{120}$ of our haemoglobin is renewed daily. Some cells have a very short life — as in the jejunal mucosa, for example, where most of the cells do not live longer than one day. There are indications that, once their protein is formed, this protein is stable for the rest of the day (3). This is certainly not the case in, for instance, the liver, where the life-span of the cell seems to be very long and the half-life of the protein *mixture* in the cells is not more than 4 days; here, therefore, we may speak of a real dynamic state.

If we come then to consider large protein masses, such as collagen, which has been estimated to constitute $1/_3$ of our body protein, estimates of renewal rate are very complicated.

NEUBERGER et al. (5), about a decade ago, came to the conclusion that collagen of the tendon in rats is probably metabolically inert, with a half-life of several hundred days. He made some very cautious suggestions with relation to collagens in other sites. However, in recent years (9), at least 3 biochemically different fractions of collagen have been recognised, of which one may even have a half-life of 2 days. Furthermore, the pregnant uterus shows rapid deposition and, upon involution, rapid removal of collagen.

Before ending our remarks on the dynamic state of proteins, attention should be drawn to the opinion of several authors that only a minor part of the body protein is rapidly renewed. This, however, does not imply that the participation of large deposits of proteins (e.g. muscle proteins) in the amino-acid pool is small. Assuming an integrated half-life of 100 days for muscle proteins, the amount of amino acids participating daily in intermediary metabolism would be 20 g.

If we now turn to the methods available for investigating protein metabolism in human beings in health or disease, the most widely used technique is the study of the nitrogen balance, based on the concept of dynamic equilibrium. In view of what I have

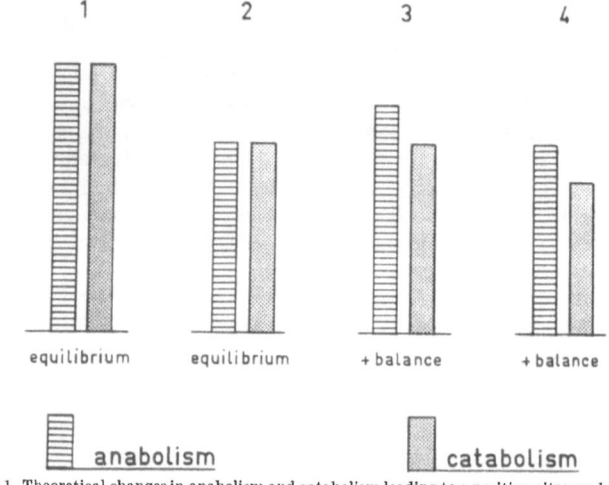

Fig. 1. Theoretical changes in anabolism and catabolism leading to a positive nitrogen balance

already said, it is clear that the information obtained with this technique will be limited. The results indicate only a difference between intake and output, being the integrated outcome of the anabolism and catabolism of many different protein pools. There is no evidence nor reason to assume that all these protein pools will move in the same direction and that the result of the balance study is representative of the trend in the body. Another relevant point is that a similar difference between anabolism and catabolism in a particular pool can be achieved at varying levels of anabolism, for example, as is shown in Fig. 1. It is probable that such explanations may account for ALLISON's classic observations (*1*) which indicate that achievement of nitrogen balance is dependent on the adaptation of the body to previous feeding periods.

This simple statement — to the effect that the body contains proteins with different renewal rates, which may be affected by disease or by active agents in a *different* way — already demonstrates the complexity of the study of protein metabolism. We should of course like to define in cases of disease *which* proteins are affected in their metabolism and whether all the proteins of the body are involved or only special ones. With some diseases we have the clinical impression that most proteins undergo changes — as, for example, in Cushing's syndrome. The big protein stores seem to become reduced: skin and muscles are atrophic and the skeleton breaks down. Experimental protein depletion, however, shows a preference for certain organs. In his able monograph, ALLISON summarised the available information on the effect of depletion as follows: "In general the data reveal that total protein and enzymes of the liver are the most labile, while the total protein and the enzymes of the brain are most resistant. The total protein and enzymes of the ventricle of the heart are more resistant than those of the kidney, spleen or skeletal muscle and almost as resistant as those of the brain."

Recently it has been shown that in mice placed on a diet containing enough calories, but no protein, the colon and other tissues atrophy markedly, but the jejunum does not. This again also shows the danger of generalisations when the tissues are not sharply defined. To explain this type of result the term "labile protein reserves" is sometimes used, an expression which to me is not clearly defined. Perhaps some clarification on this point can be achieved during our conference.

It seems to me that the example of Cushing's disease, where the skin, too, is affected in such a marked way that it is paper-thin and shows ruptures appearing as large striae, raises the question as to why this phenomenon is not seen in other disease states, such as malnutrition. This brings us to the problem of how factors responsible for changes in protein metabolism operate. *Changes* in the amount of a protein present in the body at a given moment must be dependent on the amount synthesised per unit time and the amount broken down per unit time. If we assume that both processes are slowed down, the total amount present does not necessarily change. If, however, break-down is accelerated and synthesis does not increase or does not accelerate enough, large changes in the amount of protein will occur. The second possibility might well apply in Cushing's disease. The catabolic action of corticosteroids may affect the body as a whole, and the counter-regulation of increasing synthesis may be insufficient. Studies of the half-life of albumin during treatment with corticosteroids seem to support this hypothesis. The half-life of albumin under these conditions is shortened; synthesis, however, is also increased, but not enough to neutralise the increased degradation (7).

This immediately raises a number of questions which, I hope, will be very pertinent for this Symposium. What is the condition which limits synthesis in such a way that it does not catch up with degradation? Are we able with physiological or pharmacological agents to influence a disrupted equilibrium between protein anabolism and catabolism? And furthermore, if specific protein pools are affected, do so-called anabolic agents act on these pools, or is the result only apparent in the balance study because more protein is deposited in sites which were unaffected?

In posing these questions, we are touching on problems concerning the mechanisms of protein deposition and hence on problems of tissue growth. It seems quite clear that there are at least three conditions to be fulfilled before protein formation can proceed: the presence of enough building blocks (amino acids) and calories, as well as an adequate machinery for the manufacture of proteins. Limitation of growth due to absence of one essential amino acid is a well-known phenomenon. It may also be encountered in cases where insufficient energy is provided. The third condition, the machinery, is much more difficult to visualise. It involves a whole series of extremely complicated processes, such as transfer of information from nucleus to cytoplasm, activation of amino acids, and assembling of amino acids into proteins with their special tridimensional arrangement. The output of this complicated machinery can apparently be regulated by the size as well as the activity of its constituent parts, each of which may limit the rate of production. Moreover, the machinery's output is, as I have said, dependent on energy supply and building material.

Research on the action exerted by hormones on protein metabolism (e.g. oestrogens, androgens, somatotrophic hormone, insulin) has been concerned, in turn, with each one of these main conditions: supply of amino acids, availability of energy, and construction and/or speeding up of the protein-building machine. Interpretation of the results is often difficult, because all these three main conditions involve enzymes, which are themselves proteins. These enzymes are furthermore subject to regulation within the cellular organisation. Changes in enzymatic activity per cell under the influence of a hormone may therefore be both a consequence and a cause of increased protein synthesis. MÜLLER's recent work on the action exerted by oestradiol on rat uterus is a very clear example and illustration of this problem. Oestradiol caused acceleration of the synthesis of phospholipids, ribonucleic acid, and proteins in this chronological order. When *de novo* protein synthesis was blocked with puromycin, the existing rate of synthesis of phospholipids and ribonucleic acid continued, but was not accelerated by oestradiol. MÜLLER (4) concluded that protein synthesis was stimulated by oestradiol through increased formation of a *complex* of enzymes, all of which are necessary for growth.

These considerations have a bearing on the question of how protein synthesis in *cells* may be stimulated, a topic which will be discussed today. Another aspect of the regulation of protein synthesis in the *organism* is whether hormones or drugs which act on protein metabolism do so by exerting their influence on all cells in the body or whether they show a specific affinity for certain tissues. This is a crucial question, not only for the physiologist, but also for the clinician. For example, if so-called anabolising agents are given, on which tissues do they exert their effect?

In these remarks of mine I have only superficially touched on some of the problems which we hope will be discussed at our meeting. Please consider them therefore simply as the result of a little uninhibited thinking on the part of your Chairman, with no other purpose than to weave some pattern into our programme.

References

1. ALLISON, J. B., and W. H. FITZPATRICK: In: Dietary proteins in health and disease. Charles C. Thomas, U.S.A., 1960. — 2. BORSOOK, H., and G. L. KEIGHLEY: Proc. Roy. Soc., London, Biol. Sc. 118, 488 (1935). — 3. LIPKIN, M., and H. QUASTLER: J. Clin. Invest. (U.S.A.) 41, 646 (1962). — 4. MÜLLER, G.: In: Mechanism of action of steroid hormones. Ed. by VILLEE, C. A., and L. C. ENGEL. The Pergamon Press, 1961, p. 181. — 5. NEUBERGER, A., and H. G. B. SLACK: Biochem. J. (G.B.) 53, 47 (1953). — 6. PAULING, L.: In: Significant Trends in Medical Research. Ciba Foundation Symposium. Churchill Ltd., London, 1959. — 7. ROTHSCHILD, M. A., C. S. SCHREIBER, M. ORATZ, and H. L. McGEE: J. Clin. Invest. (U.S.A.) 37, 1229 (1958). — 8. SCHOENHEIMER, R., S. RATNER, and D. RITTENBERG: J. Biol. Chem. (U.S.A.) 130, 703 (1939). — 9. SLACK, H. G. B.: Amer. J. Med. 26, 113 (1959).

Action of hormones at the cellular level

The effect of growth hormone on protein synthesis

By

A. KORNER

During the past few years I have been investigating the way in which growth hormone and other hormones control and regulate the rate of protein biosynthesis in animals. The liver of the rat, like the body as a whole, loses weight and protein content as a result of hypophysectomy and, when treatment of the animal with growth hormone reverses this trend on body-weight and protein content, the liver too is stimulated to synthesise more protein. Because of this I have used rat liver as a tissue for study in most of my work. In addition, rat liver has been one of the tissues that have been used successfully in a study of the detailed reactions which result in the biosynthesis of protein. SIEKEVITZ (1952) and ZAMECNIK and KELLER (1954) have shown, for instance, that a simple cell-free system from liver can incorporate radioactive amino acids into protein, and there is evidence suggesting that the reaction sequence which has been elucidated with the aid of this simple cell-free system reflects the processes which result in protein synthesis *in vivo* (see HOAGLAND, 1960, for review).

Some years ago (KORNER, 1958) it was shown that removal of the pituitary gland from the rat diminished the rate and extent of amino-acid incorporation into protein in the simple cell-free system from rat liver. If the hypophysectomised rats were treated with growth hormone, the diminished rate of amino-acid incorporation found in the liver cell-free system was restored towards the normal level. The cell-free system is essentially the supernatant fluid of an homogenate of rat liver centrifuged at $15,000 \times G$ and it contains a large variety of factors which are involved in protein synthesis (see Fig. 1). Amino acids enter the cell, or are synthesised inside it, and are activated by specific amino-acid-activating enzymes present in cell sap which, with the aid of ATP, form amino acyl adenylates on the surface of the activating enzymes (HOAGLAND, KELLER, and ZAMECNIK, 1956). The activated amino acid is transferred to the terminal ribose of a ribonucleic acid of low molecular weight, called soluble or transfer RNA, also present in the cell sap (HOAG-

LAND, ZAMECNIK, and STEVENSON, 1957). It is believed that there
are twenty types of soluble RNA molecules (sRNA), one for each of
the 20 amino acids found in protein. Part or all of this soluble RNA
carrying the amino acid is transferred with the aid of GTP and a
transferring enzyme (TAKANAMI, 1960) to the RNA template, where
it is deposited on a site specific for that particular amino acid
(HOAGLAND, STEVENSON, SCOTT, HECHT, and ZAMECNIK, 1958).
The amino acids form peptide bonds with one another on the
template so that a polypeptide chain is produced with a specific

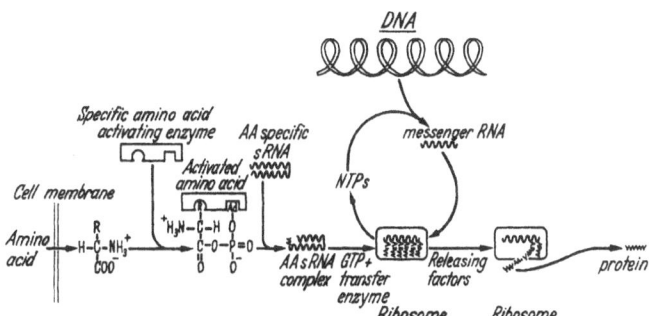

Fig. 1. Diagram showing the reactions which result in protein biosynthesis

sequence of amino acids, and the finished polypeptide is released
from the template with the aid of enzymes and other factors
(HULTIN, 1961; LAMFROM, 1961). LITTLEFIELD, KELLER, GROSS,
and ZAMECNIK (1955) have produced good evidence that the microsomal particles, ribonucleoprotein granules about 150 Å in diameter noted by PALADE and SIEKEVITZ (1956), are the actual sites
where protein is synthesised. These particles, called ribosomes by
ROBERTS (1958), contain about 50% protein and 50% RNA.

The primary amino-acid sequence of the protein is assumed to
determine the secondary and tertiary structure of the protein and,
consequently, its ability to perform its biological functions. The
amino-acid sequence is determined by the genetic information
contained in the DNA of the cell. Some people believe that the
RNA of the ribosomes is the template and that it is synthesised
under the control of part of the DNA of the cell so that each ribosome has, for the whole of its life, a template capable of synthesising
a particular polypeptide sequence in accordance with the information contained in a particular gene (see Fig. 2). More recently,
JACOB and MONOD (1961) have suggested that ribosomes are

capable of making any protein, provided that they are furnished with a messenger RNA which acts as a temporary template. The messenger RNA, in their hypothesis, is synthesised on the DNA of the gene and is, consequently, furnished with a specific nucleotide sequence which allows a unique polypeptide chain to be formed on it when it has attached itself to a ribosome. The messenger RNA is destroyed after synthesis of the polypeptide, leaving the ribosome free to accept another messenger RNA molecule.

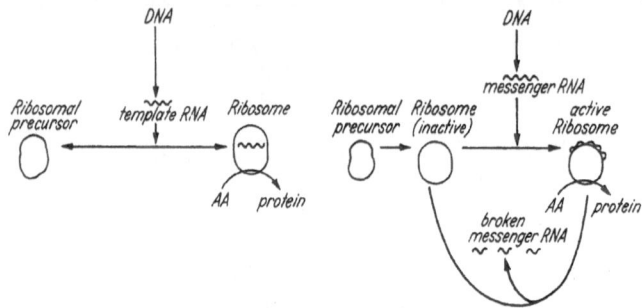

Fig. 2. The traditional and messenger hypothesis of genetic control of protein synthesis. *Traditional:* the gene makes the template of each ribosome which synthesis many molecules of the same protein. *Messenger:* the gene makes messenger RNA molecules which attach themselves to any ribosome; a polypeptide corresponding to the messenger RNA is synthesised and the messenger RNA is destroyed; the ribosomes can now accept a second messenger RNA carrying either the same message or a message different from the previous one

The new messenger RNA may contain the same message as the previous one, so that the same polypeptide chain is made, or it may contain a different message, with the result that an entirely different polypeptide chain is synthesised (see Fig. 2).

The simple experiment I carried out some years ago shows that growth hormone in some way controls the rate at which amino acids are incorporated into protein in this complex system, and attempts were made to find out which of the steps involved in these reactions was controlled and directed by growth hormone (KORNER, 1959a). The cell-free system was separated, by spinning at $100,000 \times G$, into microsomes and cell sap. Neither of these two portions of the cell-free system would incorporate amino acids into protein when incubated alone under the usual conditions, but when they were combined together again, amino-acid incorporation occurred at the same rate as in the unseparated system. When microsomes from the liver of normal rats were incubated with cell sap from the liver of hypophysectomised rats the rate of incorporation into the microsomal protein was the same as when the microsomes from normal rats were incubated with cell sap also from normal rat liver. In other words, the low incorporation into the cell-free system from the liver of hypophysectomised rats could not be accounted for by any change in the cell-sap portion and, therefore, no change had occurred, as a result of hypophysectomy, in the ability of rat liver to activate amino acids, to form amino acid sRNA complexes, or to transfer the amino acid from this complex to the template. When the microsomes from livers of hypophysectomised rats were incubated with the sap from normal rats, the same low rate of incorporation into protein was found as when the microsomes from hypophysectomised rats were incubated with cell sap from the same rats, thus confirming that growth hormone controlled the ability of liver microsomes to assemble amino acids into peptide chains.

Many experiments were carried out to make sure that these changes in the microsomes were not caused artefactually during the preparation of the system as a result, for example, of an increase in the rate of degradation of the microsomes from the liver of hypophysectomised rats due to increased activity of RNAase or cathepsin in these livers (KORNER, 1959a). Still other experiments, in which amino acids were injected into the femoral vein of rats and the incorporation of the amino acids into the protein of subcellular fractions of rat liver was studied *in vivo*, confirmed the results of the *in vitro* experiments and led inexorably to the conclusion that growth hormone can alter the metabolism of the rat so that the microsomes, isolated from liver, are changed in their ability to synthesise protein (KORNER, 1960a).

Microsome is a term applied by CLAUDE (1943) to a fraction of cells which spins down at high centrifugal force and which contains broken pieces of different sizes of the endoplasmic reticulum of liver, both the lipoprotein membranes and the ribosome granules. LITTLEFIELD's experiments (LITTLEFIELD et al., 1955) suggested that it is these granules which are the incorporating site, and so I developed methods of preparing ribosomes from rat liver in such a way that they retained their ability to incorporate amino acids into protein *in vitro* (KORNER, 1959b; 1961a). It was soon clear (KORNER, 1961b) that the changes found in the microsomes as the result of hormonal treatment of the rat could be accounted for entirely by changes in the ribosome part of the microsomes, for ribosomes from hypophysectomised rat liver incorporated less amino acid into their protein than those from normal rat liver, no matter whether they were incubated with sap from normal or from hypophysectomised rats. In addition, the counts in the normal and hypophysectomised rat-liver ribosomes bore the same relationship to each other as the counts in the normal and hypophysectomised rat-liver microsomes from which they had been prepared (see Fig. 3). It was also found that treatment of the hypophysectomised rat with growth hormone could restore the amino-acid incorporation found in ribosomes, prepared from the liver, towards the normal level.

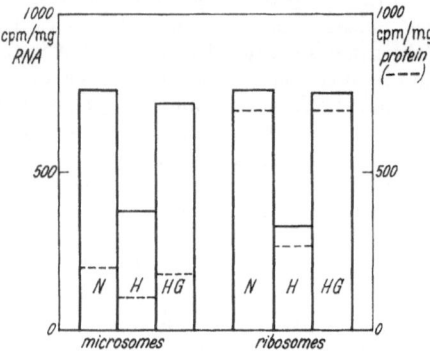

Fig. 3. Incorporation of amino acids into the protein of microsomes and ribosomes from normal (N) and hypophysectomised rats (H) and hypophysectomised rats treated with growth hormone (HG)

Incorporation of amino acids into TCA-precipitable protein is, of course, not synonymous with the biosynthesis of protein. It has not yet been possible to demonstrate net synthesis of protein in a mammalian cell-free system; and indeed the evidence of BISHOP, LEAHY, and SCHWEET (1960) might indicate that this is an impossible task, for they suggest that only the completion of started polypeptide chains is achieved in the *in vitro* system from reticulocytes. I have, however, been able to show that amino acids are incorporated into a specific and identifiable protein in the liver-ribosome *in vitro* system. At various times during an incorporation experiment, samples were removed from the mixture, and serum albumin was extracted from them by a chemical technique, followed by immunological precipitation and identification of the protein. I was able to show (KORNER, 1962) that amino-acid incorporation into serum albumin follows that into the total protein of the system, and I later found that the proportion of the counts which were incorporated into total protein appearing in albumin is the same in the normal rat-liver system and in the system from hypophysectomised rat liver (see Fig. 4). It would appear then that

growth-hormone control of the ability of the ribosomes to incorporate amino acid into protein is a reflection of the hormonal
ability to regulate the synthesis (or at least the labelling) of a real
protein.

At this point one can ask three questions:

1. Does growth hormone regulate the ability of other subcellular fractions of cells to synthesise protein?

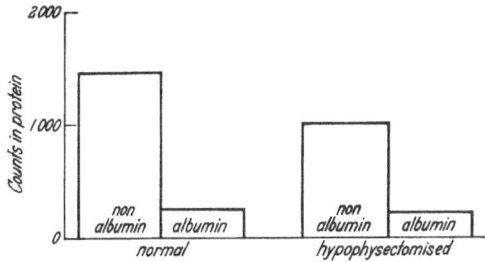

Fig. 4. Proportion of the total counts incorporated into protein in the ribosome system which
appear in albumin in preparations from normal and hypophysectomised rat liver

2. Do hormones other than growth hormone, which regulates
the rate of protein biosynthesis, do so by influencing the ability
of ribosomes to synthesise protein or do they alter the ability of the
cell sap to prepare amino acids for incorporation into protein?

3. What is the nature of the hormonally controlled change in
the rat which results in a change in the ability of ribosomes to
incorporate amino acid into protein in the cell-free system and what
is the nature of the change in the ribosomes?

I wish to concentrate on the third of these questions, but before
doing so, a little should be said about the other two. Mitochondria
of rat liver incorporate amino acid into protein to a small extent
(ROODYN, REIS, and WORK, 1961; TRUMAN and KORNER, 1962) and
it seems possible that somewhat similar changes have occurred in
the mitochondria after hypophysectomy as have occurred in the
microsomes and ribosomes (KORNER, 1959c). Much work has been
carried out to answer the second question and it has been found
that many of the hormones do influence the rate at which protein
can be synthesised in cell-free systems. Again, the biggest effect is
found on the ability of the microsomes and ribosomes to assemble
amino acids into polypeptide chains, though occasionally changes
in the ability of the cell-free system have been detected (KORNER,
1960b; 1960c; MUELLER, HERRANEN, and JERVELL, 1958).

There is good evidence that growth hormone exerts its protein anabolic effect by stimulating the secretion of insulin from the β cells of the islets of LANGERHANS and by allowing insulin, the real protein anabolic hormone, to exert its effect in stimulating protein synthesis while holding in check the hypoglycaemic effects of insulin (YOUNG, 1939; 1941; 1945; KORNER and MANCHESTER, 1960; YOUNG and KORNER, 1960; MANCHESTER and YOUNG, 1962). Using the cell-free system and the isolated ribosome system I have described, I have been able to show that ribosomes from alloxan-diabetic rats incorporate less amino acid into protein than those of normal rats (KORNER, 1960c) and that one can reverse this and, indeed, enhance incorporation above the normal level, by treating the diabetic rats with insulin. Furthermore, in support of the idea that insulin can stimulate protein synthesis in the absence of growth hormone if it is possible to preserve the animal from the hypoglycaemic effects of insulin, it has been shown that treatment of hypophysectomised rats with insulin stimulates the ability of the isolated microsomes and ribosomes of rat liver to incorporate amino acid into protein *in vitro* (KORNER, 1960d).

The third question is the one to which the rest of the paper is devoted. The yield of ribosomes which can be obtained from the liver of a hypophysectomised rat is, of course, less than that which can be obtained from the liver of a normal rat. My experiments revealed, in addition, a qualitative change in the ribosomes from rats which had been hypophysectomised, for their rate of incorporation, calculated per mg. RNA, was lower than normal. What has happened in the liver to cause this modification in the nature of the ribosome and what is the nature of this change ? Let us for a moment accept it its entirety the modern theory of protein biosynthesis including the messenger concept (see Figs 1 and 2). It is unlikely that growth hormone is affecting the gene itself, since the amino-acid sequence of proteins does not appear to change in growth-hormone depleted or growth-hormone treated animals. Growth hormone could act at the genetic level by affecting the rate at which messenger RNA is synthesised on the genes. In the theory of JACOB and MONOD (1961) it is proposed that a series of genes, each responsible for synthesising one enzyme in a metabolic series, is controlled by an operator gene (see Fig. 5). It is further suggested that the operator is in its turn negatively controlled by a repressor gene which can thus switch off the synthesis of all of the enzymes that are under the control of one operator. The repressor substance can be influenced, according to this theory, by cytoplasmic substances, so that, for example, substances known to induce enzyme formation in bacteria are substances which interfere with the repressor; similarly, repressors of enzymic activity are substances which stimulate the repressor to interfere with the operator. One could argue that growth hormone could interfere with the repressors of a large number of protein molecules, thus stimulating general protein synthesis. Another possibility is that growth hormone is acting at the cytoplasmic level by enhancing the ability of ribosomes to accept

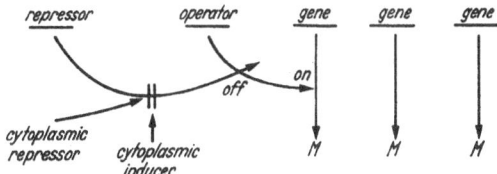

Fig. 5. Diagram showing JACOB and MONOD's operator and repressor theory

messenger RNA or by decreasing the rate at which messenger RNA is broken down in the cytoplasm (see Fig. 2). A third type of possibility, which does not necessarily depend on the acceptance of the messenger hypothesis, is that the ribosomes themselves are unstable and are constantly being broken down and resynthesised in the cell and that growth hormone stabilises the ribosomes or stimulates their synthesis. If this latter possibility is to be accepted, one must argue that ribosomes can exist in non-active forms, either as precursors or as immediate break-down products, which, in the experiments described here, are not distinguished from active ribosomes. TISSIÈRES, SCHLESSINGER, and GROS (1960) have shown that only some sizes of ribosomes from E. coli are able to synthesise protein, and our experiments suggested that larger sizes of ribosomes are the ones which incorporate best and indicated a difference in size distribution between ribosomes from normal and hypophysectomised rats (KORNER, 1961 c).

The first two of these possibilities depend upon the existence of messenger RNA in mammalian tissues and this has yet to be proved. The theory was originally propounded to explain the rapid rate at which adaptive enzymes begin to be synthesised as soon as an inducer is added to a suitable bacterial culture. The authors (JACOB and MONOD, 1961) suggest that the hypothesis can also be applied to mammalian tissues and they cite as evidence the work of LAMFROM (1961) and of KRUH, DREYFUS, and SCHAPIRA (1961). These workers incubated ribosomes from reticulocyte cells

of one species (A) with cell sap obtained from reticulocytes of a
second species (B) and examined the *in vitro* radioactive labelling
of the haemoglobins of both species. If the messenger concept is
correct and if messenger RNA is present in the cell sap, then
ribosomes of species A incubated with cell sap from species B
might produce both A and B haemoglobins. Both groups of workers
found this to be so and concluded that ribosomes of species A could
be induced to make the "wrong" type of haemoglobin if supplied
with messenger RNA of the "wrong" (i.e. B) species. Unfortunately
these experiments were not well controlled and no account was
taken of the possibility of fortuitous labelling of the wrong haemoglobin in the systems that were used. We have repeated these
experiments (Table 1) and have found that without the controls
there appears to be synthesis of haemoglobin of a type different
from that of the ribosomes which were used. If, however, the
apparent synthesis of haemoglobin in the control experiments in
which sap alone is incubated is taken into account, then there is no
evidence that a messenger RNA exists in the sap, for all of the
counts in the B haemoglobin would have come from B ribosomes
contaminating the B sap, prepared in the manner of KRUH et al.
(1961). This result does not disprove the messenger hypothesis,
but it certainly does not support it.

Table 1. *Apparent synthesis of rabbit haemoglobin on guinea-pig ribosomes*

Source of cell fraction		Counts in haemoglobin	
Ribosomes	Cell sap	Guinea-pig	Rabbit
Guinea-pig	Guinea-pig	1400	0
Guinea-pig	Rabbit	2000	1100
—	Rabbit	1200	1800

Line 1. Guinea-pig ribosomes incubated with guinea-pig cell sap produce
labelled guinea-pig haemoglobin but no label in rabbit haemoglobin.
Line 2. Guinea-pig ribosomes incubated with rabbit cell sap produce a label
in both guinea-pig and rabbit haemoglobin. It would appear that rabbit cell
sap contains information which causes guinea-pig ribosomes to synthesise
rabbit haemoglobin, but:
Line 3. Rabbit cell sap incubated alone causes labelling of both guinea-pig
and rabbit cell sap to an extent which explains the apparent synthesis of
rabbit cell sap in line 2.

WEISS (1960) discovered an enzyme in rat-liver nuclei capable
of synthesising RNA from nucleotide triphosphates which requires
the presence of DNA as a primer. The RNA so produced has a
base composition similar to the DNA of the primer, and it appears

likely that the RNA synthesised has a base sequence similar to
the base sequence of the DNA of the primer. Perhaps WEISS's
enzyme is one which synthesises messenger RNA. Experiments
were carried out in which the WEISS enzyme was incubated with
nucleotide triphosphates and DNA under conditions where RNA
could be expected to be synthesised, and this brew was then added
to the ribosome cell-free system. Stimulation of incorporation by
the WEISS RNA was noted only once, and it has not yet been
possible to repeat this finding on a number of occasions.

If ribosomes from normal rat liver carry more messenger RNA
than those from livers of hypophysectomised rats one might argue
that the latter would be more likely to accept added messenger
RNA. Experiments designed to examine this possibility have been
carried out in the following fashion. Ribosomes from normal rat
liver and from hypophysectomised rat liver were pre-incubated
in the presence of RNA extracted from various subcellular fractions
of normal liver in the hope that the RNA might be messenger
RNA and might attach itself to the ribosomes. The extent of
amino-acid incorporation into protein of the RNA-fortified ribosomes was compared with that of controls to which RNA had not
been added (Table 2). The addition of RNA to ribosomes from

Table 2. *Stimulation of incorporation of amino acids into protein by liver
ribosomes from hypophysectomised rats on the addition of RNA from normal
rat-liver ribosomes*

Source of cell fractions		Source of	CPM/mg. protein
Ribosomes	Cell sap	RNA	
Normal	Normal	---	250
Normal	Normal	Normal ribosomes	262
Hypo	Normal	—	107
Hypo	Normal	Normal ribosomes	143

normal rat liver has so far not resulted in a stimulation of amino-
acid incorporation, but ribosomes from hypophysectomised rat
liver showed a 30% stimulation of incorporation of amino acids
into proteins when RNA extracted from normal rat-liver microsomes was added to them. We believe that we have not yet
achieved optimal binding of RNA to ribosomes, for we have not
been able to detect stimulation of incorporation of phenylalanine
into protein-like material when polyuridilic acid was added, although NIRENBERG and MATTHAEI (1961) and OCHOA and his
colleagues (LENGYEL et al., 1961) have found such a stimulation
in bacterial systems.

If the messenger hypothesis is true for mammalian tissues it should be possible to detect in these tissues a newly synthesised RNA with a molecular weight smaller than that found in the ribosomes, with a rapid rate of turnover, and with a nucleotide

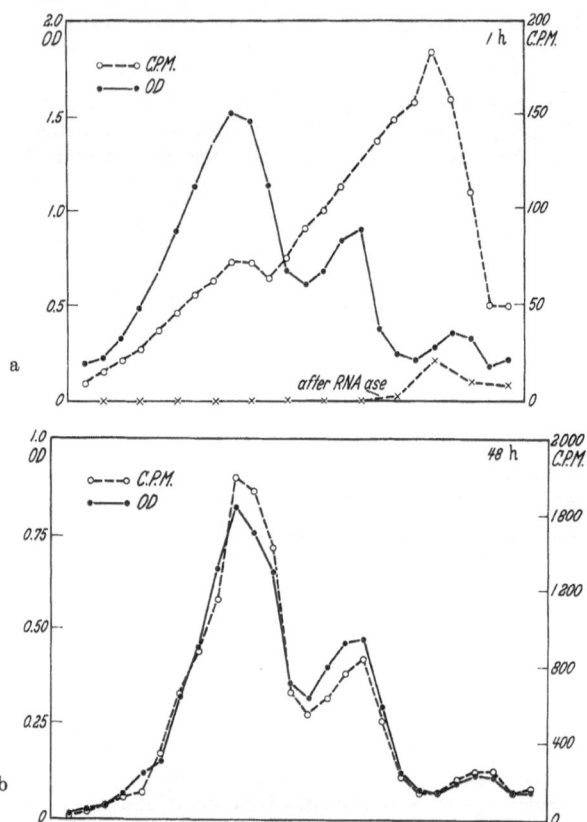

Figs 6a and b. Distribution of ^{32}P in RNA isolated from rat-liver microsomes at (a) 1 hour and (b) 48 hours after injection of radioactivity

sequence related to the nucleotide sequence of the DNA of the cell. With my student, Mr. A. J. MUNRO, I have injected radioactive phosphorus into a rat, extracted microsomal RNA at various time intervals after the injection, analysed the RNA by the sucrose gradient centrifugation technique, and studied the distribution of

the radioactivity in the RNA (see Fig. 6). We have shown that one
hour after injection of radioactive phosphorus there is a strongly
labelled fraction which runs on the gradient in a position between
that of the soluble RNA and that of the smaller of the two RNA
moieties of the ribosomes. If one examines labelling of the RNA
48 hours after the radioactive injection, all of the radioactivity is
associated with the ribosomal RNA. The peak of radioactivity
detected at 1 hour is associated with a very low amount of material
with an O. D. peak at 200 mμ., and it is therefore possible that this
material is not ribonucleic acid but is some other phosphorus-
containing material. This highly radioactive peak disappears,
however, when the material is treated with ribonuclease, and it is
therefore probable that the labelled material is indeed ribonucleic
acid. Its position on centrifugation in a sucrose gradient suggests
that its molecular weight is of the order of that of a messenger RNA
found in bacterial systems (HAYASHI and SPIEGELMAN, 1961).
Experiments are now being conducted in order to detect possible
differences in the behaviour of this newly synthesised RNA which
are controlled by growth hormone.

Summary

Growth hormone has been shown to influence the ability of a cell-free
system from rat liver to incorporate amino acids into protein. It has been
shown that the activity of the amino-acid-activating enzymes is not affected
by growth hormone, nor is the acceptor or transfer activity of the soluble
RNA influenced by this hormone. The microsomes of rat liver are, however,
under the control of growth hormone in their ability to assemble activated
amino acids into polypeptide chains.

It has been shown that the change in the microsomes brought about
by growth hormone can be accounted for by changes in the ribosomes of
the microsomes.

The messenger hypothesis is described and its applicability to mammalian
tissues is assessed on the basis of a number of experiments designed to
detect the presence of messenger RNA in rat liver. The possible point at
which growth hormone exerts its effects on protein synthesis is discussed in
terms of the messenger hypothesis and of an alternative hypothesis.

Zusammenfassung

Wachstumshormon fördert die Fähigkeit eines einfachen aus Rattenleber gewonnenen zellfreien Systems zum Einbau von Aminosäuren in
Eiweiß. Es wurde gezeigt, daß die Wirksamkeit der Aminosäuren aktivieren-
den Fermente durch Wachstumshormon nicht beeinflußt wird, und ebensowenig verändert Wachstumshormon die Aufnahme- oder Überträgerfunktion
der löslichen Ribonucleinsäure. Dagegen stehen die Mikrosomen der Rattenleber in bezug auf ihre Fähigkeit, aktivierte Aminosäuren zu Polypeptid-
ketten zusammenzufügen, unter der Kontrolle von Wachstumshormon.

2*

Für die durch Wachstumshormon in den Mikrosomen hervorgerufenen Veränderungen sind Veränderungen in den Ribosomen der Mikrosomen verantwortlich.

Die Übermittler- (messenger) Hypothese wird beschrieben und ihre Übertragbarkeit auf Säugetiergewebe auf Grund verschiedener Versuche beurteilt, die zum Ziel hatten, die Anwesenheit der die Information übermittelnden Ribonucleinsäure (messenger RNA) in der Rattenleber festzustellen. Der mögliche Angriffspunkt von Wachstumshormon in der Eiweißsynthese wird im Sinne der Übermittler-Hypothese und einer anderen Hypothese besprochen.

Résumé

L'hormone de croissance stimule la capacité d'un système simple exempt de cellules, tiré du foie du rat, à incorporer les acides aminés dans les protéines. On a pu établir que l'hormone de croissance n'influence pas l'action des ferments qui activent les acides aminés et ne modifie pas non plus les fonctions d'absorption ou de transfert de l'acide ribonucléique soluble. En revanche, les microsomes du foie du rat sont sous le contrôle de l'hormone de croissance en ce qui concerne leur capacité à grouper les acides aminés en chaînes polypeptidiques.

Les modifications provoquées par l'hormone de croissance dans les microsomes sont dues à des changements dans les ribosomes des microsomes.

L'auteur expose l'hypothèse du "messager" (substance de transmission) et discute des possibilités de l'appliquer aux tissus des mammifères en se fondant sur diverses expériences dont le but est d'établir la présence de l'acide ribonucléique transmettant l'information ("messenger RNA") dans le foie du rat. Le point d'attaque de l'hormone de croissance dans la synthèse des protéines est discuté à la lumière de l'hypothèse du "messager", ainsi que d'une autre hypothèse.

References

BISHOP, J., J. LEAHY, and R. SCHWEET: Proc. Nat. Acad. Sc. U.S. 46, 1030 (1960). — CLAUDE, A.: Science (U.S.A.) 97, 451 (1943). — HAYASHI, M., and S. SPIEGELMAN: Proc. Nat. Acad. Sc. U.S. 47, 1564 (1961). — HOAGLAND, M. B.: In: Nucleic Acids, Vol. 3. Ed. by CHARGAFF, E., and J. N. DAVIDSON. Academic Press, New York, 1960, p. 349. — HOAGLAND, M. B., E. B. KELLER, and P. C. ZAMECNIK: J. Biol. Chem. (U.S.A.) 218, 345 (1956). — HOAGLAND, M. B., P. C. ZAMECNIK, and M. L. STEPHENSON: Biochim. biophysica acta (U.S.A.) 24, 215 (1957). — HOAGLAND, M. B., M. L. STEPHENSON, J. F. SCOTT, L. I. HECHT, and P. C. ZAMECNIK: J. Biol. Chem. (U.S.A.) 231, 241 (1958). — HULTIN, T.: Biochim. biophysica acta (U.S.A.) 51, 219 (1961). — JACOB, F., and J. MONOD: J. Mol. Biol. 3, 318 (1961). — KORNER, A.: Nature (G.B.) 181, 422 (1958); — Biochem. J. (G.B.) 73, 61 (1959a); — Biochim. biophysica acta (U.S.A.) 35, 554 (1959b); — Exper. Cell Res. (U.S.A.) 18, 594 (1959c). — Biochem. J. (G.B.) 74, 462 (1960a); — J. Endocr. (G.B.) 21, 177 (1960b); — J. Endocr. (G.B.) 20, 256 (1960c); — Biochem. J. (G.B.) 74, 471 (1960d); — Biochem. J. (G.B.) 81, 168 (1961a); — Biochem. J. (G.B.) 81, 292 (1961b); — Mem. Soc. Endocr. No. 11 (1961c); — Biochem. J. (G.B.) 83, 69 (1962). — KORNER, A., and K. L. MANCHESTER: Brit. Med. Bull. 16, 233 (1960). — KRUH, J. ,J. ROSA, J. C. DREYFUS, and G. SCHAPIRA: Biochim. biophysica acta (U.S.A.) 49, 509 (1961). — LAMFROM, H.: J. Mol. Biol. 3, 241 (1961). — LENGYEL, P., J. F. SPEYER, and S. OCHOA: Proc. Nat. Acad. Sc. U.S. 47, 1936 (1961). — LITTLEFIELD, J. W., E. B. KELLER, J. GROSS, and

P. C. ZAMECNIK: J. Biol. Chem. (U.S.A.) **217**, 111 (1955). — MANCHESTER, K. L., and F. G. YOUNG: Vitamins and Horm. (U.S.A.) **19**, 95 (1962). — MUELLER, G. C., A. M. HERRANEN, and K. F. JERVELL: Recent Progr. Hormone Res. (U.S.A.) **14**, 95 (1958). — NIRENBERG, M. W., and J. H. MATTHAEI: Proc. Nat. Acad. Sc. U.S. **47**, 1588 (1961). — PALADE, G. E., and P. SIEKEVITZ: J. Biophys. Biochem. Cytol. (U.S.A.) **2**, 171 (1956). — ROBERTS, R. B.: In: Microsomal Particles in Protein Synthesis. Pergamon Press, London, 1958, p. 8. — ROODYN, D. B., P. J. REIS, and T. S. WORK: Biochem. J. (G.B.) **80**, 9 (1961). — SIEKEVITZ, P.: J. Biol. Chem. (U.S.A.) **195**, 549 (1952). — TAKANAMI, M.: Biochim. biophysica acta (U.S.A.) **39**, 318 (1960).— TISSIÈRES, A., D. SCHLESSINGER, and F. GROS: Proc. Nat. Acad. Sc. U.S. **46**, 1450 (1960). — TRUMAN, D. E. S., and A. KORNER: Biochem. J. (G.B.) **83**, 588 (1962). — WEISS, S. B.: Proc. Nat. Acad. Sc. U.S. **46**, 1020 (1960). — YOUNG, F. G.: Brit. Med. J. 1939/II, 393; — Brit. Med. J. 1941/II, 897; — Biochem. J. (G.B.) **39**, 515 (1945). — YOUNG, F. G., and A. KORNER: In: Diabetes. P. C. Hoeber Inc., New York, 1960, p. 216. — ZAMECNIK, P. C., and E. B. KELLER: J. Biol. Chem. (U.S.A.) **209**, 337 (1954).

Discussion

STAEHELIN: We have been greatly impressed by Dr. KORNER's beautiful papers, and working on similar lines we have tried to repeat some of his experiments. The effect of hypophysectomy and of subsequent growth-hormone treatment is certainly evident in the ribosomes of the cell-free system.

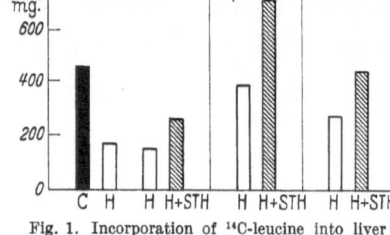

Fig. 1. Incorporation of ¹⁴C-leucine into liver ribosomes and muscles of hypophysectomised rats and hypophysectomised rats treated with growth hormone (1 mg./day for 5 days)

We too have studied the physico-chemical states of the ribosomes from normal and hypophysectomised animals. TISSIÈRES[1] has shown that *E. coli* ribosomes disaggregate in $\frac{m}{1000}$ Mg⁺⁺. Under these conditions, we have consistently found that ribosomes from hypophysectomised animals disintegrate almost completely into smaller subunits and show only a very small peak of 80 s particles, whereas more 80 s and some 110 s particles are present in normal ribosomes. It seems, therefore, that there is a change in the physico-chemical state of the ribosomes. We have also found that ribosomes from various animals can differ in their lability towards lower and higher Mg⁺⁺ concentrations measured by the incorporation of amino acids, although under optimal conditions they have similar activities.

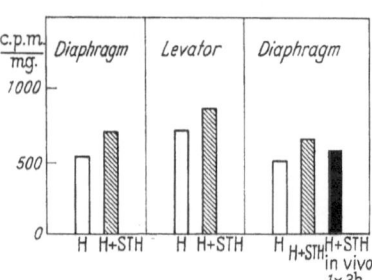

Fig. 2. Incorporation of ¹⁴C-leucine into muscles of hypophysectomised rats in the presence and absence of growth hormone (10 mcg./ml.) and after one single injection of 1 mg. growth hormone 3 hours before the experiment

The second point which I should like to discuss in relation to Dr. KORNER's paper is the effect of growth hormone on amino-acid incorporation in muscle. This form of protein synthesis is also stimulated by treatment with growth hormone. This effect was first demonstrated by KOSTYO and KNOBIL[2] on the diaphragm. We have found that the levator ani responds to growth hormone in a similar manner (Fig. 1). In muscle, this stimulation by growth hormone can also be demonstrated *in vitro*, although to a much lesser degree (Fig. 2). If growth hormone exerted a direct

[1] TISSIÈRES, A., D. SCHLESSINGER, and F. GROS: Proc. Nat. Acad. Sc. U.S. **46**, 1450 (1960).

[2] KOSTYO, J. L., and R. KNOBIL: Endocrinology (U.S.A.) **65**, 395 (1959).

effect on a messenger RNA one would expect this action to be very fast. However, upon injection of a single dose of growth hormone 3 hours prior to sacrifice of the animals, we have found only an extremely small stimulation of amino-acid incorporation as compared to the marked effect after 5 days of treatment. This seems to argue against an immediate effect on the synthesis of a messenger RNA.

I wonder whether Dr. KORNER or Dr. YOUNG have done experiments to show whether the action of growth hormone or of insulin requires the synthesis of a new RNA ? If this were the case, the hormone effect should be abolished or impaired by pre-treatment of the animals with a nucleic acid analogue, e.g. 8-azaguanine or fluorouridine. This experiment should provide an answer as to whether the synthesis of a new messenger RNA is required for hormone action.

KORNER: Regarding the first question about the sizes of ribosomes, we have found differences in the size distribution of ribosome preparations from normal and hypophysectomised animals. We find that most of the ribosomes from rat liver are not 70 s at all, unlike those from the bacterial system. In fact, we don't get anything less than 80 s, unless we take stringent measures to break the ribosomes down. We find a small peak of 80 s and then a much larger peak of about 110 s, and then smaller peaks of even higher s values. We think that the preparation from the normal rat has more of the bigger sizes of ribosomes than that from the hypophysectomised rat.

It is difficult to be sure, in these experiments, that there are not differences in the amount of magnesium attached to the ribosomes in different preparations made even in the absence of magnesium in the media used. The size of ribosomes is very dependent on the magnesium level.

Magnesium is also needed in the reactions to activate the amino acids, and it is difficult to distinguish the effect of magnesium on the activation of amino acids and on the size of the ribosomes.

As to the second point you mentioned, I take it your results were with isolated hemidiaphragms or diaphragms and not with ribosomes from them. We have been trying to obtain ribosomes from muscle in a state in which they will incorporate amino acids *in vitro* and to study the effects of hormones on the system. The difficulty is to break up the muscle in such a way that we don't also completely destroy the system.

Your third point related to time of treatment: I often give growth hormone for several days, but I have detected changes in incorporation which are statistically significant 3 hours after treatment with insulin, and within 6 hours after treatment with growth hormone.

Now you mentioned the diaphragm experiments. MANCHESTER and others[1] have shown that insulin will stimulate incorporation of radioactive carbon into protein, even if the radioactivity is given in the form not of amino acids but of carbon in some other form, so that the substrate which enters the muscle has been metabolised to amino acids and then enters protein. Yet insulin is able to stimulate protein synthesis under these conditions. Furthermore, these authors have been able to show that glucose need not be present for stimulation of amino-acid incorporation in protein. WOOL[2], who worked in Cambridge for a while, has found that if the dia-

---

[1] MANCHESTER, K. L., and M. E. KRAHL: J. Biol. Chem. (U.S.A.) **234**, 2938 (1959).

[2] WOOL, I. G.: unpublished.

phragm of the rat is incubated with insulin, there is an increase in the amount of RNA and also in the rate of labelling of RNA.

DREYFUS: Dr. KORNER has referred to the cross-experiments between guinea-pig and rabbit haemoglobin which we made with KRUH and SCHA-PIRA[1]. We certainly did not prove that the synthesis is possible, because such proof must involve dissection of the peptides and production of evidence that the amino acid has been incorporated in the right positions. But I would not agree with the factual results, since when we use rabbit microsomes and no guinea-pig system, we do not get an incorporation into guinea-pig haemoglobin — a fact which apparently contradicts what has been shown by Dr. KORNER. So I still think that our results are technically valid, but the final proof must await identification of the haemoglobin, which has not been done either by LAMFROM[2] or by us. Moreover, in the reticulocyte system we do not observe the proportionality between the synthesis of haemoglobin and the incorporation into ribosomes which has been reported by Dr. KORNER with serum albumin. On the contrary, there is often a reciprocal relationship: the more haemoglobin is produced, the less counts are found (at the end of the experiment at least) in the microsomes. On the other hand, I would confirm and agree with what Dr. KORNER says about the existence of a fast labelled RNA fraction in mammalian tissues. The same results as he obtained have also been reported by HIATT[3] with liver nuclei, by ALLFREY et al.[4] with thymus nuclei, and by my colleagues KRUH, GROS, and MARKS[5] in a mixed system of reticulocytes and leucocytes (the counts probably come from leucocytes). So I think a fast labelled fraction, which might be a messenger, exists in mammalian cells, but it is not proven and perhaps not true that it has a very short life-span, since in the reticulocyte system the messenger can act for a longer period.

LARON: I would like to ask you about the somewhat confusing problem of the relationship between insulin and growth hormone. At one point you mentioned that the cytoplasmic inducer may be acted on directly by growth hormone, whereas at another point you mentioned that growth hormone acts through the presence of insulin. On the other hand, we also know from the work of your colleagues, such as Dr. MANCHESTER, that the requirement of glucose for an insulin effect varies in different tissues. Do we have to believe that growth hormone does not act directly without insulin? Or may we also consider the possibility that there is a different relationship in various tissues?

KORNER: May I mention some results of my colleague, Dr. RANDLE, which are of interest? He and his colleagues, Dr. HALES and Dr. NEWSHOLME[6],

---

[1] KRUH. J, J. ROSA, J. C. DREYFUS, and G. SCHAPIRA: Biochem. biophysica acta (U.S.A.) **49**, 509 (1961).

[2] LAMFROM, H.: J. Mol. Biol. (U.S.A.) **3**, 241 (1961).

[3] HIATT, H. H.: Fed. Proc. (U.S.A.) **21**, 381 (1962).

[4] SIBATANI, A., S. R. DE KLOET, V. G. ALLFREY, and A. E. MIRSKI: Proc. Nat. Acad. Sc. (U.S.) **48**, 471 (1962).

[5] MARKS, P. A., C. WILLSON, J. KRUH, and F. GROS: Biochem. Biophys. Res. Commun. (U.S.A.) **8**, 9 (1962).

[6] HALES. C. N., and P. J. RANDLE: Biochem. J. (G. B.) **84**, 79 (1962).
RANDLE, P. J.: In: Disorders of Carbohydrate Metabolism. Ed. by D. M. PIKE. Pitman Press, London, 1962.
NEWSHOLME, E. A., and P. J. RANDLE: Biochem. J. (G.B.) **84**, 79 (1962).

have developed a method of measuring blood insulin very accurately and in very small amounts, and they found that if they placed themselves on a low carbohydrate diet for a few days the blood non-esterified fatty acid level was high, the blood glucose level was somewhat higher than normal, and the blood insulin level was also high. On the basis of these and other results they suggest that, when fatty acids are released from adipose tissue, they are preferentially oxidised in muscle and inhibit the oxidation of glucose by muscle. In consequence, the blood sugar rises and stimulates insulin release to restore the status quo by stimulating glucose oxidation by muscle and adipose tissue and the synthesis of fat. Suppose growth hormone acts by stimulating fatty acid release: the result will be a rise in the blood sugar level because of decreased oxidation of glucose, and this will cause insulin secretion. Hence the insulin-like action of growth hormone. If treatment with growth hormone is continued until the β-cells can no longer respond by insulin secretion, diabetes may result.

I do not know if growth hormone is itself the immediate agent for controlling protein synthesis. It may act via insulin or through some other secondary effect.

Regulation of protein synthesis by androgens and estrogens

By

J. D. Wilson

The administration of testosterone and several other steroid hormones clearly causes a marked diminution in nitrogen excretion (*3*) and results in increased protein deposition in kidney, liver, muscle, carcass, and accessory sex tissue (*4*). However, the mechanisms of the protein-anabolic action have never been fully explained. While this effect of testosterone is probably the result of an enhancement of protein synthesis, previous attempts to study this aspect of hormonal action have been complicated by two factors: first, the major enzymatic steps in the synthesis of protein have been described only in the past few years (*5*); and, second, previous attempts to demonstrate an influence of testosterone on protein synthesis in several non-sexual tissues of the rat (*6*) and mouse (*7*, *8*) have yielded effects which, although consistent, are quite small.

Several observations suggested to us that the accessory sex organs might serve as suitable tissues for an exploration of the mechanisms by which steroid hormones influence protein metabolism. The accessory sex organs are very responsive to the administration of certain of these agents, and, in fact, Scow has demonstrated that as much as 25% of the total weight gain induced by testosterone in castrated rats occurs in the secondary sex tissue (*9*). Furthermore, there is at least one report that the rat prostate rapidly and selectively concentrates testosterone-C^{14} (*10*).

Acceleration of protein synthesis *in vitro* by testosterone administration

The initial studies, therefore, were directed to an examination of the influence of testosterone on protein synthesis in these tissues. First, an attempt was made to ascertain whether testosterone administration does, in fact, influence protein synthesis *in vitro*. Immature male rats were injected with 5 mg. testosterone-proprionate per day for one to three days. The prostate and seminal vesicle were removed, sliced, and incubated with amino acid-C^{14} for one hour. The reaction was then stopped, and the proteins were purified and assayed for C^{14} in a liquid scintillation counter by previously described methods (*1*).

Table 1. *Influence of testosterone administration on protein synthesis from L-valine-C^{14} and L-tyrosine-C^{14} by slices of rat seminal vesicle*

Time of injection	Protein synthesis from:			
	L-valine-C^{14}		L-tyrosine-C^{14}	
	cpm.	cpm./mg.	cpm.	cpm./mg.
0	454	649	516	974
24 hours	2790	3204	2076	2227
48 hours	6214	3088	1850	1284
72 hours	9666	3710	4006	1659

As demonstrated in Table 1, testosterone administration for 24 hours produced a three to five-fold rise in the incorporation of both C^{14}-labeled tyrosine and valine into total protein; as shown in the adjacent columns, the specific activity of newly synthesized protein was similarly increased. Although usually the total counts per minute continued to rise in subsequent days, the maximal change in specific activity was seen within 24 hours in 50-gram rats and somewhat later in 100-gram rats.

Mechanism of protein biosynthesis

The *in vitro* demonstration of the sensitivity of protein synthesis to testosterone administration made it possible to evaluate some of the biochemical mechanisms by which this effect might be mediated. In Fig. 1 are summarized the major steps in protein synthesis. Free amino acids within the cell may arise from one of two sources. First, they may be transported into the cell from the extracellular fluid or, second, amino acids may be synthesized by the fixation of ammonia with α-ketoglutarate to form glutamic acid, which can subsequently undergo transamination to form any of the non-essential amino acids. Intracellular amino acids arising from either of these two sources are then activated in a reaction requiring ATP to form amino-acid adenylates, which are then bound to soluble ribosenucleic acid, forming the soluble RNA-amino-acid complexes. A separate and specific soluble RNA exists for each of the amino acids. In a reaction requiring guanosine triphosphate (GTP), these RNA-amino acids are then transferred into the ribosomes of the microsomes where peptide bonding occurs, resulting in the formation of ribonucleoproteins. The completed protein is subsequently stripped off the ribosome particle and released into the soluble portion of the cell.

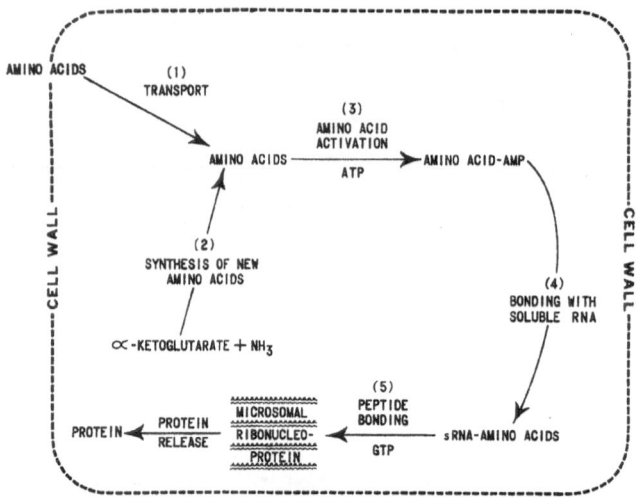

Fig. 1. The mechanism of protein biosynthesis

In an attempt to identify the precise locus of testosterone action, this pathway of synthesis in seminal-vesicle slices was studied at four critical sites; amino-acid transport (Step 1), amino-acid synthesis (Step 2), soluble RNA-amino-acid formation (Step 4), and microsomal ribonucleoprotein formation (Step 5).

Site of action of testosterone on protein synthesis in seminal vesicle

Because the studies of CHRISTIANSEN (11) and others (12) have suggested that the intracellular penetration of the non-utilizable amino-acid, α-aminoisobutyric acid, is enhanced by testosterone and by several other hormonal agents, it was of interest, first, to determine if testosterone might stimulate protein synthesis by enhancing the intracellular transport of a naturally occurring amino acid (Step 1 in protein biosynthesis).

In the experiment shown in Fig. 2 the rate of intracellular penetration of L-tyrosine-U-C^{14} was studied in slices of seminal vesicles from normal and testosterone-treated rats. Each 100 mg.

of seminal vesicle was incubated with one microcurie of tyrosine-C^{14}, and at varying time intervals the amount of free tyrosine-C^{14} within the cell was determined. In addition, at each time interval the extent of incorporation of the labeled tyrosine into protein was measured. The specific activity of intracellular tyrosine at the varying time intervals is shown by the dotted line, and the specific activity of protein is demonstrated by the solid line. Despite a

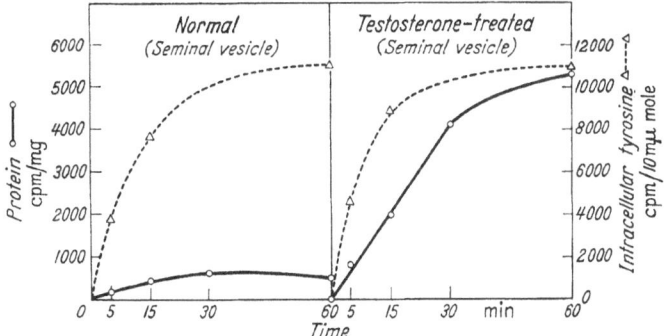

Fig. 2. Time course of transport and incorporation into protein of L-tyrosine-C^{14} by slices of rat seminal vesicle. [Reprinted with permission from the Journal of Clinical Investigation **41**, 153 (1962)]

profound difference in the specific activities of the intracellular protein-C^{14}, at no time was there a significant difference in the specific activity of the intracellular tyrosine-C^{14}. Thus, it was concluded that the anabolic effect of testosterone is not caused by an enhancement of amino-acid transport. Furthermore, subsequent experiments in which the pool size of intracellular tyrosine was measured directly have demonstrated that the free tyrosine pool is actually smaller in the slices from the testosterone-treated animal than in the normal preparation. Certainly, an increase in amino-acid-C^{14} transport may occur as a late event following testosterone administration. However, this invariably takes place after the enhancement of protein synthesis, and it seems clear, therefore, that the effects on amino-acid transport are secondary to the increased utilization of intracellular amino acids in protein synthesis.

The effect of testosterone administration on the *de novo* synthesis of amino acids (Step 2) was then examined. As shown in Table 2, acetate-2-C^{14} was incubated with slices of seminal vesicle and prostate either from testosterone-treated or normal rats, and the rate of incorporation into glutamic acid and into combined amino acids was measured. Duplicate preparations were incubated with tyrosine-C^{14} in order to determine simultaneously the rate of protein synthesis. There was no significant difference between the rates of glutamic-acid and total amino-acid synthesis before and after testosterone administration, despite the fact that in the duplicate experiment protein synthesis from tyrosine-C^{14} was increased three-fold in the testosterone-treated slices. It is clear from this that testosterone does not influence the *de novo* synthesis of amino acids. The anabolic effect of testosterone, therefore, must occur at some step subsequent to the synthesis and intracellular transport of amino acids.

Table 2. *Absence of effect of testosterone on amino-acid synthesis from acetate-2-C^{14} by rat seminal vesicle*

Treatment	Acetate-2-C^{14} converted to:	
	Glutomic acid	Total amino acid
	cpm.	cpm.
None.	7352	8519
Testosterone .	7824	8502

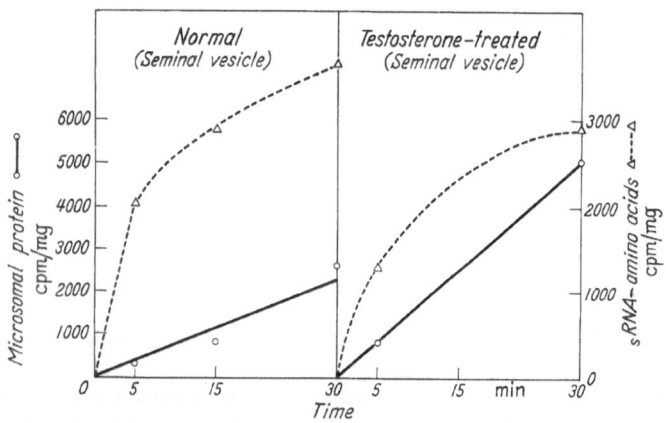

Fig. 3. Time course of incorporation of L-tyrosine-U-C^{14} into soluble RNA-amino acid and microsomal protein by slices of rat seminal vesicle. [Reprinted with permission from the Journal of Clinical Investigation 41, 153 (1962)]

Step 4 in protein biosynthesis, the formation of soluble RNA-amino acids, was then examined (Fig. 3). In this experiment, following incubation with tyrosine-C^{14}, the slices were homogenized, and the cell fractions were separated by differential ultracentri-

fugation. Soluble RNA-amino-acid complexes were purified by the
method of HOAGLAND and his co-workers (13) and analyzed both
for RNA content and for C^{14} in order to permit calculation of the
specific activity of the soluble RNA-tyrosine. In addition, cellular
fractionation allowed the determination of the rate of incorporation
of tyrosine-C^{14} into the microsomal protein. The specific activity
of soluble RNA-tyrosine at varying time intervals is shown by the
dotted line, and the specific activity of microsomal protein is
indicated by the solid line. At no time could a difference be
demonstrated in the specific activity of the soluble RNA-tyrosine,
despite the marked differences in the specific activities of the
microsomal protein between the slices from the normal and the
testosterone-treated animals.

As shown in Fig. 4, the same relationship can be demonstrated
in the seminal vesicle for L-valine-C^{14}. The specific activity of
soluble RNA-valine is shown by the dotted line, and the specific
activity of microsomal protein by the solid line. Again, although
there was no difference in the rate of soluble RNA-valine synthesis,
marked differences could be shown for the rate of microsomal protein synthesis.

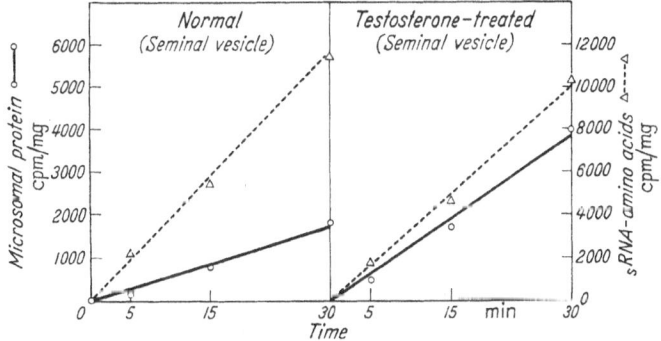

Fig. 4. Time course of incorporation of L-valine-C^{14} into soluble RNA-amino acid and microsomal protein by slices of rat seminal vesicle. [Reprinted with permission from the Journal
of Clinical Investigation **41**, 153 (1962)]

These experiments, in themselves, rule out the possibility that
any of the preceding steps in protein synthesis — either amino-
acid transport, amino-acid synthesis, amino-acid activation, or
soluble RNA-amino-acid formation — can be rate limiting in the
non-testosterone-treated tissue. They clearly indicate that testosterone must enhance one specific step in protein synthesis, the

peptide bonding of soluble RNA-amino-acid complexes to form microsomal ribonucleoprotein.

Site of action of estradiol on protein synthesis in uterus

It seemed logical, next, to determine whether this specific biochemical site of action of testosterone is limited to testosterone alone or whether the same type of action is characteristic of different steroid hormones. Therefore, the effect of estradiol administration on protein synthesis in slices of rat uterus was then examined. In these experiments, one milligram of estradiol was administered intramuscularly to immature rats 12 hours prior to death.

As shown in Fig. 5, the intracellular penetration of L-valine-C^{14} was studied in slices of uterus from normal and estradiol-treated rats. Again the concentration of intracellular tyrosine at varying

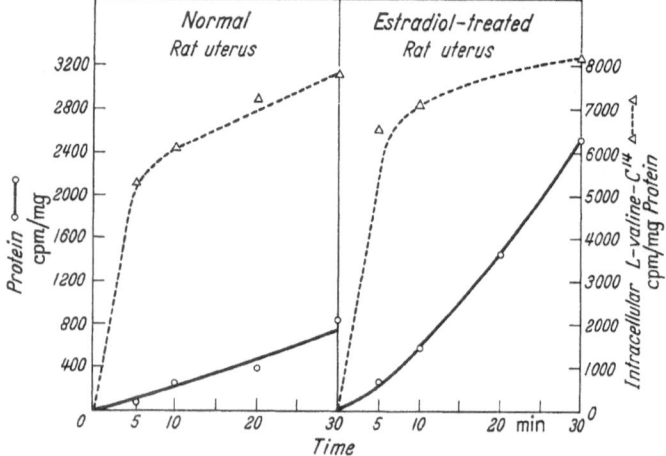

Fig. 5. Time course of transport and incorporation into protein of L-valine-U-C^{14} by slices of rat uterus

time intervals is shown by the dotted lines, and the specific activity of protein is demonstrated by the solid lines. And, as before, despite marked differences in the specific activity of the protein-C^{14} between the two preparations, at no time was there a significant difference in the concentration of the intracellular valine-C^{14}. Thus, it is clear that the acceleration of protein synthesis in rat uterus is not secondary to the enhancement of the transport of this naturally occurring amino acid.

Using techniques similar to those employed in the studies on testosterone action, Step 4 in protein synthesis, the formation of soluble RNA-amino acids, was then studied in slices of uterus from normal and estradiol-treated rats (Fig. 6). The specific activity of soluble RNA-valine at varying time intervals is shown by the dotted line, and the specific activity of microsomal protein is indicated by the solid line. As in the seminal vesicle, at no time

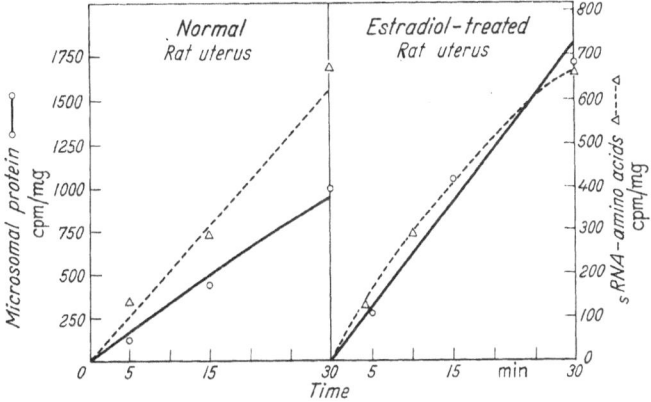

Fig. 6. Time course of incorporation of L-valine-U-C¹⁴ into soluble RNA-amino acids and microsomal protein by slices of rat uterus

was a difference demonstrated in the specific activity of soluble RNA-valine-C¹⁴ between the normal and the treated slice, despite marked difference in the specific activities of the microsomal protein. Subsequent studies have also demonstrated that estradiol does not influence the stripping off of the completed protein from the microsome. Thus, the acceleration of protein synthesis in accessory sex tissue of the rat by both testosterone and estradiol appears to be secondary to the enhancement of the same step in protein synthesis, the peptide bonding of soluble RNA-amino acids to form microsomal ribonucleoprotein.

Studies on the mechanism of the steroid-hormone effect on the peptide-bonding step in protein synthesis

While these studies have localized the hormone-sensitive step in protein synthesis, they do not furnish evidence as to the mechanism by which this effect is mediated. The formation of peptide bonds, Step 5 in protein biosynthesis, is a complex reaction requiring, in addition to the 17 or so different RNA-amino acids and the ribosome acceptor, the co-factor GTP, a soluble transfer enzyme, magnesium, and, in some preparations, sulfhydryl compounds. In order, therefore, to determine which of these factors mediates this hormonal action, it was necessary to evaluate the various components of the peptide-bonding reaction, and such a study could only be done in a cell-free system. However, repeated attempts to solubilize the protein-synthesizing system of immature rat uterus and seminal vesicle in our laboratory have been unsuccessful.

In order to solubilize a protein-synthesizing system sensitive to steroid hormone, a third tissue, the hen oviduct, was examined. BRANDT and NALBANDOV had reported in 1956 that the secretion of ovalbumen is under the control of both estrogenic and androgenic steroids (14). And it was possible to demonstrate that the administration of estradiol results in a four-fold increase in protein-C^{14} synthesis and that the administration of testosterone causes a two-fold increase in protein synthesis, without either having an effect on amino-acid transport. In some experiments there was an additive effect of the two hormones; when maximal quantities of either hormone are given, however, this additive effect is not seen. Furthermore, this increase in protein synthesis is marked within 2 hours after estradiol administration.

In order to exaggerate the effects as much as possible for the homogenate studies, in the subsequent experiments oviducts from 10-week-old pullets were studied 36 hours following the intramuscular administration of five milligrams of estradiol. Thirty to sixty oviducts were pooled for each experiment; three to one homogenates were prepared in a Dounce homogenizer using either the sucrose-electrolyte buffer as described by ZAMECNIK and KELLER (15) or the Tris-electrolyte buffer described by NIRENBERG and MATTHAEI (16). The homogenates from such preparations were centrifuged at $10,000 \times G$ and the supernatant portion containing soluble and microsomal fractions was then studied.

When such homogenates were prepared from estradiol-treated oviduct and supplemented with an ATP-generating system (creatine-phosphate plus creatine-phosphate-ATP transphosphorylase), they were capable of active protein synthesis. The further addition of either ATP or GTP, with or without an ATP-generating system, had no effect on protein synthesis in either control or treated homogenate. Thus, the effect of estradiol on protein synthesis is not mediated through an effect on GTP.

The 10,000 $\times$ G supernatant was then subfractionated by centrifugation at 104,000 $\times$ G into microsomal and soluble fractions (Table 3). The recombination of estradiol-treated microsomes and estradiol-treated soluble fraction and of control microsomes and control soluble fraction gave results similar to the unfractionated 10,000 $\times$ G supernatants. Furthermore, although the combination of control microsomes and estradiol-treated

Table 3. *Fractionation of the protein-synthesizing system of oviduct from estradiol-treated hens*

Homogenate	System				L-Valine-U-C^{14} incorporated into protein cpm./mg.
Unfractionated homogenate	Estradiol-treated				463
	Control				40
	Microsomes		Soluble fraction		
	Estradiol-treated	Control	Estradiol-treated	Control	
Fractionated homogenate	+		+		363
		+		+	45
	+		+		359
		+	+		36

soluble fraction caused no enhancement of protein synthesis over the control system, the combination of estradiol-treated microsomes and control soluble fraction resulted in as great an enhancement of protein synthesis as was seen in the complete estradiol-treated system. These findings clearly demonstrate that the site of enhancement of protein synthesis in this tissue by estradiol administration is in the microsome itself and they exclude the possibility of the involvement of any soluble factors in this effect.

Having localized this effect to the microsome, a series of experiments were performed in an attempt to determine exactly how estradiol enhances this microsomal reaction. In the experiment shown in Table 4, soluble fraction from estradiol-treated oviduct was added either to control microsomes or to microsomes from estradiol-treated oviduct. Protein synthesis in this homogenate clearly requires the presence of treated microsomes, an ATP-generating system, and soluble fraction. To demonstrate that the estradiol effect was not mediated by sulfhydryl groups, this ex-

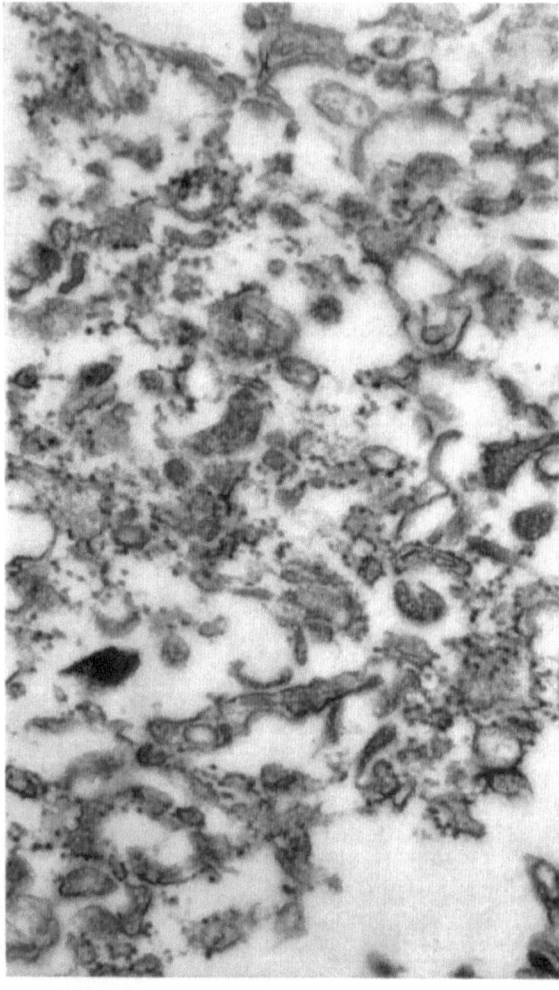

Fig. 7 a

Figs 7 a and b. Comparison by electron microscopy of the particulate fractions obtained by
a) Control;

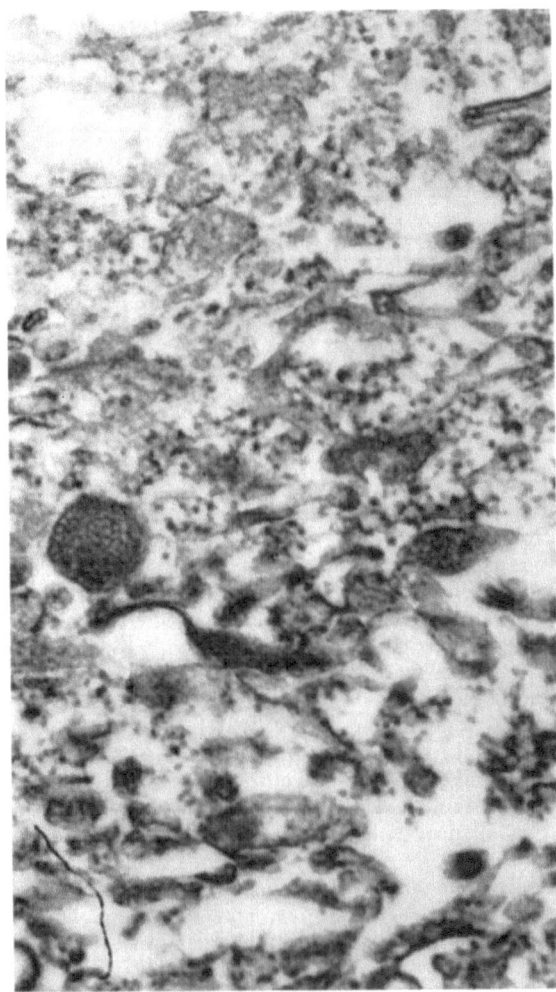

Fig. 7 b

centrifugation at 104,000 × G from control and estradiol-treated oviducts (× 36,000).
b) Estradiol-treated

periment was performed in a buffer containing 6×10^{-3} M mercaptoethanol; the difference between the two preparations is clearly not influenced by the presence of sulfhydryl groups.

Table 4. *Characteristics of the estradiol-sensitive protein synthesizing system*

| Characteristics | L-valine-C^{14} incorporated into protein: | |
	Estradiol-treated microsomes cpm./mg.	Control microsomes cpm./mg.
Complete system . . .	137	12
— ATP-gen. sys. . .	0	0
— Microsomes . . .	0	0
— Supernatant . . .	0	0
+ 0.05 μm. puromycin .	5	0
+ mRNA	205	21
+ Estradiol (1 mcg.) . .	126	21

Furthermore, the enhanced protein synthesis can still be inhibited by puromycin (which de-acylates soluble RNA-amino acids from the ribosomes). It should be pointed out that the microsome preparations from the control and treated oviduct were similar in respect to both RNA content and protein content. However, one obvious possibility for the different rates of protein synthesis between the two microsome preparations was that the estradiol-treated microsomes might contain more messenger or template RNA for the synthesis of protein. Therefore, microsomal RNA was isolated from estradiol-treated ribosomes by the method of NOMURA, HALL, and SPIEGELMAN (*17*) and added back to the control microsomes. No significant enhancement of protein synthesis was found in either system. Nor did estradiol, when added directly, enhance synthesis in the microsomes. Thus, these experiments give some evidence which tends to make it unlikely that there are two possible mechanisms for this action. The effect does not appear to be solely due to an increase in the number of microsomes, since the enhancement could be demonstrated even when the quantity of microsomes (as measured by the RNA and protein contents) was the same. Furthermore, although a negative experiment is not really conclusive, the absence of an enhancement of added microsomal RNA to the untreated system certainly does not lend support to the idea that the estradiol effect is mediated through an increased template RNA.

Finally, with the aid of a colleague, Dr. WALTER NORTON, the microsomal fraction obtained upon centrifugation of the oviduct homogenates at 104,000 × G was examined under the electron microscope (Fig. 7). These photographs were taken at about a 36,000 × magnification. The treated and control preparations appear to be very similar. In each instance the ribosome particles are numerous and are located on the ergastoplasmic membranes; these studies are in accord with the observations of DEANE and PORTER that the population density of ribosomes in cytoplasm of the mouse seminal vesicle did not change following castration (18). This approach, therefore, did not yield any clues as to how this effect might be mediated.

Thus, the experiments which have been described demonstrate unequivocally that the enhancement of protein synthesis by steroid hormones is localized in the microsomal fraction of the cell. However, it must be concluded that the exact mechanism for this effect remains unexplained.

Summary

A series of studies on the influence of testosterone and estradiol on protein synthesis have been performed in slices and homogenates of the accessory sex tissue of the rat and chicken. Evidence has been obtained in the slice experiments that the enhancement of protein synthesis which is mediated by these hormones is independent of either amino-acid transport or synthesis; indirect evidence in the slice studies and direct evidence in the homogenates has clearly demonstrated that this effect is secondary to the acceleration of a specific step in protein synthesis, the conversion of soluble ribosenucleic acid-amino acid complexes to microsomal ribonucleoprotein. Finally, fractionation of the homogenates has demonstrated that this effect is localized in the microsome itself. The ultimate mechanism by which this physiological action is mediated awaits elucidation.

Zusammenfassung

Es wurden verschiedene Untersuchungen über den Einfluß von Testosteron und von Oestriadiol auf die Eiweißsynthese an Schnitten und Homogenaten akzessorischer Sexualorgane von Ratte und Huhn durchgeführt. In den Versuchen mit Gewebsschnitten fand sich, daß die durch diese Hormone gesteigerte Eiweißsynthese unabhängig von Transport oder Synthese der Aminosäuren ist; indirekte Hinweise in den Untersuchungen an Schnitten und direkte Beobachtungen an Homogenaten haben eindeutig gezeigt, daß die vermehrte Synthese von Eiweiß Folge der Beschleunigung eines spezifischen Schrittes ist, nämlich der Umwandlung von Ribonucleinsäure-Aminosäurekomplexen in mikrosomales Ribonucleoprotein. Fraktionierung der Homogenate ergab, daß dieser Effekt in den Mikrosomen zu lokalisieren ist. Der Mechanismus, durch den diese physiologische Wirkung vermittelt wird, ist noch unbekannt.

Résumé

On a fait diverses recherches sur l'influence de la testostérone et de l'œstradiol sur la synthèse protéique en recourant à des coupes et homogénats de tissus sexuels secondaires de rats et de poulets. La méthode des coupes tissulaires a montré que la stimulation de la synthèse protéique par ces hormones est indépendante du transport ou de la synthèse des acides aminés; indirectement pour les coupes tissulaires et directement pour les homogénats, ces recherches ont montré de façon indiscutable que cet effet résulte de l'accélération d'une étape spécifique, qui est la transformation des complexes d'acide ribonucléique-acide aminé soluble en ribonucléoprotéine microsomique. Le fractionnement des homogénats a enfin prouvé que cet effet est à localiser dans le microsome lui-même. Le mécanisme d'où résulte cette action physiologique est encore inconnu.

Acknowledgements

Portions of this work have been previously published (1, 2). The investigation was supported by grants from the National Institutes of Health (A-3892) and the Institutional Grants Committee of the University of Texas Southwestern Medical School.

The author of this paper is an Established Investigator of the American Heart Association.

References

1. WILSON, J. D.: J. Clin. Invest. (U.S.A.) 41, 1953 (1962). — 2. WILSON, J. D.: Biochem. Biophys. Res. Com. 8, 175 (1962). — 3. KOCHAKIAN, C. D.: Proc. Soc. Exper. Biol. Med. (U.S.A.) 32, 1064 (1935). — 4. KOCHAKIAN, C. D.: In: A Symposium on Steroid Hormones. Ed. by E. S. GORDON. University of Wisconsin Press, Madison, 1950, p. 113. — 5. LOTTFIELD, R. B.: In: Progress in Biophysics and Biophysical Chemistry, Vol. 8. Ed. by J. A. V. BUTLER and B. KATZ. Pergamon Press, New York, 1957, p. 347. — 6. BERNELLI-ZAZZERA, A., M. BASSI, R. COMOLLI,and P. LUCCHELLI: Nature (G.B.) 182, 663 (1958). — 7. FRIEDEN, E. H., M. R. LABY, F. BATES, and N. W. LAYMAN: Endocrinology (U.S.A.) 60, 290 (1957). — 8. FRIEDEN, E. H., E. H. COHEN, and A. A. HARPER: Endocrinology (U.S.A.) 68, 862 (1961). — 9. SCOW, R. O.: Endocrinology (U.S.A.) 51, 42 (1952). — 10. GREER, D. S.: Endocrinology (U.S.A.) 64, 898 (1959). — 11. NOALL, M. W., T. R. RIGGS, L. M. WALKER, and H. N. CHRISTENSEN: Science (U.S.A.) 126, 1002 (1957).— 12. METCALF, W., and E. GROSS: Science (U.S.A.) 132, 41 (1960). — 13. HOAGLAND, M. D., M. L. STEPHENSON, J. F. SCOTT, L. I. HECHT, and P. C. ZAMECNIK: J. Biol. Chem. (U.S.A.) 231, 241 (1958). — 14. BRANDT, J. W. A., and A. V. NALBANDOV: Poultry Sc. (U.S.A.) 35, 692 (1956). — 15. ZAMECNIK, P. C., and E. B. KELLER: J. Biol. Chem. (U.S.A.) 209, 337 (1954). — 16. NIRENBERG, M. W., and J. H. MATTHAEI: Proc. Nat. Acad. Sc. U.S. 47, 1588 (1961). — 17. NOMURA, M., B. D. HALL, and S. SPIEGELMAN: J. Mol. Biol. (U.S.A.) 2, 306 (1960). — 18. DEANE, H. W., and K. PORTER: Zschr. Zellforsch. (G.) 52, 697 (1960).

Discussion

KASSENAAR: I should like to ask Dr. WILSON two things which I didn't quite understand. *Firstly*, am I right in thinking that you concluded from the experiment in which you added the RNA fraction that this had no effect? I got the impression from the slides that there was an increase of 137 to about 205 counts in the treated group and from 12 to 21 in the non-treated group. Don't you consider this difference significant? *Secondly*, in the experiments you did on the specific activity of the soluble RNA, did you express your data per unit of RNA? If so, don't you think that the amount of RNA in the microsomal fraction, as well as in the supernatant at the beginning of the incubation period, is so much different in the two groups that this might mask the results which were obtained?

WILSON: Let me answer the second question first, since that is easier. In the experiment in which soluble RNA was studied, the cell fractions were separated by differential centrifugation, so that we only measured soluble RNA. There was no difference in soluble RNA pools between the treated and the untreated tissues in our experiments, and we have done this a number of times. I did not show an experiment in which I measured pool size for soluble RNA, but we have been unable to demonstrate any difference in pool size as related either to protein or to total weight. Therefore, the values do represent a specific activity of soluble RNA amino acids. I would like to emphasise that our experiments were not performed on castrated animals.

Homogenisation of the seminal vesicle, as you and I have previously discussed, is very difficult if one attempts to produce viable homogenates. However, one can certainly grind them up completely if one is not attempting to get live preparations at the end. It is just a matter of the time of grinding. HOAGLAND et al.[1] had demonstrated that sRNA amino acids can be synthesised at 0° C, and so those reactions were stopped by the addition of large quantities of unlabelled amino acids, and then they were homogenised very carefully and separated by differential centrifugation. By this sort of technique we were unable to demonstrate any difference. The worst objection to the design of this type of experiment is that the total number of counts is so low that one always questions whether or not the specific activity calculations are meaningful or whether this RNA might be contaminated with some other fraction. I conclude from a whole series of other experiments that it probably is a valid approach, and I think that the best proof of this has been the confirmation of the results in the homogenate experiments.

The first question is much more difficult to answer. At this extremely low level of protein synthesis, increases from 12 to 21 counts per minute are probably not meaningful. I showed you the one experiment out of several which has the greatest effect, but it is impossible to get even that much

---

[1] HOAGLAND, M. B., M. L. STEPHENSON, J. F. SCOTT, L. I. HECHT, and P. C. ZAMECNIK: J. Biol. Chem. (U.S.A.) **231**, 241 (1958).

effect usually, and I doubt whether under these circumstances this difference has any significance. I have tended in these experiments to neglect differences in protein synthesis which are less than twofold, because it is quite obvious that the difference between the treated and the untreated preparations is at least a tenfold one; and in view of the fact that, even if the results were valid, one gets less than a 10% effect, I don't see how one can attach much physiological significance to such an effect. I do want to emphasise that I do not think that a negative experiment of this kind really is not meaningful physiologically.

ASCHKENASY: I should like to ask Dr. WILSON whether he has any explanation for the fact that the biochemical effects of androgens, such as stimulation of the conversion of RNA-amino acid complexes to microsomal ribonucleoproteins, are apparently limited to certain sensitive tissues and do not extend to others. Do you think that there might be some relationship between differences in tissue reactivity and differences in the availability of certain co-factors such as Mg or sulphhydryl compounds ?

WILSON: I don't know, but this is an important question, which confronts any physiological interpretation of such *in vitro* experiments. At one time I thought there was rather good evidence that the specificity of action might be due to differential rates of penetration of different hormones into tissues, particularly because GREER[1] had shown in the case of testosterone a selective concentration in the accessory sex tissues. I think one possible explanation for the effect on accessory sex tissues is that the concentration of any hormone in these tissues is likely to be higher, because the hormones can possibly reach the oviduct, the seminal vesicle, and the uterus directly from the gonads without entering into the general circulation, and the local concentration of these hormones may be much higher than one can determine in the systemic blood. But the question which you raise is obviously the reason why one should interpret with great caution any physiological implications of such an *in vitro* study.

KORNER: I was interested in the negative result you obtained, Dr. WILSON, when you added ribonucleic acid to your system. We found that great care had to be exercised not to allow the RNA to be degraded during extraction. Did you look at the size of the RNA that you extracted and added to your system ?

WILSON: We did perform some experiments on microsomal RNA in the oviduct; the Schlieren diagrams turned out very poorly, and these experiments will be repeated.

KORNER: There are three problems as I see it. Firstly, can one extract messenger RNA without destroying it ? Secondly, can one get it to fix on to the ribosomes ? Thirdly, does messenger RNA really exist in mammalian tissues ?

WILSON: Yes, and I emphasise that it is a negative experiment and may not be physiologically meaningful. We followed, I might add, the NIRENBERG and MATTHAEI modification of the NOMURA, HALL, and SPIEGELMAN technique[2] for purification of the microsomal RNA, and all one can say is that it has been unsuccessful in our hands.

[1] GREER, D. S.: Endocrinology (U.S.A.) **64**, 898 (1959).

[2] NOMURA, M., B. D. HALL, and S. SPIEGELMAN: J. Mol. Biol. (U.S.A.) **2**, 306 (1960).

KORNER: We have used the addition of polyuridilic acid and its ability to stimulate phenylalanine incorporation as a test system to see if we were getting the polynucleotide to attach itself to the ribosomes. Unfortunately, we have not yet been successful in obtaining stimulation of phenylalanine incorporation in the liver system with polyuridilic acid. I wonder if you have tried this in your system?

WILSON: No, I have not. It seems to me, since one has localised the influence of both growth hormone and steroid hormones to very nearly the same reaction of protein synthesis, that the question of whether or not messenger RNA exists in mammalian cells becomes a very crucial one. If you don't mind my asking you a question too, since you reintroduced this problem, it was not clear to me exactly what your present views are as to whether or not there is a messenger RNA system in mammalian cells. I had always thought previously that one of the best pieces of evidence against the messenger hypothesis in mammalian cells is the fact that mammalian homogenates, so far as I am aware, have never been reported to be sensitive to inhibition by chloramphenicol. The oviduct homogenates are not sensitive to inhibition by chloramphenicol, and I feel that the best evidence is that chloramphenicol interferes with the joining of messenger RNA to the ribosome. Consequently, I had always assumed that the evidence for a free messenger RNA in the mammalian system was very poor, but I was impressed by ARNSTEIN's[1] report in last week's Nature that there is a messenger RNA system for rabbit reticulocytes and that polyuridilic acid has a phenomenal effect on phenylalanine incorporation in that system. Exactly what is your present view on this matter?

KORNER: My verdict at the moment on the existence of messenger RNA in mammalian tissue is: "not proven". We are working hard to see if we can either prove or disprove it. Some of our results support the messenger theory, others do not. This theory was originally postulated for bacterial tissues, where the bacterium has to be able to adapt and start making new types of protein very rapidly. With mammalian tissues this adaptability is less important, so it is quite likely, as Dr. DREYFUS mentioned earlier, that the messenger RNA, if it exists in mammalian tissues, is much more stable than has been postulated to be the case in bacterial tissue. Now, if messenger RNA is more stable in mammalian tissues, it might be that when one prepares ribosomes, there are still enough living messengers stuck on to the ribosomes to carry out a little protein synthesis in the test-tube. Whether messenger RNA, floating about in the sap, can be attached to the ribosome, is, I think, the crucial question. You have mentioned ARNSTEIN's results on polyuridilic acid mammalian ribosomes. We have not, so far, been able to get polyuridilic acid to stick on to liver ribosomes and stimulate incorporation, but this may be a question of conditions. In the experiments I described, where RNA from liver ribosomes stimulates incorporation of amino acid into the ribosomes from hypophysectomised rats, we had to work quite hard to get the RNA to stick on. We heated the RNA and then added the ribosomes in quite a high magnesium concentration — much higher than is optimum for incorporation — and then we pre-incubated for ten minutes or so to give the RNA a chance to stick on to the ribosomes. I don't know whether all these precautions are important, but they appeared to work.

[1] ARNSTEIN, H. R. V., R. A. COX, and J. A. HUNT: Nature (G.B.) **194**, 1042 (1962).

WILSON: It is not only a semantic difference to say that the messenger is more stable in the mammalian systems, because the very definition of messenger is that it is free RNA, and if one says that it is more stable or that it is already in the ribosome, then what one really has is ribosome synthesis and not free messenger synthesis.

KORNER: I think there may be an eventual synthesis between these two opposing doctrines. The exponents of the messenger hypothesis are already modifying their original view that each messenger RNA is destroyed as soon as one polypeptide chain is synthesised on it. It is agreed that the messenger RNA, if it exists in mammalian cells, must be more stable than this. Either a compromise will be agreed upon between the permanent-template exponents and the "messengerists" or we must accept that an important difference exists in this regard between mammalian and bacterial cells.

Histological aspects of the action of androgens and oestrogens

By

J. A. SZIRMAI

I. Introduction

For generations endocrinologists have been puzzled by the sensitivity of certain organs, tissues, or cells to various hormones. At present we are still unable to explain the basis of this specific responsiveness and to define in scientifically acceptable terms what we really mean by "secondary sex characteristics", "target organs", "end organs", or "sexual tissues". Although the study of the mode of action of various hormones and of the mechanism of the response at the cellular level might provide us with a wealth of analytical data, this approach leaves at present unanswered the crucial questions concerning specific sensitivity.

The great diversity of the types of response of closely related tissues to the same hormone, or the analogous effects of various hormones on seemingly unrelated structures, the marked differences in sensitivity of homologous tissues in different species, and the interdependent influences exerted by the entire endocrine system on these reactions — all these aspects present an intricate but fragmentary mosaic which resists any attempt at facile generalisations. Yet generalisations will have to be made if we are to find a common denominator for the effects of a given hormone and thus to arrive at a better understanding of its action(s) and function(s). Generalisations — in spite of their obvious dangers — might help to reveal gaps in our knowledge or point out discrepancies between observed facts, but they might also reconcile seemingly contradictory and paradoxical observations.

Considering the action of hormones at the tissue level — rather than at the level of cells or organs — already implies a generalisation, since it involves selecting certain aspects and neglecting others. Nevertheless, this histological approach has certain advantages: it furnishes sufficient details concerning the cellular reactions involved, while being universal enough to demonstrate the extent to which the various tissues are participating in the reaction of the whole organism to a hormone. Furthermore, examination of the hormonal response of tissue systems and extension of these observations to various species of vertebrate, reveals certain facts concerning the histogenesis of the responsive tissues, which might give some clue to the determination of the specific responsiveness at an early stage of embryonic development.

The tissues considered below — certain connective tissues, epithelia, and muscle — belong to what would generally be designated as "target" tissues. However, this survey of their response to androgenic or oestrogenic hormones will not lead to a useful definition of the above term. On the contrary, the great variations in the degree of sensitivity will clearly illustrate the difficulty of distinguishing between a "target" and a "non-target" tissue. In spite of this, the term "target" will be used, but merely in an operational sense to indicate a relatively high degree of sensitivity in a tissue to a given hormone.

II. Connective tissues

1. **Comb and other head furnishings in birds.** One of the first recognised targets in the history of endocrinology was a connective tissue organ, the rooster comb, which was shown to depend on the presence of the testis by BERTHOLD in 1849 (*13*). His remarkable experiment, involving castration of six young roosters with resulting regression of the comb and its restoration by testicular implants, remained unnoticed for more than 60 years. Similarly, the description of a so-called mucoid connective tissue (*"Schleimgewebe"*) in the rooster comb by VIRCHOW in 1851 did not receive any attention for at least the same period of time (*131*). Mainly through the studies of PÉZARD (*102, 103*) and of CHAMPY and co-workers (*26—31*) these two observations have been rediscovered and linked together, forming the basis for the use of the capon or chick comb in the bioassay of androgens. This practical use prompted many investigations into the factors influencing the comb response (for review, see *23, 41, 100, 103, 128*). The dependence on androgens of many other head furnishings was also demonstrated, including the wattles and ear lobes in roosters (*28, 123*), the eye patches in pheasants (*27, 31, 107*), the corunculated skin areas in turkeys, the wattles in the guineafowl, and many other skin formations in various birds (*49, 79, 101*).

Early histological studies by CHAMPY and KRITCH (*29*) indicated that the connective tissue of the dermal layer of the comb represents the structure responding to androgenic stimulation, and this has been confirmed by several subsequent investigations (*47, 52, 57, 80, 123, 124, 125, 126*). On the basis of these studies the histological

features of this tissue can be summarised as follows: in the comb of
the normal adult rooster the dermal layer (also designated mucoid
or intermediate layer) has a considerable thickness (up to 10 or
15 mm. in the White Leghorn breed) and consists of oedematous,
loose connective tissue. The thin fibrils of the collagen network and
the fibroblast-like cells are widely separated by large extracellular
spaces, filled with a viscous and highly water-soluble ground
substance (Fig. 1 a). Application of selective stains in histological
studies suggested that this substance was of a polysaccharidic
nature, giving the tissue an appearance in many respects similar to
that of the so-called Wharton jelly of the umbilical cord. Castration
will result in the gradual disappearance of the extracellular spaces,
condensation of the fine collagen fibrils into dense bundles, and
complete loss of the stainability of the ground substance (Fig. 1 b).
The reversibility of these changes is illustrated by restoration of the
normal structure following androgen treatment in capons: the
shrunken comb starts to increase in size in one or two days, owing
mainly to the accumulation of fluid in the dermal layer, the collagen
fibres split up into thin elementary fibrils, and the extracellular
space becomes filled with increasing amounts of ground substance.
The fibroblasts show some enlargement and perhaps increased
basophilia; cell divisions were observed in the connective tissue of
the comb in androgen-treated chicks (*80*), but not in the case of
treated adult capons (*126*). Concomitant changes in the course of
treatment with androgens include increased vascularisation,
ectodermal proliferation, and decreased thickness of the keratin
layer — the latter in contrast to the thick horny layer covering
the capon comb.

Further elucidation of the nature of the comb reaction by
means of histochemical and chemical studies was provided with a
firm basis by the work of Boas (*14*), who demonstrated the presence
of large amounts of hyaluronic acid in the rooster comb. This
suggested that this mucopolysaccharide might be the main component of the ground substance in the mucoid layer and be
responsible for its staining characteristics, and also that accumulation of this compound might represent an essential feature of the
androgenic action. Combined histochemical studies utilising the
metachromatic staining of mucopolysaccharides by cationic dyes
and determinations of the total hexosamine content of the tissue
(as an approximate measure of the mucopolysaccharide content)
showed good agreement in both testosterone-treated cockerels (*113*)
and testosterone-treated capons (*124, 125, 126*). In the latter studies
it was shown, indeed, that the very low hexosamine content of the

capon comb corresponds to the absence of metachromatic staining
in the dermal layer, and that the gradual increase of the hexosamine

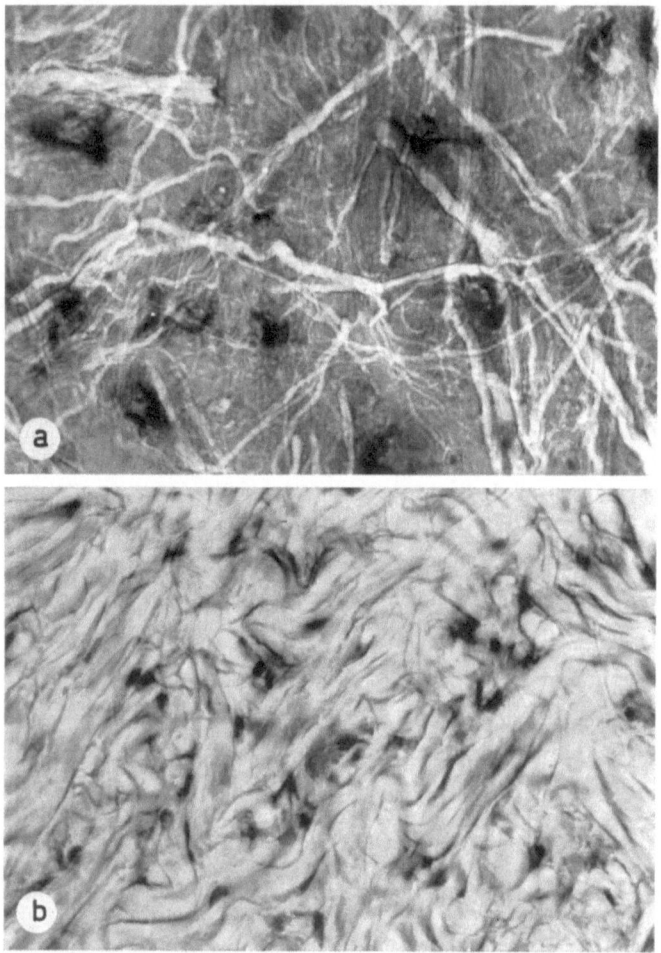

Fig. 1. (a) Structure of the intermediate layer of the comb of a normal adult rooster. Unfixed frozen section, stained with azure A to show the (dark) interfibrillar ground substance,
magnified approx. 400 ×. (b) Structure of the corresponding area of the comb in the capon,
showing coarse collagen bundles and absence of interfibrillar ground substance. Same technique and magnification as in (a). [From Szirmai (128), with permission of the Acad. Press
Inc., New York]

content of the comb, reaching or even exceeding the normal level, can be related to the gradual development of metachromasia in the extracellular space. The wattles, histologically identical with the comb, react in the same way but to a lesser extent, whereas the ear lobes show only traces of metachromatic staining in the very dense dermal layer and no significant increase in hexosamine content (Fig. 2).

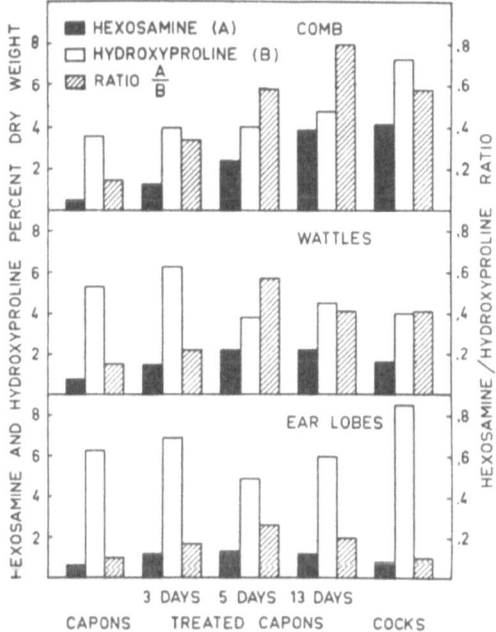

Fig. 2. Hexosamine and hydroxyproline content of the comb, wattles, and ear lobes in testosterone-treated capons (daily dose: 2 mg. testosterone-propionate). For comparison, the corresponding values in untreated capons and normal roosters are given. (From SZIRMAI (126), with permission of the Society for Experimental Biology and Medicine)

Detailed chemical studies provided further proof that hyaluronic acid is the principal compound involved in this reaction. Hyaluronic acid and chondroitin sulphate B (dermatan sulphate) were shown to be present in a ratio of 15 : 1 in extracts of epithelium-free comb tissue (114). In the dissected mucoid layer of the rooster comb, chondroitin sulphate was present in a concentration of 0.1 to 2.0% of the total mucopolysaccharide (11, 42); in the latter studies some evidence was obtained indicating that the chondroitin sulphate might be structurally bound to the collagen fibrils and not present at all in the interfibrillar ground substance. Finally, the total hyaluronic acid has been quantitatively isolated and determined in a large series of growing cockerels, in testosterone-treated capons, and in regressing capons, i. e. after cessation of hormone treatment (*10, 43*). In all these cases a nearly linear correlation was found between the weight or size of the comb and its hyaluronic acid content (Fig. 3). Bearing in mind the practically linear response of the comb size to androgens, as apparent from the many examples from bioassay work, the above data would indeed mean that accumulation of hyaluronic acid in the comb and the other head furnishings is the principal result of the androgenic action on these tissues. The high water content of the comb (up to 90—95%) and its dependence on androgens should be mentioned in this connection (*125, 129*); the positive correlation found between the hyaluronic acid and water content (*43, 126*) seems to indicate the significance of this mucopolysaccharide for the binding of water in these tissues.

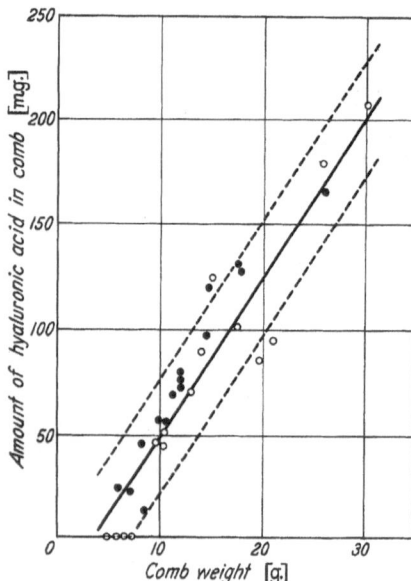

Fig. 3. Correlation between the total hyaluronic acid content and the comb weight in White Leghorn capons aged 9 months (○) and 21 months (●). Determinations were made after regression periods of 3 to 24 days, following 15 days of treatment with testosterone-propionate (1 mg./day). The dotted lines indicate the standard deviation of the calculated regression line ($r = 0.94$). [From DOYLE, SZIRMAI, and DE TYSSONSK (*43*)]

Once the accumulation of hyaluronic acid in the connective tissue of the comb had been established as the principal result of androgenic action, the obvious step was to investigate its source of origin. Cytological studies at the light-microscope level failed to supply any evidence that the cellular components participate in this process (*52, 53, 127*). Isotope studies using C^{14}-labelled glucose revealed that this can be utilised as a direct precursor and is rapidly incorporated into the hyaluronic acid of the comb (*9, 10, 12*).

The biological half-life of the hyaluronic acid in the comb of adult White Leghorn roosters was found in these studies to be about 4 days; the rate of incorporation was highest in a series of growing cockerels at the age of 3 to 4 months, which is the period of maximal growth rate of the comb, whereas testosterone treatment markedly increased the specific activity of the hyaluronic acid in both normal and castrated animals. Autoradiographic studies showed that C^{14} is, in fact, located in the intermediate layer (*12*), and studies using incubation of comb slices under tissue-culture conditions in media containing C^{14}-glucose indicated that the cells of the connective tissue of the comb are capable of synthesising hyaluronic acid (*8*). Although the above studies did not reveal the site at which androgens influence the biosynthetic pathways involved, they suggest that local stimulation of the biosynthesis of hyaluronic acid is the chief biochemical parameter of this characteristic hormone-induced reaction of connective tissue.

Although androgens are evidently the causative agent of the comb response, a great many factors have been recognised to influence or modify its extent considerably. Such factors involve effects of other endocrine organs, such as the adrenal, pituitary, and thyroid glands, food intake, vitamins, effects of fatigue, lighting conditions, and seasonal variations, in addition to factors such as differences in the sensitivity of various breeds (for review, see *128*). Though such influences might ultimately be reflected in the metabolism of hyaluronic acid, it appears unlikely that they represent direct effects. One exception is perhaps the antagonistic action of oestrogens (*2, 15, 20, 72, 86, 87*), which might be interpreted as a direct effect at the target site (*81, 85*).

2. Sexual skin in monkeys. Another connective tissue, similar to that of the rooster comb in structure and reactivity, is present in the areas of the perineal skin in several genera of primates (baboon, mandril, mangabey, macaque, and other monkeys and apes). Detailed studies of the anatomy and structural variations of the sexual skin in relation to the reproductive cycle have been made (*1, 46, 54, 59, 136, 139*); the histological, histochemical, and biochemical aspects of this tissue reaction have been reviewed in greater detail elsewhere (*128*).

The development of the sexual skin is a periodic phenomenon related to the menstrual cycle: the swelling occurs in the follicular phase and subsides in the luteal phase. Histologically, the swollen sexual skin is characterised by an oedematous dermal layer, consisting of a loose network of collagen fibrils separated by large amounts of extracellular material, with positive histochemical tests for mucopolysaccharides (*4, 5, 36, 44*). The belief that this tissue is analogous to the rooster comb has also been substantiated by chemical studies — e. g. the early investigations into the chemical properties of the so-called exudate, obtained by squeezing and extracting chopped sexual skin, which indicated the polysaccharidic nature of this material (*25, 74, 97*). Further identification of the mucopolysaccharides of the "active" sexual skin in various species (baboon, rhesus monkey, pig-tailed monkey) showed the main component to be hyaluronic acid, with small admixtures of chondroitin sulphate(s) (*109, 110*); a positive correlation between the hyaluronic acid content and water content of the sexual skin was also observed in these studies.

In the "inactive" sexual skin, i. e. in the luteal phase or in the prepuberal or castrated animal, the amount of extracellular ground substance is markedly reduced or altogether absent, the dermal layer becomes coarse, and there is a relative increase in the number of cells per unit area. This process again is analogous to what can be observed in the head furnishings of birds. Owing to the large amounts of tissue involved in this reaction of the sexual skin, some effects can be observed on the whole organism, such as fluctuation in the body-weight parallel to the water retention in the phase of swelling and increased volume of the urine upon subsidence (*74*). Changes in the circulating mass of plasma proteins have also been observed and interpreted as indicating that plasma proteins become lodged in the intercellular fluid of the sexual skin during the follicular phase of the cycle (*33, 34*).

The morphological and biochemical parameters of the response of the sexual skin seem to be identical to those of the rooster comb. A puzzling difference is the causative effect of oestrogens in the case of the sexual skin (*1, 136*) and the clearly antagonistic action of progestational and androgenic steroids (*53, 54*). This is true not only for the female, which normally possesses the prerogative of the sexual skin, but also for the male, in which it is lacking but can be produced by oestrogen treatment (*5, 53, 139*). One should recall in this connection that the comb in fowls is maintained by androgens and inhibited by oestrogens in both sexes.

3. Other "steroid-sensitive" connective tissues. The comb in roosters and the sexual skin in monkeys are two illustrative examples of the reaction of target connective tissues with an entirely analogous reaction pattern, but with two distinct steroid hormones as the stimulating agent. This makes it impossible to qualify such reactions as typically androgenic or oestrogenic. While one may wonder whether these two steroids act directly and in the same way in the chain of events resulting in the increased biosynthesis of hyaluronic acid in these connective tissues, all that one can do is to assume a trigger effect on the part of one or other of these hormones and designate such tissues in a very general way as "steroid-sensitive", in contrast to similar tissues without such sensitivity (*124, 128*). These seemingly paradoxical facts appear less surprising as more and more similar tissues are recognised to react in an identical way to either androgens or oestrogens, independently of the sex of the subject.

Such target connective tissues are widely distributed in the animal kingdom and occur either as parts of the integument or as components of the reproductive organs. Examples of the first group are the incubation patches of the abdominal skin in several birds (*7, 49, 76*), the dorsal crest and cloacal labia in salamanders (*3, 21, 30, 88*), various areas of the skin in fishes and amphibia (*40, 78*), and the skin around the external genitals in various mammals (*23*). To the second group belong connective tissues of the male urogenital tract, such as the stroma of the seminal vesicle, vas deferens, prostate, and urethra (*22, 23, 60, 105, 133*), and parts of the female reproductive organs, such as the stroma of the oviduct in fishes and amphibia (*40*), birds (*16, 100*), rodents (*117*) and various domestic animals (*35*), as well as the stroma of the uterus and partly also of the vagina in a variety of animals (*23, 35, 82*). In all these cases the connective tissue is normally influenced either permanently or periodically either by oestrogens (in some cases progesterone and prolactin increase the oestrogen effect) or by androgens, resulting in a development of what can be described histologically as a mucoid connective tissue. Withdrawal of the hormone, castration, or seasonal inactivity results in the reversal of this structure to the dense, compact, fibrillar connective tissue.

4. **Location and histogenesis of the "steroid-sensitive" connective tissues.** There is an analogy not only in the structural response of the target connective tissues briefly mentioned above (for a more extensive discussion, see *128*), but also in their location in certain areas of the body, which reveals a systematic pattern (Fig. 4). In the skin, the sensitive areas are located in the facial and cranial region, whereas another group is situated around the external genitals. Furthermore, this target type of connective tissue may be found in the urogenital tract of both the male and female. The schematic representation of these tissues given in Fig. 4 does not by any means indicate that in all vertebrate species these areas exhibit the same degree of sensitivity or that they have to occur simultaneously in a given species. It serves merely to point out the

remarkable location of such targets — if present — at the caudal and cranial parts of the body, surrounding the main body openings, and further that connective tissues of all reproductive glands or ducts in both sexes can exhibit such reactivity.

Several observations suggest that the sensitivity of these areas to a given steroid hormone is located in the tissue itself, as illustrated by the results of transplantation experiments in roosters (24, 73) and monkeys (6, 139). In most cases the sensitive areas are sharply demarcated. Considerations of the histogenetic origin of the tissues

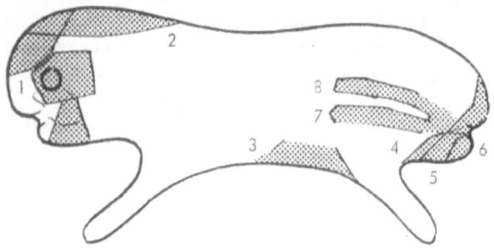

Fig. 4. Diagram showing the location of "steroid-sensitive" connective tissues in various vertebrates. [Modified from SZIRMAI (124)]. 1. Head furnishings (birds); 2. Dorsal crest (amphibia); 3. Incubation patches (birds); 4. Copulatory organs (birds, mammals); 5. Scrotum, vulva, cloaca (various vertebrates); 6. Sexual skin (monkeys); 7. Female genital tract (various vertebrates); 8. Male genital tract (various vertebrates)

involved would suggest, first of all, that target connective tissues are found in all parts of the Wolffian and Mullerian duct derivatives as well as in the mesodermal parts of the original sinus urogenitalis (see also Fig. 4). Concerning the skin, ectodermal parts are involved around their invaginations into the body cavities and their surroundings. Although a single observation cannot substantiate a hypothesis to any extent, it is remarkable that such skin areas, namely around the face and the external genitals, show peculiar structural changes in early embryogenesis in the human (120, 121). Their extension gradually decreases during later embryonic life, and only parts of the orbit, lips, and vulva or scrotum retain structural differences at birth in both ectoderm and connective tissue. The variability observed in the response of the sexual skin in monkeys, such as its extension to the orbital parts and face following oestrogen overdosage (6), or generalisation of the oestrogen response of this type in genetically abnormal mice (118), are examples of expansion over the borderline of sensitivity. Normally the "non-target" areas of homologous tissue do not react to either hormone with the typical response: compared with the markedly increased hexosamine content of the comb, there is no such increase in the remaining skin of the testosterone-treated chick or capon (*113, 124*). The sensitivity of these target tissues to steroid hormones, as well as their degree of responsiveness, seem to be determined in the course of their histogenesis, although other factors might modify the extent of the final response.

III. Epithelia

1. Male reproductive organs. Besides the capon comb, employed in early endocrinological studies, other targets have also been utilised for bioassay. Such examples of glandular tissues, where the main response is due to structural changes of the epithelium, are the seminal vesicle or the prostate of rodents, shown to depend in structure and secretory activity on the presence of gonads or androgenic stimulation (*23,41,84,105*). A recent review by PRICE and WILLIAMS-ASHMAN (*105*) summarises in admirable detail the structural changes of these organs upon administration of androgens and oestrogens, and reference is made to this work for further data. In general, it would appear that these glands respond — due allowance being made for great variations in species sensivity, age, and other influences — to androgens with development of the secretory apparatus and maintenance of the secretory process. In histological terms, this means the presence of a columnar type of epithelium with recognisable signs of the secretory process, such as secretory granules, a well-developed Golgi apparatus, often oriented towards the apical surface, and marked cytoplasmic basophilia. Castration leads not only to cessation of the secretory activity but also to a marked regression of the tissue structure: the epithelial lining becomes low, the cells flat or cuboidal, the secretory granules disappear, and there is marked reduction or loss of cytoplasmic basophilia and other histochemical characteristics associated with secretion.

In recent years further structural details have been revealed through the use of the electron microscope, and studies have been made on the normal ultrastructure of these glands in various laboratory animals such as the rat (*17, 58, 115, 130*), mouse (*18, 19, 39*), or dog (*116*). In general these studies (for review, see *105*) shed light on several details associated with the secretory function: microvillous surface structure, extensive Golgi complex, in some instances clearly associated with the secreted material, and in most cases an elaborate system of rough-surfaced endoplasmic reticulum with a dense ribosomal population in various parts of the epithelial cell. Castration affects these organelles in most of the cases investigated, and results mainly in changes involving the Golgi complex and the

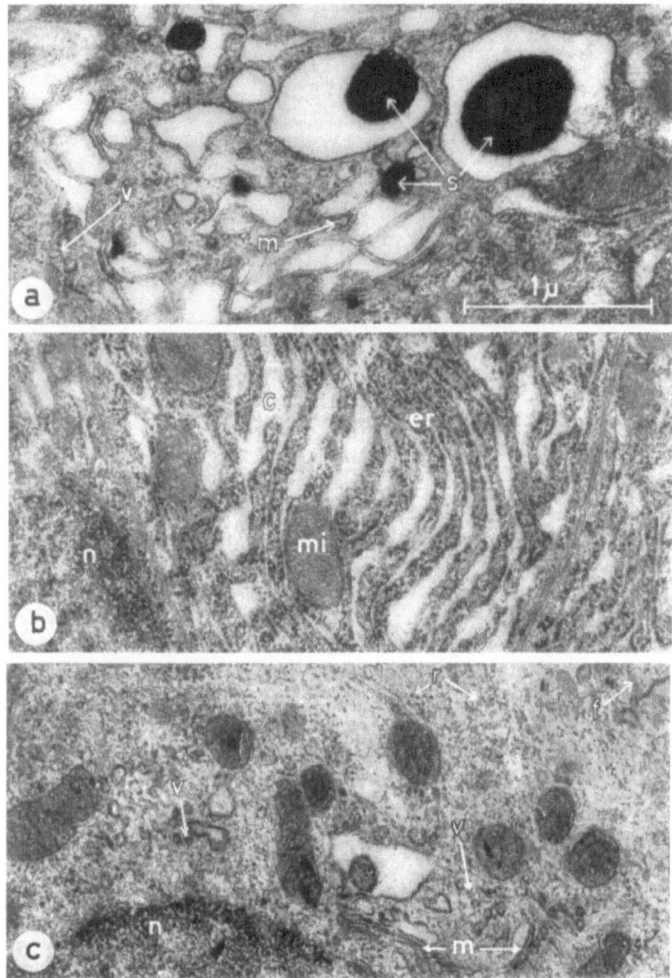

Fig. 5. (a) Seminal vesicle epithelium from a normal rat (4 months old). Supranuclear region of the cell. Arrows indicate secretory granules (s), Golgi membranes (m), and small Golgi vesicles (v). (b) As in (a), perinuclear region, showing rough-surfaced endoplasmic reticulum (er) with cisternae (c) and some mitochondria (mi). Nucleus = n. Magnification as in (a). (c) Seminal vesicle of a rat (same age), castrated at 7 weeks. Apical portion of a cell with part of the free surface (f) and of the nucleus (n). Note involution of the Golgi membranes and vesicles (m, v), disorganised endoplasmic reticulum, and clusters of free ribosomes (r). Magnification as in (a). [From Szirmai and van der Linde (130)]

endoplasmic reticulum. Early changes have been observed within 24 hours of castration, and these include reduction of the microvilli, collapse of the endoplasmic reticulum sacs and of the Golgi vesicles, and reduction — to a varying extent — of the ribosome particles. These changes will obviously vary greatly depending on the organ,

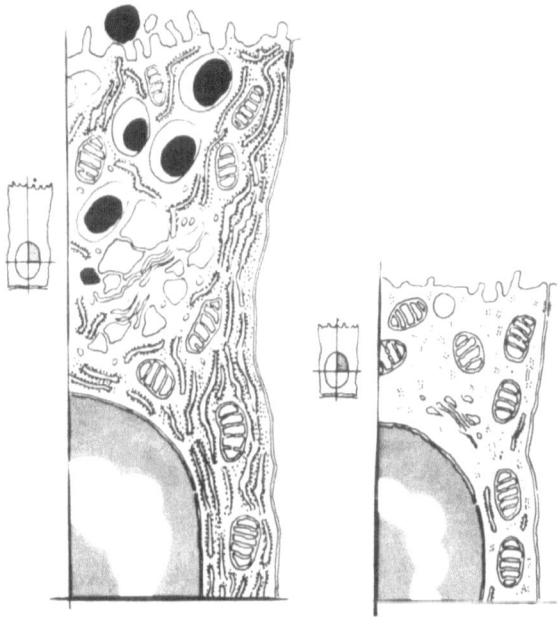

Fig. 6. Diagram showing the main features of the ultrastructure of the seminal vesicle epithelium in the normal rat (left) and its changes upon castration (right), based on the electron-microscopic findings illustrated in Fig. 5

species, time of castration and its duration, as well as on other conditions. As an illustration of what nevertheless might be the principal feature of the androgenic effect, some details of the ultrastructure of the rat seminal vesicle in the normal and castrated state are shown in Fig. 5. A summary of these changes is given in a diagram (Fig. 6), which may serve as a general illustration showing the reversible changes of these epithelia as determined by the presence or absence of androgens.

The practical application, in bioassay work, of the changes in weight or in histological structure of these reproductive glands upon androgenic stimulation provided many data on additional factors influencing this reaction. The scope of this review does not permit us to give a detailed description of them, and mention will be made only of the general antagonistic effect of oestrogens. Oestrogens suppress or prevent the secretory process and the structural development of the secretory apparatus of these epithelia, and can lead to an atrophy similar to that produced by castration, or they may even produce metaplastic changes of the squamous type to be discussed below (*22, 23, 71, 105*).

2. Female reproductive organs. Whereas oestrogens seem to have — as a broad generalisation — an inhibiting action on the epithelia of the male reproductive glands (exceptions, such as the secretory type of response to oestrogens in the prostate of the cat [119] have been noted), oestrogens in general exert a stimulant effect on the glandular epithelia of the female genital tract. This applies to the glandular parts of the oviduct in practically all vertebrates, where the secretory activity is induced and maintained chiefly by oestrogens, although progesterone and in some cases androgens may act synergistically (for fishes and amphibia, see *40, 55*, for birds *64, 111*, for rodents and domestic animals *23, 35*). Similar changes are produced, mainly owing to oestrogenic stimulation, on the uterine epithelium in many species, including human beings, and appear to be in several aspects analogous to the androgen-induced types of secretory response. This is true not only at the level of the light microscope, but also with respect to the ultrastructure. As an example we would quote the studies of Nilsson (*89—94*) on the effect of castration and oestrogen treatment on the uterine epithelium in mice, demonstrating changes in the Golgi complex, endoplasmic reticulum, and cell surface. These alterations are practically identical to those observed in the male organs. This type of response, conveniently called "glandular" (*137*) or secretory, might also extend to areas of the cervix or the upper part of the vagina. This development differs from the response of the male organs in its periodical fluctuations with the reproductive cycle; the alternating influence of progestational or androgenic steroids might counteract the oestrogenic effect or modify the final response by exerting a genuine action.

If we examine the above changes from a histological point of view, we arrive at the conclusion that an analogous differentiation — or rather modulation — of the epithelia of the reproductive glands can be obtained by either androgens or oestrogens, with some degree of relationship to sex: androgens produce the glandular response mainly in the male organs, whereas oestrogens act in the same way on female structures. This general statement is complicated not only by the synergistic action of androgenic or progestational steroids, but also by the other types of response which oestrogens can elicit in structures belonging to the urogenital tract. A different response of this kind has been designated as the "squamous" type (*137*) and is apparently confined to the vagina and parts of its internal and external surroundings in a variety of mammals (*99, 135*). This response has also been utilised in bioassay work and represents a different effect, at first glance opposite to the glandular type described above. This apparent divergency is questionable in view of some general biological considerations of the epithelial structure and function. The squamous type of response is nothing but a somewhat different secretory process and, as such, equivalent to the glandular type. Considering that both involve biosynthetic processes (specific proteins and other compounds in the glandular type, keratin in the squamous type), this would imply similarities of the secretory machinery in these cells. Although they show some distinct features as regards endoplasmic reticulum and free ribosomes, the "secretory" and "retaining" types of epithelial cell (*83*) are fundamentally equivalent. Similarly equivalent might be the seemingly divergent actions of a hormone, and one has to accept that certain sensitive epithelia respond to oestrogens by preference with a glandular reaction, while others favour the squamous type. Both responses can in many cases be counteracted by androgens and depend on various endocrine factors, as well as, to a considerable extent, on age and species. Differences in the degree of response can also be observed, such as the most marked squamous reaction of the lower vagina in comparison to the gradually decreasing extent of keratinisation at more distant areas (*137*).

3. Other ectodermal and glandular epithelia. In addition to the epithelia of the male and female reproductive system, whose response to both androgens and oestrogens has been examined above, there are many epithelial structures outside these organs which react in a strikingly similar way. Examples of such reactions are the keratinised structures, recognised as secondary sex characteristics, on the head of fishes, reptiles, and amphibia (*40, 95*), probably controlled by androgens, the ectodermal glands in various areas of the integument (fingers, thigh, abdomen, cloaca) in amphibia and reptiles (*40, 95*), or various epithelial structures of the mammary gland, reacting to both androgens or oestrogens and exhibiting either the glandular or the squamous type of reaction (*23, 70*). All these structures can be lumped together under the heading of secondary sex characteristics. Seemingly much less sex-related epithelial structures have been observed to exhibit certain degrees of response to steroids, mainly to androgens — such as the serous tubules of the submaxillary gland in mice and rats (*48, 56, 65, 75*) and various kidney structures. The latter include the so-called sexual segment of the mesonephros in fishes, amphibia, and reptiles (*40, 106*); probably the (metanephric) kidney elements, showing hypertrophy upon treatment, should be mentioned as well in this respect, although the response itself in this case cannot properly be qualified as glandular. Whereas in most cases these reactions, whether of the squamous or of the glandular type, are induced by androgens, there are examples of oestrogenic effects leading to keratinisation outside the genital region (*137*).

4. **General features of the androgenic and oestrogenic action, location, and histogenesis of the "steroid-sensitive" epithelia.** When attempting to summarise under an integral concept the manifold effects of androgens and oestrogens on epithelial tissues, one has to consider the intrinsic capacity of the epithelial cell to modulate in several directions — the squamous and glandular being only two of the possible reactions — together with the location of these partic- ular tissues and their histogenetic origin. The two main types of response can be represented in an oversimplified way (Fig. 7, upper part), leaving undecided whether such modulations are entirely reversible in both directions or can develop only from an indifferent germinal cell. Considering such modulations with respect to androgens, it would appear that they lead in the majority of cases to the glandular type of reaction, although occasionally a squamous response is also elicited. Oestrogens do stimulate the same develop- ment in many structures, but result more frequently in a squamous response. The fact that in many cases a probably direct antagonism of these two hormones is mutually present can, for many examples, be traced from the direction of the arrows in Fig. 7. The synergistic action of androgens with oestrogens on certain structures, such as the oviduct or other derivatives of the Mullerian duct, is also noticeable in this diagram.

The diagram opposite (Fig. 7) was partially inspired by the con- siderations of ZUCKERMAN (*137, 138*), who noted that the squamous or glandular type of response to oestrogens is principally restricted to two regions: derivatives of the Mullerian duct (oviduct, uterus, cervix) respond with the glandular type, whereas derivatives of the urogenital sinus (vagina and part of the external genitals) favour the squamous reaction. This seems to be valid for derivatives of these embryonic structures in both sexes, shown in a diagram in Fig. 8. Considering that the prostate is derived from the urogenital sinus, this will explain the squamous response of its epithelium to prolonged treatment with oestrogens, observed in several rodents (see *23*). These considerations also support the view that the vagina is derived rather from the urogenital sinus than from the Mullerian

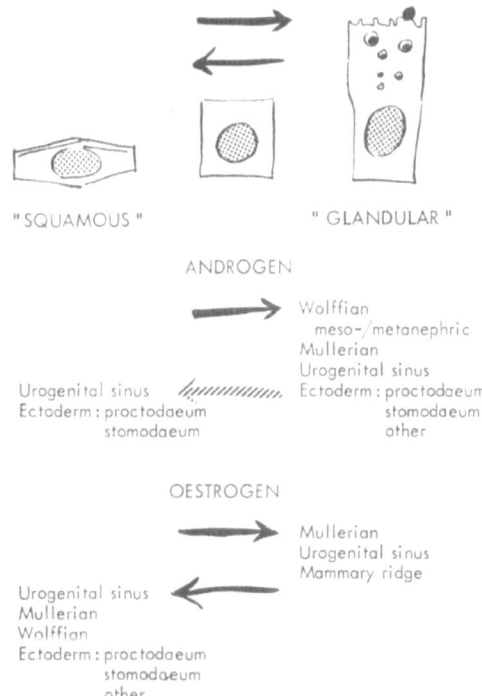

Fig. 7. Schematic representation of the two principal epithelial modulations (the squamous and glandular type) resulting from androgen or oestrogen treatment, with the histogenetic derivation of the responsive tissues

duct, in contrast to earlier opinions (*134, 138*). The suggestions of ZUCKERMAN (*137, 138*), concerning oestrogen-sensitive tissues only, can easily be supplemented by similar indications regarding androgens. It will then appear that the glandular type of response is found in derivatives of the Wolffian duct (vas deferens, seminal vesicle) and also in derivatives of the urogenital sinus (prostate) or of the Mullerian duct (oviduct, uterus), the androgens acting in the latter case at least synergistically, rather than antagonistically, with oestrogens. This "amphisexual" nature of the Mullerian duct and its derivatives has been noted in several studies in various vertebrates; in many cases its responsiveness to one or other steroid also depends on species, age, or critical stage of development (77). That mesonephric derivatives, such as the sexual segment of the

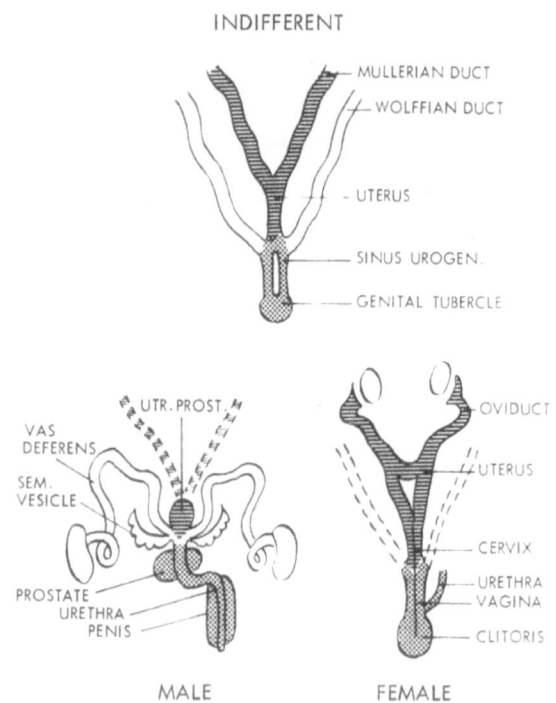

Fig. 8. Diagram showing the histogenesis of the male and female reproductive organs from their indifferent embryonic precursors. [Modified from PONSE (104)]

kidney in lower vertebrates, exhibit sensitivity for androgens will not be surprising in view of these considerations, and the same might apply also to the responsiveness of metanephric structures.

The reactivity of the integumental ectoderm constitutes a more complex problem from the point of view of histogenesis. Whereas the reproductive organs described above are derived from the two closely related gonaducts and their regions of contact with the neighbouring urogenital sinus (see Figs 8 and 9), many of the other responsive epithelia have no relationship to these regions. It appears, however, as already noted by ZUCKERMAN (137) and as also found for "steroid-sensitive" connective tissues (124), that the caudal and cranial regions — in addition to the mammary gland — are the principal sites of occurrence of epithelial targets. These areas are derived from the ectodermal parts invaginating towards the blind endings of the cloacal and oral cavities, i. e. the proctodaeal and stomodaeal ectoderm. In both areas, exact demarcation between the entodermal or mesodermal derivatives is difficult, as illustrated by the always doubtful margin

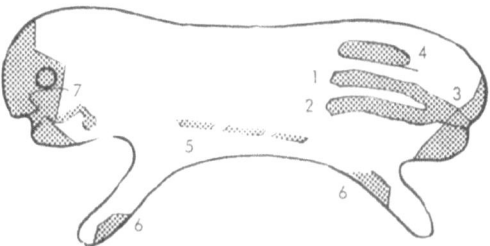

Fig. 9. Diagram showing the location of "steroid-sensitive" epithelia in various vertebrates, with their histogenetic derivation. 1. Mullerian duct derivatives; 2. Wolffian duct derivatives; 3. Derivatives of urogenital sinus and proctodaeal ectoderm; 4. Mesonephros and metanephros; 5. Mammary ridge derivatives; 6. Various ectodermal glands; 7. Derivatives of stomodaeal ectoderm

at the vagina-cervix junction. The endocrinological evidence, i. e. the squamous response to oestrogens, could in the case of the vagina be utilised to refute earlier anatomical and embryological dogmas regarding the Mullerian origin of this organ: it is in fact probably a Mullerian structure only in the anlage, becoming gradually covered by the invading epithelium of the urogenital sinus and possibly even the surrounding ectoderm. Such events illustrate the difficulties of these histogenetic inferences and motivate some doubts about the accepted rigidity in our concept of homologies of the various reproductive organs, especially when faced with paradoxical actions of hormones.

The summary of the above data in a diagram, representing the location in various vertebrates of what could be called "steroid-sensitive" epithelia, is merely an attempt to find a common denominator for the sensitivity of these targets (Fig. 9). The analogy in location and certain correspondences in histogenetic origin would indicate that further investigations in this area might throw some light on how the responsiveness of these epithelia is determined. That they are in many respects located similarly to target connective tissues (see Fig. 4) might be another argument in support of a common histogenesis. Again, this schematic presentation should by no means be taken as a final picture: the variations in sensitivity from species to species and in the various areas are of course completely neglected in such a diagram, although their existence is obvious from a great number of data.

IV. Muscle

1. Effect of androgens on skeletal muscle. Sexual dimorphism in skeletal muscle is one of the earliest recognised secondary sex characteristics, having first been noted in frogs and in many amphibian species (*45, 51, 61, 66, 95, 96, 108*). In several frogs and toads the musculature of the forelimb in the male undergoes marked hypertrophy in the breeding season, which gradually subsides afterwards. The phenomenon of the "clasping reflex", peculiar to the mating habit of these animals, seems to be associated with this tissue response. Anatomically, the muscles involved are mainly the flexors of the ventral arm and shoulder as well as some pectoral muscles — in fact those used in performing the mating act. The muscles affected are characterised histologically by the fact that the muscle fibres are of larger dimensions and the myofibrils have an increased diameter, as shown in early studies by JASIENSKI (*62*). Castration prevents these muscles from becoming hypertrophic in the breeding season (*96*) and they remain comparable to those of the female. Remarkably enough, no studies on the effect of androgen treatment seem to have been carried out in frogs.

Other instances of sexual dimorphism of muscles have been observed, but few studies have been published on the subject. JASIENSKI (*62*) compared the dorso-scapular muscle in the bull, steer, and cow: although he describes his observations as being common slaughter-house experience, his histological study appears to be the first published account of this muscular hypertrophy evidently conditioned by androgens. Casual observations of a similar sexual dimorphism in deer (carrying heavy antlers), the seasonal fluctuations in the size of the neck musculature in ungulated horn-bearing animals (*61*), the robust pectoral and shoulder musculature of the male cat, and possibly the more highly developed pectoral and shoulder musculature in man (*112*) illustrate that certain biological subjects are less susceptible to scientific study than others. In any case, these observations appear to have been less inspiring in endocrinological studies than the discovery of the "sexual muscles" in laboratory rodents like the guinea-pig (*98*) or the rat (*132*). Many studies have been performed in these animals on the stimulating effect of androgens upon the hypertrophy of especially sensitive muscles like the M. levator ani in the rat or the M. masseter, M. temporalis, and M. retractor penis in the guinea-pig (for review, see *112*). The histological aspects of such muscles in

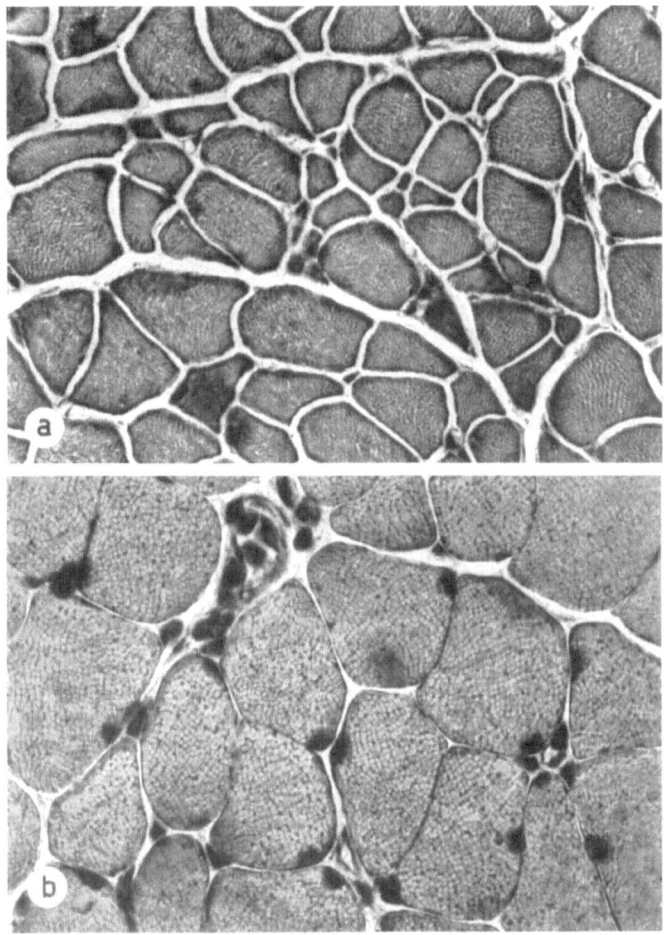

Fig. 10. (a) Structure of the M. levator ani in the castrated rat (8 months old, castrated at the age of 7 weeks). Unfixed frozen section, Weigert's haematoxylin, magnified approx. 400 ×. (b) Structure of the M. levator ani after treatment of castrated rats (as above) with testosterone-propionate (daily dose of 250 mcg. for 6 days). Technique and magnification as in (a). (From KASSENAAR and SZIRMAI, unpublished)

castrated animals and following androgenic treatment are illustrated in Fig. 10. This shows that the principal features of the response to androgen consist of an increase in the diameter of the muscle fibres as well as of the myofibrils.

Some controversy seems to exist as to the extent to which various skeletal muscles participate in this reaction. Illustrative in this respect are the data on the guinea-pig, where castration leads to a marked atrophy of the three muscles mentioned, but subsequently also affects a great number of muscles in the neck, forearm, shoulder, and even the heart and diaphragm. A similar gradation in the response to androgens can be observed also by varying the dose or as a result of the influence of other conditions, such as the nutritional state (67, 68). The data presented make it possible to distinguish a gradient of responsiveness in the skeletal musculature of the guinea-pig, which is quite different from the situation in the rat. In the latter species, the M. levator ani and other muscles of the perineal complex show an extreme degree of reactivity, whereas the majority of the other skeletal muscles, although undoubtedly reacting, do so as a rule to a limited extent and in proportion to the total body-weight (69). Since body-weight is determined to such a large extent by the weight of the skeletal muscles, such comparisons seem to be of limited value. In fact, as remarked by RUSSEL and WILHELMI (112), our ideas concerning the myotrophic and anabolic action of androgens might have been influenced by the accidental use of the rat in many of these investigations: the musculature of the rat is perhaps too well developed to react differentially. The graded responsiveness such as occurs in the guinea-pig or in the frog might be in fact a more universal pattern. In the case of the frog, considerable variations in sensitivity can be deduced from weight measurements, indicating for certain muscles a male/female weight ratio of over 13, and decreasing for other muscles to about 1.3 or less (108). The fact that the M. gastrocnemius of the frog, which is by no means a secondary sex characteristic, is also significantly heavier and has a higher working capacity in the male than in the female (50) bears out this point.

The graded response of certain skeletal muscles seems to be controlled in all cases by androgens, which might account for the generally larger body-weight in the male than in the female in many species, although the contrary might occur as well. Other parts of the skeleton might also be involved in the formation of several male secondary sex characteristics. The heavier bones of the forearm of the male frog and the seasonal fluctuations in the size of the crista medialis humeri, serving as an insertion for the markedly responsive

M. flexor carpi radialis (38, 45), indicate that more than muscle only
might be affected. This could also be the case with the hypertrophied hindlimb of the male alligator and crocodile (40), where
the extent of muscle participation is not clear.

2. **Effect of oestrogens on smooth muscle.** The response of the
skeletal muscle in all the cases studied seems to be due to the action
of androgens. By contrast, there are many examples of oestrogens
exerting an analogous effect on smooth musculature occurring in
different parts of the reproductive organs. The muscular hypertrophy of the uterine wall as a result of oestrogenic stimulation is
too well known to be discussed in detail here (for data, see 37).
Similar muscular hypertrophy has been observed in many cases in
the oviduct (23) and also in the muscular parts of the male reproductive organs, such as the prostate, seminal vesicle, or coagulating
gland (23, 32, 63, 71, 105). Many other instances of what has been
described as "fibromuscular hypertrophy" in early endocrinological
studies might be accounted for by similar reactions on the part of
smooth muscle.

3. **Some general features of androgenic and oestrogenic action
on muscle.** In biochemical terms, the effects described above can
probably be translated as a stimulation of the biosynthesis of

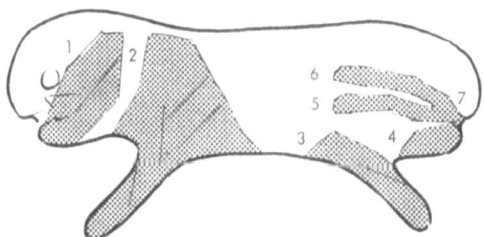

Fig. 11. Diagram showing the location of some "steroid-sensitive" muscles in various vertebrates. 1. Oral and facial muscles (guinea-pig); 2. Forelimb and shoulder muscles (frog,
guinea-pig, bull); 3. Hindlimb muscles (reptilia); 4. Muscles of copulatory organs (various
vertebrates); 5. Mullerian duct derivatives (fishes, amphibia, mammals); 6. Wolffian duct
derivatives (mammals); 7. Derivatives of urogenital sinus (amphibia, rodents)

specific muscle proteins, and it appears difficult to distinguish
categorically between a typical androgenic or oestrogenic action in
this respect. Anatomically, these effects seem to be distinct: the
smooth musculature, responsive to oestrogens, is confined to the
urogenital tract, derived from both embryonic gonaducts and the
urogenital sinus (Fig. 11). The androgen-sensitive skeletal muscles
can be localised less accurately, owing not only to the limited

amount of information available but also to the presence of obvious
gradients in sensitivity. Such gradients make it extremely difficult,
if not impossible, to decide which muscle of the frog or the guinea-
pig should be considered as a "target" muscle and which not. The
sketchy representation of some responsive skeletal muscles in the
diagram shown in Fig. 11 has consequently hardly any other aim
than to indicate the need to collect further data. It does suggest,
however, that certain areas are favoured, namely the hypaxial
musculature of the trunk and limbs. Whereas histogenetic considerations can be partly satisfied by the knowledge that the
oestrogen-sensitive muscles are derived from the embryonic
gonaducts, little can be said about the embryonic development and
histogenesis of the androgen-sensitive skeletal muscle. The observation that, for example, in amphibia the responsive muscles are all
hypaxial, occupying a ventral position on the forelimb, makes one
wonder whether the origin of this musculature and its derivation
from certain parts of the prospective limb could determine its
reactivity. The studies of SWETT (*122*) on the contribution of various
parts of the primitive limb disc to the forelimb musculature in
Amblyostoma, and his finding of a distinct spatial relationship of
the ventral and dorsal prospective parts to the pronephric tissues,
might give some motive for such speculations. In any case, these
considerations indicate that tracing the histogenetic origin of these,
as well as of other steroid-sensitive tissues, might constitute a reward-
ing approach in future studies.

Summary

On the basis of the observations summarised and systematised in a
fragmentary way in these considerations, several general remarks can be
made concerning the action of androgens and oestrogens at the tissue level.
Androgens behave as a stimulant or trigger of certain structural changes,
involving synthetic processes such as the formation of mucopolysaccharides
in target connective tissues, development of the secretory apparatus, and
maintenance of the secretory process in target epithelia, or increased formation of muscle proteins in especially sensitive parts of the body musculature.
Oestrogens are capable of inducing practically the same types of reaction, in
some cases in homologous tissues, in others, as in the case of smooth muscle,
in anatomically distinct tissue systems. Only in a few instances do androgens
and oestrogens act in the same way on homologous or identical structures; in
most cases their effects are mutually antagonistic.

Both androgens and oestrogens produce these reactions with a variable
degree of intensity in tissues of anatomically distinct and demarcated loca-
tions, with great variations in respect of species, age, and other endocrine or
nutritional conditions. There is a certain systematic pattern recognisable in
the anatomical location and distribution of these steroid-sensitive tissues
throughout the various vertebrate species, which is suggestive of a common

origin and determination of sensitivity in early ontogenesis. In many cases, a given sensitive structure is responsive to both types of hormone, be it in the sense of stimulation of the same type of response or of suppressing or antagonistic effects.

The gradients of sensitivity within the reactive tissues suggest that a definition of the term "target" is not feasible on this basis, and also that terms like "secondary sex characteristic" or "sexual tissue" are rather meaningless in view of the lack of relationship between the type of effective hormone and the sex of the subject. One can only recognise in a general way the sensitivity of certain tissues, tissue systems, or organs to certain steroid hormones: the type of response will not only depend on the kind of hormone, androgen or oestrogen, but will be influenced by interdependent effects of the endocrine balance and to a great extent by additional conditions relative to general metabolism, nutritional state, and age.

Most of the effects described, brought about by either androgens or oestrogens, involve stimulation of biosynthetic processes and can be considered as anabolic. Not only the degree of sensitivity and the type of response, but also the amount of tissue involved in respect of total body mass, will determine how far these reactions contribute to a generally recognisable anabolic effect on the part of these hormones. The fact that skeletal muscle makes up a relatively large proportion of total body-weight might account for the great contribution of this target tissue to the so-called general anabolic effect of androgens. However, all other target tissues will contribute to the grand sum, which can only be visualised as being derived from all the tissue reactions described — and many others as well. One should not forget that it is this sum which is measured by the rather undiscriminating parameter of nitrogen retention, and that the concept of the general anabolic effect of certain steroids is based on measuring in this way the sum of their manifold actions on the tissues of the body.

Zusammenfassung

Die in den vorstehenden Ausführungen fragmentarisch und vereinfacht zusammengefaßten Beobachtungen erlauben einige allgemeine Bemerkungen über die Wirkung von Androgenen und Oestrogenen auf die Gewebe. Androgene regen bestimmte strukturelle Veränderungen an oder lösen sie aus; dazu gehören synthetische Prozesse wie die Bildung von Mucopolysacchariden im Bindegewebe der Erfolgsorgane, die Ausbildung des sekretorischen Apparates in den betroffenen Epithelien und die Aufrechterhaltung der Sekretion oder die vermehrte Bildung von Muskeleiweißen in den besonders empfindlichen Teilen der Skelettmuskulatur. Oestrogene verursachen gleichartige Reaktionen nicht nur in homologen Geweben, sondern auch in anatomisch davon abweichenden Gewebssystemen, wie z. B. der glatten Muskulatur. Nur gelegentlich wirken Androgene und Oestrogene in derselben Weise auf homologe oder gleiche Strukturen, in den meisten Fällen antagonisieren sich ihre Effekte gegenseitig.

Sowohl Androgene als auch Oestrogene rufen diese Reaktionen in verschiedener Intensität in anatomisch voneinander differenzierten und gegeneinander abgegrenzten Geweben hervor, wobei große Unterschiede im Hinblick auf Tierart, Alter und andere endokrine oder ernährungsmäßig bedingte Verhältnisse bestehen. Die anatomische Lokalisierung und Verteilung der steroidempfindlichen Gewebe folgt einem bestimmten systematischen Schema, das durch die gesamte Reihe der Wirbeltiere erkennbar ist und das für einen gemeinsamen Ursprung und eine Festlegung der Empfindlichkeit während

der frühen Entwicklung spricht. In vielen Fällen reagiert ein bestimmtes emp-
findliches Gewebe auf beide Typen von Hormonen, sei es im Sinne einer För-
derung einer gleichartigen Reaktion oder eines hemmenden bzw. antago-
nistischen Effektes.

Die Empfindlichkeitsgradienten in einem reaktionsfähigen Gewebe sprechen dafür, daß der Ausdruck „Erfolgsorgan" in diesem Zusammenhang
nicht zweckmäßig ist, und daß auch Bezeichnungen wie „sekundäre Ge-
schlechtsmerkmale" oder „Sexualgewebe" wenig aussagen im Hinblick auf
die fehlenden Beziehungen zwischen dem Typus des wirksamen Hormons und
dem Geschlecht. Es ist lediglich generell anzugeben, daß bestimmte Gewebe,
Gewebssysteme oder Organe gegenüber bestimmten Hormonen reaktions-
bereit sind. Der Typus der Reaktionen hängt nicht nur von der Art des
Hormons ab, Androgen oder Oestrogen, sondern wird auch durch davon
unabhängige Einflüsse mitbestimmt, wie das endokrine Gleichgewicht oder
zusätzliche Bedingungen, die sich aus der allgemeinen Stoffwechsellage, dem
Ernährungszustand und dem Alter ergeben.

Die meisten der beschriebenen Wirkungen, seien sie durch Androgene
oder Oestrogene hervorgerufen, führen zu einer Anregung biosynthetischer
Vorgänge und können somit als anabol charakterisiert werden. Nicht nur das
Ausmaß und die Art der Reaktion, sondern auch die davon betroffenen Men-
gen von Geweben im Verhältnis zur gesamten Körpermasse bestimmen, in-
wieweit sich daraus ein allgemein erkennbarer anaboler Effekt der genannten
Hormone ergibt. Der relativ große Anteil der Skelettmuskulatur am gesamten Körpergewicht erklärt den erheblichen Beitrag, den dieses Erfolgsorgan
an den sogenannten allgemein anabolen Effekt der Androgene leistet. Alle
anderen Erfolgsorgane sind jedoch ebenfalls an der Gesamtwirkung beteiligt,
die nur unter Berücksichtigung aller beschriebenen und zahlreicher anderer
Gewebsreaktionen voll verständlich ist. Es ist darauf hinzuweisen, daß diese
Summe von Effekten durch den an sich wenig aufschlußreichen Parameter
der Stickstoffretention erfaßt wird, und daß die Vorstellung von einer allgemein anabolen Wirkung einzelner Steroide darauf beruht, daß auf diese
Weise die Summe ihrer mannigfaltigen Auswirkungen auf die Körpergewebe
gemessen wird.

Résumé

Des observations fragmentaires et simplifiées permettent quelques
remarques générales sur l'action des androgènes et des oestrogènes sur les
tissus. Les androgènes stimulent ou déclenchent certaines modifications
structurelles, parmi lesquelles on doit citer les processus de synthèse tels
que la formation des mucopolysaccharides dans le tissu conjonctif des
organes effecteurs, le développement de l'appareil sécréteur et le maintien
de la sécrétion dans les épithéliums intéressés ou la formation accrue de protéines musculaires dans les parties spécialement sensibles de la musculature
squelettique. Les oestrogènes déterminent des réactions du même genre non
seulement dans les tissus homologues, mais encore dans des systèmes tissulaires anatomiquement différents, par exemple ceux de la musculature lisse.
Androgènes et oestrogènes n'agissent qu'occasionnellement de la même
façon sur des structures homologues ou semblables; dans la plupart des cas,
leurs effets sont antagonistes.

Tant les androgènes que les oestrogènes provoquent ces réactions à des
intensités diverses dans des tissus anatomiquement différents et de localisation distincte; selon l'espèce animale, l'âge et d'autres conditions endo-
criniennes ou alimentaires, on trouve des différences considérables. La
localisation anatomique et la distribution de ces tissus sensibles aux stéroïdes

se font selon un certain schéma que l'on retrouve à travers toute la série des vertébrés, ce qui est en faveur d'une origine commune et d'une sensibilité déterminée tout au début de l'ontogénèse. Dans de nombreux cas, un tissu sensible réagit aux deux types d'hormones, soit par un renforcement d'une réponse de même type, soit par une inhibition ou un effet antagoniste.

Les "gradients de sensibilité" constatés dans un tissu qui réagit indiquent que l'expression "organe effecteur" (target organ) ne convient pas ici, et que des termes tels que "caractères sexuels secondaires" ou "tissus sexuels" ne disent pas grand'chose étant donné l'absence de rapport entre le type de l'hormone active et le sexe du sujet. On peut simplement reconnaître une tendance générale dans la sensibilité de certains tissus, systèmes de tissus ou organes à diverses hormones. Le type de réaction dépend non seulement de la nature de l'hormone, androgène ou oestrogène, mais encore de facteurs intriqués tels que l'équilibre endocrinien, ou encore d'autres conditions résultant du métabolisme général, de l'état de nutrition, de l'âge.

La plupart des actions décrites, qu'elles soient attribuables aux androgènes ou aux oestrogènes, aboutissent à une stimulation des processus de biosynthèse et peuvent par conséquent être qualifiées d'anaboliques. Non seulement la nature et l'intensité de la réaction, mais aussi la quantité de tissus intéressés par rapport à la masse totale de l'organisme déterminent jusqu'à quel point ces hormones auront un effet anabolisant général décelable. La part assez considérable que représente la musculature squelettique dans le poids total explique le rôle important de cet organe effecteur dans l'effet anabolisant général des androgènes. Tous les autres organes effecteurs participent cependant aussi à l'effet d'ensemble; ce dernier n'est pleinement compréhensible qu'en tenant compte de toutes les réactions tissulaires décrites, et de beaucoup d'autres encore. Il ne faut pas oublier que ce qu'en registre le paramètre "rétention azotée", peu concluant par lui-même, est cette somme d'effets; l'idée que certains stéroïdes ont une action anabolique générale se base sur le fait qu'on mesure ainsi la somme de leurs multiples effets sur les tissus de l'organisme.

References

1. ALLEN, E.: Contr. Embryol. Carnegie Inst. Washington **19**, 1 (1927). — 2. ANASTASSIADIS, P. A., W. A. MAW, and R. H. COMMON: Canad. J. Biochem. Physiol. **33**, 627 (1955). — 3. ARON, M.: Arch. biol. (Fr.), **31**, 1 (1924). — 4. AYKROYD, O. E., and S. ZUCKERMAN: J. Physiol. (G. B.) **94**, 13 (1938). — 5. BACHMAN, C., J. B. COLLIP, and H. SELYE: Proc. Roy. Soc., London. Biol. Sc. **117**, 16 (1935). — 6. BACHMAN, C., J. B. COLLIP, and H. SELYE: Proc. Soc. Exper. Biol. Med. (U.S.A.) **33**, 549 (1936). — 7. BAILEY, R. E.: Condor (U.S.A.) **54**, 121 (1952). — 8. BALAZS, E. A.: personal communication. — 9. BALAZS, E. A., P. H. MARS, and J. A. SZIRMAI: Abstr. in: J. Amer. Chem. Soc. 1955, p. 46C. — 10. BALAZS, E. A., P. H. MARS, and J. A. SZIRMAI: unpublished. — 11. BALAZS, E. A., and J. A. SZIRMAI: J. Histochem. Cytochem. (U.S.A.) **6**, 416 (1958). — 12. BALAZS, E. A., J. A. SZIRMAI, and G. BERGENDAHL: J. Biophys. Biochem. Cytol. (U.S.A.) **5**, 319 (1959). — 13. BERTHOLD, A. A.: Arch. Anat. Physiol. (G.) **2**, 42 (1849). — 14. BOAS, N. F.: J. Biol. Chem. (U.S.A.) **181**, 573 (1949). — 15. BOAS, N. F., and A. W. LUDWIG: Endocrinology (U.S.A.) **46**, 299 (1950). — 16. BOLTON, W.: J. Endocr. (G.B.) **9**, 440 (1953). — 17. BRANDES, D., and D. P. GROTH: Exper. Cell Res. (U.S.A.) **23**, 159 (1961). — 18. BRANDES, D., and A. PORTELA: J. Biophys. Biochem. Cytol. (U.S.A.) **7**, 505 (1960). — 19. BRANDES, D., and A. PORTELA: J. Biophys. Biochem. Cytol. (U.S.A.) **7**, 511 (1960). —

20. BRENEMAN, W. R.: Endocrinology (U.S.A.) **30**, 609 (1942). — 21. BRES-CA, G.: Roux' Arch. Entw. mech (G.) **29**, 403 (1910). — 22. BURROWS, H.: Proc. Roy. Soc., London, Biol. Sc. **118**, 485 (1935). — 23. BURROWS, H.: Biological Actions of Sex Hormones. Univ. Press., Cambridge, 1949. — 24. CARIDROIT, F., and A. PÉZARD: Compt. rend. Soc. biol. (Fr.) **95**, 296 (1926). — 25. CHAIN, E., and E. S. DUTHIE: Brit. J. Exper. Pathol. **21**, 324 (1940). — 26. CHAMPY, C.: Compt. rend. Soc. biol. (Fr.) **94**, 311 (1926). — 27. CHAMPY, C.: Compt. rend. Soc. biol. (Fr.) **122**, 550 (1936). — 28. CHAMPY, C., H. BULLIARD, N. KRITCH, and M. L. DEMAY: Arch. anat. microsc. (Fr.) **27**, 301 (1931). — 29. CHAMPY, C., and N. KRITCH: Compt. rend. Soc. biol. (Fr.) **92**, 683 (1925). — 30. CHAMPY, C., and N. KRITCH: Arch. morph. gén. (Fr.) **25**, 1 (1926). — 31. CHAMPY, C., N. KRITCH, and A. LLOMBART: Compt. rend. Ass. anat. (Fr.) **24**, 120 (1929). — 32. CHEVREL-BODIN, M. L., and D. LEROY: Ann. endocr. (Fr.) **2**, 226 (1941). — 33. COHEN, S.: J. Endocr. (G.B.) **12**, 196 (1955). — 34. COHEN, S.: Biochem. J. (G.B.) **64**, 286 (1956). — 35. COLE, H. H., and P. T. CUPPS: (Eds.) Reproduction in Domestic Animals, Vol. 1. Acad. Press Inc., New York/London, 1959. — 36. COLLINGS, M. R.: Anat. Rec. (U.S.A.) **33**, 271 (1926). — 37. CSAPO, A.: Ann. N. Y. Acad. Sc. **75**, 790 (1959). — 38. DAUWART, A.: Arch. mikrosk. Anat. (G.) **103**, 504 (1924). — 39. DEANE, H. W., and K. R. PORTER: Zschr. Zellforsch. (G.) **52**, 697 (1960). — 40. DODD, J. M.: In: Marshall's Physiology of Reproduction. Ed. by A. S. PARKER. Vol. 1., Part 2. Longmans Green & Co. Ltd., London, 1960. — 41. DORFMAN, R. J., and R. A. SHIPLEY: Androgens. John Wiley, New York, 1956. — 42. DOYLE, J., and J. A. SZIRMAI: Biochim. biophys. acta (Neths) **50**, 582 (1961). — 43. DOYLE, J., J. A. SZIRMAI, and E. R. DE TYSSONSK: in preparation. — 44. DURAN-REYNALS, F., H. BUNTING, and G. VAN WAGENEN: Ann. N. Y. Acad. Sc. **52**, 1006 (1950). — 45. ECKER, A.: Die Anatomie des Frosches. Friedrich Vieweg & Sohn, Brunswick, 1864. — 46. ECKSTEIN, P., and S. ZUCKERMAN: In: Marshall's Physiology of Reproduction. Ed. by A. S. PARKER. Vol. 1., Part 1. Longmans Green & Co. Ltd., London, 1956. — 47. ELKNER, A., and P. SLONIMSKI: Bull. histol. appl. (Fr.) **4**, 263 (1927). — 48. FRANTZ, M. J., and A. KIRSCHBAUM: Cancer Res. (U.S.A.) **9**, 257 (1949). — 49. FREUND, L.: Zool. Anz. (G.) Suppl. **2**, 153 (1926). — 50. GAULE, J.: Pflügers Arch. Physiol. (G.) **83**, 83 (1901). — 51. GAUPP, E.: Anatomie des Frosches.3rd Ed. Friedrich Vieweg & Sohn, Brunswick, 1896. — 52. GARRAULT, H.: Arch. anat. microsc. (Fr.) **30**, 5 (1934). — 53. GILBERT, C.: S. Afr. J. Med. Sc. **9**, 125 (1944). — 54. GILLMAN, J., and C. GILBERT: S. Afr. J. Med. Sc. **11**, 1 (1946). — 55. GORBMAN, A.: Proc. Soc. Exper. Biol. Med. (U.S.A.) **42**, 811 (1939). — 56. GRAD, B., and C. P. LEBLOND: Endocrinology (U.S.A.) **45**, 250 (1949). — 57. HARDESTY, M.: Amer. J. Anat. **47**, 277 (1931). — 58. HARKIN, J. C.: Endocrinology (U.S.A.) **60**, 185 (1956). — 59. HARTMAN, C. G.: Contr. Embryol. Carnegie Inst. Washington **23**, 1 (1932). — 60. HERINGA, G. C., and S. E. DE JONGH: Zschr. Zellforsch. (G.) **21**, 629 (1934). — 61. HOWELL, A. B.: Copeia (U.S.A.) **4**, 188 (1935). — 62. JA-SIENSKI, J.: Compt. rend. Soc. biol. (Fr.) **101**, 533 (1929). — 63. JONGH, S. E. DE: Acta brevia Neerl. physiol. pharmacol. **3**, 112 (1933). — 64. JUHN, M., and R. G. GUSTAVSON: J. Exper. Zool. (U.S.A.) **56**, 31 (1930).—65. JUNQUEIRA, L. C., A. FAJER, M. RABINOVITCH, and L. FRANKENTHAL: J. Cellul. Comp. Physiol. (U.S.A.) **34**, 129 (1949). — 66. KÄNDLER, R.: Jena. Zschr. Med. Naturw. **60**, 175 (1924). — 67. KOCHAKIAN, C. D., and D. COCKRELL: Proc. Soc. Exper. Biol. Med. (U.S.A.) **97**, 148 (1958). — 68. KOCHAKIAN, C. D., C. TILLOTSON, and J. AUSTIN: Endocrinology (U.S.A.) **60**, 144 (1956). — 69. KOCHAKIAN, C. D., C. TILLOTSON, and G. L. ENDAHL: Endocrinology (U.S.A.) **58**, 226 (1956). — 70. KON, S. K., and A. T. COWIE: (Eds.) Milk: the

mammary gland and its secretion. Vol. 1. Acad. Press Inc., New York/London, 1961. — 71. KORENCHEVSKY, V., and M. DENNISON: J. Path. Bact. (G.B.) 41, 323 (1935). — 72. KOSIN, I. L., and S. S. MUNRO: Endocrinology (U.S.A.) 30, 102 (1942). — 73. KOZELKA, A. W.: J. Hered. (U.S.A.) 20, 3 (1929). — 74. KROHN, P. L., and S. ZUCKERMAN: J. Physiol. (G.B.) 88, 369 (1937). — 75. LACASSAGNE, A.: Compt. rend. Soc. biol. (Fr.) 133, 227 (1940). — 76. LANGE, B.: Jb. Morph. mikrosk. Anat. (G.) 59, 601 (1928). — 77. LEATHEM, J. H., and R. C. WOLF: Mem. Soc. Endocr. No. 4. Univ. Press., Cambridge 1955, p. 220. — 78. LEYDIG, F.: Biol. Zbl. (G.) 12, 205 (1892). — 79. LORENZ, F. W.: In: Reproduction in Domestic Animals. Vol. 2. Ed. by COLE, H. H., and P. T. CUPPS. Acad. Press Inc., New York/London, 1959. — 80. LUDWIG, A. W., and N. F. BOAS: Endocrinology (U.S.A.) 46, 291 (1950). — 81. MARTIN, J. E., J. H. GRAVES, and F. C. DOHAN: Amer. J. Vet. Res. 16, 141 (1955). — 82. McKAY, D. G.: Amer. J. Obstet. Gynec. 59, 875 (1950). — 83. MERCER, E. H.: Keratin and Keratinization. Pergamon Press, Oxford, 1961. — 84. MOORE, C. R., W. HUGHES, and T. F. GALLAGHER: Amer. J. Anat. 45, 109 (1930). — 85. MORATÓ-MANARO, J., and A. ALBRIEUX: Endocrinology (U.S.A.) 24, 518 (1939). — 86. MÜHLBOCK, O.: Acta brevia Neerl. physiol. pharmacol. 8, 50 (1938). — 87. MÜHLBOCK, O.: Acta brevia Neerl. physiol. pharmacol. 8, 142 (1938). — 88. NAKAMURA, T.: Compt. rend. Soc. biol. (Fr.) 96, 524 (1927). — 89. NILSSON, O.: J. Ultrastructure Res. 1, 375 (1958). — 90. NILSSON, O.: J. Ultrastructure Res. 2, 73 (1958). — 91. NILSSON, O.: J. Ultrastructure Res. 2, 185 (1958). — 92. NILSSON, O.: J. Ultrastructure Res. 2, 331 (1959). — 93. NILSSON, O.: J. Ultrastructure Res. 2, 342 (1959). — 94. NILSSON, O.: J. Ultrastructure Res. 2, 373 (1959). — 95. NOBLE, G. K.: The Biology of Amphibia. McGraw-Hill Book Co., New York, 1931. — 96. NUSSBAUM, M.: Erg. Anat. (G.) 15, 39 (1905). — 97. OGSTON, A. G., J. S. L. PHILPOT, and S. ZUCKERMAN: J. Endocr. (G.B.) 1, 231 (1939). — 98. PAPANICOLAOU, G. N., and E. A. FALK: Science (U.S.A.) 87, 238 (1938). — 99. PAPANICOLAOU, G. N., H. F. TRAUT, and A. A. MARCHETTI: The Epithelia of Woman's Reproductive Organs. The Commonwealth Fund, New York, 1948. — 100. PARKES, A. S., and C. W. EMMENS: Vitamins and Horm. (U.S.A.) 2, 361 (1944). — 101. PARKES, A. S., and A. J. MARSHALL: In: Marshall's Physiology of Reproduction. Ed. by A. S. PARKER. Vol. 1, Part 2. Longmans Green & Co. Ltd., London, 1960. — 102. PÉZARD, A.: Compt. rend. Acad. Sc. (Fr.) 153, 1027 (1911). — 103. PÉZARD, A.: Erg. Physiol. (G.) 27, 552 (1928). — 104. PONSE, K.: Rev. suisse zool. 55, 477 (1948). — 105. PRICE, D., and H. G. WILLIAMS-ASHMAN: In: Sex and Internal Secretions. Ed. by YOUNG, W. C., and G. W. CORNER. Vol. 1. The Williams & Wilkins Co., Baltimore, 1961. — 106. REGAUD, C., and A. POLICARD: Compt. rend. Soc. biol. (Fr.) 55, 216 (1903). — 107. RÉGNIER, M. V.: Compt. rend. Soc. biol. (Fr.) 95, 171 (1926). — 108. REY, M. P: Ann. endocr. (Fr.) 8, 35 (1947). — 109. RIENITS, K. G.: Biochem. J. (G. B.) 48, 58 P (1951). — 110. RIENITS, K. G.: Biochem. J. (G. B.) 74, 27 (1960). — 111. ROMANOFF, A. L., and A. J. ROMANOFF: The Avian Egg. Wiley, New York, 1949. — 112. RUSSEL, J. A., and A. E. WILHELMI: In: The Structure and Function of Muscle. Ed. by G. H. BOURNE. Vol. 2. Acad. Press. Inc., New York/London, 1960. — 113. SCHILLER, S., E. P. BENDITT, and A. DORFMAN: Endocrinology (U.S.A.) 50, 504 (1952). — 114. SCHILLER, S., and A. DORFMAN: Proc. Soc. Exper. Biol. Med. (U.S.A.) 92, 100 (1956). — 115. SCHRODT, G. R.: J. Ultrastructure Res. 5, 485 (1961). — 116. SEAMAN, A., and M. WINELL: Anat. Rec. (U.S.A.) 139, 272 (1961). — 117. SELYE, H.: Anat. Rec. (U.S.A.) 76, 145 (1940). — 118. SELYE, H.: Arch. Dermat. Syph. (U.S.A.) 50, 261 (1944). — 119. STARKEY, W. F., and J. H. LEATHEM: Anat.

Rec. (U.S.A.) **75**, 85 (1939). — 120. STEINER, K.: Arch. Dermat. Syph. (G.) **157**, 446 (1929). — 121. STEINER. K,: Arch. Dermat. Syph. (G.) **162**, 577 (1931). — 122. SWETT, F. H.: J. Exper. Zool. (U.S.A.) **37**, 207 (1923). — 123. SZIRMAI, J. A.: Anat. Rec. (U.S.A.) **105**, 337 (1949). — 124. SZIRMAI, J. A.: Thesis, University of Amsterdam, 1954. — 125. SZIRMAI, J. A.: J. Histochem. Cytochem. (U.S.A.) **4**, 96 (1956). — 126. SZIRMAI, J. A.: Proc. Soc. Exper. Biol. Med. (U.S.A.) **93**, 92 (1956). — 127. SZIRMAI, J. A.: Acta endocr. (Den.) **25**, 225 (1957). — 128. SZIRMAI, J. A.: In: The Amino Sugars. Ed. by JEANLOZ, R.W., and E. BALAZS. Acad. Press. Inc., New York (in press). — 129. SZIRMAI, J. A., and E. A. BALAZS: Acta histochem. (G.) Suppl. **1**, 56 (1958). — 130. SZIRMAI, J., A. and P. C. VAN DER LINDE: In: Electron Microscopy. Ed. by S. S. BREESE, Jr., Vol. 2. Acad. Press Inc., New York/London, 1962. — 131. VIRCHOW, R.: Verh. Phys.-Med. Ges. Würzburg **2**, 314 (1851). — 132. WAINMAN, P., and G. C. SHIPOUNOFF: Endocrinology (U.S.A.) **29**, 975 (1941). — 133. WEIDE, U.: Thesis, University of Munich, 1952. — 134. WELLS, L. J.: Ann. N. Y. Acad. Sc. **83**, 80 (1959). — 135. WHITELOCK, O. v. S.: (Ed.) The vagina. Ann. N. Y. Acad. Sc. **83**, 77 (1959). — 136. ZUCKERMAN, S.: Proc. Zool. Soc. London **2**, 691 (1930). — 137. ZUCKER. MAN, S.: Biol. Rev. (G.B.) **15**, 231 (1940). — 138. ZUCKERMAN, S.: Arch. anat. microsc. (Fr.) **39**, 608 (1950). — 139. ZUCKERMAN, S., G. VAN WAGENEN, and R. H. GARDINER: Proc. Zool. Soc. London **108**, 385 (1938).

Discussion

QUERIDO: I think Dr. SZIRMAI has touched on some extremely important problems concerning the specificity of tissue reactions, which, of course, in the final analysis must be dependent on nuclear properties. This poses the question as to whether these complicated reactions result from stimulation of a well-organised complex of reactions — starting possibly in the nucleus — or from enhancement of certain steps in the metabolism of the cell.

TANNER: One of the tissues most reactive to different sex hormones, at least in the primates, is cartilage. The cartilage in the shoulder-girdle in many primates, including man, reacts specifically at puberty to stimulation with androgens, and the Y-shaped cartilage in the acetabulum reacts specifically to oestrogens. There are great species differences in this; for example, in the chimpanzee there is not a great deal of sexual dimorphism, whereas in the gorilla and in the rhesus there is. Cartilage provides an instructive example, because, as far as I know, the androgen reaction and the oestrogen reaction in shoulder cartilage look the same under ordinary histological examination. I don't know if there have been any transplantation experiments, but I would certainly presume that these reactivities are fixed into the cell at quite an early stage of differentiation.

DESAULLES: I have been very interested by the paper presented by Dr. SZIRMAI, but with regard to the comments made by Dr. TANNER I feel that the epiphyseal cartilage of the long bones of the rat is a sensitive target organ by which to study the response to steroid treatment during sexual development. GARDNER and PFEIFFER[1], LICHTWITZ et al.[2], and BAISSET et al.[3] have shown that, in immature rats of both sexes, growth of the epiphyseal plate of long bones is stimulated by androgens as well as by oestrogens; after sexual maturity is reached, however, androgens — but not oestrogens — are active in males, whereas in females oestrogens are active and androgens much less so, though not completely inactive. The contrast is so sharp that there must certainly be a marked change in sensitivity at the time of sexual maturation which could not be observed before.

TANNER: This of course is not true to the same extent in the primate, because here you have this differential sensitivity before puberty.

LEATHEM: There are two questions I would like to ask. Firstly, have you looked at the influence of oestrogen on the seminal vesicle, since it is markedly stimulated by oestrogen? In fact, weight-wise, the degree of stimulation is just as great as in response to androgen on a dose-for dose basis. Secondly, there are also quite sharp species differences for different rodents

---

[1] GARDNER, W. U., and C. A. PFEIFFER: Physiol. Rev. (U.S.A.) **23**, 139 (1943).
[2] LICHTWITZ, A., G. THIÉRY, R. PARLIER, and M. DELAVILLE: Sem. hôp. Paris **27**, 247 (1951).
[3] BAISSET, A., L. DOUSTE-BLAZY, P. MONTASTRUC, H. PLANEL, and J. VIRENQUE: Sem. hôp. Paris **29**, 63 (1953).

in response to androgens; and if you compare a series of androgens you can find, for example, that a decreased response to testosterone-propionate can be seen with rat, mouse, and hamster. You might actually reverse that sequence, however, by injecting another androgen. Do you have any suggestions about this ?

SZIRMAI: We have no experience with oestrogens. But such studies have been reported by others[1] concerning the effect of oestradiol on the epithelium of the rat prostate, in which it clearly acts as an antagonist to androgens. The effect is similar to that of castration in that it produces regression and collapse of the endoplasmic reticulum.

LEATHEM: May I interrupt you just to say that it is very clear that these two tissues (seminal vesicle and ventral prostate) respond quite differently to oestrogen. The seminal vesicle is stimulated and the prostate is inhibited.

SZIRMAI: The question of species differences is certainly a very complicated one and can give rise to many difficulties in interpretation. Dr. WILSON mentioned the experiments of DEANE and PORTER[2] which suggest that castration does not have any marked effect on the ribosomes in the seminal vesicles of mice. In contrast, our results in the rat following castration indicate very evident changes in the ribosomes in this same organ. When discussing these controversial results with Dr. DEANE some time ago we could not arrive at any reasonable explanation. One possibility might be that there is a difference in the results of castration in rats and mice, a possibility for which some evidence is offered by the work of HOWARD[3] on the X-zone of the adrenals. Another explanation might be the age at which castration is performed (in our case 6—7 weeks in rats, in Dr. DEANE's case about 3 months in mice), especially since, as shown by PRICE[4], the androgen sensitivity of the accessory organs differs considerably depending upon age. We have repeated the experiments of Dr. DEANE with mice of the same age and under identical experimental conditions, and so far we have been able fully to confirm her data: there is certainly no striking loss of ribosomes a few weeks after castration, carried out at the age of 3 months, although the whole endoplasmic reticulum seems to change somewhat and to present a disorganised appearance. We now have studies in progress on the influence of age and duration of castration with both species.

ASCHKENASY: Dr. SZIRMAI has some reason to claim that the anabolic action of androgens on muscle represents the most important cause of the general anabolic effect of these hormones. Nevertheless, there is no definite parallelism between the two actions: the general anabolic effect, which is responsible for the increase in body-weight and for the nitrogen retention, is, in male rats, only of short duration; moreover, it is observed only in castrated or undernourished males, but is absent in normal male rats[5].

On the other hand, the stimulating action exerted by androgens on the accessory sex organs, on certain muscles, and also on other specific tissues and organs is a long-term action, and it does not entirely depend on the absence of the testes or on the nutritional status. It may therefore be assumed

[1] GROTH and BRANDES: J. Ultrastructure Res. 4, 166 (1960).

[2] Zschr. Zellforsch. (G.) 52, 697 (1960).

[3] Amer. J. Anat. 65, 105 (1939).

[4] Physiol. Zool. (U.S.A.) 20, 213 (1947).

[5] KOCHAKIAN, C. D., and J. A. WEBSTER: Endocrinology (U.S.A.) 63, 737 (1958).

that the most constant effect of the androgens is not a general nitrogen retention, but a specific anatomic distribution of proteins: whereas certain tissues or organs (accessory sex organs, some muscles, kidneys, bone marrow, salivary glands, etc.) are subject to an anabolic effect, some — including especially the thymus — are inhibited, and others are not influenced at all by androgens.

SZIRMAI: I find it very difficult to comment on this, not knowing anything about the general anabolic action. Nor do I quite appreciate the recently so popular emphasis on the cellular mechanism of action, since I never thought of anything but a direct action of a hormone on the cells and tissues of the body. In my paper I illustrated the graded response of only a few tissue systems to various steroid hormones. Some of these responses are so striking that macroscopic inspection or a quick glance through the microscope reveals them immediately; others are hard to discover at all, as could be illustrated by many examples from the history of endocrinology. To take but one example, the biosynthesis of hyaluronic acid in the capon comb can certainly be described as an anabolic process. The capon comb, weighing about 2 g., will respond to a total dose of 15 mg. testosterone-propionate with a weight increase to about 25 g. Of this, perhaps 23 g. will be water, the total dry mass increasing only slightly. Only about 100—150 mg. of the increased organic mass are accounted for by hyaluronic acid, but the amount of nitrogen "spared" for the formation of this compound would certainly not be discovered by a method based on measurement of nitrogen retention in a very indiscriminate fashion. At the same time, many other tissues may contribute to the overall effect, depending on their degree of sensitivity, the kind and intensity of their response, and their total amount. This could apply to connective-tissue derivatives like bone, tendons, and skin, which constitute such a large proportion of the body. Collagen formation is greatly influenced both by oestrogens, for example in the uterus, and by androgens, as in the tendons, musculature, and skin of the capon[1]. Obviously the muscular response is quantitatively very important, but it is also very insufficiently documented; I am not aware, for example, of any data concerning a gradient of responsiveness of the muscles in chicken. I feel, however, that we are not justified in assuming a general anabolic effect merely because we do not have sufficient insight into the extent of the participation of the various tissues in the final sum of effects; the reason why the contribution of the musculature is so important is probably simply because there is so much muscle in the body.

---

[1] HERRICK, E. H.: Anat. Rec. (U.S.A.) **93**, 145 (1945).

Introduction to the General Discussion:

The importance of experimental conditions for the demonstration of hormone effects *in vitro* and *in vivo*

By

W. DIRSCHERL

Though it may seem remarkable that there should be such a large number of hormones displaying protein-anabolic effects, the reason is obvious when one bears in mind the importance of protein synthesis in all living creatures. From a chemical point of view, however, it is not easy to understand why hormones which are as chemically different as the protein hormones STH and insulin and, on the other hand, steroid hormones, should all show the same action. But, is it really the same action? It is possible that the various protein-anabolic hormones may influence different steps of protein synthesis. Hence the importance and value of investigating their mode of action in detail, as has been done in such an excellent manner by Dr. KORNER in the case of growth hormone and by Dr. WILSON in that of testosterone and oestradiol. Besides differences in mode of action, one also has to consider differences as far as the target organs are concerned. For example, growth of the sex organs seems to be influenced only by the sex hormones, but not by growth hormone or by insulin. To take another example, the glucocorticoids, which generally have an anti-anabolic effect on protein metabolism, can exert a protein-anabolic action in the liver. This has been extensively investigated for the induction of certain liver enzymes. In our institute, Dr. OTTO (1962) studied the influence of glucocorticoids on liver glutamic-pyruvic-transaminase in experiments *in vivo* on young rats. The transaminase was purified several hundredfold and finally chromatographed on ion-exchange cellulose. Compared with controls, the active fractions showed an increased protein content, together with increased enzymatic activity.

After these more general remarks, I should like to refer to those experiments which demonstrate the influence of hormones on protein synthesis as measured by the incorporation of labelled amino acids into the proteins of certain organs.

We can distinguish here between 3 types of experiment:

1. The true *in vivo* experiment: the animals are first pre-treated with hormone and later the amino acid is injected; the rate of incorporation of the labelled amino acid is then measured.

2. The partial *in vivo* experiment: after pre-treatment with hormone, the animals are sacrificed, and the incorporation of labelled amino acid is studied *in vitro*.

3. The true *in vitro* experiment: here, the animals are not pre-treated, but the incorporation of labelled amino acid is investigated *in vitro* in the presence of the hormone.

Regarding the first type of experiment, it has been found that, in rats and guinea-pigs, testosterone or its propionate increase the rate of incorporation of amino acids (e.g. glycine or leucine) into the seminal vesicles, the prostate, and/or certain muscles.

As to the second type of experiment, we have just heard two fine papers dealing with studies of this category. In addition, other investigators have found the following: in the case of mice pre-treated with testosterone or its propionate, the incorporation of ^{14}C-glycine was stimulated in kidney homogenate or kidney slices. Similar results have been obtained with regenerating rat liver.

If this enhancement of protein synthesis as revealed by experiments of types 1 and 2 can be traced back to an immediate effect of the anabolic hormones used, then it should be possible to demonstrate this effect in the true *in vitro* experiment.

Not many investigations of this kind have been undertaken. According to KIT and BARRON (1953), the incorporation of labelled glycine, phenylalanine, and acetate into the proteins of lymphatic cells decreases in response to 17-hydroxycorticosterone or 11-dehydro-17-hydroxycorticosterone employed mainly in concentrations of 100 mcg./ml. This corresponds well to the picture of glucocorticoids as protein-anti-anabolic agents. In similar fashion, HAUSCHILDT and GROSSMAN (1953) showed that testosterone, oestradiol, desoxycorticosterone, and cortisone — in amounts of 50—300 mcg./ml. — inhibit the incorporation of ^{14}C-glycine into the proteins of rat-liver slices. In experiments with human-liver slices, testosterone generally inhibited incorporation, although in 3 out of 13 cases the rate of incorporation was actually increased. This predominantly inhibitory effect is inconsistent with the anabolic activity of this steroid.

In 1955 I pointed out that, in experiments *in vitro*, increases in incorporation may be observed more frequently if the experimental conditions are changed — for example, if smaller hormone concentrations are used. Metabolic processes and enzymatic reactions can be either inhibited or enhanced by compounds generally known in German as "*Effektoren*", i.e. which act either as inhibitors or as activators. In our investigations we gained the impression that one and the same steroid could be both an inhibitor or an activator,

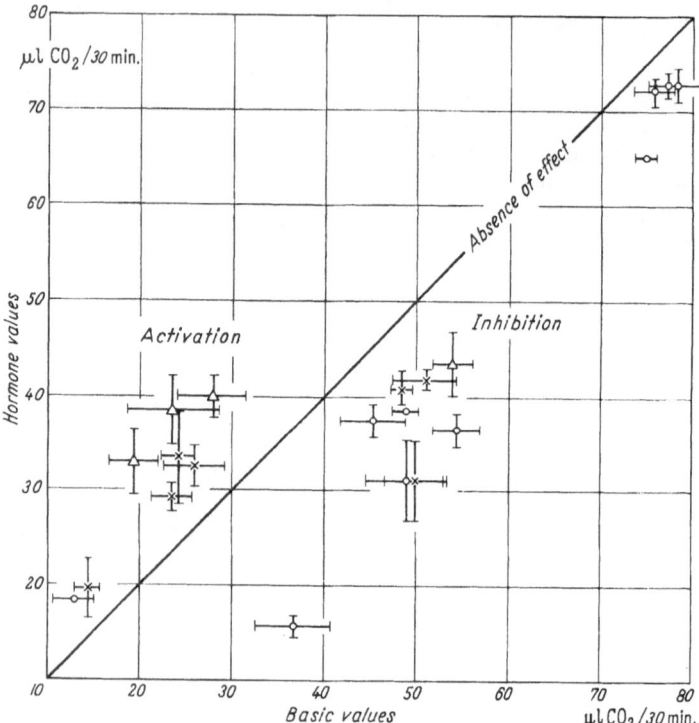

Fig. 1. Effect of desoxycorticosterone on the activity of hexokinase from EHRLICH's mouse-ascites cells (autolysed by toluene). Concentration of desoxycorticosterone (mcg./ml.): △ 25; × 50; ○ 100. From DIRSCHERL and PROJAHN: Zschr. Vitamin-Hormon-Ferm. forsch., in press (1962)

depending on the experimental conditions, and I should just like to mention a few examples.

The degradation of hyaluronic acid by hyaluronidase was activated by cortisone below pH 6 but inhibited above pH 6 (DIRSCHERL and KRÜSKEMPER, 1952).

The effect exerted by desoxycorticosterone acetate on the oxidation of octanoate to aceto-acetic acid by rat-liver mitochondria depended upon the concentration of desoxycorticosterone acetate employed: small amounts caused both activation and inhibition, whereas somewhat larger amounts caused inhibition only (DIRSCHERL and MÜLLER, 1953). Other investigators, too, have made similar observations. According to JONES and WADE (1953), the ATPase of rat-liver homogenates can be either activated or inhibited by progesterone, depending once again on the concentration used.

Even the enzyme concentration, however, may be of importance in determining whether the hormonal effect will be positive or negative. For example, the glycolytic action of rabbit-muscle extracts could be influenced in different ways by one and the same amount of desoxycorticosterone: activation was observed with small enzyme concentrations (0.26%), and inhibition with larger ones (0.64%) (DIRSCHERL and OTTO, 1953). The hexokinase activity of ascites cells was also modified by desoxycorticosterone and cortisone in somewhat the same levelling manner (DIRSCHERL and PROJAHN, 1962; Fig. 1). For this phenomenon we use the term "nivellierender Effekt".

I should like to mention just one more example, although here the "effector" involved was not a steroid hormone but an amino acid. Since amino acids are more readily soluble in water, this meant that we were able to use a wider range of concentrations in order to obtain a solid basis for our kinetic studies. We investigated the effect of amino acids (particularly histidine and leucine) on the de-amination of D-alanine by D-amino-acid oxidase (DIRSCHERL and MOSEBACH, 1957). One and the same amount of leucine caused activation with small enzyme concentrations corresponding to a low velocity of reaction (Fig. 2), and inhibition with larger enzyme concentrations corresponding to a higher velocity. This can be expressed by the following equation:

$$\Delta v = k'v - k''v^2$$

in which v is the velocity, Δv is the change in velocity, and k' and k'' are constants. Δv is positive (signifying activation) if the first term of the equation is larger than the second; this is the case with a low velocity. Δv becomes negative (signifying inhibition) if the second term becomes larger than the first; this is the case with a high velocity.

I am convinced that there are many instances in which an inhibitor can activate, and vice versa. Quite often, however, only inhibition is actually observed, because the experimental conditions are not sufficiently varied.

After this interlude I should now like to come to our *in vitro* experiments dealing with the influence of anabolic steroid hormones on protein synthesis.

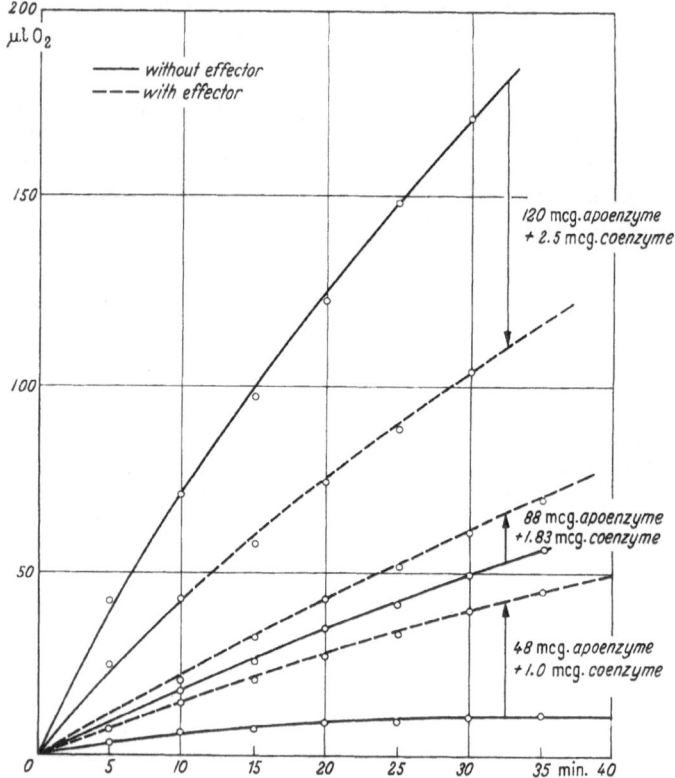

Fig. 2. Effect of L-leucine on the degradation of D-alanine by D-amino-acid oxidase in various enzyme concentrations. 37° C, m/15 phosphate buffer pH 7.2, m/100 D-alanine, m/40 L-leucine; air. From DIRSCHERL: Proc. of the Fourth International Congress of Biochemistry, Vienna, 1958. Symposium IV: Biochemistry of Steroids. Pergamon Press, London, 1959, p. 123, 136

First we studied the formation of a dipeptide, in this case p-aminohippuric acid from p-aminobenzoic acid and glycine, by rat-liver homogenates in the presence of testosterone or desoxy-corticosterone (1—200 mcg./ml.). We observed only inhibition, the degree of which increased as the concentration of hormone rose, but we did at least discover that both steroids, employed in very small amounts, can influence the formation of a dipeptide.

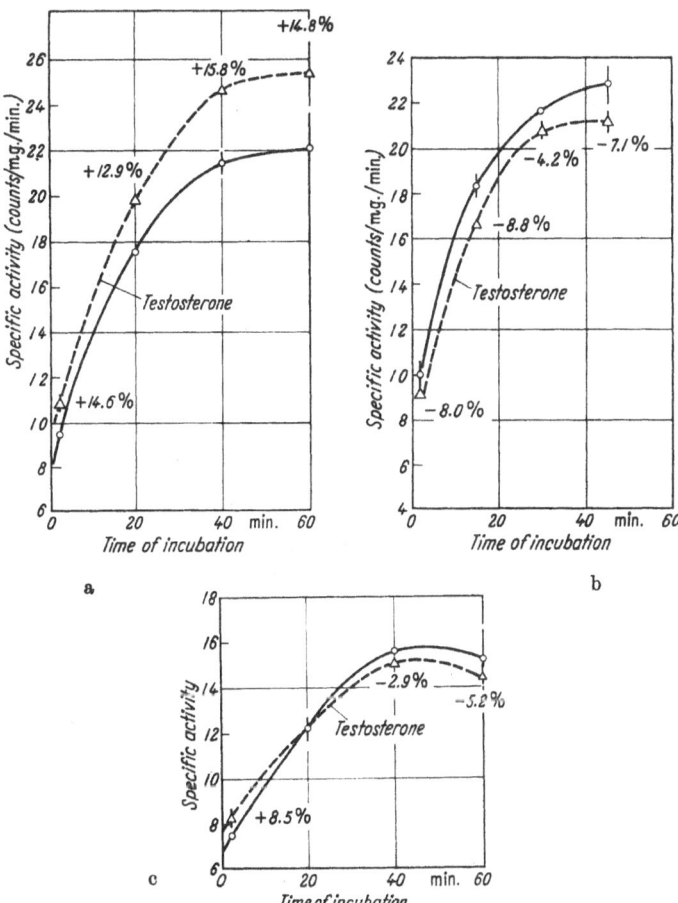

Fig. 3. Varying effects of 10 mcg. testosterone/3 ml (interrupted line), on the rate of incorporation of 0.3 μC ^{14}C-glycine-(U) into proteins of mouse-liver homogenate. a) activation; b) inhibition, c) biphasic reaction

Later we switched to incorporation experiments, which were conducted by Dr. STREHLOW (1962). We examined the incorporation of ^{14}C-glycine into the proteins of mouse-liver homogenate under

various conditions, with particular emphasis on very short incorporation times. Even during the very first few minutes, rapid
incorporation occurs, which comes to a halt after about 1 hour. The
effect of added testosterone was studied at various times and with
various concentrations: 3 mcg. testosterone per 3 ml. did not exert
any effect on the rate of incorporation, whereas 10 or 100 mcg. per
3 ml. did influence incorporation, i.e. in two different directions.
In response to 10 mcg. testosterone there was for 60 minutes either
only activation (Fig. 3a) or only inhibition (Fig. 3b), or, alternatively, early activation followed by inhibition (Fig. 3c). In response
to 100 or 200 mcg. testosterone, we observed activation for the first
2—3 minutes, followed by inhibition (Fig. 4).

This means that for the first time we were able to demonstrate
activation of the incorporation of amino acids by adding testosterone. That other investigators have not made similar observations may perhaps be explained by the fact that they did not
perform such short-term experiments.

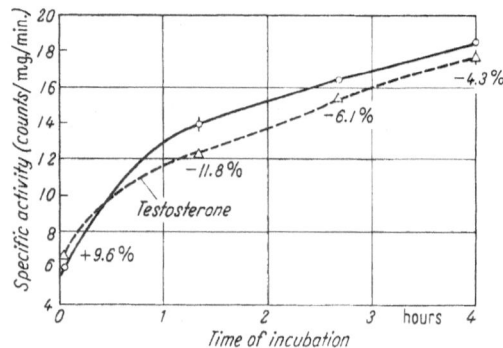

Fig. 4. Effect of 100 mcg. testosterone/3 ml. on the rate of incorporation of 0.3 μC ^{14}C-glycine-
(U) into proteins of mouse-liver homogenate

It is interesting to note that the early incorporation shows a
temperature coefficient of 1 between 0° C and 27° C, and of 1.1
between 27° C and 37° C. This would seem to suggest that the process is more of a physical one. Whether incorporation under the
influence of testosterone or incorporation at later times shows the
same temperature coefficient has not yet been investigated.

Similar results were obtained with mouse-kidney homogenate.
The incorporation of glycine was activated by testosterone after
2—3 minutes.

All these experiments to which I have just referred raise a number of questions that still have to be answered.

Finally, I should like to discuss the problem as to why it is that the so-called anabolic steroids show relatively little androgenic

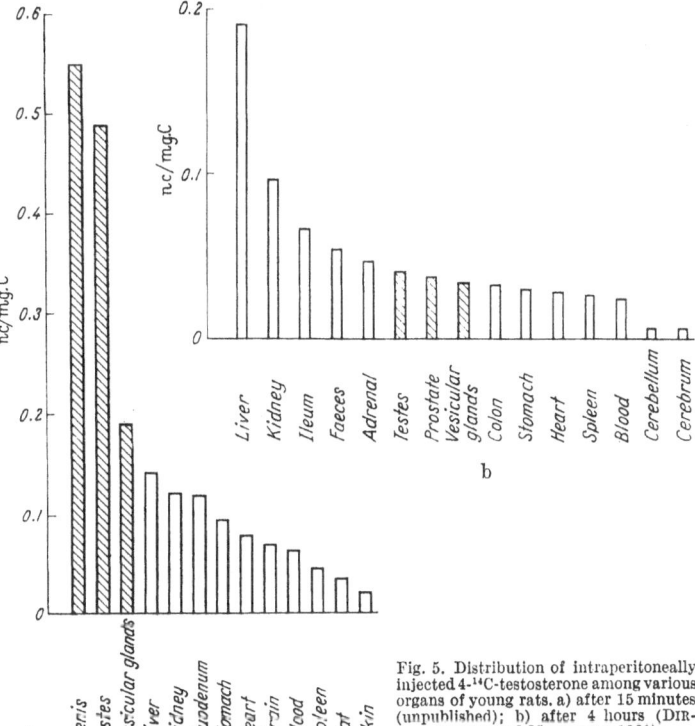

Fig. 5. Distribution of intraperitoneally injected 4-^{14}C-testosterone among various organs of young rats. a) after 15 minutes (unpublished); b) after 4 hours (DIRSCHERL and MOSEBACH, 1961)

activity. The ratio of anabolic to androgenic activity can be evaluated by means of the Hershberger test, in which growth of the levator ani muscle is compared with that of the seminal vesicles. It seemed important and useful to study the distribution of radioactive-labelled testosterone and of an anabolic steroid such as 1-methyl-^{14}C-Δ^1-5α-androsten-17β-ol-3-one-17β-acetate by injecting them into rats and determining the radioactivity in the various organs, including especially the levator ani muscle and the seminal vesicles. These experiments were performed by

Dr. MOSEBACH. Young rats were sacrificed 15 minutes after injection of the steroid into the jugular vein, and the specific activities (^{14}C/total C) in the organs were then determined. It was found that with testosterone the specific activity in the seminal vesicles was about five times as high as in the levator ani, whereas with the 1-methyl compound there was only 1.4 times as much activity. This finding is consonant with the fact that the 1-methyl compound displays a weaker androgenic effect. I believe that the anabolic effect of testosterone is the basis for its androgenic activity. Androgenic activity can be expected of a compound only if it enters the sex glands in sufficient quantity.

Figs 5a and 5b show the distribution of intraperitoneally injected ^{14}C-testosterone among various organs of young rats. After 15 minutes an accumulation of radioactivity is apparent in the sex organs; after 4 hours the concentration in these organs is reduced.

References

DIRSCHERL, W.: Erg. Physiol. (G.) **48**, 112 and 167 (1955). — DIRSCHERL, W., and H.-L. KRÜSKEMPER: Biochem. Zschr. (G.) **323**, 1 (1952). — DIRSCHERL, W., and K.-O. MOSEBACH: Liebigs Ann. Chem. (G.) **604**, 75 (1957). — DIRSCHERL, W., and K.-O. MOSEBACH: Acta endocr. (Den.) **36**, 115 (1961). — DIRSCHERL, W., and K.-O. MOSEBACH: unpublished data (1962). — DIRSCHERL, W., and T. MÜLLER: quoted from DIRSCHERL, W. (1955). — DIRSCHERL, W., and K. OTTO: Biochem. Zschr. (G.) **324**, 172 (1953). — DIRSCHERL, W., and K. OTTO: Acta endocr. (Den.) **25**, 64 (1957). — DIRSCHERL, W., and G. PROJAHN: Zschr. Vitamin-Hormon-Ferm. forsch. (Austria), in press (1962). — HAUSCHILDT, J. D., and C. M. GROSSMAN: Endocrinology (U.S.A.) **53**, 306 (1953). — JONES, Jr., H. W., and R. WADE: Science (U.S.A.) **118**, 103 (1953). — KIT, S., and E. S. G. BARRON: Endocrinology (U.S.A.) **52**, 1 (1953). — OTTO, K.: unpublished data (1962). — STREHLOW, J.: Thesis, Bonn, 1962.

General Discussion

WILSON: I think Dr. KORNER and I are in essential agreement that the messenger hypothesis is a logical hypothesis and that there are two possibilities: either there is a basic difference between the bacterial system and the mammalian system or they will turn out to be the same and a compromise will be made in the pure theory. The question is not yet quite settled and it's still open to a great deal of investigation.

RABEN: Dr. SZIRMAI's pictures were impressive in showing structural changes in the microsomal pattern, and I can't help wondering whether Dr. KORNER and Dr. WILSON paid perhaps too little attention to the rather gross change in microsomes in terms of numbers of microsomes and perhaps of the actual integrity of the microsomes themselves. I would like to ask

Dr. KORNER, for example, if all the isolated ribosomes have the same activity, if there is any information as to whether there are active and inactive ribosomes among the collection that are isolated, and also whether the number of ribosomes is affected by growth hormone.

KORNER: I don't know if there are gross changes in the endoplasmic reticulum of the liver after hypophysectomy. But the fact that I can detect changes in the isolated ribosomes, i. e. the ribonucleoprotein particles divorced from their lipoprotein membranes, suggests that STH changes the ribosome part of the endoplasmic reticulum. The yield of liver ribosomes from hypophysectomised rats is less than from the liver of normal rats. Nevertheless, per gramme of liver, as far as I can judge, the yield of ribosomes is about the same. The important thing is that the protein-synthesising activity of ribosomes per mg. RNA is affected by growth hormone, and this must mean that there are less active ribosomes in the preparations from hypophysectomised rats than in those from normal rats. It is suggested on the basis of work with bacterial systems (TISSIÈRES et al.)[1] that only about 5% of the ribosomes in any preparation are able to synthesise protein. The authors argue that only those ribosomes which have messenger RNA are able to synthesise protein. Clearly there are in our preparations particles which look like ribosomes from a physico-chemical point of view, but which are not ribosomes if a ribosome is defined as a particle which can incorporate amino acids into protein. The proportion of these active ribosomes in the ribosome preparations from normal and hypophysectomised rat liver will differ from each other. We are trying to find out how they differ and what differences exist in the proportions of the active ribosomes, but we have not yet succeeded.

WILSON: The difference between Dr. SZIRMAI's preparation and mine probably is that I have not used castrated animals compared with treated animals. I used non-castrated animals compared with animals given hormone. And we have always performed our studies at much earlier times than indicated on the slides he showed. Now we have done a whole series of electron-microscopic studies on the effect of oestradiol on the hen oviduct, and although we could demonstrate an effect on protein synthesis 2 hours after the injection of oestradiol, there was no difference in different experiments in the ergasto-plasmic membrane or in the ribosomes at this time interval. Now to be sure, at the end of 36 hours, as one would perhaps predict from previous studies, there is more endoplasmic reticulum involved in this increase in the cyto-plasm. But the disorganisation which Dr. SZIRMAI showed in his tissues is not present in the non-castrated immature organ. However, to get around this possibility, we assayed the microsome content in the preparation, as measured by RNA and protein content, and we were still able to demonstrate these ten-fold differences between the treated and non-treated preparations.

TUCHMANN-DUPLESSIS: With reference to the first very stimulating paper on protein synthesis and growth hormone I would like to make two general remarks. One is concerned with postnatal life and the other with prenatal life. There is no doubt that growth hormone is one of the keystones of protein synthesis. However, there is not always a close correlation between the amount of growth hormone and the rate of growth. It has been shown that

---

[1] TISSIÈRES, A., D. SCHLESSINGER, and F. GROS: Proc. Nat. Acad. Sc. U.S. **46**, 1450 (1960).

the growth-hormone content of the human pituitary is just the same in infants, adults, and old people. This raises the question: is growth determined only by the amount of growth hormone or also by the sensitivity of tissues to growth hormone ?

The second question is concerned with the influence of growth hormone on protein synthesis in foetal life. We have very good evidence that a foetus deprived of its pituitary will have just the same rate of growth as a normal foetus. On the other hand, many experiments have been done in which growth hormone was injected into the pregnant mother. No difference in the rate of growth of the foetuses in such cases has been shown. Thus, as far as prenatal growth and protein synthesis are concerned, it seems that they are not regulated by growth hormone. It is curious to see that in the period at which protein synthesis is so intense the main mechanism of hormonal regulation, i.e. growth hormone, does not seem to be involved.

QUERIDO: Is there anybody who would like to venture an answer to Dr. TUCHMANN-DUPLESSIS' question ?

LARON: If one finds the same amount of growth hormone in the pituitary at different ages, this does not imply that the secretion is identical at these various ages. It may be quite different, as we know from other glands. One fact of interest in this connection is that we produce growth hormone today from pituitaries of all ages. So I would say that one of the most important effects of growth hormone, which I am sure Dr. RABEN will discuss, is not its growth-promoting effect but presumably its many other basic metabolic effects.

LEATHEM: I just want to ask Dr. KORNER if he would comment about RNase and hypophysectomy. I'm afraid I don't remember whether you said it was altered or that it was not altered. I am interested in a relationship to protein depletion.

KORNER: It is known that the ribosomes contain ribonuclease, and obviously if in differently treated rats one finds a difference in the ability of ribosomes to incorporate amino acids into protein, one of the things one has to watch is that in some way activation of ribonuclease has not occurred. We did a lot of experiments to look at the rate at which the ribosomes lose their ability to synthesise protein when incubated under conditions where ribonuclease would be expected to be active. The rates at which ribosomes from hypophysectomised and normal rats lost their ability to synthesise protein paralleled each other. — Now I may answer Dr. TUCHMANN-DUPLESSIS' question. When I get discouraged about discovering how growth hormone influences ribosomes, I go to the rat-room and I look at hypophysectomised rats and normal rats, and the palpable difference between them persuades me that growth hormone does control growth.

ARIAS: Attractive as it always is to combine morphological and biochemical studies, I think one should not place a great deal of emphasis on the number of ribosomal particles seen on electron-microscopic examination of pellets obtained by differential ultracentrifugation. When combined radioautography and electron microscopy become perfected to the extent of permitting direct examination of amino-acid incorporation into protein in tissue sections, the morphologist may be able to clarify the important biological question of whether all ribosomes are active in protein synthesis.

SZIRMAI: I certainly would agree with the comment of Dr. ARIAS. One has to realise that the electron-microscopic image reveals dense particles in tissue sections which we call ribosomes and which we suppose to contain RNA only on circumstantial evidence. In the biochemist's test-tube, ribosomes are particles in which the presence of RNA can be demonstrated and its biochemical activity assessed. Although the two may be identical, electron microscopy reveals nothing about the state of activity of the particles, nor about the other forms of cytoplasmic RNA which we know is present in the cell and participates in protein synthesis. Differences in the biochemical activity of the ribosome or the microsome fraction are obviously not detectable with present-day electron microscopy, and conflicting data on the electron-microscopic appearance of the endoplasmic reticulum or the ribosomal particles in tissue sections should not be considered as necessarily discordant with biochemical data.

Factors influencing protein metabolism in the organism

Protein malnutrition and replenishment with protein in man and animals

By

J. C. WATERLOW

The subject covered by this title is a wide one with a very large literature. All I can do is to put forward a point of view derived from the work of our Unit on malnourished children in the West Indies. In quoting from other work I shall be deliberately selective, which I think is justified in a Symposium of this kind, where there is an opportunity for discussion and contradiction.

The patients we are concerned with are from 6 months to 2 years old. There is evidence that, even within this narrow range, age influences the course of the malnutrition (GARROW, 1962). The younger the child, the worse the outlook. The protein requirement of the child is much higher than that of the adult. This is not entirely accounted for by the extra nitrogen laid down during growth; there are also differences in metabolism — a higher basal metabolic rate and a higher endogenous nitrogen excretion. Whether these differences are ones of degree only, or of kind, we do not know. But, since they exist, one must be cautious about applying to adults ideas derived from work on infants.

The clinical picture of severe protein malnutrition

A typical patient is a baby 1 year old weighing 5 kg. — that is, half the normal weight for its age. The child is apathetic and miserable, with extreme muscular wasting, although subcutaneous fat may be fairly well preserved. Other common findings are oedema, an enlarged fatty liver, changes in the skin and mucosae, and depigmentation of the hair. This is the clinical picture to which the name kwashiorkor has been given. The salient biochemical features are a reduction in the plasma levels of total protein, albumin, and many enzymes (WATERLOW, CRAVIOTO, and STEPHEN, 1960); loss of potassium, magnesium, and phosphorus — that is, the intracellular minerals (HANSEN, 1956; SMITH and WATERLOW, 1960; MONTGOMERY, 1961); and retention of sodium and chloride. The mortality in advanced cases is from 10 to 20%.

Some of the associated changes, e. g. lesions of the skin and mucosae, probably result from vitamin deficiency. In Jamaica severe anaemia in malnourished infants is usually caused by lack of folic acid, and responds dramatically to this vitamin (BACK and MacIVER, 1960). These features are, however, very variable.

The main elements in the dietary pattern are:

1. Deficiency of protein. There is strong evidence that this is a deficiency of total protein, and not of one of the essential amino acids.

2. Some degree of calorie imbalance. If the intake of calories as well as that of protein is very low, the result is total starvation or marasmus. Usually in practice the diet provides calories, chiefly in the form of carbohydrate, in relative excess compared with the low level of protein. This imbalance seems to promote some of the specific features of kwashiorkor, particularly fatty liver and oedema, which statistically tend to occur together (WATERLOW, BRAS, and DEPASS, 1957). In fact, this association between fatty liver, oedema, and calorie excess was recorded more than 70 years ago by the Austrian paediatrician CZERNY, under the name *Mehlnährschaden* ("flour-feeding injury"). In spite of this long history, the underlying nature of the relationship is not understood, because it has proved unexpectedly difficult to reproduce experimentally the full picture of human kwashiorkor. Attempts to do so with the rat have on the whole failed; the pig may prove to be a better experimental animal (HEARD et al., 1958).

In practice the clinical picture usually falls somewhere between the two extremes of kwashiorkor and marasmus. In what follows I shall make no distinction, but for the sake of simplicity shall consider them all as cases of protein malnutrition or protein depletion. This certainly does not mean that one can ignore the effect of calorie intake on protein metabolism, but as a first approach it is easier to consider these two aspects of the problem separately.

The composition of the protein-depleted body

The malnourished child, half the normal weight for its age, is not just a miniature edition of the normal; its composition is abnormal in many ways.

Whole body. The body contains an excess of water, which is only to be expected if the child is oedematous. Usually, however,

the total body water is still high even after all detectable oedema
has disappeared (Table 1). This means that the reduction in body
solids is greater than would be expected from the deficit in body-
weight. So far it has not been possible to measure how much of the
solids is made up by fat and how much by protein and minerals, but
work along these lines is being pursued in Jamaica.

Table 1. *Total body water of malnourished infants as % of body-weight, at
different stages of recovery*

Stage	No. of subjects	Mean body water $\pm$ SEM
In presence of oedema	13	84.5 ± 1.4
Within 35 days of loss of oedema	18	73.1 ± 1.5
On recovery (more than 50 days after loss of oedema)	7	62.6 ± 2.3

Data from Smith, R. (1960).

For some reason, rats seem to differ from human beings in that
a low-protein diet usually does not cause water retention. Even
in severely depleted rats water content and nitrogen content per
unit body-weights are normal (Widdowson and McCance, 1957;
Stanier, 1957; Wallace et al., 1956).

Organs. The fact that the total protein concentration of the
body may be normal is deceptive; if we think in terms of *amounts*
rather than concentrations it is im-
mediately apparent that the *distribution* of protein in the body is abnormal. Addis and his co-workers
showed that in rats kept for a short
time on a protein-free diet the main
loss of protein was from the liver —
an observation confirmed many times
since (Addis, Poo, and Lew, 1936).
Table 2 shows measurements made
by Kerpel-Fronius in Hungary on
infants dying of malnutrition (Kerpel-Fronius and Frank, 1949). This
well illustrates the two extremes:
the brain, whose weight is main-

Table 2. *Body composition in
atrophic infants*

Part of body	% of normal weight
Whole body .	52
Brain	90
Kidneys . .	80
Heart	60
Fat	5
Skeleton. . .	85
Muscle . . .	30
Surface area .	70

Data from Kerpel-Fronius
and Frank (1949).

tained at all costs, and muscle, which bears the brunt of the loss.
Muscle forms the largest protein reservoir of the body; therefore,
if we are concerned with quantities, it needs particular attention.

The extent to which muscle is depleted in malnutrition is not always realised. Creatinine output has long been regarded as an index of muscle mass. Table 3 shows that in a series of patients

Table 3. *Daily creatinine output by malnourished babies at intervals during recovery*

Days after admission to hospital	Creatinine output/24 hours	
	mg./cm. height	mg./kg. weight
Less than 20	0.79	9.4
20—39	1.04	11.9
40—59	1.48	14.3
More than 60	1.86	15.0
Normal: 1 year (STEARNS, 1958)		13.5
2 years (STEARNS, 1958)		16.3

in Jamaica the output was reduced to one-third of normal. The loss of muscle substance has been demonstrated even more dramatically by MONTGOMERY (1962) by measurement of the number and size of muscle bundles (Table 4).

Table 4. *Number and cross-sectional area of fibres in sartorius muscle*

Patient	No. of fibres × 1000	Area of fibre bundles sq. mm.
32 weeks premature foetus	64	2.2
Full-term still-birth	101	9.8
Well-nourished infant, 13 months	128	28.3
Severely malnourished infant, 12 months		
(body weight 35% of normal)	60	2.1
Adult, 74 years	114	133

From MONTGOMERY, R. D. — Muscle morphology in infantile protein malnutrition (to be published).

Tissues and cells. Depletion not only affects differently different organs within the body, but also different tissues within an organ, and different proteins within a cell. I can but give a few examples. Collagen, once formed, is a very stable substance. In malnourished muscle, because of wasting of muscle fibres, there is a relative increase in collagen content, both chemically (Table 5) and microscopically. It would, of course, be quite wrong to call this a fibrosis. Another very stable substance is DNA, the amount of which per cell is determined genetically and not by environmental influences.

For this reason it is useful as a basis of reference when, as in the living human subject, only a sample of an organ is available for

Table 5. *Collagen content of muscle in rats on a low-protein diet, and after refeeding*

Treatment	Weight of paired gastrocnemii (g.)	Total collagen N (mg.)	Collagen N (% of total N)
Weaned	0.146	0.09	3.0
Normal diet, 30 days	2.23	1.53	3.1
'Jamaican' diet[1], 28 days . . .	0.26	1.44	34.1
'Jamaican' diet, then refed 10 days.	0.72	2.76	21.1
'Jamaican' diet, then refed 20 days.	1.56	2.91	10.5

[1] Contains 6% protein.

analysis, the total weight of the organ being unknown. Table 6 shows results obtained on human liver and muscle, using DNA as reference base. If it is assumed that the DNA content per cell is constant, these figures show that in malnutrition each cell on the average had lost 30—40% of its protein (WATERLOW and WEISZ, 1956).

Table 6. *Nitrogen content of biopsy samples of liver and muscle from malnourished infants, related to DNA — phosphorus (N/DNAP)*

	N/DNAP, mg./mg.	
	Before treatment	After treatment
Liver	49	83
Muscle . . .	237	343

There is evidence that this loss of protein is not uniform. The attempt by LUCK (1936) 25 years ago to show differential loss of protein from the liver in starvation was not successful, because only rather crude methods of protein fractionation were available. More recently KAPLANSKY (1959) has shown by electrophoretic separation that there is a differential loss of liver proteins. DICKERSON's extremely interesting studies on muscle suggest a fairly uniform loss of sarcoplasmic and myofibrillar proteins (DICKERSON and CABAK, 1962). All this work, however, is very much in its early stages.

In a parenchymatous organ such as the liver it is probable that enzymes account for a very large part of each cell's complement of proteins. Enzymatic activity is a convenient method of distinguishing specific proteins. There has been a great deal of experimental work, summarised by KNOX, AUERBACH, and LIN (1956), on the enzymatic activity of tissues in different nutritional states. Undoubtedly there is some differential loss of enzymes, though with no very obvious relation to functional activity. A very few studies have been done on man. Assuming that activities are proportional to amounts, the results summarised in Table 7 indicate that in malnutrition some enzyme proteins are reduced more than others.

Table 7. *The activity of some enzymes in biopsy specimens from the livers of malnourished infants before and after treatment*

Enzyme and units	Before treatment	After treatment
I. Cholinesterase Q_{CO_2}	6.8	8.9
Dehydrogenases Q_{O_2}		
lactic	54	52
malic	237	173
glutamic	6.5	8.4
Transaminase Q_{CO_2}	299	280
Cytochrome reductase Q_{O_2}	38	31
Succinoxidase Q_{O_2}	21	27
II. Xanthine oxidase, μmoles/g. protein/hr	2.6	6.9
D-amino-acid oxidase, μmoles/g. protein/hr	150	432
Glycolic acid oxidase, μmoles/g. protein/hr	471	506
DPNH-dehydrogenase, mmoles/g. protein/hr	18.5	15.4
Malic dehydrogenase, mmoles/g. protein/hr	116	106
Transaminase, mmoles/g. protein/hr . .	82	71
III. Catalase, ml. O_2/mg. N/hr	3.2	1.4
Cholinesterase units/mg. wet weight/hr	0.06	0.3
Alkaline phosphatase, mcg. P/mg. wet weight/hr	4.13	2.12

Data from: I. WATERLOW and PATRICK (1954). II. BURCH et al. (1957). III. MUKHERJEE and SARKAR (1958).

All these changes can conveniently be summarised in one phrase: *alterations in pattern*. In protein depletion there is an alteration in the pattern of protein distribution in the body at the level of the organ, the tissue, and the cell. Quite clearly, however, there is no single abnormal pattern, but a progression of changes which are dependent upon the severity of the protein deficiency and its duration, and probably also upon the supply of calories. Some results of WIDDOWSON and MCCANCE (1956) are an excellent illustration of how these factors may affect the pattern of nitrogen loss (Table 8).

I do not know of any detailed experimental studies which show
how the pattern changes with time as the depletion progresses. With
human patients it is almost impossible to study this, but one can
more easily observe the reverse sequence — that of repletion.
There are many observations which indicate that the process of
repletion, like that of depletion, is not uniform. In the malnourished
infant plasma proteins and enzymes rise very rapidly after treatment is begun (DEAN and SCHWARTZ, 1953). The duodenal enzymes
are also very quickly restored (THOMPSON and TROWELL, 1952).

Table 8. *Nitrogen content of liver and body in undernourished and starved rats*

	$^o/_o$ of control		
	Body-weight	N in liver	N in rest of body
Rats acutely starved	73	61	88
Rats chronically undernourished	70	68	72

Data from WIDDOWSON, E. M., and R. A. McCANCE (1956).

On the other hand, body-weight increases more slowly, even when
allowance is made for the fact that loss of excess water tends to
mask gains in tissue. We have found that, after a satisfactory food
intake has been established, it may still take 3—4 weeks before
there is a significant increase in creatinine output, suggesting a lag
in the laying down of new muscle. The same thing is seen in
experiments with rats: on repletion after a low-protein diet, the
liver gains protein very rapidly at first, but the rate of gain quickly
falls off. In muscle the sequence is the opposite: a lag period,
followed by growth which is slow at first, and faster later on
(MENDES and WATERLOW, 1958).

In an earlier review (WATERLOW, CRAVIOTO, and STEPHEN, 1960)
it was suggested that the proteins of the body might be divided
into two broad categories — mobile and fixed. In the normal body
it seems from the data collected by DARROW and HELLERSTEIN
(1958) that the proportions would be about 60% mobile and 40%
fixed. It may be more useful to distinguish 3 groups, according
to the response to depletion and repletion:

1. Proteins which are little affected, e. g. of brain and some
extracellular proteins (collagen).

2. Proteins which are lost rapidly at first, then more slowly,
e. g. of liver.

3. Proteins which are lost slowly at first, then more rapidly,
e. g. of muscle.

These groups are not sharply divided; the differences are only ones of degree (with the possible exception of brain). One explanation for the differences in behaviour of the proteins in these broad groups is that they depend upon differences in the rates of turnover. So far we have been considering the changes that result from depletion and its reversal from a purely static point of view. To understand more clearly how they are brought about, one must take into account the dynamic state of the body proteins.

Dynamic changes

The changes that have been described in the composition and pattern of tissues and cells must result from changes, relative or absolute, in rates of synthesis or break-down of their proteins. Measurement of these rates is difficult, and in human subjects even more difficult than in animals. Most types of metabolic study, whether with isotopes or by simple balance methods, take some time, during which it is not possible to keep the subject depleted. He must be treated with food, and therefore is not in a steady state. Nevertheless, I shall refer mainly to work on human subjects, as there are others better qualified to discuss the animal experiments.

Catabolism. Measurements of catabolic rate that can be accepted as valid have been made only on plasma proteins. In animals the results are not very clear-cut, but they do suggest that on a low-protein diet the rate of catabolism is reduced, and the half-life lengthened (STEINBOCK and TARVER, 1953; JEFFAY and WINZLER, 1958). In the malnourished human infant it has been found by PICOU, working in our Unit, that the fractional rate of albumin break-down is greatly decreased (PICOU and WATERLOW, 1962). Albumin labelled with I^{131} was injected soon after admission to hospital, when the child was severely depleted, and again just before discharge when the nutritional state was good. The results are shown in Table 9. In this work the break-down rates were calculated from the urinary excretion of iodide. No valid deductions can be drawn from the rate of decay of radioactivity in the plasma, because, particularly during the initial study period, the albumin pool is increasing in size as a result of net synthesis. This has the effect of increasing the slope of the plasma decay curve and artificially shortening the half-life. The table shows the discrepancy between results calculated from measurements on plasma and on urine. It is probably for this reason that GITLIN and his co-workers in Mexico, in a study based on plasma decay rates, found no difference in albumin half-life between malnourished and normal infants (GITLIN, CRAVIOTO, FRENK, MONTANO, RAMOS-GALVAN, GOMEZ, and JANEWAY, 1958).

Preliminary studies by PICOU show that the catabolic rate of γ-globulin is also reduced in the malnourished child. It is tempting to suppose that the slowing of break-down rate demonstrated for these two proteins is an adaptation to shortage of supplies. If

Table 9. *Turnover rate and half-life of plasma albumin labelled with I^{131} in malnourished and recovered infants*

	Malnourished	Recovered
Fractional turnover rate, ...	3.5	5.9
% of albumin pool per day	± 0.24	± 0.38
Absolute turnover rate, ...	1.41	2.78
g. per day	± 0.13	± 0.28
Half-life, from urinary	24.6	13.1
excretion, days	± 2.5	± 0.7
Half-life from plasma	10.2	10.4
slope, days	± 0.7	± 0.7

Mean, ± SEM.

tissue proteins behave in the same way, this clearly represents a compensatory reaction of great importance. There is as yet no direct evidence of changes in the rate of tissue-protein break-down; there is, however, indirect evidence of the existence of a fairly substantial pool of protein, perhaps mainly in the liver, whose turnover rate is close to that of plasma proteins, both in the normal and in the depleted animal (GARROW, 1959).

Synthesis. Even the severely malnourished child appears to retain a remarkable capacity for protein synthesis. This statement naturally applies only to those patients who survive and recover. It may well be that in others a point of no return is reached, where protein synthesis fails and the patient dies, but in this situation systematic investigation is almost impossible. Synthesis in the patient who survives is shown by the rapid increase in concentrations of plasma proteins, duodenal enzymes, etc., and by the large retentions of nitrogen found in balance measurements. These, however, are only indications of net synthesis; what we would like to have are measurements of absolute rates of synthesis during depletion and recovery.

Attempts to make such measurements with labelled amino acids are notoriously difficult to interpret. In none of the nutritional studies with which I am familiar have all the data been obtained that are necessary for a rigorous kinetic analysis. When a difference is found between a normal and a malnourished subject, the problem that arises is to distinguish between two kinds of effect: those that may be called physical or statistical, and those that represent real biological differences. The first group are often regarded simply as hindrances to a proper interpretation of the experiment, but I think they have some significance. I shall try to illustrate this by some rather naive examples, mainly from my own work and that of my colleagues.

1. When a labelled amino acid is injected, more label appears in the plasma proteins in the depleted than in the normal subject (GARROW, 1957; 1959). Without knowledge of the specific activity of the precursor and the pool size of the plasma proteins, no conclusions can be drawn about rates of synthesis. Nevertheless, the fact remains that a greater proportion of the precursor has been utilised.

2. When plasma proteins are labelled by injecting or feeding a labelled amino acid, the apparent half-life is longer than the true half-life, because of re-utilisation of labelled material liberated from other tissues. The apparent half-life is longer, sometimes much longer, in the depleted than in the normal organism (Table 10) (GARROW, 1959; YUILE, LUCAS, OLSON, and SHAPIRO, 1959). This indicates a greater degree of re-utilisation, that is, of economy.

Table 10. *Effect of diet on the apparent half-life of plasma proteins*

	Apparent half-life, days	
	Normal diet	Protein-free diet
Dogs, C^{14}-lysine (YUILE et al., 1959)	14	>45
Dogs, S^{35} methionine (GARROW, 1959) . . .	14.5	24

3. When labelled methionine was injected in normal and depleted rats, and the extent of incorporation measured in different organs at various time intervals, in the depleted as compared with the normal rats there was more incorporation in the liver and internal organs, and less in the muscle, carcass, and skin (WATERLOW, 1959). This was originally interpreted as evidence of a concentration of synthesis in the essential organs at the expense of the less essential, but the effect could result simply from a greater dilution in the normal animal of amino acids coming to the liver by unlabelled material from the food.

In a recent repetition of these experiments with C^{14}-lysine, in which measurements were made of the amount and specific activity of the free amino acid, preliminary results suggest that there is a real decrease in the rate of incorporation in muscle.

Of all these experiments it may be said that there is no evidence of real changes in rates of synthesis or break-down. All the findings that suggest increased incorporation may be explained on the reasonable assumption of a decrease in amount and an increase in specific activity of the free amino-acid precursor. This, however, does not mean that there is no biological interest attached. All these changes are evidence of a greater economy of utilisation of available material. McFARLANE has made an interesting development of this point: analysis was made by a computer of a model in which a precursor pool was supposed to be in communication with two other pools, one with a high and one with a low rate of turnover (McFARLANE, WATERLOW, and GARROW, 1960). If the precursor pool contracts in size, the proportion of material taken by the high-activity pool increases, and that of the low-activity pool decreases. If this purely statistical effect occurs in the living organism, it would result in increased incorporation in one organ at the expense of another, and ultimately in a change in the distribution of protein.

Protein metabolism presents a supreme example of a physiological equilibrium. It is remarkable that with so many proteins turning over at different rates, the net result in the normal subject should not only be nitrogen balance, but presumably a constant composition of different parts of the body. In depletion this equilibrium is disturbed. The first line of defence is a physical one — a kind of self-regulatory mechanism which leads to greater economy of utilisation and some redistribution of nitrogen. The fact that these changes are physically determined does not destroy their biological significance. An analogy may be drawn with acid-base regulation; the first line of defence is the physico-chemical buffering by bicarbonate and haemoglobin, which precedes regulation by respiratory centre and kidney.

This, which I have called a buffering mechanism, must surely be supplemented by biological reactions of a homoeostatic nature. Examples have already been given of what seem to be real changes in rates of synthesis and break-down. We know very little about how these rates are normally regulated. Hormones, particularly steroid hormones, which form the main subject of this Symposium, undoubtedly play an important role, though I think that in the field of endocrines, as in that of nutrition, there are few experiments from which precise deductions can be drawn. As an exception, the effects of ACTH, cortisol, and thyroxin on the rate of albumin catabolism are well documented, although how they are brought about ist not known.

I believe that the part played by insulin may also be very important, at least in relation to the nutritional situation that I have described in children. As YOUNG and his co-workers have shown, insulin *in vitro* promotes the uptake of amino acids by muscle (MANCHESTER and YOUNG, 1958). MUNRO found that *in vivo* insulin causes a gain of nitrogen by muscle, but a loss from liver, while cortisone has the opposite effect (MUNRO, 1956; MUNRO, BLACK, and THOMSON, 1959). Perhaps this action of insulin may explain in part the apparently harmful effect of a high carbohydrate intake in protein-depleted children.

In ending, I should like to emphasise that the characteristics of protein depletion in adults, as seen for example in war-time famine oedema, seem in some ways to be different from those in infants. From the metabolic point of view, one major difference is in the rates of turnover. In the normal adult the albumin half-life is twice as long as in the child. The overall rate of turnover in the adult has been calculated to be of the order of 5—10 g. protein per kg. per day — i. e. about 5 times the intake from food (SAN PIETRO and RITTENBERG, 1953; MAURER, 1960). Our preliminary observations suggest that it may be double this in infants. Perhaps related to this difference in turnover rate is the finding of CHRISTENSEN and his co-workers that in the infant there is a much greater concentration difference of free amino acids between cells and extracellular fluid; the authors call this effect "tissue amino-acid hunger" (CHRISTENSEN, THOMPSON, MARKEL, and SIDKY, 1958). Compared with the hungry infant, the adult probably has less capacity for synthesis, but he is also likely to be more economical under stress.

I hope that this contribution may do something to bridge the gap, which is larger than it ought to be, between those working on protein metabolism and nutrition in infants, mostly overseas, and those studying metabolic problems of adults in well-developed countries, where malnutrition is usually secondary to an endocrine or other disturbance.

Summary

The protein requirement and protein turnover in children are much higher than in adults, not only because of the extra nitrogen laid down during growth, but also because of differences in metabolic rate. The clinical picture of severe protein malnutrition in children (kwashiorkor, flour-feeding injury or "*Mehlnährschaden*") differs from that seen in adults. No satisfactory animal-experimental model for human kwashiorkor or similar conditions has yet been found.

In the protein-depleted child, the composition of the body is abnormal in many respects; the protein depletion does not affect all organs and tissues to the same extent, nor are all cellular proteins affected in the same way. Similarly, the process of protein repletion, like that of depletion, is not uniform in all the organs. A distinction can be drawn between three groups of proteins, according to their response to depletion and repletion: a) proteins which are little affected; b) proteins which are lost rapidly at first, and then more slowly, e.g. of the liver; c) proteins which are lost slowly at first, and then more rapidly, e.g. of muscle. These groups are not sharply divided, however, the differences being only ones of degree.

On a low-protein diet the rate of protein catabolism is reduced and the rate of albumin and globulin break-down decreases. Amino-acid incorporation appears to be favoured in some organs, e.g. the liver, at the expense of others, e.g. muscle. Very little is known about how these processes are normally regulated, but hormones (cortisol, thyroxin, insulin) appear to play a role in this connection.

Zusammenfassung

Eiweißbedarf und Eiweißumsatz des Kindes sind wesentlich größer als beim Erwachsenen, nicht nur wegen des erhöhten Eiweißansatzes, sondern auch als Folge einer anderen Stoffwechselrate. Das klinische Bild des schweren Eiweißmangels beim Kind (Kwashiorkor, Mehlnährschaden) weicht von dem des Erwachsenen ab. Bisher ist kein tierexperimentelles Äquivalent zum kindlichen Eiweißmangel bekannt.

Die Zusammensetzung des Körpers beim Eiweiß-unterernährten Kind unterscheidet sich qualitativ von der Norm; der Eiweißverlust betrifft weder alle Organe und Gewebe in gleichem Maße, noch sind alle Zellproteine in gleicher Weise betroffen. Ebensowenig verläuft der Vorgang der Wiederauffüllung mit Eiweiß in allen Organen einheitlich. Prinzipiell lassen sich drei Gruppen von Körpereiweißen im Hinblick auf ihr Verhalten bei Mangel und Wiederzufuhr unterscheiden: a) Proteine, die durch exogenes Eiweiß wenig beeinflußt werden; b) Proteine, die bei Eiweißmangel zunächst rasch, dann langsam abnehmen (Leber); c) Proteine, die erst langsam, dann schnell abnehmen (Muskel). Dabei finden sich alle Übergänge zwischen diesen Gruppen.

Bei niedrigem Eiweißangebot ist der Katabolismus der Eiweiße eingeschränkt, der Abbau von Albuminen und γ-Globulinen vermindert. Bei Wiederzufuhr scheint der Einbau von Aminosäuren in einzelnen Organen, z. B. der Leber, begünstigt zu sein auf Kosten anderer, z. B. Muskel. Über die Regulation dieser Vorgänge ist wenig bekannt, doch ist anzunehmen, daß Hormone (Cortisol, Thyroxin, Insulin) daran beteiligt sind.

Résumé

Les besoins et le "turnover" des protéines chez l'enfant sont beaucoup plus importants que chez l'adulte, non seulement parce qu'il y a une assimilation accrue de protéines, mais encore parce que le métabolisme de base est plus élevé. Le tableau clinique de l'insuffisance protéique grave chez l'enfant (kwashiorkor, dystrophie farineuse) diffère de celui de l'adulte. Jusqu'à présent, on ne connaît pas chez l'animal d'équivalent expérimental de la carence protéique infantile.

La composition du corps chez l'enfant sous-alimenté en protéines est qualitativement différente de la normale; la déplétion protéique n'atteint pas tous les organes et tissus dans la même mesure; toutes les protéines cellulaires non plus ne sont pas touchées de la même façon. De même, le processus de récupération en protéines ne se déroule pas de façon uniforme dans tous les organes. En principe, selon leur comportement en cas de carence et de réadministration, on distingue 3 groupes de protéines dans le corps: a) Protéines peu influencées par les apports protéiques exogènes; b) Protéines diminuant d'abord rapidement puis lentement en cas de carence protéique (foie); c) Protéines diminuant d'abord lentement puis rapidement (muscle). Entre ces groupes, on trouve tous les types intermédiaires.

Si l'on réduit l'apport de protéines, leur catabolisme diminue, de même que la dégradation des albumines et des globulines γ. L'incorporation des acides aminés semble favorisée dans certains organes, par exemple le foie, aux dépens d'autres, par exemple le muscle. On ne sait pas grand'chose sur la régulation de ces processus, mais on peut admettre que des hormones (cortisol, thyroxine, insuline) y participent.

References

ADDIS, T., L. J. POO, and W. LEW: J. Biol. Chem. (U.S.A.) **115**, 117 (1936). — BACK, E. H., and J. MACIVER: Arch. Dis. Childh. (G.B.) **35**, 134 (1959). — BURCH, H. B., G. ARROYAVE, R. SCHWARTZ, A. M. PADILLA, M. BÉHAR, F. VITERI, and N. S. SCRIMSHAW: J. Clin. Invest. (U.S.A.) **36**, 1579 (1957). — CHRISTENSEN, H. N., D. H. THOMPSON, S. MARKEL, and M. SIDKY: Proc. Soc. Exper. Biol. Med. (U.S.A.) **99**, 780 (1958). — DARROW, D. C., and S. HELLERSTEIN: Physiol. Rev. (U.S.A.) **38**, 114 (1958). — DEAN, R. F. A., and R. SCHWARTZ: Brit. J. Nutr. **7**, 131 (1953). — DICKERSON, J. W. T., and V. CABAK: Proc. Nutr. Soc. (G.B.) **21**, 3 (1962). — GARROW, J. S.: J. Clin. Invest. (U.S.A.) **38**, 1241 (1959); — S³⁵-Methionine uptake in protein-depleted Jamaican children. In: Amino-acid Malnutrition: XIIIth Annual Protein Conference, New Brunswick, U.S.A. Rutgers University Press, 1957, p. 14. — To be published (1962). — GITLIN, D., J. CRAVIOTO, S. FRENK, E. L. MONTANO, R. RAMOS-GALVÁN, F. GOMEZ, and C. A. JANEWAY: J. Clin. Invest. (U.S.A.) **37**, 682 (1958). — HANSEN, J. D. L.: S. Afr. J. Laborat. Clin. Med. **2**, 206 (1956). — HEARD, C. R. C., B. S. PLATT, and R. J. C. STEWART: Proc. Nutr. Soc. (G.B.) **17**, 41 (1958). — JEFFAY, H., and R. J. WINZLER: J. Biol. Chem. (U.S.A.) **231**, 111 (1958). — KAPLANSKY, S. Y.: Clinical Chem. (U.S.A.) **5**, 186 (1959). — KERPEL-FRONIUS, E., and K. FRANK: Ann. paediatr. (Switz.) **173**, 321 (1949). — KNOX, W. E., V. H. AUERBACH, and E. C. C. LIN: Physiol. Rev. (U.S.A.) **36**, 164 (1956). — LUCK, J. M.: J. Biol. Chem. (U.S.A.) **115**, 491 (1936). — McFARLANE, A. S., J. C. WATERLOW, and J. S. GARROW: International Atomic Energy Agency. Proceedings of a Conference in Bangkok, 1960. Vienna, 1962. — MANCHESTER, K. L., and F. G. YOUNG: Biochem. J. (G.B.) **70**, 363 (1958). — MAURER, W.: In: Dynamik des Eiweißes. 10. Colloquium der Gesellschaft für Physiol. Chem. Springer-Verlag, Berlin-Göttingen-Heidelberg, 1960. — MENDES, C. B., and J. C. WATERLOW: Brit. J. Nutr. **12**, 74 (1958). — MONTGOMERY, R. D.: Lancet (G.B.) 1960/I, 74; — In press (1962). — MUKHERJEE, K. L., and N. K. SARKAR: Brit. J. Nutr. **12**, 1 (1958). — MUNRO, H. N.: Scott. Med. J. **1**, 285 (1956). — MUNRO, H. N., J. G. BLACK, and W. S. C. THOMSON: Brit. J. Nutr. **13**, 475 (1959). — PICOU, D., and J. C. WATERLOW: Clin. Sc. (G.B.) **22**, 459 (1962). — SAN PIETRO, A., and D. RITTENBERG: J. Biol. Chem. (U.S.A.) **201**, 457 (1953). — SMITH, R., and J. C. WATERLOW: Lancet (G.B.) 1960/I, 147. — STANIER, M. W.: Brit. J. Nutr. **11**, 206 (1957). — STEARNS, G.,

K. J. Newman, J. B. McKinley, and P. C. Jeans: Ann. N. Y. Acad. Sc. **69**, 857 (1958). — Steinbock, H. L., and H. Tarver: J. Biol. Chem. (U.S.A.) **209**, 127 (1954). — Thompson, M. D., and H. C. Trowell: Lancet (G.B.) 1952/I, 1031. — Wallace, W. M., W. B. Weil, and A. Taylor: The effect of variable protein and mineral intake upon the body composition of the growing animal. Contribution presented at meeting of American Pediatric Society, May, 1956 (cyclostyled). — Waterlow, J. C.: Nature (G. B.) **184**, 1875 (1959). — Waterlow, J. C., G. Bras, and E. Depass: J. Trop. Pediatr. (G.B.) **2**, 189 (1957). — Waterlow, J. C., J. Cravioto, and J. M. L. Stephen: Advances Protein Chem. **15**, 131 (1960). — Waterlow, J. C., and S. J. Patrick: Ann. N. Y. Acad. Sc. **57**, 750 (1954). — Waterlow, J. C., and T. Weisz: J. Clin. Invest. (U.S.A.) **35**, 346 (1956). — Widdowson, E. M., and R. A. McCance: Brit. J. Nutr. **10**, 363 (1956). — Brit. J. Nutr. **11**, 198 (1957). — Yuile, C. L., F. V. Lucas, J. P. Olson, and A. B. Shapiro: J. Exper. Med. (U.S.A.) **109**, 173 (1959).

Discussion

ASCHKENASY: I should like to make two remarks concerning Dr. WATER-
LOW's paper. The first is related to the difference found in the water content
of tissues in protein-starved rats, which is normal, an in children with
kwashiorkor, in whom it is increased. I should like to suggest the possibility
that the difference may be due to the association in human subjects of some
other alimentary deficiencies, such as deficiencies in methyl groups and/or
in vitamin B_{12}. I would remind you that choline deprivation has been shown
to induce oedema in protein-starved rats[1]. My second remark concerns the
hormonal control of the tissular distribution of proteins in protein malnutrition. I should like to draw attention to some special aspects of such control
in the rat. First of all, the very early atrophy of the accessory sex organs in
protein starvation is obviously due to inhibition of androgen production.
On the other hand, the characteristic involution of thymus and spleen may
be partly due to a relative hyperactivity of the adrenals, which seems to
exist at least at the beginning of malnutrition. In protein-depleted rats,
adrenalectomy prevents the involution of these organs, as well as the decrease
in the blood lymphocyte levels[2].

Finally, hormonal disturbances certainly play an important role in the
development of protein-starvation anaemia. Thus, the well-known priority
of erythropoiesis in protein deprivation seems to be conditioned, at least
in part, by the functional activity of the thyroid gland; the removal of the
latter considerably enhances the anaemia, whereas thyroxin injections
significantly delay the onset of anaemia[3].

WATERLOW: As far as your first point about choline and B_{12} is concerned,
Dr. ASCHKENASY, the fatty liver naturally has led us to think that choline
deficiency may be important. All I can say is that neither choline nor B_{12}
have the slightest effect in treatment. Blood levels of B_{12} do not seem to be
particularly low in these patients. They are very often deficient in folic acid,
but that is another matter. As for the question of the hormonal control of
protein loss and protein repletion, I came here to learn and have nothing to
offer; we know almost nothing about the hormonal status of these children.
At post-mortem the adrenals are quite markedly hypotrophic. A few measurements of aldosterone output have been made in relation to oedema and
electrolyte changes; we expected it to be increased, but in fact it was decreased. Thyroid activity appears to be more or less normal, judging from
Dr. MONTGOMERY's[4] results in iodine-uptake studies, but otherwise I am
afraid I have very little to contribute on this point.

FRASER: May I ask if that means that you have not looked at the
pituitaries?

[1] ALEXANDER, H. P., and H. E. SAUBERLICH: J. Nutrit. (U.S.A.) **61**, 329
(1957).
[2] ASCHKENASY, A.: Rev. belge path. **24**, 365 (1955).
[3] ASCHKENASY, A.: Sang (Fr.) **25**, 15 (1954).
[4] MONTGOMERY, R. D.: in press (J. Clin. Path. [G. B.] 1962)

WATERLOW: No, I have not done so personally. I have not heard that the pathologists found anything interesting in the pituitaries as far as ordinary histological methods go.

RUNDLES: Some years ago, GILLMAN[1] reported from South Africa some interesting studies on nutritional deficiency, pellagra, and cirrhosis in infants. I believe there were two striking features that may or may not pertain to your patients. There was first a striking increase in iron in the liver, and secondly they observed a remarkably effective therapeutic result from the administration of defatted hog stomach. I suppose in modern parlance the latter might be the result of protein replacement or folic-acid therapy.

WATERLOW: The iron accumulation seems to be peculiar to South Africa and some parts of West Africa. I expect Dr. HOFFENBERG from South Africa could help us about that. It certainly does not occur in kwashiorkor in other parts of the world. As far as the hog stomach is concerned, we were familiar with GILLMAN's work, but it is of course extremely difficult to do comparative trials and to compare the efficiency of different kinds of protein; even so, I think later experience has not confirmed GILLMAN's claim that hog's stomach is any better than milk.

HOFFENBERG: I have no wish to become involved in the controversy regarding the origin of the excessive hepatic iron, i. e. whether it is derived from simple dietary-iron overload or whether there is a primary hepatic cellular disturbance — possibly due to malnutrition — with secondary iron deposition. I believe that this question has not finally been answered. In the Endocrine Laboratories in Cape Town, an investigation of adrenal function in kwashiorkor was made and showed no significant alteration, as measured by 17-ketosteroid and 17-hydroxysteroid excretion and by ACTH stimulation[2]. This investigation is now being extended to include studies of pituitary function using Metopirone.

I am pleased that Dr. WATERLOW stressed the difficulty in determining absolute rates of synthesis, which is considerable. I would also like to emphasise the need for maintenance of a steady state during isotope studies of this sort. Some earlier publications reached fallacious conclusions because of altered plasma-albumin specific activity resulting from lack of a steady state over the period of testing. In the Department of Paediatrics at the University of Cape Town, 4 infants with kwashiorkor have been studied with labelled proteins. In all cases the albumin turnover rate was extremely slow and returned to normal levels after refeeding. In addition, a very high turnover rate of globulin was found in these infants[3].

SCHREIER: I should like to say only a few words about the results we obtained in studies on malnourished rats and rabbits[4]. We had an opportunity of studying the metabolism of C^{14}-labelled lysine in protein, carbohydrate, and fat-depleted rats and in newborn rabbits. In agreement with the results of Dr. WATERLOW, we found that a sort of shift takes place: the incorporation rate of C^{14}-lysine into liver and kidney proteins increases in particular, whereas the incorporation into haemoglobin, myosin, etc. decreases markedly.

[1] GILLMAN, T., and J. GILLMAN: Arch. Int. Med. (U.S.A.) **76**, 63 (1945). GILLMAN, J., and T. GILLMAN: Arch. Path. (U.S.A) **40**, 239 (1945).
[2] LURIE, A. O., and W. P. U. JACKSON: Clin. Sc. (G. B.) **22**, 259 (1962).
[3] PURVES, L.: personal communication (1962).
[4] SCHREIER, K.: Dtsch. Zschr. Verdauungskrkh. **20**, 80 (1960).

Dr. WATERLOW has already made the necessary critical comments on this problem. Of course, there are big differences between absolute starvation and protein depletion as regards their influence on the incorporation of labelled amino acids. I would only add that the catabolism of lysine apparently decreased. The C^{14} in the expired air was about 50% in the protein-depleted animals as compared with the normal controls. The "free" activity of the amino-acid pool in the liver of these animals was increased.

WATERLOW: Yes, I am familiar with the experiments to which Dr. SCHREIER referred. We injected S^{35}-methionine into protein-depleted and normal rats, which were killed at various intervals, and measured the distribution of the dose in different organs. The results showed that in the depleted animal at all time intervals more of the dose was recovered from the liver and internal organs and less from the carcass, muscle, and skin (Fig. 1).

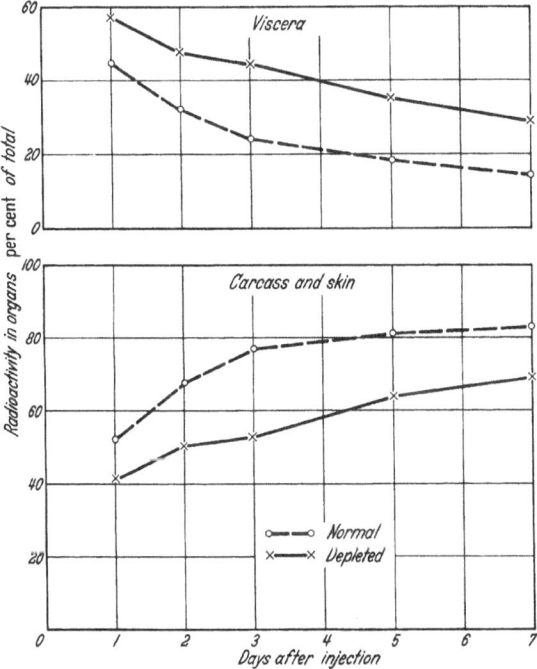

Fig. 1. The percentage of total body radioactivity in the viscera and in the carcass and skin of normal and protein-depleted rats after injection of S^{35}-methionine

I interpreted this, as I think Dr. SCHREIER does, as a kind of concentration of synthesis in essential organs at the expense of the less essential. This, I feel, was a naive interpretation. The results could be explained, at least in theory, on the basis that in the normal animal there is a greater dilution of labelled amino acid by non-labelled amino acid coming in from the food. However, Dr. STEPHEN and I have recently been repeating some of this work with labelled lysine. We are now measuring the specific activity of the free lysine in the tissues, and the preliminary results do suggest a real decreased rate of incorporation into muscle, because the free amino-acid specific activity is high in the presence of relatively low specific activity in the muscle protein. So I think there is a real change here, but how it is brought about, I don't know.

QUERIDO: Before ending up, I should like to put just one other question to Dr. WATERLOW. Do you actually do away with the expression of labile protein reserves?

WATERLOW: Entirely, yes.

QUERIDO: Would you explain them on the basis of the turnover rate?

WATERLOW: Yes, that would be my feeling.

The bearing of the plane of nutrition on growth and endocrine development

By

R. A. McCance and E. M. Widdowson

We have been studying for some time the effect of varying the plane of nutrition on development and subsequent growth, and for this purpose we have used rats, pigs, and cockerels. The experiments are not complete, but results have been obtained which indicate that the endocrine system is involved in ways which are not yet understood.

In our experiments the young rats in two litters born on the same day were mixed. Three, taken at random, were given to one mother and the remainder, preferably 15 to 18, to the other. Both mothers usually accepted the young and reared them. At three weeks of age the rats suckled in groups of 3 were two to four times the weight of those suckled in groups of 15. They were, moreover, a correspondingly greater length. The animals were all weaned at 3 weeks and afterwards given unlimited amounts of the stock diet. In spite of this, the small ones continued to grow more slowly than the large ones, and the difference in size continued to increase (WIDDOWSON and McCANCE, 1960). The failure of a young animal to catch up with its fellows or littermates for a long time, and indeed its tendency to get further and further behind for months after a nutritional handicap *before* birth, is a common experience with many species. The average growth curves given by McCANCE and WIDDOWSON (1962) demonstrate what may happen in rats when the growth rates only begin to diverge *after* birth. The imprint of the differential food intakes during suckling persisted throughout the animals' lives, and those that were small when they were 3 weeks old never attained the stature and weight of the others in spite of unlimited food (WIDDOWSON and KENNEDY, 1962).

The body-weights are not the only interesting features of these experiments. WIDDOWSON and McCANCE (1960) and DICKERSON and WIDDOWSON (1960) studied the influence of the rate of growth on the composition of the animals and on their physiological development in point of time. Some features — the appearance of the teeth and the opening of the eyes, for example — were very little affected by the plane of nutrition during suckling and appeared at the normal chronological age (Outhouse and Mendel, 1932—1933). Others, such as the onset of puberty and the desire to explore new surroundings (Lát et al., 1961), were accelerated by a high rate of growth, and in these particular experiments seemed to be linked with the attainment of a definite size. This may have been fortuitous so far as puberty was concerned, for, as will be shown later, by varying the plane of nutrition it is possible to dissociate the rate of sexual development from the rate of somatic growth and to produce signs of puberty when the animals are still quite small.

The outstanding questions posed by these experiments seem to be: (1) To what extent does the plane of nutrition during the first three weeks of life bring about its effects through the operations of the growth hormone ? Perhaps not at all if one can judge from the age at which growth ceases in pituitary dwarfs and from what we know about growth rates *in utero*, where nutrition *per se* seems to dominate the situation (McCance, 1962). (2) Why do the animals which have been suckled in large groups not attain the same stature as the others ? The epiphyses of a rat do not fuse even in adult life, and the animal probably stops growing because it ceases, after a certain time, to respond to the available growth hormone. There is no evidence that the amount of hormone to be found in the pituitary falls off with advancing age in the rat more than it does in other species in which the epiphyses do fuse (Baird et al., 1952; Armstrong and Hansel, 1956; Baker et al., 1956; Solomon and Greep, 1958; Bowman, 1961), but failure of response offers an explanation as to why a slow rate of growth for any reason should lead to permanent stunting in this animal, and Fábry and Hrůza (1956) have shown that in intermittent starvation the administration of growth hormone can prevent it (Hrůza and Fabry, 1957). If "failure of response" can be equated with maturation of the bones, we have some evidence that this may underlie the failure to attain full stature. Fig. 1, which is taken from Dickerson and Widdowson (1960), shows the forepaws of three rats. A is the forepaw of a rat 7 days old which had been suckled in a large group and grown slowly. B is the forepaw of a 7-day-old animal which had grown fast. It is larger, as might have been expected, but it is also clearly more mature as judged by epiphyseal development. C is the forepaw of a rat 21 days old which had grown slowly and which at that age had only attained the weight of the animal 7 days old which had been growing fast. Paws B and C are the same size, but C is much more mature as judged by the usual criteria, and this was confirmed by finding more calcium in the body as a whole and in the femur, which was analysed by itself. At any given size, therefore, the bones of the animals suckled in large groups were more mature, and they may have

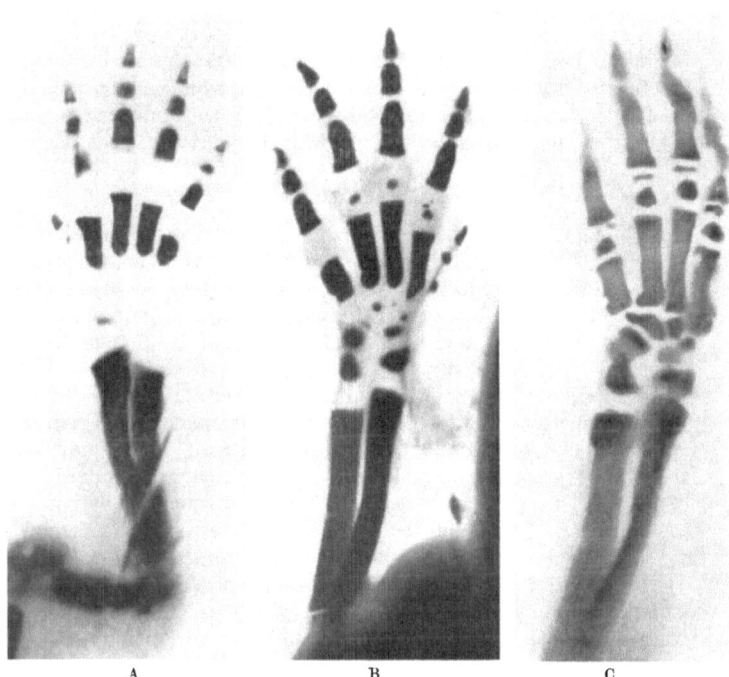

A   B   C

Fig. 1. The development of the forepaw. A. Forepaw of slow-growing 7-day-old rat ($\times$ 5); B. Forepaw of fast-growing 7-day-old rat ($\times$ 5); C. Forepaw of slow-growing 21-day-old rat ($\times$ 5)

become sufficiently mature by the time the animals were 300 g. in weight to stop growing any further. In man, undernutrition holds up both maturation and growth, but the latter more than the former (ACHESON and HEWITT, 1954; HEWITT et al., 1955; ACHESON and MACINTYRE, 1958; ACHESON and FOWLER, 1962).

A check in the rate of growth later in life did not reduce the final stature of the animals (MCCANCE, 1962). Rats which were growing fast were undernourished from 9 weeks of age till they weighed only as much at 12 weeks as those which had grown slowly all their lives because they had obtained less food during suckling. The rats held back from 9—12 weeks at once gained weight at great speed on being given unlimited food and quickly rejoined their fellows which had never been checked. It will probably be found that these results are true in a general way of all species, although they may vary considerably in detail from one to another; but there has not yet been enough controlled work done on larger animals to be quite sure. From the work on rats carried out by ourselves and others it would appear that the younger the animal the more serious a nutritional set-back will be, and permanent effects from undernutrition during foetal life are a distinct possibility.

Cockerels and pigs

If Rhode Island Red cockerels are fed on very restricted amounts of food on which they will grow splendidly if given enough of it, they can easily be maintained below a weight of 180 g. for 6 months, in which time brood mates will grow to 4 kg. or thereabouts. Newborn pigs weigh about 1 kg. and can be held below 6 or 7 kg. in just the same way till they are a year old. Normal pigs weigh 120 to 150 kg. at this age. The undernourished animals are very thin and they must be kept in a warm environment, for even at ambient temperatures of 21° C they have subnormal body temperatures. Some growth takes place in spite of the undernutrition, particularly in organs with strong anabolic tendencies. The bones grow a little and, although the long bones appear osteoporotic and the cortex thin, this cortex has a high calcium/collagen ratio (Dickerson and McCance, 1961). In some instances the development of one organ seems to have a high priority at the appropriate age. The teeth are a good example of this and continue to develop, although there is not room for them in the mandible and the maxilla. In other instances development seems to be hormone-directed, and into this category come the testicles, which have been known for some time to grow faster than the rest of the body in undernourished rats (Siperstein, 1921).

When the pigs and cockerels were given unlimited food they began to grow at speeds which were not far short of those at which they would have grown 6 months or a year before, and the body temperatures soon rose to normal. This growth was accomplished on the usual amounts of food (cf. Mendés and Waterlow, 1958), and presumably normal amounts of growth hormone were made available as soon as full nutrition was restored. These animals seem to differ from infants with kwashiorkor, who seldom gain weight in the early phases of recovery unless their calorie intakes have risen well above the normal (WATERLOW, 1961). This renewed growth of the pigs and cockerels involved expansion of the soft tissues and extensive remodelling of the bones. New bone began to be laid down by the growth cartilage and at the periosteum, but for some time the old acellular and overcalcified cortical bone of the undernourished animals remained unresolved. X-rays showed this surviving rudiment of undernourished cortical bone inside the bones of partially rehabilitated cockerels (PRATT and McCANCE, 1961). In a few months the histological appearances of the bones became normal, but the cockerels have never attained their expected stature or weight, and the epiphyses closed when the bones were shorter than those of their brood mates (PRATT and McCANCE, 1961). This is probably what will happen to the pigs, but we are not yet quite sure, and some interesting issues are involved.

The skull and jaws also underwent complete remodelling and assumed what appeared to be the right size and proportions for the animal. This allowed the impacted molars to disentangle themselves and appear in the mouth, but when they did so the 3rd molars, which had been laid down during the period of undernutrition, were still impacted and only some two-thirds' the expected size. While, therefore, most of the changes brought about by undernutrition were of a temporary nature and reversible, others were permanent (McCANCE et al., 1961).

After the female pigs had been undernourished for a year, the vulva was larger than it should be in an animal of the same weight, and the ovaries sometimes contained large cystic follicles which were quite out of place in such a small animal. When the animals were rehabilitated the vulva grew relatively much faster than the rest of the body and became very conspicuous (McCANCE, 1960). It is thought that this rapid growth is likely to be a response to oestrogens produced in the cystic ovary. The testicles also became too large for the size of the animal at a similar stage of rehabilitation. This was primarily due to a gradual enlargement of the tubules during undernutrition, for the interstitial cells, which are very numerous in the early weeks of a pig's life, lost their cytoplasm quite early, although some of the nuclei may have persisted. When the animals were rehabilitated after a year of undernutrition, the tubules underwent rapid expansion and maturation, and interstitial cells later reappeared. The tubules had a diameter of 150—180 μ after 2—3 months of rehabilitation and contained abundant sperm, although the animals still only weighed 35 kg. Normal animals were about 60 kg. in weight before their testicles had reached a similar stage of development.

Hitherto the emphasis has always been on the delayed sexual development of the undernourished child or animal (Ellis, 1945; Butler et al., 1945) and these differential effects have not been studied. At the right chronological age for their maturation, however, some development of the sexual glands evidently takes place under what would appear to be the influence of the follicle-stimulating hormone of the pituitary (Asdell, 1957) — even if the undernutrition is very severe — and organs in the female such as the vulva, which respond to the oestrogens produced, begin to grow faster than the rest of the body. When, moreover, rehabilitation is practised at the right moment the growth priorities of these secondary target organs allow them to become most conspicuous. Testicles, large for the weight of the body, have been recorded before in both undernourished rats (Winters et al., 1927; Jackson and Smith, 1931; Smith, 1931; Clarke and Smith, 1938) and goats (Wilson, 1960). Siperstein (1921) made a detailed histological study of the effect of undernutrition and rehabilitation on the testicles of the rat. The physiological side of this has never been investigated, but there are many facets of it to be explored. From the histological appearances of the ovaries and testes it looks as though undernutrition made the pituitary secrete the follicle-stimulating rather than the luteinising hormone. Some such effect as this may have accounted for the gynaecomastia which was often described in male prisoners of war in 1945—46 (Keys et al., 1950; McCance et al., 1951).

Puberty in human beings puts an end to bone growth after an initial spurt. The latter is probably due to androgens in both sexes produced by the adrenal glands (Wilkins, 1957), and in males also by the testicles. The maturation of the bones and the cessation of growth are attributed primarily to oestrogens, but androgens will do the same, and even the male has some oestrogens in his tissues (Seckel, 1946). In normal children, growth after puberty varies inversely as the growth before (Gallagher and Seltzer, 1961), and precocious puberty, whatever its first effects, ultimately leads to dwarfism in both sexes (Seckel, 1950; Jolly, 1955; Ferrier et al., 1961; McGeorge and O'Connor, 1961). Failure of bones, which rehabilitation makes quite normal histologically, to reach their proper length after prolonged undernutrition may be secondary to a "puberty" which is certainly precocious for the size of the animal although admittedly not so for its age. This problem can probably be solved by the undernutrition and rehabilitation of castrates and the assay and administration of hormones. Its complete solution, however, may take some time and require the study of primates (TANNER, 1962). The role of the adrenals, moreover, which are always enlarged in undernourished pigs, will certainly have to be considered at the same time. Undernutrition has been spoken of as causing "pseudohypophysectomy" (MANN, 1960), but if the follicle-stimulating hormone of the pituitary initiates the development of the sexual glands of these pigs, and through them the secondary target organs, the pituitary must have a high metabolic priority which enables it to do this relatively normally with respect to time in spite of the general hold-up due to the undernutrition.

Summary

In rats, overfeeding or underfeeding during the first 3 weeks after birth results in an increased or decreased rate of growth. If all are then transferred to an unlimited diet, the animals undernourished during the first few weeks of life never attain the size and weight of their fast-growing littermates. At any given size, however, the bones of the slow-growing rats are more mature. If fast-growing rats are later placed on an inadequate diet and their weight reduced to that of the slow-growing ones, they quickly make good the deficit when given unlimited food again. The younger the animal, the more serious the nutritional set-back is, and permanent effects from undernutrition during foetal life are a distinct possibility.

The growth of cockerels and young pigs can be greatly retarded by giving them very restricted amounts of food on which they would have grown splendidly if given enough of it. Although the long bones appear osteoporotic and the cortex thin, this cortex has a high calcium/collagen ratio. The development of some organs, e. g. the teeth, goes forward slowly, but not in the normal fashion. The sexual organs develop and may become larger than normal for the size of the body. This may be due to the pituitary producing an abnormal ratio of follicle-stimulating to luteinising hormone.

Zusammenfassung

An Ratten führt Über- oder Unterernährung während der ersten 3 Wochen nach der Geburt zu beschleunigtem oder verlangsamtem Wachstum. Werden die Tiere anschließend auf eine unlimitierte Diät umgesetzt, so erreichen die während der ersten Lebenswochen unterernährten Tiere weder Länge noch Gewicht ihrer schnell wachsenden Geschwistertiere des gleichen Wurfes. Bei jeder Körpergröße zeigen jedoch die Knochen der langsam wachsenden Ratten einen höheren Reifungsgrad. Erhalten schnell wachsende Ratten später eine ungenügende Nahrung, so daß ihr Gewicht bis zu dem ihrer langsam wachsenden Geschwister absinkt, so gleichen sie diesen Verlust rasch aus, wenn sie wieder unbegrenzt Nahrung erhalten. Je jünger das Tier ist, desto schwerwiegender sind die Folgen einer auch nur vorübergehenden Unterernährung, die zu Dauerschäden führt, wenn sie in der Foetalperiode eintritt.

Junge Hähne und junge Schweine können durch quantitative Begrenzung einer qualitativ vollwertigen Ernährung erheblich im Wachstum zurückgehalten werden. Die Röhrenknochen erscheinen osteoporotisch und haben eine dünne, jedoch übercalcifizierte Cortex mit hohem Calcium/Collagen-Verhältnis. Einzelne Organe, z. B. die Zähne, entwickeln sich langsam, aber

8*

nicht in normaler Weise. Die Sexualorgane können sich stärker entwickeln als der Körpergröße entspricht. Dafür kann möglicherweise ein verändertes Verhältnis von follikelstimulierendem zu luteinisierendem Hormon verantwortlich sein.

Résumé

Si l'on suralimente ou sous-alimente des rats pendant les 3 premières semaines de leur vie, on accélère ou ralentit leur croissance. Si on les nourrit ensuite tous normalement, les animaux sous-alimentés durant les premières semaines de leur vie n'atteignent jamais le poids ni la longueur des rats plus grands et bien nourris dès leur naissance. Pour n'importe quelle taille cependant, les os des animaux sous-alimentés sont à un stade de maturation plus avancé. Si les rats qui se sont développés rapidement reçoivent plus tard une alimentation insuffisante réduisant leur poids à celui des rats qui se développent lentement, ils récupèrent rapidement le déficit dès que les restrictions alimentaires cessent. Plus l'animal est jeune, plus les conséquences d'une sous-alimentation sont graves; si elle intervient à la période foetale, elle peut entraîner des lésions durables.

La croissance de jeunes coqs et de jeunes porcs peut être considérablement retardée si on leur donne des quantités très réduites de la nourriture qui leur aurait assuré un développement pleinement satisfaisant, si elle avait été administrée en quantité suffisante. Bien que les os longs semblent atteints d'ostéoporose et que la corticale paraisse mince, cette dernière présente un taux calcium/collagène élevé. Le développement de certains organes, par exemple des dents, progresse lentement. Les organes sexuels se développent, mais pas de façon normale, et peuvent devenir trop grands pour les dimensions du corps. Cela pourrait être dû au fait que l'hypophyse produit l'hormone folliculo-stimulante et l'hormone de lutéinisation dans une proportion anormale.

References

Acheson, R. M.: In: Symposia of the Society for the Study of Human Biology, Vol. 3. Human Growth (Ed. by J. M. Tanner). Pergamon Press, London (1960). — Acheson, R. M., and G. B. Fowler: Nature (G.B.) (In press) (1962). — Acheson, R. M., and D. Hewitt: Brit. J. Prev. Soc. Med. 8, 59 (1954). — Acheson, R. M., and M. N. MacIntyre: Brit. J. Exper. Path. 39, 37 (1958). — Armstrong, D. R., and W. Hansel: J. Anim. Sc. (U.S.A.) 15, 640 (1956). — Asdell, S. A.: In: Progress in the Physiology of Farm Animals, Vol. 3 (Ed. by J. Hammond). Butterworth, London 1957, p. 743. — Baird, D. M., A. V. Nalbandov, and H. W. Norton: J. Anim. Sc. (U.S.A.) 11, 292 (1952). — Baker, B., R. Hollandbeck, H. W. Norton, and A. V. Nalbandov: J. Anim. Sc. (U.S.A.) 15, 407 (1956). — Bowman, R. H.: Nature (G.B.) 192, 976 (1961). — Butler, A. M. J., M. Ruffin, M. M. Sniffen, and M. E. Wickson: N. England J. Med. 233, 639 (1945). — Clarke, M. F., and A. H. Smith: J. Nutrit. (U.S.A.) 15, 245 (1938). — Dickerson, J. W. T., and R. A. McCance: Brit. J. Nutr. 15, 567 (1961). — Dickerson, J. W. T., and E. M. Widdowson: Proc. Roy. Soc., London, Biol. Sc. 152, 207 (1960). — Ellis, R. W. B.: Arch. Dis. Childh. (G.B.) 20, 97 (1945). — Fábry, P., and Z. Hrůza: Physiol. Bohemoslov. 5, 10 (1956). — Ferrier, P., T. H. Shepard, and E. K. Smith: Pediatrics (U.S.A.) 28, 258 (1961). — Gallagher, J. R., and C. C. Seltzer: Pediatrics (U.S.A.) 27, 984 (1961). — Hewitt, D., C. K. Westropp, and R. M. Acheson: Brit. J. Prev. Soc. Med. 9, 179 (1955). — Hrůza, Z., and P. Fábry: Gerontologia (Switz.) 1, 279 (1957). — Jackson,

C. M., and D. E. SMITH: Amer. J. Physiol. **97**, 146 (1931). — JOLLY, H.: Sexual Precocity. Blackwell, Oxford, 1955. — KEYS, A., J. BROŽEK, A. HENSCHEL, O. MICKELSEN, and H. L. TAYLOR: The Biology of Human Starvation. University of Minnesota Press, Minneapolis, 1950. — LÁT, J., E. M. WIDDOWSON, and R. A. McCANCE: Proc. Roy. Soc., London, Biol. Sc. **153**, 347 (1960). — McCANCE, R. A.: Brit. J. Nutr. **14**, 59 (1960); — Food, growth and time. Lancet (G.B.) (In press) (1962). — McCANCE, R. A., R. F. A. DEAN, and A. M. BARRETT: Spec. Rep. Ser. Med. Res. Coun. (G.B.) No. **275**, p. 135 (1951).—McCANCE, R. A., E. H. R. FORD, and W. A. B. BROWN: Brit. J. Nutr. **15**, 213 (1961). — McCANCE, R. A., and E. M. WIDDOWSON: Nutrition and growth. Proc. Roy. Soc., London, Biol. Sc. (In press) (1962). — McGEORGE, M., and D. V. CONNOR: Arch. Dis. Childh. (G.B.) **36**, 439 (1961). — MANN, T.: Proc. Nutr. Soc. (G.B.) **19**, 15 (1960). — MENDÉS, C. B., and J. C. WATERLOW: Brit. J. Nutr. **12**, 74 (1958). — OUTHOUSE, J., and L. B. MENDEL: J. Exper. Zool. (U.S.A.) **64**, 257 (1932—33). — PRATT, C. W. M., and R. A. McCANCE: Brit. J. Nutr. **15**, 121 (1961). — SECKEL, H. P. G.: Med. Clin. North America **30**, 183 (1946); — Amer. J. Dis. Child. **79**, 278 (1950). — SIPERSTEIN: D. M.: Anat. Rec. (U.S.A.) **20**, 355 (1921). — SMITH, A. H.: J. Nutrit. (U.S.A.) **4**, 427 (1931). — SOLOMON, J., and R. O. GREEP: Proc. Soc. Exper. Biol. Med. (U.S.A.) **99**, 725 (1958). — TANNER, J. M.: personal communication. (1962). — WATERLOW, J. C.: J. Trop. Pediatr. (G.B.) **7**, 16 (1961). — WIDDOWSON, E. M., and G. C. KENNEDY: Proc. Roy. Soc., London, Biol. Sc. **156**, 96 (1962). — WIDDOWSON, E. M., and R. A. McCANCE: Proc. Roy. Soc. London, Biol. Sc. **152**, 188 (1960). — WILKINS, L.: The Diagnosis and Treatment of Endocrine Disorders in Childhood and Adolescence. Blackwell, Oxford, 1957. — WILSON, P. N.: J. Agric. Sc. **54**, 105 (1960). — WINTERS, J. C., A. H. SMITH, and L. B. MENDEL: Amer. J. Physiol. **80**, 576 (1927).

Discussion

SCHREIER: I am sure that paediatricians — and especially those interested in obesity — will be very grateful to Dr. McCANCE for having demonstrated that overnutrition accelerates bone age in animals. This is a finding which we have ample opportunity of observing in our obese children (particularly girls) suffering from what has been termed "adipose gigantism", i.e. a form of obesity occurring in pre-puberty. I should like to ask Dr. McCANCE one question in this connection. What do you think about the influence of overnutrition on the enzymes? Isn't it possible that not only growth hormone is responsible for this "adipose gigantism" and for the effects of overnutrition in animals, but also the excessively high content of substrates for the enzymes?

McCANCE: I am not able to answer your question about enzymes; it is rather a difficult one. We are all familiar with the fact that the overnourished child tends to be a tall child, but whether growth hormone is involved or not, I do not know.

WATERLOW: In relation to your remarks about chronological age, I wonder if you can explain an observation that Dr. MONTGOMERY of our unit has made in children. He measured their total oxygen uptake at various stages after recovery; it is rather low at first and then rapidly rises to a plateau, whereas the body-weight goes up more slowly. The height of this plateau corresponds to the oxygen uptake that a normal child should have at the same age. In other words, in the malnourished child the total uptake is high in relation to its weight, but not in relation to its age. Can you elucidate this at all?

McCANCE: Dr. MONTGOMERY thinks that his results are largely due to the metabolism of the brain — doesn't he? — for this is an important part of the total metabolism of the malnourished infant.

WATERLOW: Yes, I think that is one factor.

McCANCE: In pigs we do not get this result, but of course in pigs the brain represents a very much smaller portion of the weight of the animal.

PRADER: Dr. McCANCE, you did not mention the bone maturation or the bone age of your pigs. In dwarfed or undernourished children we usually see that the onset of puberty is more closely correlated with bone maturation than with chronological age or with the size or the patient. I got the impression from your slides that the bone maturation in your undernourished pigs was not as retarded as the size of the animals. This would explain why they reach puberty before their size corresponds to the usual size at puberty. My first question is therefore whether you find a correlation between bone maturation and sexual development in your pigs.

My second question concerns your demonstration that rats and pigs show a very different growth pattern after they have been undernourished during the early period of their life. Would you say that in this respect man is rather like a pig or like a rat? If I understood you correctly, the human baby is rather like a pig in this respect. I mean that an undernourished baby will show compensatory growth as soon as it gets enough food and will soon reach its normal growth channel, whereas your rats do not show this "catch-up" growth. It would be most interesting to know whether the puberty and bone age of your pigs are correlated.

MCCANCE: We have not tried to correlate bone age with sexual development, and I am therefore grateful for your suggestion. Small babies, as you know, tend to remain small children, at any rate until puberty; but the infant which has received a nutritional set-back early in life certainly has a much better chance of reaching the height predicted for him genetically than has the rat.

BAUER: One thing strikes me about the x-ray you showed in your Fig. 1. In the undernourished animals, the shaft doesn't look quite as slender as in the normal controls. It almost looks as if this animal had hypoparathyroidism or lacked vitamin D.

MCCANCE: The bones of the pigs do change in shape, but I don't think we can relate this to parathyroid activity. Each bone seems to react rather differently, and in cockerels the corresponding bones may behave differently again.

TANNER: One question which always arises in work on different species in relation to growth is what phases of growth are physiologically comparable in the two species, or how the two can be made equivalent in "time". This is a very difficult problem, and I must make it clear that this figure (Fig. 1 a—f, p. 120) is only one of the many possible approaches. In the figure are plotted the rates of growth in weight of several species. In the rat, and also in the mouse, birth is very early in the curve; in the guinea-pig it is at about the peak of weight velocity, as in man and other primates. The second peak is peculiar to primates, the manner of whose growth differs very considerably from that of non-primate species.

I want to ask Dr. McCANCE what the curve for the pig looks like; my guess is that it is nearer that of the guinea-pig than the rat.

MCCANCE: Yes.

TANNER: If you use this sort of system (or several others for that matter) for equating the chronological ages, then the rat is born exceedingly early as compared with the human, with the guinea-pig, and with many other species. So the time at which the rat is subjected to malnutrition in Dr. McCANCE's quite beautiful method corresponds to a time somewhere in the human which is well in the foetal period. Quite recently, BARRACLOUGH[1] at Cambridge and GEOFFRY HARRIS[2] have made the very interesting observation that if at 5 days after birth in the rat, which is still on the rising portion of the curve, you give one single injection of oestrogen to the male rat, or of androgen to the female rat, you cause what must be lasting organisational changes in the hypothalamus. I would like to suggest that this is also what Dr. McCANCE has done by malnourishment here at this early time, i.e. that he has caused a lasting structural change in the hypothalamus, in the area concerned with regulation of the food intake, and that this is the reason for the very persistent change.

---

[1] BARRACLOUGH, C. A.: personal communication.
[2] HARRIS, G.: personal communication.

MCCANCE: The pig is much more advanced at birth than the rat. It is in some ways comparable with the human. It has very fair temperature regulation, whereas the rat has practically none. But so far as rate of growth is concerned, the pig is comparable with the rat. The rat doubles its birthweight in 5 days and the pig can do this in 6 or 7 days.

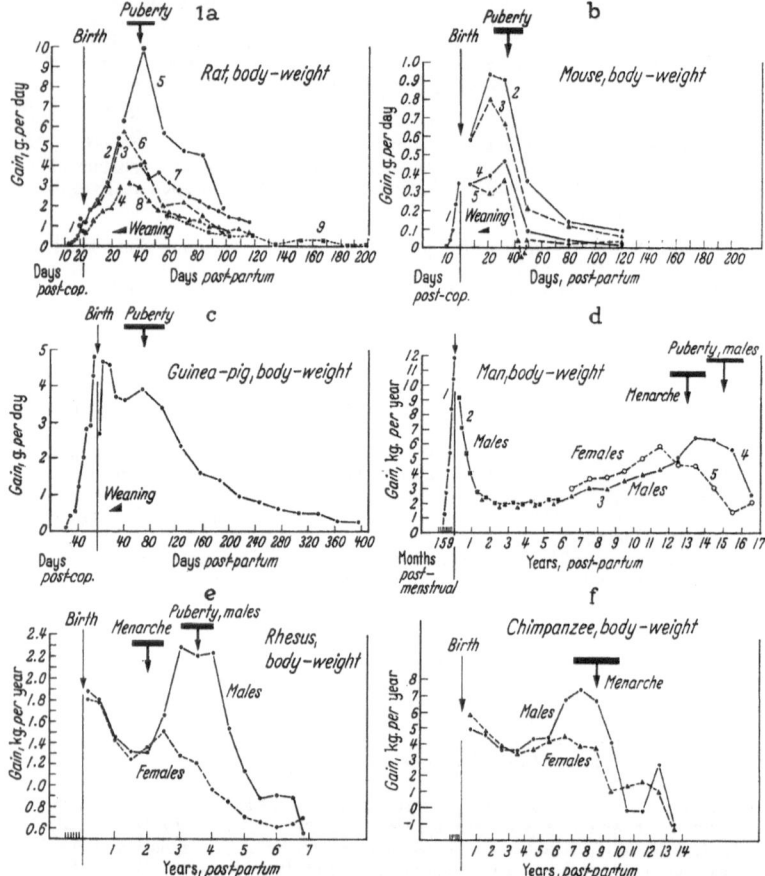

Fig. 1a—f. Rates of growth in weight of six species (from: "Growth at Adolescence", 2nd Edition. Blackwell's Sc. Publ., Oxford, 1962, where the sources of data used are given)

Metabolic effects of growth hormone in man

By

M. S. Raben, P. R. Minton, M. L. Mitchell, and H. Juarez-Penalva

Since it became evident several years ago that human and simian growth hormone, in contrast to bovine and porcine, were biologically active in man, the metabolic effects of human growth hormone in man have been extensively studied. While in general the results have been similar to those previously seen in animals, the metabolic responsiveness of man to small doses of human growth hormone has been impressive, and the consistency of the human response has been particularly gratifying in view of the many earlier abortive efforts to obtain such results.

The balance studies of BECK, PEARSON, HENNEMAN, BERGENSTAL, IKKOS and their associates established that human growth hormone caused retention of protoplasmic components in man (1). Nitrogen, phosphorus, and potassium were regularly retained, and sodium less consistently. Calcium balance was sometimes positive and other times not, but in either case the urinary loss of calcium was increased by growth hormone during these relatively brief periods of study. Serum calcium did not change. Based on a recent tabulation of published cases by FORBES et al. (2), the average increase in nitrogen retention with HGH during balance studies, disregarding the variations in dose and length of study, was 1.9 grams per day for 18 hypopituitary subjects and 2.4 grams per day for 5 normal subjects.

Other changes noted after growth hormone were enlargement of the chloride space and increased renal function as measured by the clearance of inulin, of creatinine, and of para-aminohippurate and by the maximal tubular excretion of para-aminohippurate. There was a rapid fall in blood urea nitrogen and, with continued treatment, a slow rise of serum phosphorus and alkaline phosphatase. The serum "sulfation factor" in man, like that in the rat, was shown to be controlled by growth hormone.

Recently a new finding has been added to the list by DULL et al. (3) and by JASIN et al. (4), namely that urinary excretion of hydroxyproline is increased by growth hormone, probably reflecting an alteration in the metabolism of collagen, the only protein containing hydroxyproline. Previous work had shown that soluble collagen, the presumed precursor of fibrous collagen, was present in larger amounts in growing animals. If urinary hydroxyproline, which is almost entirely in peptide form, is largely derived from the soluble collagen, the amount may be related to the abundance of this component. Children were found to excrete relatively more hydroxyproline than adults, and an acromegalic more than normal individuals, and the administration of HGH to hypopituitary subjects increased the excretion. Since urinary hydroxyproline was also high in hyperparathyroidism, PAGET's disease, and hyperthyroidism, the excretion may be related to the rate of formation of new collagenous tissue, whether this is part of growth or only of a more rapid turnover of this tissue.

Effects on metabolism of fat and carbohydrate

Human growth hormone causes in man a rise in free fatty acids in the plasma beginning about two hours after its administration. The rise is preceded by a transient fall of both fatty acids and glucose occurring about half an hour after intravenous injection. The fall was not attributable to insulin in the one study in which it was measured (5). The increase in fatty acids is substantial and prolonged, indicating an increased rate of delivery of free fatty acids by the adipose tissue to the blood (6). Several years of study have failed to clarify how growth hormone produces this increased mobilization of fat. Growth hormone appears to differ from corticotropin and epinephrine, which are lipolytic on direct contact with isolated adipose tissue, apparently by causing a rapid enhancement of lipase activity (7). Growth-hormone preparations are active *in vitro* only in doses so large as to suggest that traces of contaminants are the effective ingredients.

The *in vivo* effect of growth hormone on fatty acids has been regularly reproduced and generally accepted, but the physiological and pathological significance of this action remains speculative. DE BODO (8) has recently found that dogs receiving 1 mg./kg./day of bovine growth hormone show only a transiently increased mobilization and oxidation of fatty acids while maintaining a persistently increased delivery and use of glucose. The increased turnover of free fatty acids was evident within 3 hours, but lasted only 3 to 5 days, whereas the turnover of glucose was increased only after 24 hours, but was still rapid with continued treatment after 6 to 8 days; $1/_{25}$ of this dose of growth hormone was enough to raise plasma fatty acids after 3 hours, but had no effect on the turnover of glucose. The source of the additional glucose with the larger dose of growth hormone in the face of the protein-sparing action of the growth hormone is mysterious unless the animals consumed more food, which might also tend to inhibit the mobilization of fat, or unless less of the food was converted to fat. If dogs do fail to maintain an increased output of fatty acids in response to this dose of growth hormone, it may be that their secretion of insulin is sufficiently enhanced to prevent it. It is indeed possible that the dose of 1 mg./kg./day is a sub-diabetogenic dose for the very reason that it is insufficient to maintain a high concentration of plasma free fatty acids which might be a prerequisite for the development of diabetes. The much greater sensitivity of the hypophysectomized diabetic patient and the partially pancreatectomized dog to the ketogenic and hyperglycemic effects of growth hormone, and of the alloxan-treated dog to fatty acid mobilization (9), indicates the efficacy of pancreatic insular function in opposing these effects.

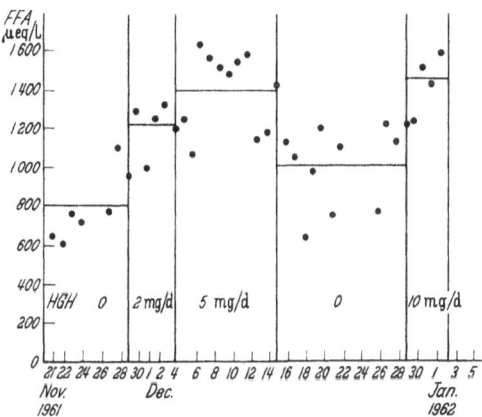

Fig. 1. Effect of HGH, administered intramuscularly daily at 6:00 p.m., on free fatty acids in the serum [μeq/l] drawn in the morning before breakfast. The horizontal bars indicate the average value for each period. Subject (♂ 71 yrs) had decreased glucose tolerance, but required no treatment

Relevant to the problem of whether prolonged excess of growth hormone can cause persistently excessive fat mobilization is the finding of RABINOWITZ and ZIERLER (10) that the output of fatty acids from the forearm, determined by arteriovenous difference using a superficial vein, was three times as high in acromegalics as in normal subjects. It had previously been noted that the value for plasma fatty acids in acromegalics was normal, though failing to rise after injected HGH, suggesting that the plasma value may

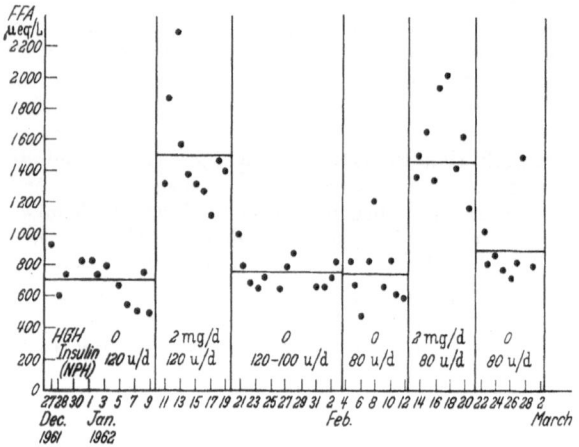

Fig. 2. Effect of HGH on fatty acids. HGH administered and FFA determined as in Fig. 1. Subject (♂ 62 yrs) was a diabetic patient receiving insulin as indicated

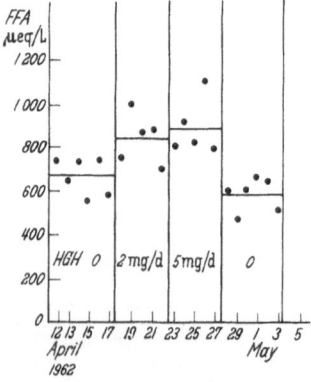

Fig. 3. Effect of HGH on fatty acids. HGH administered and FFA determined as in Fig. 1. Subject (♀ 19 yrs) was a normal individual

be a better indication of a change than of the absolute rate of fatty-acid output.

We have observed in one subject with diminished glucose tolerance, and in one subject with diabetes treated with insulin, a persistent increase in fatty acids with daily injections of HGH. The fatty acids were measured before breakfast about 14 hours after the injection of HGH. Figs 1 and 2 illustrate this response, lasting for as many as 16 days, the longest period tested. A normal subject showed only a small effect when tested in this way (Fig. 3). It is known from previous experience that these doses of HGH (2 to 5 mg.) produce a considerable rise in fatty acids in fasted normal subjects 4 hours after the injection, but a single injection of 4 mg. HGH raised fatty acids for less than 24 hours in normal subjects, although for longer than 24 hours in hypopituitary individuals.

The patient with diminished glucose tolerance (Fig. 1) had no change in blood glucose during treatment with growth hormone, but in the diabetic patient of Fig. 2, the treatment increased blood sugar as well as fatty acids. In the normal subject, the fasting blood sugar was unchanged, but the rate of disappearance of intravenously administered glucose may have diminished somewhat. From our experience with a small number of diabetic patients it appears that sensitivity to both the hyperglycemic and the fat-mobilizing effects varies considerably among subjects. Other investigators have documented the extreme sensitivity of severe diabetics who have been hypophysectomized, particularly to the ketogenic effect, but also to the hyperglycemic effect of growth hormone. An impressive example is the hypophysectomized juvenile-type diabetic studied by PEARSON who, following a single intramuscular injection of 1 mg. HGH, developed hyperglycemia and severe ketosis and acidosis lasting 72 hours (11). The hypophysectomized diabetic has had his need for insulin reduced and is therefore receiving at the time of testing a smaller amount of insulin than is an equally severe diabetic with an intact pituitary. The greater sensitivity to growth hormone supports the view that it is the elimination of growth hormone that at least partly accounts for the amelioration of diabetes after hypophysectomy. It is under such circumstances that the full potency of growth hormone in causing metabolic effects is revealed. Non-diabetic pituitary dwarfs and short normal children tolerate chronic treatment with growth-promoting doses of HGH for several years without change in blood sugar, and larger doses (of the order of 5 to 10 mg./day) have little effect in normal adults beyond slightly decreasing glucose tolerance in some and, more commonly, lessening sensitivity to insulin.

Several experiences suggest that HGH may be clinically helpful at times in counteracting hypoglycemia. Two children with retarded growth and troublesome hypoglycemia are said to have grown well and maintained more normal blood sugars when treated with growth hormone (12, 13). Two patients with malignant insulinomas were kept free from severe hypoglycemia with HGH (14). One of these patients, who had been diabetic before developing the tumor, became hyperglycemic with treatment. The failure of HGH to improve hypoglycemia in one infant has been reported.

Postulated consequences of excessive fat mobilization

The list of effects of growth hormone on the metabolism of fat and carbohydrate, some of which have now been seen in man, includes increased output of fatty acids by the adipose tissue, increased fat in the liver and in the blood, ketosis, storage of glycogen in the heart and in skeletal muscle, resistance to insulin, and finally diabetes. It seems reasonable now to attempt to explain all of the other effects of growth hormone on fat and carbohydrate as consequences of the excessive mobilization of fatty acids. With more fatty acid supplied to the liver, more would be esterified and stored, and as triglyceride accumulated it would be released into the blood in greater amounts. In addition to increased esterification and storage, more fatty acid might be degraded in the liver to acetyl-coenzyme A, which could be the source of increased ketone production. Ketone formation would be abetted by a diminished ability of the liver after growth hormone treatment (15) to re-synthesize the acetate to long-chain acid. The decreased synthesis of fatty acids by such livers may, indeed, itself be a consequence of greater provision of fatty acids, just as synthesis in the livers of animals fed on a high-fat diet is reduced. Tissues other than the liver would have available an abundance of fatty acids and also more ketone bodies, and if these substrates were oxidized in preference to glucose, more of the glucose entering the cells would be stored as glycogen. The restricted oxidation of glucose, along with the reduced ability to synthesize fat, would also predispose to the development of hyperglycemia. In these circumstances, insulin would have little effect on the oxidation of glucose, but would correct the aberrant metabolic state at its source by reducing the output of fatty acids from adipose tissue.

Synthesis of fatty acids is reduced after growth hormone not only in the liver, but also, as recently shown by GOODMAN (16), in the adipose tissue. It is possible that in the adipose tissue reduced synthesis is related to the mechanism leading to increased fatty-acid output rather than as a consequence of it.

The proposed sequence of events requires that a substantial portion of the body tissues oxidize fatty acids and ketones preferentially. Thus far we have examined only one tissue. In agreement with the findings of SHIPP, OPIE, and CHALLONER (17), and of WILLIAMSON and KREBS (18), the isolated perfused rat heart oxidized either palmitate or acetoacetate in preference to glucose. The production of $C^{14}O_2$ during a 45-minute retro-aortic perfusion was measured with either glucose-C^{14} (uniformly labelled), palmitate-1-C^{14} or acetoacetate-3-C^{14} as the radioactive substrate, and the effect of competing unlabelled substrate on the $C^{14}O_2$ formation was determined. Fig. 4 shows the effect of 1 mM. palmitate on the oxidation of glucose. With a concentration of glucose of 100 mg.%, palmitate reduced CO_2 production from glucose by 80%, and with 200 mg.% glucose plus insulin (10 milliunit/ml.), by 51%. Similarly, in Fig. 5, 4 mM. acetoacetate decreased oxidation of glucose by 75% and by 88%. The uptake of glucose by the heart was not changed in these experiments by palmitate or acetoacetate, but a smaller fraction of the glucose removed from the medium was oxidized. Conversely, with either palmitate or acetoacetate as the labelled substrate, the addition of glucose had little effect on production of $C^{14}O_2$ (Figs 6 and 7). It is of incidental interest that acetoacetate was oxidized in preference to palmitate, and that the uptake of palmitate was reduced with acetoacetate present.

If skeletal muscle resembles the heart in its use of substrates, the influence on total body metabolism would be considerable and the likelihood of the proposed consequences of excessive fat mobilization would be increased. Several studies with isolated rat diaphragm have shown that palmitate is oxidized in greater amounts as the concentration is raised and that glucose is usually ineffective in suppressing its oxidation (12, 20, 21), indicating at least to that extent similarity to the heart.

Stimulation of growth

The relation of the metabolic changes caused by growth hormone to the distinctive and characterizing action of the hormone in promoting growth is not readily explained. It is possible that the abundant provision of calories to the tissues is in itself an anabolic stimulus, growth hormone acting to divert calories away from the adipose tissue through mobilization and perhaps also inhibition of formation of fat. Other studies have tended to relate the anabolic action of the hormone more directly to an effect on protein metabolism, showing an influence on amino-acid transport and on ribosomal synthesis of protein (22, 23). Studies in man have not helped to clarify this aspect of the subject, but have shown that, in addition to exhibiting the metabolic effects which have been enumerated, man can be made to grow with parenterally administered growth hormone (24).

Human growth hormone satisfactory for clinical use has been prepared from human pituitaries obtained at routine autopsy and stored for many months either in acetone or frozen. It has been administered for long periods without untoward reaction, either local, general, metabolic or immunological. The preparation made

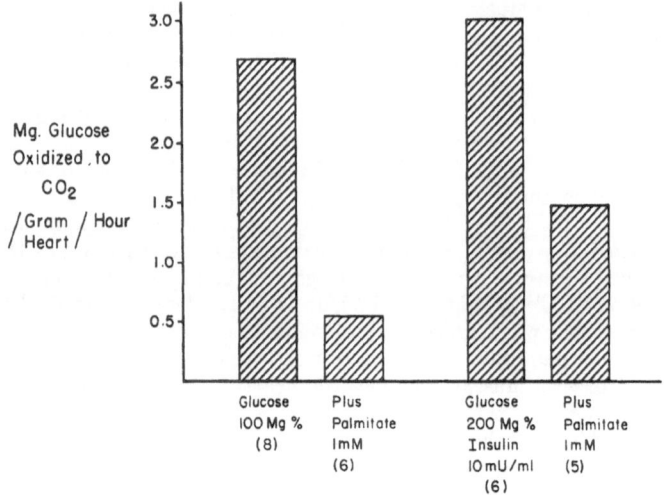

Fig. 4. Effect of palmitate on the oxidation of glucose to CO_2 during 45-minute perfusion of isolated rat hearts, calculated from $C^{14}O_2$ produced from glucose-U-C^{14} (*26*)

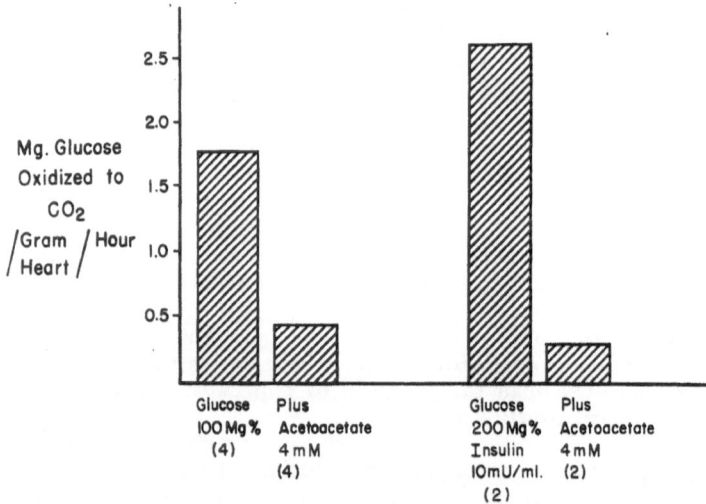

Fig. 5. Effect of acetoacetate on the oxidation of glucose-U-C^{14} to CO_2 in perfused rat hearts (*26*)

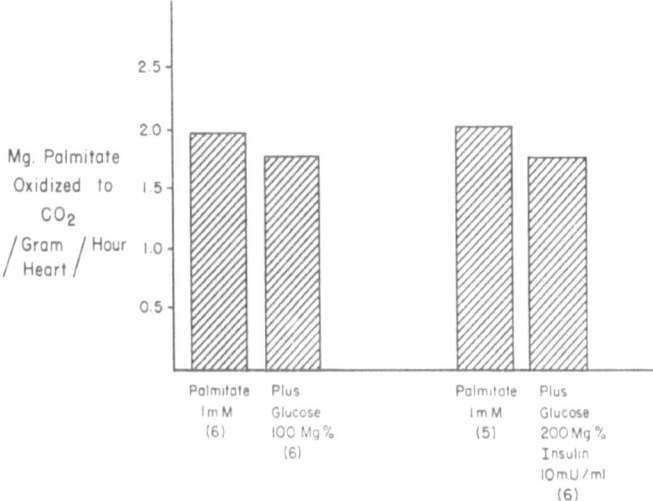

Fig. 6. Oxidation of palmitate-1-C[14] by perfused rat hearts with and without glucose in the medium (26)

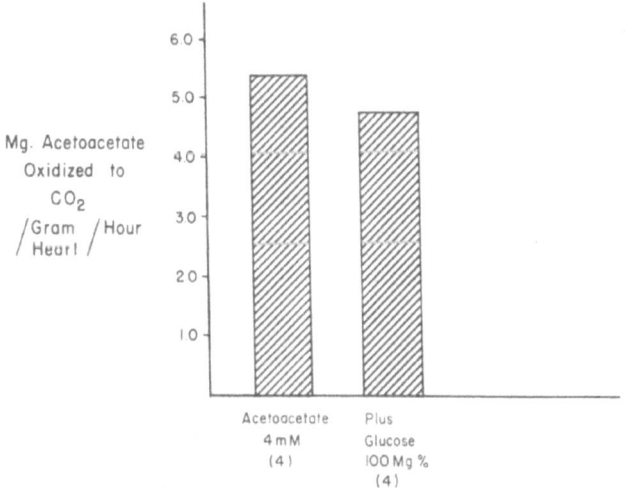

Fig. 7. Oxidation of acetoacetate-3-C[14] by perfused rat hearts in the presence and absence of glucose (26)

by the glacial acetic acid extraction method (25), which has been the most extensively used, has not caused other pituitary-hormonal effects in pituitary dwarfs and has been used in normal short children without preventing the normal onset and progression of puberty. In three pituitary dwarfs, doses of 2 to 4 mg. three times a week continued to stimulate growth for periods of 5, $3^1/_2$, and $2^1/_2$ years, resulting in increases in height of 10.6, 9.75, and 6.75 inches (27, 25, and 17 cm.), respectively. The growth was proportionally distributed, maintaining the proper relations of torso to limbs. The hormone did not seem to hasten the maturation of bones. There was some tendency for the growth rate to become less rapid with continued treatment, but it was possible to maintain a rate at least as fast as that of the normal child for two years without increasing the dose of hormone.

Seven children who were extremely short, but apparently normal otherwise, were treated for periods of up to two years with doses of 1 to 4 mg. three times a week, and in each case the growth rate exceeded the pre-treatment rate. One "bird-headed" dwarf also responded well. Four patients out of a total of 17 responded poorly or not at all to HGH. One was a case of ovarian dysgenesis, two were girls, aged 17 and 21, with uncertain diagnoses and bone ages over 14 years, and the other was an incompletely matured 19-year-old male with a bone age of 15.

It is likely that human growth hormone will eventually be a substance of considerable clinical usefulness, but thus far it has been shown to be clearly of value only in the treatment of pituitary dwarfism. Although it has also been shown to accelerate growth in short normal children, it will require a longer evaluation to know whether such children will be substantially benefited.

Summary

In man, human growth hormone (HGH) causes retention of nitrogen, phosphorus, and potassium, enlargement of the chloride space, enhancement of renal function, and an increase in the excretion of hydroxyproline which probably reflects an alteration in the metabolism of collagen. Injection of HGH leads to a rise in the free fatty acid concentration in the plasma, although the mechanism and significance of this response is still unknown.

In diabetic patients, sensitivity to the hyperglycemic and fat-mobilizing effects of HGH varies considerably from one individual to another. Severe diabetics who have been hypophysectomized are extremely sensitive, particularly to the ketogenic effect, but also to the hyperglycemic effect of HGH. The elimination of HGH may at least partly account for the amelioration of diabetes after hypophysectomy. Non-diabetic pituitary dwarfs and short normal children tolerate chronic treatment with growth-promoting doses of HGH for several years without change in blood sugar, while on the other hand HGH may at times be helpful in counteracting hypoglycemia.

Increased mobilization of fatty acids may lead to an accumulation of triglyceride in the liver and to an increase in the degradation of acetyl-coenzyme A in the liver, which, in turn, could be the source of increased ketone production. At the same time, synthesis of fatty acids is reduced both in the liver and in the adipose tissue. It would appear that a substantial portion of the body tissues oxidize fatty acids and ketones preferentially. The isolated perfused rat heart oxidizes either palmitate or acetoacetate in preference to glucose. Both palmitate and acetoacetate diminish the oxidation of glucose, also in the presence of insulin, whereas glucose does not interfere with fatty-acid oxidation.

The relation between the metabolic changes caused by HGH and the action of this hormone in promoting growth cannot be readily explained. It is possible that the abundant provision of calories to the tissues is in itself an anabolic stimulus or that HGH has a more direct effect on protein metabolism by influencing amino-acid transport and ribosomal synthesis of protein.

Zusammenfassung

Menschliches Wachstumshormon (HGH) führt beim Menschen zu Retention von Stickstoff, Phosphor und Kalium, einer Zunahme des Chloridraumes, gesteigerter Nierenfunktion, sowie zu vermehrter Ausscheidung von Hydroxy-prolin, die möglicherweise Folge eines Eingreifens in den Kollagenstoffwechsel ist. Injektion von HGH ruft einen Anstieg der Konzentration von freien Fettsäuren im Plasma hervor, jedoch sind Mechanismus und Bedeutung dieses Vorganges nicht bekannt.

Bei diabetischen Patienten bestehen große individuelle Unterschiede gegenüber der hyperglykämischen und der fettmobilisierenden Wirkung von HGH. Schwere Diabetiker sind nach Hypophysektomie besonders empfindlich und reagieren mit starker Hyperglykämie und Ketonkörper-bildung. Möglicherweise beruht die Besserung eines Diabetes nach Hypophysektomie auf Ausschaltung des Wachstumshormons. Hypophysäre Zwerge ohne Diabetes und im Wachstum zurückgebliebene, sonst aber normale Kinder vertragen HGH-Injektionen während Jahren ohne wesentliche Beeinflussung des Blutzuckers. Andererseits kann HGH bei Hypoglykämie günstig wirken.

Die vermehrte Mobilisierung von Fettsäuren hat eine Ansammlung von Triglyceriden in der Leber zur Folge sowie Abbau zu Acetyl-Coenzym A, das seinerseits Ursache vermehrter Ketonbildung sein kann. Gleichzeitig nimmt die Fähigkeit zur Synthese von Fettsäuren ab, und zwar sowohl in der Leber als auch im Fettgewebe. Verschiedene Gewebe oxydieren vorzugsweise Fettsäuren und Ketone. Das isolierte Rattenherz oxydiert Palmitat oder Acetoacetat bevorzugt im Vergleich zu Glukose. Sowohl Palmitat als auch Acetoacetat vermindern die Glukoseoxydation, auch in Anwesenheit von Insulin, während andererseits Glukose die Fettsäureoxydation nicht beeinträchtigt.

Die Beziehungen der verschiedenen durch HGH hervorgerufenen Stoffwechseländerungen zur Wachstumswirkung sind nicht geklärt. Möglicherweise wirkt die reichliche kalorische Versorgung der Gewebe bereits als anaboler Reiz, oder der Eiweißstoffwechsel wird durch Einwirkung auf den Aminosäurentransport und die Eiweißsynthese in den Ribosomen beeinflußt.

Résumé

L'hormone de croissance humaine (HGH) détermine chez l'homme une rétention d'azote, de phosphore et de potassium, une augmentation de "l'espace de chlorure", une intensification de la fonction rénale, ainsi qu'une augmentation de l'excrétion d'hydroxyproline, qui est peut-être due à une intervention dans le métabolisme du collagène. L'injection d'HGH augmente la concentration des acides gras libres dans le plasma, mais on ignore encore le mécanisme et la signification de ce phénomène.

Chez le diabétique, on trouve de grandes différences individuelles en ce qui concerne l'action hyperglycémiante et la mobilisation des graisses par l'HGH. Les diabétiques graves sont particulièrement sensibles après hypophysectomie; ils réagissent par une forte hyperglycémie et la formation de corps cétoniques. Il se pourrait que l'atténuation du diabète après l'hypophysectomie provienne de la suppression de l'hormone de croissance. Les nains hypophysaires non diabétiques et les enfants qui tardent à se développer mais sont par ailleurs normaux supportent les injections d'HGH pendant des années sans modifications notables de leur glycémie. D'autre part, l'HGH peut exercer une influence favorable en cas d'hypoglycémie.

La mobilisation accrue d'acides gras détermine une accumulation des triglycérides dans le foie, ainsi que leur dégradation en coenzyme A acétylé, qui peut être à son tour la cause d'une formation accrue de corps cétoniques. Simultanément, la faculté de synthétiser les acides gras diminue, aussi bien dans le foie que dans le tissu adipeux. Divers tissus oxydent de préférence les acides gras et les cétones. Le coeur isolé de rat oxyde le palmitate ou l'acétyl-acétate plutôt que le glucose. Tant le palmitate que l'acéto-acétate diminuent l'oxydation du glucose, même en présence d'insuline, tandis que le glucose n'a pas d'effet sur l'oxydation des acides gras.

Les rapports entre les diverses modifications du métabolisme provoquées par l'HGH et l'action sur la croissance sont inconnus. Il se pourrait que la provision calorique abondante dans les tissus soit déjà un stimulus anabolisant ou que l'action sur le transport des acides aminés et sur la synthèse des protéines dans les ribosomes influence le métabolisme des protéines.

Acknowledgements

This work was supported by Research Grants A-1567 and A-3729, and Graduate Training Grant HTS 5391, from the National Institutes of Health, U. S. Public Health Service.

We are greatly indebted to Mrs. Vera Grinbergs for her invaluable assistance in these studies.

References

1. Raben, M. S.: Recent Progr. Hormone Res. (U.S.A.) 15, 71 (1959). — 2. Forbes, Anne P., J. G. Jacobsen, E. L. Carroll, and M. M. Pechet: Metabolism (U.S.A.) 11, 56 (1962). — 3. Dull, T. A., L. Causing, and P. H. Henneman: J. Clin. Invest. (U.S.A.) 41, 1355 (1962). — 4. Jasin, H. E., C. Fink, D. Smiley, and M. Ziff: J. Clin. Invest. (U.S.A.) 41, 1368 (1962). — 5. Zahnd, G. R., J. Steinke, and A. E. Renold: Proc. Soc. Exper. Biol. Med. (U.S.A.) 105, 455 (1960). — 6. Raben, M. S., and C. H. Hollenberg: J. Clin. Invest. (U.S.A.) 38, 484 (1959). — 7. Hollenberg, C. H., M. S. Raben, and E. B. Astwood: Endocrinology (U.S.A.) 68, 589 (1961). — 8. Winkler, B., R. Steele, N. Altszuler, A. Dunn, and R. C. de Bodo: Fed. Proc. (U.S.A.) 21, 198 (1962). — 9. McKenzie, J. M., and M. S. Raben: unpublished observations. — 10. Rabinowitz, D., and K. L. Zierler: Abstracts of 44th Meeting of the Endocrine Society, Chicago, 1962. — 11. Pearson, O. H., J. M. Dominguez, E. Greenberg, and B. S. Ray:

Transact. Ass. Amer. Physicians **73**, 217 (1960). — 12. HORWITH, M.: personal communication. — 13. GREENBERG, A. J.: personal communication. — 14. MAHON, W. A., M. L. MITCHELL, J. STEINKE, and M. S. RABEN: Abstracts of 44th Meeting of the Endocrine Society, Chicago, 1962. — 15. GREENBAUM, A. L., and R. F. GLASCOCK: Biochem. J. (G.B.) **67**, 360 (1957). — 16. GOODMAN, H. M.: to be published. — 17. SHIPP, J. C., L. H. OPIE, and D. CHALLONER: Nature (G.B.) **189**, 1018 (1961). — 18. WILLIAMSON, J. R., and H. A. KREBS: Biochem. J. (G.B.) **80**, 540 (1961). — 19. FRITZ, I. B., D. G. DAVIS, R. H. HOLTROP, and H. DUNDEE: Amer. J. Physiol. **194**, 379 (1958.) — 20. SCHWARTZMAN, L. F., and J. BROWN: Amer. J. Physiol. **199**, 235 (1960). — 21. EATON, P., and D. STEINBERG: J. Lip. Res. **2**, 376 (1961). — 22. KOSTYO, J. L., and E. KNOBIL: Endocrinology (U.S.A.) **65**, 525 (1959). — 23. KORNER, A.: Biochem. J. (G.B.) **81**, 292 (1961). — 24. RABEN, M. S.: N. England J. Med. **266**, 82 (1962). — 25. RABEN, M. S.: Science (U.S.A.) **125**, 883 (1957). — 26. MINTON, P. R., and M. S. RABEN: J. Clin. Invest. (U.S.A.) **41**, 1385 (1962).

Discussion

IKKOS: From what Dr. RABEN has said it is obvious that changes in fat metabolism are considered to be a very important step in the overall action of growth hormone. I would therefore like to ask him if any experiments have been done in which growth hormone was given to humans with lipoatrophic diabetes, i.e. subjects with virtually no fat tissue.

Dr. RABEN mentioned that there was an increase in the lactic acid during the heart perfusion experiments and he wondered whether this was due to hypoxia. In this connection, I would like to add that we have studied lactic acid and pyruvic acid concentrations in the peripheral blood during constant glucose loading in patients before and after administration of growth hormone. These studies showed an increase in the lactic and pyruvic acid concentration after growth hormone.

Since from our data we were unable to decide whether the increase in lactic and pyruvic acid was due to an increase in glucose break-down or to overloading of the metabolic pathway below pyruvic acid — probably by products of fat metabolism — we studied glucose turnover in a few cases by means of C^{14}-labelled glucose. These studies showed an increased — or at least an unchanged — glucose break-down at a time when the blood glucose level was increased and the glucose turnover decreased. These results bring us to the question as to how to define the criteria on the basis of which we should be justified in describing the action of a hormone as "diabetogenic". Finally, I should like to suggest that the results obtained in acute experiments might not necessarily reflect the physiological effects of the hormone. Dr. RABEN mentioned that administration of HGH to pituitary dwarfs caused among other things a rise in the free fatty acid and phosphatase level. Now, if these two variables are measured in cases involving a spontaneous chronic excess of growth hormone — i.e. in acromegalics — one finds normal values.

RABEN: I did point out with reference to the normal values obtained in acromegalic patients that the free fatty acids may not in fact be an indication of the rate of mobilisation. As for the lack of visible, measurable effects when growth hormone is administered, I have pointed out, too, that one can give growth-promoting doses for as long as five years without any change in blood sugar or in most other serum components; this must depend on the body's ability to balance things off, because the normal individual tolerates relatively large doses with little change, whereas, in the absence of pancreatic insular function, even small doses will cause severe acidosis. To interpret your observation of increased oxidation of glucose, one would have to know the entire situation at the moment; one would have to know how much insulin is being secreted, since the insulin would be counteracting some effects of the growth hormone. But, as for the source of the increased gluconeogenesis that has been reported, I am not at all certain about this. With reference to your statement about lactic acid, I think that DOROTHY HENNEMAN[1] found very little change in lactate. I believe she noted some increase in citrate production

---

[1] HENNEMAN, D. H., and P. H. HENNEMAN: J. Clin. Invest. (U.S.A.) **39**, 1239 (1960).

and rather small changes in pyruvate and lactate. Do you find larger changes than she ?

IKKOS: In experiments with HGH in ten normal subjects we got significant increases in lactate and pyruvate in the three cases in which the glucose tolerance decreased most.

YOUNG: I was very interested to see the rise in the plasma insulin activity some hours after the administration of growth hormone, and I wondered what the activity was some days later in these experiments. Did you do these estimations, Dr. RABEN ?

RABEN: In the experiments from ALBERT RENOLD's[1] laboratory, I think that those values were the only ones measured.

YOUNG: May I ask a further question ? If I understand correctly, Dr. RABEN, you interpret the mechanism of action of growth hormone in depressing insulin sensitivity as the result of an interference with the utilisation of carbohydrate by plasma free fatty acids or ketone bodies formed under the influence of growth hormone. Such circulating metabolites might interfere with the utilisation of glucose by muscle through the mechanism suggested by NEWSHOLME and RANDLE[2] and so depress the sensitivity of the animal to the action of insulin in promoting the utilisation of carbohydrate. Growth hormone would thus exert this action indirectly, through its influence on fat metabolism, rather than by a more direct interference with the action of insulin. That, as you say, is your provisional interpretation, and I may say that this view is also adopted by many of us in the Department of Biochemistry at Cambridge.

And now, finally, what is your view about the suggestion of BORNSTEIN[3] that it is possible to separate from growth hormone two polypeptides which have a direct action on metabolism, one inhibiting the action of insulin, and the other having an insulin-like anabolic action ? Have you any evidence from your experiments to dissociate the anabolic action of growth hormone from its effects on carbohydrate metabolism, on fat mobilisation, or on any of the other actions that you have been studying ?

RABEN: Before I say more, I should mention, if it was not obvious in my paper, that I have recently been more influenced by Dr. YOUNG's thinking than I was several years ago.

In response to the question about separating out substances from growth hormone, i.e. substances which display part of its action, we have had no indication of such a possibility. Now, in the recent work with starch-gel electrophoresis that has been published by FERGUSON[4] in Australia, and which has also been published by BARRET, FRIESEN, and ASTWOOD[5] and perhaps others by now, the human growth hormone has shown up on starch-gel electrophoresis as several bands rather than one, and this of course

[1] ZAHND, G. R., J. STEINKE, and A. E. RENOLD: Proc. Soc. Exper. Biol. Med. (U.S.A.) **105**, 455 (1960).
[2] NEWSHOLME, E. A., and P. J. RANDLE: Biochem. J. (G.B.) **80**, 655 (1961); **83**, 387 (1962).
[3] BORNSTEIN, J., and D. HYDE: Nature (G.B.) **187**, 125 (1960).
[4] FERGUSON, K. A., and A. L. C. WALLACE: Nature (G.B.) **190**, 632 (1961).
[5] BARRET, R. J., H. FRIESEN, and E. B. ASTWOOD: J. Biol. Chem. (U.S.A.) **237**, 432 (1962).

suggested the possibility of separation. Human growth hormone has displayed a couple of peculiarities, I might say; it does appear to be multiple, but thus far (these experiments are not final by any means) each of the bands has seemed to be immunologically identical and biologically identical or at least biologically very close. The only activity besides the usual activity expected from a growth-hormone preparation has been the prolactin-like activity, which has been found in everybody's — human and monkey — growth hormone, and this is an activity of about 10% of the potency of purified ovine prolactin. In terms of small molecules and metabolic effects, of course, the number of examples is increasing. ACTH, for instance, mimics a number of effects of growth hormone, and yet the particular type of activity — the particular mechanism of activity — seems to be different. In fatty acid mobilisation ACTH has a direct effect — a quick effect on adipose tissue which growth hormone has not. Growth hormone is impressive *in vivo*, whereas ACTH is very weak *in vivo* in mobilising fatty acids. Studies have been made of synthetic peptides patterned after the beginning of the ACTH molecule, which were prepared by KLAUS HOFMANN[1], i.e. molecules with the first ten amino acids of ACTH or the first 13. When the molecule contains 13 amino acids, it is almost identical with what is known as alpha-MSH or alpha-intermedin. The molecule first acquires its fat-mobilising activity with a length of 10 amino acids; this activity increases about 10-fold when a length of 13 amino acids is reached, but, as measured in the rat, the 13-amino-acid molecule is still only about 1/1000 as active as corticotropin. Yet in the rabbit the molecule is at least as active, and possibly a little more active, than corticotropin. Then there is the substance which RUDMAN[2] described, and the peptides which ASTWOOD[3] purified, which seem to be quite separate from corticotropin and separate from growth hormone. This type of substance is also remarkably fat-mobilising in the rabbit, but it is inert in the rat and probably in man. So there are accumulating a number of possible peptides that could be contributing to a variety of metabolic effects, but there is no evidence that any of these have come from the initial growth-hormone molecule.

QUERIDO: May I ask a question, Dr. YOUNG? I, too, was aware of BORNSTEIN's observations concerning the separation of growth hormone into three polypeptides, which seems extremely interesting. Could you add anything on this subject?

YOUNG: Dr. BORNSTEIN has kindly sent us samples of his anabolic peptide material for testing in our laboratory, but neither Dr. KORNER nor Dr. MANCHESTER have yet found *in vitro* activity of the sort we were looking for. Dr. MANCHESTER is now on his way to Australia to spend some months in Dr. BORNSTEIN's laboratory in the hope that the causes of these discrepancies can be ascertained.

---

[1] HOFMANN, K., H. YAJIMA, and E. T. SCHWARTZ: J. Amer. Chem. Soc. **82**, 3732 (1960).

[2] RUDMAN, D., M. B. REID, F. SEIDMAN, M. DiGIROLAMO, A. R. WERTHEIM, and S. BERN: Endocrinology (U.S.A.) **68**, 273 (1961).

[3] ASTWOOD, E. B., R. J. BARRET, and H. FRIESEN: Proc. Nat. Acad. Sc. U.S. **47**, 1525 (1961).

Nitrogen balance studies: the relation of protein metabolism to excessive shifts of sodium and potassium

By

P. S. BLOM, J. DE GRAEFF, A. A. H. KASSENAAR,
and H. A. SONNEVELDT

Before a true comparison between two substances acting on protein metabolism can be made by studying nitrogen balance we should make sure that the simple sequence of the experiments or that other variables inadvertently introduced are not distorting our results. This is very difficult to accomplish in the human subject and it may well be the reason why a clear impression of dose-response relations for any given substance is still lacking. It was ALLISON who stressed the importance in nitrogen balance studies of the influence exerted by the level of nutrition in the *preceding* period. Anyone working in this field should be continuously aware of this when interpreting his results. Among other influences affecting the nitrogen balance, the effect of the level of caloric supply and the composition of the non-protein moiety in it deserve particular mention. Interest in these subjects dates back as far as a century ago. In 1951, MUNRO gave a very extensive review of earlier work in this field. Of the more recent studies undertaken, reference should be made to the raft experiments of the Massachusetts General Hospital group led by GAMBLE, and to the work of PASSMORE. In our country, WOLTHUIS has very recently made a contribution to this topic.

Judging it advisable to obtain some thorough, first-hand knowledge of the effects of such simple variables as calories and the relative amounts of carbohydrate and fat, we carried out balance studies while refeeding two patients with anorexia nervosa and during a weight-reduction programme in an obese subject.

We realise that none of our subjects offered a normal physiological status as starting point. As, however, some of the correlations to be reported might be exaggerated distortions of normal mechanisms, they are perhaps worthy of closer scrutiny, particularly since some of our results are common both to the patients with anorexia nervosa and to the one suffering from obesity.

Fig. 1 gives an outline of the experimental set-up in both cases of anorexia nervosa. S. was the more severely depleted of the two patients. Food was administered throughout the balance studies in 6 equal portions per day by gastric tube. All the proteins supplied were derived from milk. The non-protein moiety always consisted of both carbohydrate and fat, the minimum supply of these foodstuffs being 180 g. carbohydrate and 36 g. fat. Changes in caloric supply and changes in protein supply were never made at the same time. Fig. 1 also indicates one of our findings to which I shall be

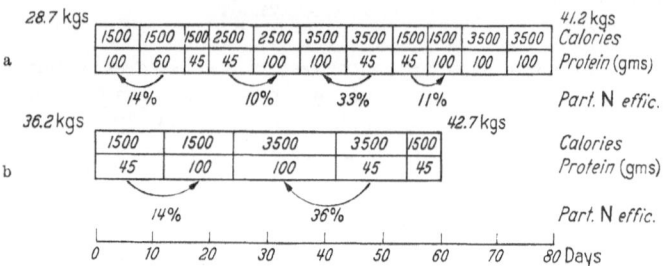

Fig. 1. Refeeding programme of 2 patients with anorexia nervosa. 1a. S ♀ 17 yrs; 1b. de W ♀ 21 yrs. Also shown are: weight at start and end-point of balance studies and partial nitrogen efficiency as calculated from results (omitting the first two days after a change in diet)

referring later, namely the nitrogen efficiency. This was calculated as a *partial* efficiency, i. e. the efficiency with which the body responds to an increment of protein supply at a constant level of caloric supply. The calculations were made after discarding the results obtained on the first two days after a change in diet. The effect of caloric supply is evident. The objection that the 1,500 cal. level could have been a sub-maintenance level can be met by referring to Figs 2 and 3, which show that de W. was then in excellent equilibrium as to weight and nitrogen balance, while S. even gained on both at this level. The interesting fact that, even at the 2,500 cal. level, S. did not improve in N efficiency should be noted. Although I shall later be mentioning an important difference between S. and de W., their N efficiency showed a remarkable degree of conformity under comparable circumstances.

Fig. 2 shows the cumulatively charted data of S. for her weight, nitrogen, and potassium. The ordinates were chosen in accordance with the relation in muscle protoplasm. There is a rather steadily rising curve of N accumulation, which only poorly reflects the differences in N efficiency as calculated by us. N accumulation is to a large extent independent of weight changes, but the potassium curve runs remarkably parallel to the weight curve. This means that the potassium balance deviated considerably from the predictions based on nitrogen.

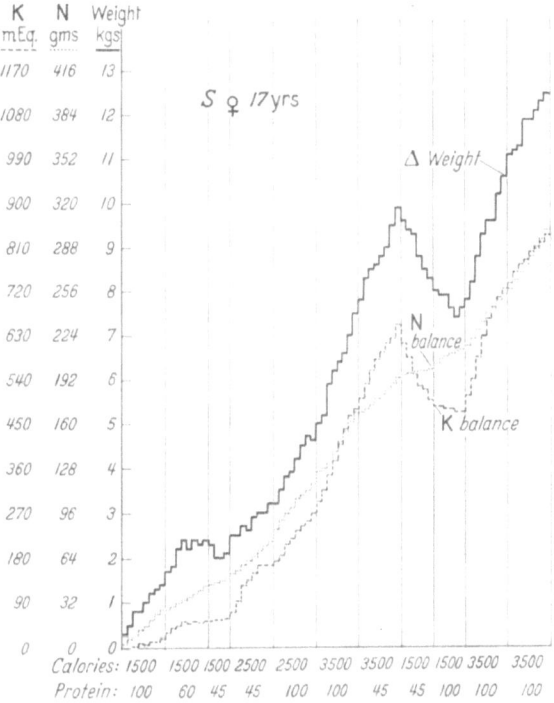

Fig. 2. Anorexia nervosa. Cumulative changes in weight, N balance, and K balance. Ordinates have been chosen in accordance with their relation in muscle protoplasm. For time relations, see Fig. 1. Total length of study 80 days

The same remarks apply to patient de W. as regards the curves for weight, nitrogen, and potassium shown in Fig. 3. Here, again, the potassium balance follows the weight curve more closely than the nitrogen balance.

These charts suggested to us that there might be an interesting correlation between the surplus accumulations and losses of potassium (and perhaps sodium) and the nitrogen efficiency. To study this hypothetical relationship, charts were plotted showing cumu-

latively all the *excess* movements of Na and K, i.e. all Na and
K movements diverging from what could be predicted from the
N balance data using the factors of ALBRIGHT and REIFENSTEIN for
muscle protoplasm (Figs 4—7).

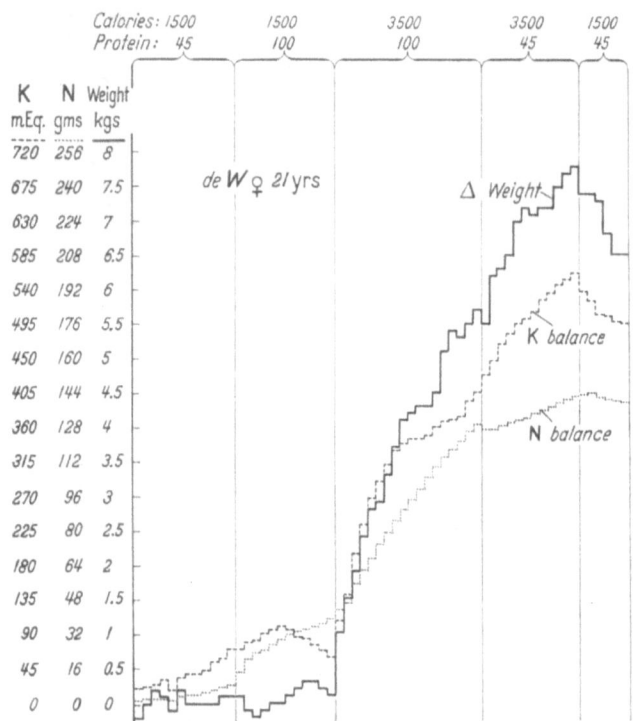

Fig. 3. Anorexia nervosa. Cumulative changes in weight, N balance, and K balance.
Compare with Fig. 2. Total length of study 60 days

Fig. 4 shows the excess movements of potassium in S. At the
start of the balance study, this patient evidently corrected a
potassium surplus existing at that moment and remained afterwards for several periods at a fairly constant level. The poor N
efficiency found at the 2,500 cal. level (see Fig. 1) correlates with a
rather steady state in the potassium sector at that point. As soon
as 3,500 cal. are given, an appreciable surplus retention of potassium occurs, which may account for 1.3 kg. of body-weight. This

correlates with a significant gain in N efficiency. As soon as we return to the 1,500 cal. level, N efficiency returns to the 11% level, and Fig. 4 shows that the potassium then also reverts to its former level. Restoration of the 3,500 cal. supply enables the patient to regain most of her potassium surplus. There is a suggestion in this curve that surplus potassium retention is not only related to caloric supply but is greatest in response to a low-protein, high-caloric supply.

Fig. 5 affords confirmation of these observations in the case of patient de W. Here again, potassium retention correlates with caloric supply, and again it is most marked in response to a limited protein supply.

Excess movements of potassium, as compared with the nitrogen balance data, have already been noted by ALBRIGHT and REIFEN-STEIN, who in their cases had good reasons for correlating them with changes in the glycogen stores. At the start of our experiments, both S. and de W. were severely depleted, and glycogen may well have accounted for part of their excess potassium shifts. But, even so, it is not clear how glycogen could explain the considerable second gain in potassium which occurred when, after the patient had for many days been on a 3,500 cal. supply, the protein content of her diet was lowered.

A deviation of potassium from calculated estimates has been found by several investigators during the action of human growth hormone and by McSWINEY and PRUNTY during the action of an anabolic steroid. In both kinds of experiment an increase in N efficiency occurs under the influence of the exogenous hormone. Most authors mention this fact without further comment. BERGEN-STAL and LIPSETT suggest that the protein formed might be used for protoplasm other than muscle protoplasm, which may possibly display a different K/N relationship. It is difficult to visualise how the very appreciable and rapid K movements in our cases could be explained by shifts in the selection of protoplasm formed, partic-ularly since the levels of caloric and protein supply play such an important role in this connection.

Let us now turn to the sodium shifts observed in both cases of anorexia nervosa. Figs 6 and 7 again show *excess* movements, which exceed the limited allowances in the Albright-Reifenstein factors for an increase in extracellular fluid in response to nitrogen retention. In S. the shifts are most impressive. She immediately starts by retaining some Na and by forming a new kind of base-line, but afterwards very marked shifts are noted, which again are closely correlated with the caloric supply and perhaps to some

extent with the protein content in it. In S. each of these shifts
could account for a body-weight change of about 2 kg. At the time,
the patient was still extremely thin, and this appreciable gain in

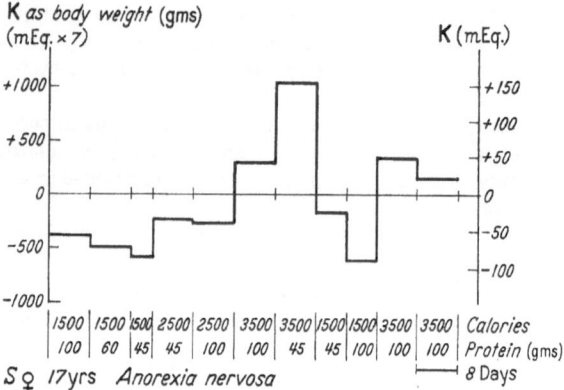

Fig. 4. Cumulative K divergence from nitrogen-balance data (using factors of ALBRIGHT and
REIFENSTEIN)

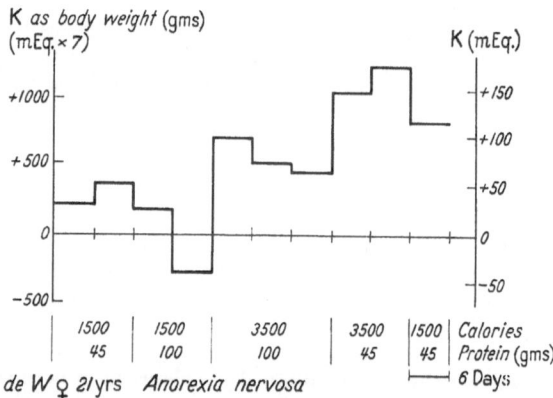

Fig. 5. Cumulative K divergence from nitrogen-balance data (using factors of ALBRIGHT
and REIFENSTEIN)

what at first sight was presumably extracellular fluid was reflected
neither in the formation of oedema nor in signs of haemodilution.
At this point I should stress once again that the supply of Na and K
was strictly maintained at a constant level throughout the entire

experiment. De W. shows the same shifts, albeit on a less impressive scale. These observations made us reluctant to regard such

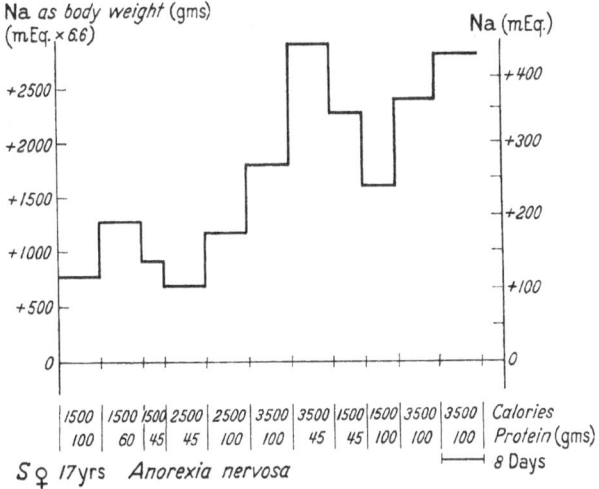

Fig. 6. Cumulative Na divergence from nitrogen-balance data (using factors of ALBRIGHT and REIFENSTEIN)

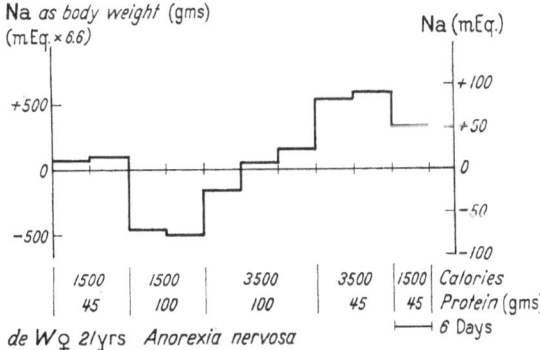

Fig. 7. Cumulative Na divergence from nitrogen-balance data (using factors of ALBRIGHT and REIFENSTEIN)

sodium shifts as representing rapid changes in extracellular fluid. This same phenomenon had already been noted some 20 years ago, and GAMBLE commented on it in his Harvey lecture of 1946. In 4 experiments on his subject E. B., he noted a considerable loss of Na in response to complete fasting for 6 days. This loss, which he still ascribed to extracellular fluid, could be prevented to the extent of about 50% by administering a daily supply of 100 g. of glucose, but not by giving 4.5 g. of NaCl daily. The supply of both glucose and NaCl abolished all sodium loss.

We think that his experiments, and the comparable effects of caloric supply in our patients, strongly suggest that the Na shifts observed derive to a considerable extent from intracellular Na. If so, the question may be asked whether these Na shifts are connected either with glycogen formation or in some way with the effect which calories have on nitrogen metabolism. Neither our own experiments as reported so far nor GAMBLE's data provide a satisfactory answer to this question.

Before studying this problem in a different experimental set-up, it seems appropriate to complete the story of the two anorexia nervosa patients. On the assumption that all Na and K shifts, including the *excess* movements, are accompanied by appropriate water shifts to preserve a constant osmotic level, it is possible to calculate a *weight divergence*. Such a weight divergence implies either a loss or gain of fat or a departure from the presumed principle of a constant osmotic level. A cumulative curve of this weight divergence in S. indicated no weight changes of any importance after a small correction at the start of the study (presumably a correction of tonicity). Thus, all the weight changes observed could be accounted for in terms of the N, Na, and K balances. This would mean that the patient accumulated practically no fat at all during the study. The only alternative possibility might be that she retained Na or K, or both, at a hyperosmotic level. The same weight-discrepancy curve for de W., however, shows a definite and regular weight gain in response to a high-caloric supply, and there are no serious objections to ascribing this to fat. There is hence an important difference between the two patients. Although their N efficiency was remarkably identical at different levels of caloric supply, the more severely depleted subject (S.) had a very poor *caloric* efficiency (which was calculated to have risen gradually from 4% to 30% during the experiment), whereas the caloric efficiency of de W. was far better and ranged around 50% throughout the study. It is not known, however, whether this poor caloric efficiency in S. resulted from a serious depletion of enzymes and co-factors important for fat formation or from excessive activity on the part of endogenous growth hormone.

We next studied the effect of carbohydrate on excess movements of Na and K, for which purpose we investigated a case of obesity. Several recent reports on carbohydrate versus fat feeding in obesity have stressed the conspicuous weight gains which may occur when refeeding carbohydrate after a period of restriction, a fact which suggested to us that electrolyte shifts might be involved.

Interest in this effect of carbohydrates on weight was revived when in 1956 KECKWICK reported on some of his experiments. Although he thought that the results he obtained were due to better caloric efficiency on carbohydrate feeding, later reports of PILKINGTON and OLESEN made his hypothesis untenable.

In our study the obese subject received a constant caloric and protein supply of 1,020 cal. and 60 g. protein. The Na and K content of the diet were constant. The composition of the non-protein moiety in this diet was varied for experimental purposes. After one period (6 days) on a fairly normal mixture of carbohydrate and fat, the patient's carbohydrate supply was severely restricted (25 g.). In the third period we refed carbohydrate, and from the fourth period onwards the patient was again subjected to very protracted carbohydrate restriction (30 days), in order to reveal, if possible, adaptation phenomena. The balance study was concluded with one period of carbohydrate refeeding. The patient's weight curve (not shown here) indicates quite a regular loss of weight, except for two definite weight "humps" in the third and last period, when carbohydrate had been reinstituted following restriction. Her cumulative N balance during the study is shown in Figs 8 and 9, superimposed on the Na and K curves. As we did not alter the protein supply, it was not possible to make a regular calculation of the partial N efficiency, because this calls for changes in the protein supply. When, however, in response to our unvarying protein supply, the N balance shows a significant change in a positive direction, this must denote an increase in N efficiency. Like other subjects whom we studied under a restricted carbohydrate supply and in accordance with earlier reports, V. suffered a nitrogen loss during the experiment; but her nitrogen balance was soon restored to equilibrium again, although it was some time before she regained her lost proteins. During the last 6 days of carbohydrate restriction, the pattern suddenly changed and she showed a clearly positive balance, despite the fact that no change had been made in her diet. This was the first time we had encountered an adaptation of this kind in a subject on a restricted carbohydrate intake, but it was also the first time that we had succeeded in keeping a patient

on this abnormal diet over such a long period. The fact must be
stressed that this change in nitrogen balance was accompanied by a
weight loss (1,850 g. in 6 days), whereas in the following period,
when we refed carbohydrate and the nitrogen balance was again
definitely positive, the patient gained 700 g. in weight in 6 days. We
may suppose, then, that during this balance study V. showed an
increase in N efficiency, especially in the last two 6-day periods (the
first when still on carbohydrate restriction, the second during

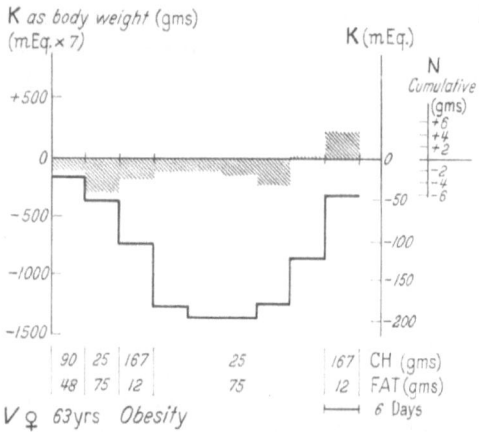

Fig. 8. Reducing diet (1,020 cal. and 60 g. protein). Total length of study 54 days. Cumulative K divergence from nitrogen-balance data (using factors of ALBRIGHT and REIFENSTEIN).
Hatched area shows cumulative nitrogen balance with independentor dinate

carbohydrate refeeding). On the other hand, the effect of carbohydrate refeeding on the nitrogen balance at the start of the
experiment was only slight. At that time, the patient was only
depleted during the 6 days of the preceding period.

Figs 8 and 9 show the excess movements of K and Na in this
study, charted in the same way as in the previous studies. The
superimposed cumulative N balance has been included merely so
that the shifts can be compared qualitatively; the ordinate scale
has no connection whatsoever with the electrolyte scales. It should
be stressed again that the electrolyte shifts shown are *excess*
movements beyond nitrogen allowances.

At the start there is quite an appreciable excess loss of potassium, which levels off somewhat later than the nitrogen blance at
about 200 mEq.; this could account for about 1.5 kg. of body-

weight. As soon as nitrogen is retained, however, potassium is also retained in excessive amounts. This phenomenon is already clearly evident when the nitrogen balance becomes positive during the period of carbohydrate restriction, and it is then almost as marked as in the following period when carbohydrate was being refed (and when the nitrogen balance was again positive). On the other hand,

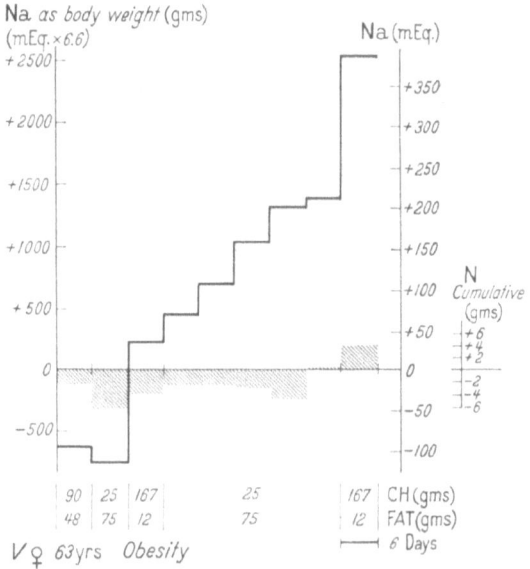

Fig. 9. Cumulative Na divergence from nitrogen-balance data (using factors of ALBRIGHT and REIFENSTEIN). Compare with Fig. 8

when carbohydrate feeding was reinstituted in the beginning of the experiment, after only a short period of depletion the slight effect on nitrogen balance was not accompanied by any parallel shift of excess K.

Fig. 9 shows how the excess sodium shifts behave differently. Here, both periods of carbohydrate refeeding are accompanied by an appreciable excess retention of Na. During the long period of carbohydrate restriction there is a fairly constant gain in body sodium. Although we cannot rule out the possibility of a systematic error in our Na determinations (for example, in our food analysis), a more satisfactory explanation might be the retention of extra-cellular fluid because of obesity with varicose veins. Ankle oedema

10*

was clearly discernible. Despite such technical or clinical difficulties, the fact remains that there were very conspicuous Na-excess shifts on refeeding carbohydrates. It should be noted that, whereas considerable K-excess gains occurred during both periods of nitrogen retention, the sodium excess was not related to the nitrogen balance when carbohydrate was still being withheld; instead, the sodium excess was confined to *both* carbohydrate refeeding periods, even when the nitrogen balance responded feebly at the start of the experiment.

Finally, I should mention briefly that the curve of weight divergence calculated from the actual N, Na, and K balances showed a remarkably constant loss of weight in most of the periods of carbohydrate restriction. The only exception was an extra weight loss at the moment when the nitrogen balance became positive at the end of the study. Absence of an expected loss had already been noted in the first carbohydrate refeeding period. I will not attempt to explain these findings. It may be that the deviations represent changes in the loss of fat (in other words, perhaps we provoked changes in caloric efficiency), or alternatively they may simply be evidence of changes of tonicity in the electrolyte stores.

To sum up, then, the cases described here indicate how nitrogen efficiency, and with it the nitrogen balance, can be influenced by changes in caloric supply or carbohydrate supply. There seems to be a mechanism by which the body can adapt itself to a minimal supply of carbohydrates and regain a good level of N efficiency.

The changes in N efficiency are accompanied rather regularly by excess shifts of Na and K. The Na shifts may perhaps to some extent derive from intracellular Na. While excess shifts of K tend to accompany a rise in N efficiency and are seldom lacking under such conditions, the extra shifts of Na seem to be more closely correlated with an abundance of calories or of carbohydrates (even when these variables have no clear influence on nitrogen efficiency).

Summary

In balance studies undertaken in patients with anorexia nervosa during refeeding, as well as in an obese subject on a weight-reduction diet, an investigation was made of the sodium and potassium shifts exceeding those which could be predicted from nitrogen-balance data.

The results indicate that appreciable extra shifts of sodium are correlated with the abundance of the caloric and/or carbohydrate supply.

The extra shifts of potassium are more closely correlated with the partial nitrogen efficiency than with the caloric or carbohydrate supply.

It is suggested that the sodium shifts are to a considerable extent accounted for by gains or losses of intracellular sodium.

Zusammenfassung

In Bilanzuntersuchungen bei Patienten mit Anorexia nervosa während der Auffütterung und bei einem Patienten mit Fettsucht während Nahrungsbeschränkung wurden die Natrium- und Kalium-Verschiebungen untersucht. Sie sind größer, als auf Grund der Stickstoffbilanzen zu erwarten wäre.

Es ergab sich, daß erhebliche zusätzliche Verschiebungen von Natrium mit überschüssiger Kalorien- und/oder Kohlenhydratzufuhr einhergehen.

Die zusätzlichen Verschiebungen von Kalium sind enger mit dem partiellen Stickstoffwirkungsgrad (der Wirkungsgrad, mit dem der Organismus auf eine Eiweißzulage bei konstanter Kalorienzufuhr reagiert) als mit der kalorischen oder Kohlenhydratzufuhr korreliert. Es wird angenommen, daß die Natriumverschiebungen zu einem erheblichen Teil auf Zunahme oder Abnahme von intracellulärem Natrium zurückzuführen sind.

Résumé

Chez des sujets atteints d'anorexie mentale en cours de réalimentation et chez un obèse soumis à un régime restrictif on a étudié les bilans du sodium et du potassium en comparaison avec les bilans de l'azote. L'attention portait surtout sur les dépassements des bilans des électrolytes par rapport à ce qu'on aurait pu attendre sur la base des bilans de l'azote. Les études ont montré que ces dépassements pour le sodium coïncidaient avec l'abondance de calories ou d'hydrates de carbone.

Les transferts excessifs du potassium ont un rapport plus étroit avec l' «efficacité partielle» de l'assimilation de l'azote. Cette efficacité partielle se définit par la quantité relative de l'azote utilisé dans la synthèse de protéines lors d'une augmentation de l'apport des protéines quand l'apport calorigène reste constant.

La coïncidence des excès de potassium avec l'apport calorique ou glucidique est moins constante. L'auteur suppose que les variations excessives du sodium sont attribuables pour une bonne part à l'augmentation ou la diminution du sodium intracellulaire.

Acknowledgement

The expenses of this research were in part defrayed by grants from the National Health Organisation T.N.O., The Hague.

References

ALLISON, J. B.: Fed. Proc. (U.S.A.) **10**, 676 (1951). — BERGENSTAL, D. M., and M. B. LIPSETT: J. Clin. Endocr. (U.S.A.) **20**, 1427 (1960). — GAMBLE, J. L.: Harvey Lect. (U.S.A.) **42**, 247 (1946). — KECKWICK, A., and G. L. S. PAWAN: Lancet (G. B.) **1956/II**, 155. — MCSWINEY, R. R., and F. T. G. PRUNTY: J. Endocr. (G.B.) **16**, 28 (1957). — MUNRO, H. N.: Physiol. Rev. (U.S.A.) **31**, 449 (1951). — OLESEN, E. S., and F. QUAADE: Lancet (G.B.) **1960/I**, 1048. — PASSMORE, R.: Nutr. et Dieta (Switz.) **3**, 1 (1961). — PILKINGTON, T. R. E., H. GAINSBOROUGH, V. M. ROSENOER, and M. CAREY: Lancet (G.B.) **1960/I**, 856. — REIFENSTEIN, Jr., E. C., F. ALBRIGHT, and S. L. WELLS: J. Clin. Endocr. (U.S.A.) **5**, 367 (1945). — WOLTHUIS, F. H.: Thesis, Amsterdam (1962).

Studies of protein metabolism in humans with the aid of N¹⁵-labelled glycine

By

A. HAAK, A. A. H. KASSENAAR, and A. QUERIDO

The concept of the dynamic state is essential for the interpretation of studies of protein metabolism. Obviously, a considerably broader insight can be obtained into the status of protein metabolism in an individual when, in addition to information on the net result of protein synthesis and break-down, data are also available on the magnitude of each.

A widely used method for studying protein metabolism in patients is evaluation of the nitrogen (N) balance. The N-balance study furnishes information on the *difference* between daily overall protein anabolism and catabolism. Studies on the excretion of N¹⁵ following the administration of N¹⁵-labelled amino acids have been used to obtain information about the level of protein synthesis. In the studies to be reported here the method described by SAN PIETRO and RITTENBERG was employed (5, 7).

Material and methods

Four patients, three with acromegaly and one with panhypopituitarism, were studied under metabolic ward conditions. Food intake and exercise were constant in all four cases.

The three patients suffering from acromegaly (B. V., male, 50 years; Z. W., male, 55 years; N. B., female, 42 years) were in an active stage of the illness. In all three cases the sella turcica was enlarged. The main symptoms were: hyperhidrosis, fatigue, and acroparaesthesia. In our opinion, one of the crucial signs of active acromegaly is swelling of the soft tissues, especially of the hands; this sign was present in all three cases (2). None of the patients showed clinical or laboratory signs of anterior-pituitary insufficiency (except patient N. B. who developed signs of hypopituitarism postoperatively).

Patient S. (male, 53 years) was seen in the out-patients department. He presented, 17 years after the removal of a chromophobe adenoma, a classic picture of panhypopituitarism (P.B.I.

2.5 mcg.%; gonadotrophin excretion in the urine less than 6 I.U.; urinary 17-ketosteroid excretion 0.9 mg./day; disturbed water tolerance). A dose of N^{15}-labelled glycine with approximately 30 atom % excess was administered intravenously in the morning before breakfast. Urine was collected at 20-minute intervals for 4—5 hours after the injection. The patients were given an additional $1^1/_2$—$2^1/_2$ litres of water to drink, so as to ensure a sufficient diuresis. Throughout the next 43 hours, urine collections were made at longer and longer intervals, ranging from 1—24 hours. In the various urine fractions the N^{15} content of the urea was determined according to the methods of SPRINSON and RITTENBERG (8). The data are expressed as atom % excess. Total amounts of nitrogen, urea, and creatinine were measured in the 24-hour urine collections. The level of N^{15} in the urea was plotted against time. From this curve the following data could be derived:

1. Height of maximum isotope concentration (Cu)
2. Time at which this maximum was reached (t_{max})
3. Slope of the declining part of the curve.

Results and discussion

Figs 1 and 2 show the results of the studies on the three patients with *acromegaly*. The excretion curves of the urinary N^{15} urea in the first two patients are essentially similar. The curves are extremely flat, with a low maximum, and the time at which this maximum is reached cannot be determined within reasonable limits. In the third patient, two studies were performed before hypophysectomy, and one study after operation. In this case a

Table 1. *Data on excretion of N^{15} in urinary urea after a test dose of N^{15}-labelled glycine (Patient N. B., female, 42 years, acromegaly)*

	Before hypophysectomy		After hypophysectomy
Cu*	0.110	0.099	0.176
t_{max}**	210	230	140
λ_0***	3.95	3.75	3.90

Cu* = Height of maximum N^{15} concentration in the urinary urea (atom % excess)

t_{max}** = Time in which this maximum is reached (minutes)

λ_0*** = Dose of N^{15}-glycine administered (mEq.)

remarkable reversal of the curve can be seen after operation: the maximum became higher and was reached earlier. The data derived from the curve of the third patient are listed in Table 1. The values of the two curves plotted prior to operation show the reproducibility (t_{max} 210' and 230'; height of maximum N[15] concentration 0.110 atom % excess and 0.99 atom % excess). Post-operatively, the maximum isotope concentration became higher (0.176) and was reached sooner (140').

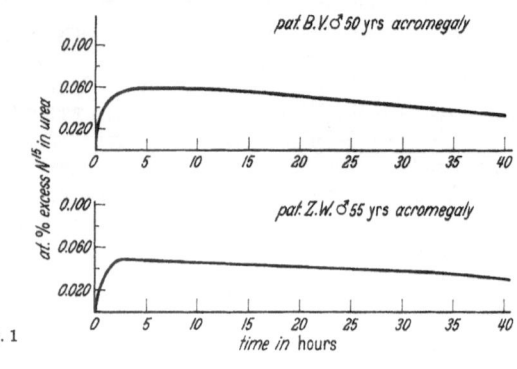

Fig. 1

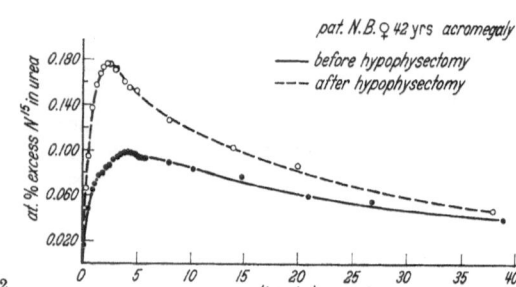

Fig. 2

Figs 1 and 2. Excretion of N[15] in urinary urea after a test dose of N[15]-labelled glycine

It may be mentioned that post-operatively the patient received a substitution dose of 37,5 mg. cortisone daily and Pitressin, but no thyroid. Later, it was evident that the patient had become hypothyroid (P.B.I. 1.7 mcg. % on the day of the N[15]-glycine study). From the data in the literature on the subject it would seem unlikely that the changes observed in the N[15] excretion pattern were due to the hypothyroid state (3).

A number of studies were performed before, during, and after administration of human growth hormone in a male patient, aged 53 years, suffering from *panhypopituitarism* (4). The patient was receiving substitution therapy consisting of 100 mg. desiccated thyroid and 37,5 mg. cortisone daily.

Administration of growth hormone[1] produced the usual metabolic effects, such as decrease in blood urea, retention of nitrogen, and increase in body-weight. Fig. 3 shows two curves of urinary N^{15} urea excretion, one before and one during administration of

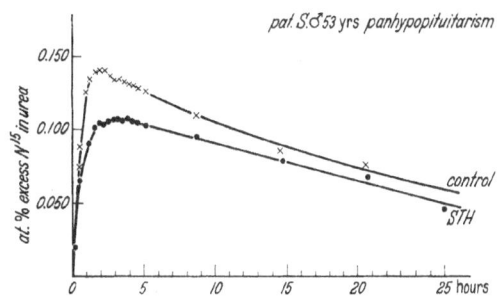

Fig. 3. Excretion of N^{15} in urinary urea after a test dose of N^{15}-labelled glycine

10 mg. human growth hormone per day. The graphs obtained during growth-hormone administration displayed the same characteristics as those of patients with active acromegaly. The maximum became lower and was reached later.

Table 2. *Data on excretion of N^{15} in urinary urea after a test dose of N^{15}-labelled glycine before, during, and after administration of STH (10 mg./day)* *(Patient S., male, 53 years, panhypopituitarism)*

Day of experiment	12	19	33	48	72
Days of STH administration			12	27	
Days after STH administration					22
Cu^{*}	0.132	0.100	0.092	0.097	0.128
t_{max}^{**}	115	130	190	180	90
λ_0^{***}	3.71	3.27	3.39	3.25	3.47

Cu^{*} = Height of maximum N^{15} concentration in the urinary urea (atom % excess)
t_{max}^{**} = Time in which this maximum is reached (minutes)
λ_0^{***} = Dose of N^{15}-glycine administered (mEq.)

Table 2 contains the results of five studies performed on the patient with panhypopituitarism.

---

[1] The growth hormone administered was prepared by the RABEN method and kindly supplied by N. V. Organon (Netherlands).

During growth-hormone treatment, t_{max} increases and the height of the maximum is slightly lower.

A comparison of the values recorded in the two experiments before the administration of growth hormone shows that fairly good reproducible data can be obtained in studies of this kind.

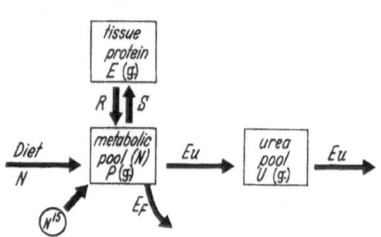

Fig. 4. Model system for the calculation of protein synthesis

San Pietro and Rittenberg devised a method based on a theoretical model (Fig. 4) which might make it possible to arrive at an approximation of the magnitude of daily overall protein synthesis (7).

Three compartments are postulated: the metabolic nitrogen pool (P), the tissue proteins, and the urea pool (through which urea is eliminated in the urine). It is assumed that the N^{15}-glycine administered is either used for protein synthesis or is excreted in the form of urea. By measuring the N^{15} content of the urinary urea and by estimating the urea pool, data may be obtained on:

1. The size of the metabolic pool (P) (in grammes of N)
2. The turnover rate of this metabolic pool (B)
3. The amount of protein synthesised per day (S) (in grammes of N).

We are fully aware of the assumptions that have to be made in calculations of this kind; however, it is our opinion that under certain conditions these calculations may give valuable information. These conditions are: the data should be reproducible in the same patient under the same conditions, and the results obtained should not be considered as absolute data.

Table 3. *Parameters of protein metabolism before and after hypophysectomy (Patient N. B., female, 42 years, acromegaly)*

	Before hypophysectomy		After hypophysectomy
B^*	16.7	16.0	31.5
P^{**}	3.9	3.6	1.4
S^{***}	55.5	53.2	34.2

B^* = Rate constant for the turnover of the metabolic pool
P^{**} = Size of metabolic pool for nitrogen (in g. of N)
S^{***} = Amount of protein which is synthesised per 24 hrs (in g. of N)

Table 3 summarises the results of these calculations, based on studies of a female patient before and after hypophysectomy. The data show that after hypophysectomy the size of the amino-acid pool (P) decreases, with a factor of 2.5, while the turnover rate of this pool is doubled. The amount of protein synthesised per day is decreased after the operation.

The reversed pattern of changes as seen in this patient was observed in the patient with panhypopituitarism during the administration of growth hormone. These data are presented in Table 4. During the administration of growth hormone, the turn-

Table 4. *Parameters of protein metabolism before, during, and after STH administration (10 mg./day) (Patient S., male, 53 years, panhypopituitarism)*

Day of experiment	12	19	33	48	72
Days of STH administration		12	27		
Days after STH administration					22
B^*	45	40	23	25	61
P^{**}	1.0	1.2	2.6	2.2	0.9
S^{***}	36	37	48	43	37

B^* = Rate constant for the turnover of the metabolic pool
P^{**} = Size of metabolic pool for nitrogen (in g. of N)
S^{***} = Amount of protein which is synthesised per 24 hrs (in g. of N)

over rate of the metabolic pool was decreased, while the size of the pool and the rate of protein synthesis were increased. The finding that the size of the amino-acid pool increases in response to the administration of growth hormone tallies well with various observations reported in the literature. BARTLETT et al. observed that the free amino-acid content of the Tibialis anticus in rats increases following administration of growth hormone (*1*). There are several observations in the literature which indicate that the plasma amino-acid level decreases during the administration of growth hormone. It therefore seems justifiable to conclude that, in response to growth-hormone stimulation, the concentration *gradient* for amino acids between the intracellular and extracellular space is increased.

The same conclusion was reached by RIGGS and co-workers, who found that growth hormone increases the uptake of the non-metabolisable amino acid alpha-aminoisobutyric acid (A.I.B.) by several tissues, while the plasma concentration is decreased (*9*).

Recently, DAUGHADAY's group found that the distribution volume of intravenously administered A.I.B. in patients with panhypopituitarism increases in size during the administration of human growth hormone (*6*).

Summary

Studies of the excretion of N[15]-labelled urea following the administration of N[15]-labelled glycine revealed that growth hormone exerts very significant effects on protein metabolism. Using the method of San Pietro and Rittenberg to calculate the results of these studies, it was found that, in response to growth-hormone stimulation, the metabolic pool for nitrogen and the rate of protein synthesis are increased, while the turnover rate of the metabolic pool is decreased. These results seem to be in accordance with observations in the literature, which were obtained with different techniques.

Zusammenfassung

Bei drei Patienten mit Akromegalie und einem Patienten mit Panhypopituitarismus wurde versucht, mit Hilfe von N[15]-markiertem Glykokoll Auskunft über den Stickstoffumsatz auf Grund der Ausscheidung von N[15]-markiertem Harnstoff zu erhalten. Aus unseren Befunden ist zu schließen, daß sich auf Grund von Änderungen der Ausscheidung von markiertem Harnstoff deutliche Einflüsse von Wachstumshormon auf den Eiweißstoffwechsel erfassen lassen. Werden unsere Resultate nach der Methode von San Pietro und Rittenberg berechnet, so findet sich unter der Einwirkung von Wachstumshormon eine Zunahme des gesamten am Stoffwechsel beteiligten Stickstoffes und der Eiweißsynthese, während der tägliche Gesamtumsatz von Eiweiß vermindert ist. Während der Gabe von Wachstumshormon nimmt der Konzentrationsgradient für Aminosäuren zwischen intra- und extrazellulärem Raum zu. Diese Befunde stehen in Übereinstimmung mit anderen veröffentlichten Beobachtungen, die mit verschiedenen Methoden gewonnen wurden.

Résumé

Chez trois sujets atteints d'acromégalie et dans un cas de panhypopituitarisme, on a cherché à se faire une idée du métabolisme de l'azote en donnant du glycocolle marqué avec du N[15] et en déterminant ensuite l'élimination de l'urée ainsi marquée. Les modifications relevées dans l'élimination d'urée marquée permettent effectivement de conclure que l'hormone de croissance influence nettement le métabolisme des protéines. Si on applique la méthode de San Pietro et Rittenberg aux résultats obtenus, on trouve que l'hormone de croissance détermine une augmentation de la réserve azotée et de la synthèse des protéines, tandis que le catabolisme protidique quotidien diminue. Pendant le traitement à l'hormone de croissance, le gradient de concentration pour les acides aminés entre l'espace intracellulaire et l'espace extracellulaire augmente. Ces constatations corroborent d'autres résultats déjà publiés et obtenus par des méthodes différentes.

Acknowledgements

The work reported was made possible by a grant from the Netherlands Organisation for Pure Research (Z.W.O.). The expert help of Miss A. Kouwenhoven, chemical engineer, and Miss M. A. M. Schuurs, dietician, is gratefully acknowledged.

References

1. BARTLETT, P. D.: In: Hypophyseal Growth Hormone, Nature and Actions. Ed. by R. W. SMITH, O. H. GAEBLER, and C. N. H. LONG. McGraw-Hill, New York, 204, 259 (1955). — 2. BRINKERINK, P. C.: Acromegaly. Thesis. Leyden University, 1961. — 3. CRISPELL, K. R., W. PARSON, and G. HOLLI-FIELD: J. Clin. Invest. (U.S.A.) 35, 164 (1956). — 4. HAAK, A., A. A. H. KASSENAAR, and A. QUERIDO: Studies on overall protein metabolism with the aid of N^{15}-labelled glycine in a case of panhypopituitarism. In preparation. — 5. KASSENAAR, A. A. H., J. DE GRAEFF, and A. T. KOUWENHOVEN: Metabolism (U.S.A.) 9, 831 (1960). — 6. KIPNIS, D. M., M. GROSS, and W. DAUGHADAY: J. Clin. Invest. (U.S.A.) 40, 1054 (1961). — 7. SAN PIETRO, A., and D. RITTENBERG: J. Biol. Chem. (U.S.A.) 201, 457 (1953). — 8. SPRIN-SON, D. B., and D. RITTENBERG: J. Biol. Chem. (U.S.A.) 180, 707 and 715 (1949). — 9. RIGGS, T. R., and L. M. WALKER: J. Biol. Chem. (U.S.A.) 235, 3603 (1960).

Discussion

TREMOLIÈRES: After Dr. BLOM's very interesting observations I want to mention one group of facts[1] that I think can be correlated with his findings. We were also struck by the problem of the different effect of carbohydrate and fats in obese people, and we studied the utilisation of carbohydrates by measuring the respiratory quotient and respiratory exchanges before and after administration of carbohydrate (1 g./kg.). We found very clearly that, when an obese person was on a reduction diet and was then given carbohydrate at this level, it did not raise his respiratory quotient — which seemed to suggest that he does not burn up the carbohydrate but stores it directly in some part of his body. You know that WERTHEIMER[2] and his group have shown that high storage of carbohydrate occurs in the fatty tissues in certain types of obesity. We therefore think that the ability of the obese tissue to store carbohydrates can be related in this connection to the discrepancy of weight and the storage of water, potassium, and sodium. At first we thought that this was simply a characteristic of obese people, but when we studied severely undernourished people by the same procedure, we have observed the same type of response as in obese patients on a reducing diet. When these undernourished patients are refed, however, you have a normal rise in the respiratory quotient after glucose, and when they are highly fed, they make fat from their carbohydrate, the respiratory quotient rising above 1. So I think that these facts can be correlated with the discrepancy you have observed between the nitrogen and potassium balances. There is an ability to store glucose or glycogen in the fat reserves which is related to the quality and quantity of the food ingested, and this may explain the retention of water and electrolytes in obese or undernourished patients; it may perhaps also have something to do with the ability of obese subjects to conserve calories in situations of emotional stress and to avoid ketosis upon caloric reduction.

BLOM: I very much agree, of course, with these remarks. It is difficult to say that I am sure what happens with this sodium and with this potassium, but I visualise for myself that they somehow cover anionic components in the metabolic pool which arise anew after calories are refed following reduction of calories or after carbohydrates are refed following reduction of carbohydrates.

QUERIDO: You mean that this might account for the excess movement of the sodium and potassium?

BLOM: Yes.

WATERLOW: I was very interested, Dr. BLOM, that your results agree with what we find in our patients, namely, that the high carbohydrate intake

---

[1] MARTINEAU, B., and J. TREMOLIÈRES: Effet d'une surcharge glucidique sur les échanges respiratoires dans l'obésité. — Nutr. et Dieta (Switz.) (to be published).

[2] WERTHEIMER, E., and R. SHAPIRO: Physiol. Rev. (U.S.A.) **28**, 451 (1948).

seems to promote the retention of sodium. And as you know, the old German paediatricians used to talk about the water-retaining effect of carbohydrates. I would like to know whether you have made measurements of body water in the course of your studies.

BLOM: I am sorry, we did not; we just relied on our sodium and potassium balances. We have not measured the body water by a direct method.

WATERLOW: I would like to ask Dr. HAAK a question. I quite agree that in measurements of this kind on human beings it's necessary to make certain assumptions, and that if you get reproducible results and are comparing results in the same patient, then these are reasonable assumptions. However, I find your results so surprising, in view of what was said this morning, that I should like to see whether there is any other explanation. I think I am right in saying that the synthesis rate S is practically the same as the product of the turnover rate and the pool size. You find an increased synthesis rate with growth-hormone administration, which is produced entirely as a result of an increase in the pool size, even though there is a decrease in the turnover rate. This seems quite contrary to the *in vitro* results reported this morning. Therefore I suppose one must question the assumptions. I believe the calculations are based on the time at which the maximum specific activity of urea is reached, i.e. about 3 hours. This assumes that, in the production of N^{15} urea, the preceding reactions — which are largely transaminations, I think — go very quickly in relation to this time interval. BISHOP[1] and his group have shown that this is not so; in this time interval you may, or you may not, get the same concentrations of N^{15} in the ammonia and amino-acid mixture as in the urea. BISHOP has therefore questioned this part of the assumptions on which the calculations are based. I wonder whether you have any measurements of the N^{15} concentration in the urinary ammonia which might be of importance in interpreting the results.

HAAK: If I may be allowed to, I should prefer to pass this buck to Dr. KASSENAAR!

KASSENAAR: The objections raised by Dr. WATERLOW are of course obvious and very pertinent ones. We fully agree with him that the assumptions which are used in the calculation of B, P, and S are certainly not completely correct and are even partially not true. The San Pietro-Rittenberg model of overall protein metabolism is an oversimplification, of which we have to be aware when performing studies of this kind. However, as stressed already by Dr. HAAK, we feel that if two conditions in one patient are compared, which differ as far as we can judge only by one external, controlled parameter, useful information on the kind of changes occurring can be obtained. Therefore we do not think, for example, that the S which is calculated expresses exactly the amount of protein synthesised per day, but we do think that, if we compare the S in two conditions, the direction in which S does change can be determined.

WATERLOW: I entirely agree with your approach. We have to make assumptions in investigations on human patients. I am not a mathematician, but I believe that P and B are not calculated independently. I think this is an extraordinarily interesting piece of work, but I wonder if you have in mind any way of independently verifying the conclusions.

---

[1] HSIEN, W., and C. W. BISHOP: J. Applied Physiol. (U.S.A.) **14**, 1 (1959).

KASSENAAR: You are right that the turnover rate and the pool size are not calculated completely independently, since in the calculation of the pool size the B is used. However, another independent parameter, namely, the maximal isotope concentration in the urinary urea, is also used. You will have seen in the patient who was studied before and after hypophysectomy that not only the time of the maximum isotope concentration, which determined the B, changes, but that also the height of the maximum isotope concentration undergoes very definite changes.

WATERLOW: You continued the urine collections for 43 hours, and you can get some information from the cumulative excretion, but you didn't say anything about this.

KASSENAAR: During the later period, the recirculation of the isotope coming from labelled protein will be much more pronounced and will complicate the calculations even further. If there were no recirculation of isotope, we think that the declining part would represent the turnover rate of the urea pool. Would you agree with that?

WATERLOW: Yes, I do.

KASSENAAR: However, we calculate from the declining part of the curve a much longer half-life for urea than after injecting labelled urea, which means in our opinion that during this period the recirculation of isotope is a serious complication.

QUERIDO: I think there is one important point that has been raised by Dr. BLOM which deserves attention. At the beginning of his paper he made a statement on the possibility of using the nitrogen balance on a quantitative basis for the assessment of agents acting on nitrogen metabolism; but he dismissed this as a possibility for which there is no proof; moreover, in view of the influence exerted by the preceding periods, i.e. by the condition of the patient before the experiment, he did not expect to get good results in this way. I would like very much to hear from the audience whether anybody has any different views on this point, because, as you know, in assessing so-called anabolic agents the nitrogen balance in monkeys is nowadays also being used for comparing the activity of compounds.

FRASER: I wasn't really going to make any references to work that is just being initiated now, but I think I ought to at this point, because the work in question has some bearing on what has been presented this afternoon on the action of growth hormone on protein metabolism. Dr. JEEJEEBHOY in our department has been measuring plasma-protein turnover with I^{131}-labelled albumin, and I suppose what happens to that may have some bearing on the sort of measurements that have been made with N^{15}-labelled glycine. We have made some measurements on patients with acromegaly before and after treatment and some on one hypopituitary subject before and after administration of growth hormone. The preliminary findings suggest that the most striking effect is on the catabolism of the protein and not on its anabolism. The net result as between the anabolic and the catabolic rates, calculated by conventional procedures, is of course that you have more retention of protein. But both rates go down after growth hormone; only the catabolism goes down lower.

Introduction to the General Discussion:

Homoeostasis in protein metabolism

By

J. TREMOLIÈRES

Factors influencing protein metabolism in the organism must be classified as nutritional and hormonal. On the one hand, these two groups of factors are integrated by the nervous system, which regulates the food intake by means of what we call "appetite". On the other hand, they are integrated in each cell, where fundamental protein structures are maintained by a flow starting with the anabolic processes and ending with what we refer to as catabolism.

This system is so perfectly homoeostatic that none of its components can be studied without reference to the others; for example, an hormonal effect cannot be accurately defined unless one is acquainted with the nutritional situation and with the other hormonal factors involved. We may postulate that nervous regulation and the target systems in the cells are well defined when the hormonal and nutritional factors are known.

The processes responsible for maintaining homoeostasis in this system are probably no better known and still as puzzling as when ATWATER (1) wrote 62 years ago:

"A man may live and work and maintain bodily equilibrium on either a higher or lower nitrogen level or energy level. One essential question is: What level is most advantageous?"

The facts already reported at this meeting show that we are still at the descriptive and analytical stage. I would therefore propose that we try to classify those homoeostatic processes and to justify such a classification by reference to some facts.

I suggest that we consider these homoeostatic processes:

1. At the highly integrated level of the organism;
2. At the level of the biochemical anatomy of the cells of a given tissue;
3. At the level of the biochemical physiology of a cell.

1. Adjustment of protein metabolism to ingesta at the level of the whole organism

Dr. McCance has already shown how the development of the pig varies in response to different diets administered from birth onwards. Dr. Waterlow has dealt with the more complex processes of adjustment to undernutrition. To complete the picture I would remind you of the work of Parizkova (5) who observed that, when rats are accustomed to consuming an excess of calories with a high-fat diet from the time of weaning onwards, they remain big eaters for the rest of their lives, even if the type of diet changes. This rather seems to correspond to the symmetrical pattern observed by Dr. McCance.

We have conducted investigations on rats which were accustomed to intermittent, partial fasting (—40% for 7 days) interrupted by periods of feeding at 100% for 7 days. When this was repeated five times it was found that, in response to total fasting, the animals lose less nitrogen and less weight. They utilise their fat more quickly.

Table 1. *Effect of total fasting on rats previously kept on a continuous diet and then subjected to 5 periods of partial fasting (— 40% for 7 days) interrupted by 7-day periods of rehabilitation (100%)*
(From M. Morin, J. Tremolières, and R. Jacquot)

Previous mode of feeding	Days of total fasting	Caloric expenditure/ rat/day	N losses mg./rat/day	R. Q.	Weight losses (g./rat/day)
Continuous	1	41	171	0.73	13
	2	39	163	0.73	9
	3	36.3	152	0.71	9
	4	35.6	147	0.70	8.3
	5	31	123	0.70	7.6
Discontinuous	1	29	112	0.70	9
	2	28	107	0.70	7
	3	27	100	0.71	7
	4	26	95	0.70	6.6
	5	24.6	90	0.70	5

Table 2 shows that food utilisation is nearly twice as efficient after fasting.

Rats accustomed to fasting in this manner show a reduction in their caloric losses of about 15% per 100 g. The old rat is not capable of such adaptability.

Thus, the amount of calories and nitrogen ingested at a particular period of life, together with the rhythm of feeding, are able to produce homoeostatic reactions which serve to maintain the nitrogen balance in the organism in accordance with the differing

Table 2. *Weight gain and food yield in rats kept on a continuous diet (control) or on a diet with alternating periods of fasting (five periods of —40% for 7 days) and rehabilitation (100% for 7 days)*
(From M. MORIN, J. TREMOLIÈRES, and R. JACQUOT)

Group	Diet	Weight gain (g./rat)	Ingested dry (g./rat)	Weight gain per gramme of dry ingested feed (g.)
Control	Continuous	18.5	113	0.16
Rehabilitation .	First rehabilitation	36	98	0.37
Rehabilitation .	Average of five periods of rehabilitation	28	94	0.30

requirements of each situation. This general adjustment of the whole organism is the result of the integration of fractional adjustments at the level of the tissues and cells.

2. Adjustment at the tissue level

a) **Changes in the distribution of the protein mass among different organs.** As the effects of growth hormone and so-called anabolic steroids have been studied carefully elsewhere, I would like here merely to mention the homoeostatic mechanism involved in the action of cortisone in adrenalectomised rats at the tissue level. This would appear to suggest a general pattern for the action of steroid hormonal factors on protein metabolism. Protein synthesis takes place in the target organs, together with increased catabolism elsewhere, producing either energy or sparing nitrogen or amino acids. The nitrogen balance is the resultant of this transfer.

In the case of cortisone in rats, the target organ is the liver, in which net protein synthesis occurs. At the same time, lymphoid tissues are catabolised, the mobilisation of muscular protein depending on the diet (6). The deposition of protein in the liver is preceded by nucleic acid synthesis (cytoplasmic RNA; cf. Fig. 2).

Another example of this type of transfer is found in pregnancy. Pregnant rats on a restricted protein intake, having a total nitrogen retention of 295 mg., store 499 mg. nitrogen in the foetus. There is a transfer from the mother to the foetus, the nitrogen balance always remaining positive (7).

Table 3. *Variation of some hepatic biochemical constituents (per 100 g. body-weight) in adrenalectomised rats * under the influence of cortisone (5 mg./24hrs)*

	Liver weight (g.)	N (mg.)	Nucleic P (mg.)	Glycogen (mg.)	Lipids (mg.)
Controls	3.55 ± 0.11	128 ± 1	4.85 ± 1	20 ± 4	166 ± 18
Cortisone 2 to 5 days	4.75 ± 0.39	153 ± 4	6.51 ± 1.3	155 ± 10	178 ± 28
Percent change . .	+ 34	+ 19.5	+ 34	+ 674	+ 7
Controls	3.2 ± 0.09	115 ± 1.3	4.28 ± 0.4	4 ± 2.5	114 ± 19
Cortisone 6 to 12 days	4.8 ± 0.31	154 ± 3.2	5.24 ± 1.1	191 ± 67	150 ± 18
Percent change . .	+ 50	+ 34	+ 22.4	+ 4700	+ 31

* Groups of 8 to 10 rats.

The reported values are mean values with probable error on the mean.

The mean nitrogen balance during the 6—12 days period was — 370 mg. N/100 g. body-weight. It indicates a 19% reduction in the nitrogen mass of the rat, the weight loss being about —9.3 g./100 g. body-weight with the 18% protein diet used.

b) Changes in protein mass within the cells. SZIRMAI has nicely illustrated the hormonal control of microsome by means of electron microscopy, and WATERLOW has dealt with nutritional control in relation to certain enzymes. With regard to this nutritional and hormonal control of the biochemical anatomy of cells, I would merely add a reference to some experiments which we conducted on rats. One group of animals was fed for 12 months on a diet with 36% of calories from protein and 21% from fat, and the other group on a diet with 23% of calories from protein and 66% from fat; the high-fat diet, which leads to atherosclerosis of the arteries, also produces in the liver and heart a reduction in RNA and in nitrogen related to DNA — in other words, a reduction in the active protein mass within the cell (2).

Fig. 1, from a study by FRAYSSINET (3) indicates the physiological significance of the relation of $\frac{cytoplasmic\ RNA}{DNA}$ to $\frac{nuclear\ RNA}{DNA}$. Total fasting reduces both cytoplasmic and nuclear RNA, whereas protein fasting reduces only cytoplasmic RNA. On the other hand, cortisone increases cytoplasmic RNA to a greater extent than nuclear RNA.

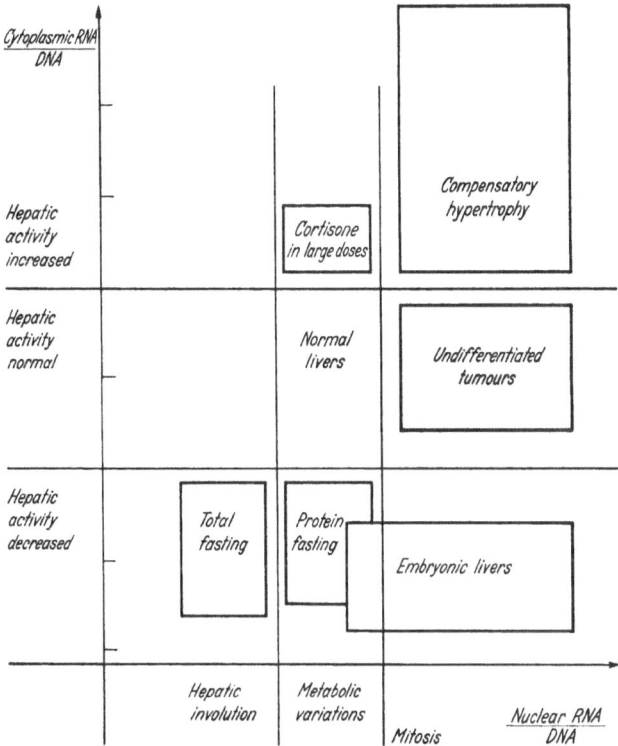

Fig. 1. Effect of diet (total fasting and protein fasting) and of cortisone on the distribution of RNA between cytoplasm and nucleus. From C. FRAYSSINET

3. Homoeostasis at the level of cellular physiology

I should like to take one example to illustrate this type of reaction. Like many other investigators, we have noted that the biological utilisation of an amino acid is greatly dependent on the previous caloric and nitrogen intake. DE GASQUET and LOWY (9) have reported that the ratio of ATP/AMP was able to regulate the availability of glutamate, amino acid oxidation, and trans-reamination in kidney homogenates. This is one mechanism, among many others, which may determine the biological value of a mixture of amino acids. It offers an example of regulation by a multi-enzymatic system, which may help to explain the way in which the organism as a whole adjusts itself to different protein intakes.

To sum up, there is a constant interplay between hormonal and nutritional factors influencing metabolism. These interactions have to be considered simultaneously in relation to the physiological functions of the organism as a whole, to the biochemical anatomy of the various organs, and to the physiology of the cell.

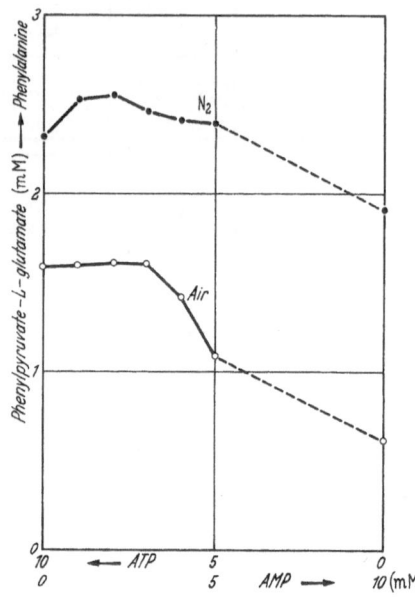

Fig. 2. Effect of ATP/AMP ratio (with a stable total concentration of ATP + AMP = 10 mM.) on phenylpyruvate-L-glutamate transamination to phenylalanine under air or nitrogen (kidney homogenate). From P. DE GASQUET and R. LOWY

I suggest that this general concept may provide a framework which can perhaps help us to establish some sort of order in the data relating to factors influencing protein metabolism. It does, at all events, underline the fact that no hormonal effect on protein metabolism can be satisfactorily studied unless one defines the nutrient intake — present and past — and unless one investigates the effect in question at three levels, i.e. in the organism as a whole, in the organs, and in the cells.

References

1. ATWATER, W., and F. BENEDICT: U.S. Dept. Agricult. Office of Experiments Station. Bull. 109. Experiments on the metabolism of matter

and energy in the human body, p. 130. 1898—1900. — 2. BRIGAND, L., R. LOWY, and J. TREMOLIÈRES: Effet d'un régime hyperlipidique sur la composition tissulaire chez le rat. Symposium Dijon 1962 (to be published later). — 3. FRAYSSINET, C.: Etude chez le rat de l'hypertrophie compensatrice du foie après ablation partielle. Thèse Sciences. Paris 1962. — 4. MORIN, M., J. TREMOLIÈRES, J. ABRAHAM, O. CHAMPIGNY, and R. JACQUOT: Compt. rend. Acad. sc. (Fr.) 252, 3142 (1961). — 5. PARIZKOVA, J.: Nutr. et Dieta (Switz.) 3, 236 (1961). — 6. TREMOLIÈRES, J., and R. DERACHE: Ann. endocr. (Fr.) 15, 827 (1954). — 7. TREMOLIÈRES, J., and R. JACQUOT: Compt. rend. Acad. sc. (Fr.) 240, 235 (1955). — 8. TREMOLIÈRES, J., M. BRUNAUD, and R. JACQUOT: Biochemistry of lipids. Pergamon Press, 1960, p, 156—161. — 9. DE GASQUET, P., and R. LOWY: Effet des nucléotides adényliques sur l'inversion in vitro des D-aminoacides. III. Influence de l'AMP et de l'ATP sur la transamination pyruvate-L-glutamate dans des homogénats de rein de rat. Enzymologia (Neths) (in press). IV. Etude de la transréamination après désamination oxydative des dérivés D, dans des homogénats de rein de rat. Enzymologia (Neths) (to be published later).

General Discussion

STAEHELIN: I should like to ask Dr. RABEN one question concerning the ketogenic effect of growth hormone. Do you have any idea about the time relationship between free fatty acid release and ketogenesis ? There have been several reports on lipolytic factors — including, for instance, the one which appeared in the Lancet about three years ago I think — in which this ketogenesis is described as occurring with much lower doses than those required to produce a release of free fatty acid. So I wonder whether you think that ketogenesis is only a consequence of the free fatty acid release or whether there is more to it than that.

RABEN: I don't know whether it's just the consequence of the free fatty acid release or not, but I am not aware of any evidence that ketosis can be produced in man with a smaller dose than is required to mobilise free fatty acids. In general, the ketosis has not been prominent except in patients who have been hypophysectomised diabetics. I think that Dr. IKKOS and Dr. GEMZELL[1] were among the authors who reported a normal patient who had a one-day ketotic response, but I think that in normal patients this is unusual. It seems conceivable that the combination of increased mobilisation of fatty acids plus a diminished ability of the liver to resynthesise long-chain fatty acids could together account for the ketosis.

LARON: I should also like to refer to Dr. RABEN's paper. We have recently studied in a child the metabolic effect of the electrophoretic fractions of human growth hormone[2] and we have found that all the fractions are active as regards nitrogen, urea, and potassium retention, but that all the fractions also induced sodium and water retention — and this brings me to my question. Are we really dealing with one hormone or with several forms of human growth hormone, or are these fractions made by us by the procedure used in preparing the growth hormone ? Furthermore, how can the water

---

[1] IKKOS, D., R. LUFT, and C. A. GEMZELL: Lancet (G.B.) 1958/I, 720.
[2] LARON, Z., S. ASSA, and R. MENASHE: submitted for publication.

retention be explained ? Is it caused by a contaminant ? I don't think it is, because how could we then account for the fact that all the fractions had this water and sodium retaining effect ? We also observed that, on long-term treatment with human growth hormone, a progressive increase in weight was later followed by a weight loss. It seems more probable that this weight loss was due to loss of water rather than to loss of nitrogen.

RABEN: I think that originally in animal studies increased body water was thought to be due to growth hormone, and similar findings emerged from studies on man. I don't know of any evidence that it's due to a contaminant. There was a suggestion in a few studies with human growth hormone that aldosterone excretion was increased, but that certainly has not been a frequent finding. Since there are so many things that could affect the body-weight, and since the patients were presumably not on a constant diet, I couldn't really begin to try to explain this question of the body-weight. I presume the patients were not on a constant diet ?

LARON: Yes, it was a more or less constant diet. But even some slight changes in the dietary intake would not seem to explain the weight loss extending over several weeks.

RABEN: I just don't know. If you follow up pituitary dwarfs treated over a long time, you find of course that their weight increases as time goes on. However, the weight changes are not constant from period to period. They may continue to grow during some periods, when the weight has gone down, and at other times, when it has gone up. But over the years the weight certainly goes up in the same way as in the case of a normally growing child. In rat studies, there has long been evidence that the immediate effect of growth hormone is to produce a weight gain which may be excessive as compared with the subsequent days. And it was seen in studies during the 1930's that, if a rat is treated for one or two days and treatment is then stopped, all the weight that it takes on will be lost, and the longer the rat is treated, the greater the percentage of the weight that is retained. It's only when a rat has been treated for a couple of days that it will lose all the weight gained.

IKKOS: I think one clinical observation might help to clarify Dr. LARON's question as to whether the water retention observed in the experiment with growth hormone is due to the hormone itself or to some contaminant. If one measures extracellular volume in patients with acromegaly, where one has no evidence to suspect a contaminant in the growth hormone, one always finds a 20-30% increase in the extracellular fluid compartment, and this would seem to support the view that it is the growth hormone *per se* which causes the increase in the volume of the extracellular fluid compartment.

SCHREIER: I should like to make two very short comments. Isn't growth as such necessarily connected with an increased water content ? If one determines the body fluid of premature infants and young babies, one finds a very high water content, which decreases over the years until growth stops. Secondly, if we discuss here the factors which influence protein metabolism, we should not forget age. We have done some research in respect to this problem and we have found that during the postnatal period not only is protein synthesis increased, but the half-life of the soluble proteins is also very considerably shortened. We injected labelled albumin into young rabbits, for instance, and found that its half-life is about 50% of the half-life in adult controls, and we got the same results in rabbits with labelled amino acid in the prenatal and postnatal period. We found that in very young rabbits the half-life of several proteins separated by paper electrophoresis, etc. was something like 10% of the half-life in the adult rabbit. I won't go into details on this subject, because there isn't time. But I would like to suggest that in all studies on protein metabolism we should always consider the age of the animals we use.

PRADER: Dr. RABEN, you mentioned that some patients with hypoglycaemia were improved by growth hormone and that some of them had an insulin-producing tumour. Could you define a little more closely which types of hypoglycaemia respond to growth hormone and which do not? We have given growth hormone to one patient with glycogen-storage disease of the classical type involving severe hypoglycaemia. There was no improvement in the hypoglycaemia and no improvement in the unresponsiveness to glucagon.

RABEN: I am afraid I can't define the cases that will or won't responds CRIGLER[1] has reported one infant with severe hypoglycaemia, and in thi patient growth hormone didn't seem to make any difference at all. Two patients treated at New York Hospital seemed to be improved. In the two patients with insulinomas, the severe hypoglycaemia seemed to be brought under control, but I don't think the type of case is well enough defined at the moment for us to be able to predict what the response will be.

---

[1] CRIGLER, J. E., J. A. KNAPP, and J. CHAGNON: Amer. J. Dis. Child. **96**, 432 (1958).

Evaluation and mode of action of anabolic steroids

Differentiation of action of various anabolic steroids

By

P. A. DESAULLES and C. KRÄHENBÜHL

The role of androgen derivatives in stimulating growth, maturation, and maintenance of the protein stores of the organism is well known. However, quite apart from their marked metabolic properties, these steroids possess androgenic, progestational, and antigonadotrophic effects, the relative intensity of which varies from one group of androgen derivatives to another. For many years intensive efforts have been made to characterise properly and to separate as completely as possible the metabolic and sexual endocrine properties of these steroids.

Since anabolism is a process which affects essentially the maintenance or the restoration of the protein balance, the principle underlying most methods of investigation is to induce experimentally a disturbance in this balance and then to try and correct it more or less completely by means of treatment. In testing anabolic steroids, all other factors which act on protein synthesis, whether hormonal or nutritional, should be kept as constant as possible. As the evaluation of the sexual endocrine properties of these steroids — considered in this context to be side effects — is at least as important as that of their specific anabolic functions, both types of action should be evaluated simultaneously in the same animal, not only in terms of the dose used but also in terms of the duration of administration; the duration of administration must be taken into account because the response of different parameters to a stimulation does not necessarily show the same rapidity of onset. The end point of the test chosen should be restoration of the impaired function to the values observed before the disturbance was produced. By measuring our results in terms of restitution, we are thus able to express the response of different functions in a more physiological manner (1). This requirement eliminates those substances which are characterised more by a lack of androgenic activity than by a complete anabolic action on muscle growth. In this short study, we shall confine ourselves to comparing the form and intensity of the myotrophic and androgenic effects measured simultaneously in the same animal in terms of the weight gain of particular organs; nevertheless, the relationship between the factors outlined above is valid for all the parameters of activity which are likely to be involved in the evaluation of these substances.

Method

All experiments were performed on rats obtained from the same source and belonging to the same local breed. Housing and nutritional factors were as uniform as possible, ambient temperature and atmospheric humidity were kept constant, and the rats were given standardised food.

The experiments were performed chiefly on near-adult male rats aged about 4 months. We used: normal, castrated, castrated thyroidectomised, castrated adrenalectomised, and castrated hypophysectomised rats. Treatment, beginning 15 days after castration, or 2 days after subsequent operation, was continued for 15 and 25 days, respectively. Steroids were administered at different dosage levels by daily parenteral (s. c.) injection in sesame oil.

Autopsies were performed on the 3rd, 6th, 9th, 15th, and 25th day following the beginning of the experiment. Groups of 6—10 rats were used for testing each dosage level.

The anabolic response was measured by reference to the weight of the levator ani muscle, and the androgenic effect by reference to the weight of the seminal vesicles; in adult rats the seminal vesicles are known to be somewhat more sensitive to androgens than the ventral prostate, whereas the opposite is true in juvenile rats (2). The weight gain of these organs in castrated treated animals was compared with the difference between the organ weight of castrated controls and normal controls, the effect being expressed in per cent of restitution to normal organ weight.

This method enables one to make a direct comparison between anabolic activity (degree of restitution of the levator ani muscle weight) and androgenic activity (degree of restitution of the seminal vesicle weight).

As an optimal anabolic action we selected the effect exerted by that dose of a substance which restores the weight of the levator ani muscle to normal control values (100%). The optimal androgenic action was taken to be the effect exerted by that dose of a substance which restores the weight of the empty seminal vesicles to normal control values (100%). (For further details on method, see 1, 3, 4.)

In our study we compared the effects of testosterone, 1-dehydro-testosterone, 1-dehydro-5α-dihydro-testosterone, 19-nortestosterone, and their 17α-methyl derivatives.

Results

The use of castrated animals makes it possible to evaluate androgenic and anabolic properties simultaneously. It has been known since the time of Bouin and Ancel (5) that castration impairs growth and weight gain in rats.

Sources of error due to the animal's natural growth in the course of the experiment can be largely avoided by using adult animals (6, 7). Fig. 1 shows the weight of the levator ani muscle and of the seminal vesicles at different time intervals following castration in young and adult male rats of our laboratory stock kept under identical conditions. In young animals the levator ani continues

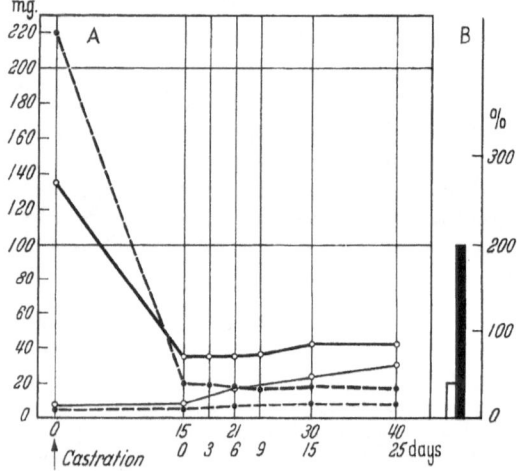

Fig. 1 A. Comparison of changes in the weights of the levator ani muscle and seminal vesicles after castration in immature and adult rats. Immature rats: Lev. ani (mg.) ——————;
Sem. ves. (mg.) — — —. Adult rats: Lev. ani (mg.) ————; Sem. ves. (mg.) — — —.
B. Relative weight gain of castrated rats after 25 days. Immature rats ▆, adult rats ▢

to grow, whereas in the adult animal, after castration and involution, the weight of the muscle remains constant. The weight of the seminal vesicles, on the other hand, remains nearly constant in castrated juvenile rats and, after involution, in adult castrated animals.

From these observations it appears that the use of adult animals offers the advantage of ensuring more constant control values over the whole duration of the test. It is also apparent that the organs studied undergo complete involution — as regards weight — within 15 days of castration; further prolongation of the interval between castration and the beginning of the experiments does not lead to any appreciable change in these values.

If we analyse the effects of the different substances tested under these conditions, it is immediately apparent that in the case of testosterone, irrespective of the dosage levels and over the whole duration of the experiment, no dissociation whatsoever between anabolic and adrogenic response appears (Figs 2—5, Table 1) (4). At a dosage level of 1 mg./kg./day s.c., both anabolic and androgenic responses are virtually complete and, as such, nearly optimal after 25 days. At a dosage level of 10 mg./kg./day s.c., the response is much more rapid, complete restitution being obtained in about 12 days for both organs. By contrast, all other steroids show a marked dissociation between anabolic and androgenic response as a function of dose or duration of administration.

Some, such as 1-dehydro-testosterone, are somewhat less active than testosterone on a weight-for-weight basis. At a dosage level of 1 mg./kg./day s.c., an incipient anabolic response begins to show up on the 25th day; at a dosage level of 10 mg./kg./day s.c., the anabolic response is nearly complete by the 15th day, the concomitant androgenic response remaining entirely subliminal at the two dosage levels.

19-nortestosterone, displaying about the same activity as testosterone on a weight basis, shows at a dosage level of 1 mg./kg./day s.c., a complete anabolic response after 25 days of treatment, the androgenic response remaining negligible. At a dosage level of 10 mg./kg./day s.c., the anabolic response is also more rapid, being complete in about 15 days. The androgenic response—no longer subliminal—amounts to 40% of the norm.

1-dehydro-5α-dihydro-testosterone, on the other hand, is about twice as active as testosterone on a weight-for-weight basis, but displays properties which are essentially similar to those of 19-nortestosterone, although its concomitant androgenic effects are more pronounced.

Since it is known that, among other possibilities, substitution in the 17α position produces qualitative and quantitative modifications in the properties of steroids, we have studied the influence exerted by the introduction of a 17α-methyl radical on the properties of the steroids mentioned above (Figs 6—9, Table 2).

17α-methyl-testosterone is somewhat less active than testosterone on a weight-for-weight basis but, although it shows to a greater degree than testosterone a dissociation between androgenic and myotrophic actions, this dissociation is not complete. 17α-methyl-1-dehydro-testosterone is about as active as 1-dehydro-testosterone on a weight-for-weight basis. At a dosage level of 3 mg./kg./day s. c., the anabolic response is complete before the 25th day, the androgenic response being subliminal. At a level of 10 mg./kg./day s. c., the anabolic response is practically complete by the 15th day, while the androgenic response amounts to about 20—25% of restitution and is therefore still liminal, not exceeding 40% even on the 25th day. At this dosage level, 17α-methyl-1-dehydro-testosterone shows about the same degree of dissociation between myotrophic and androgenic response as 1-dehydro-testosterone. At 100 mg./kg./day s. c., the response is very rapid, restitution of both anabolic and androgenic organs being complete by the 9th day. No dissociation between anabolic and androgenic effects is possible at

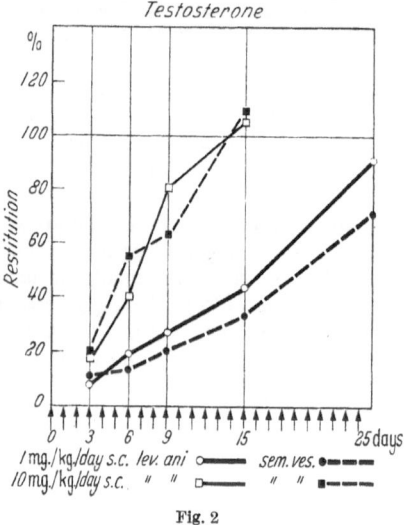

Fig. 2

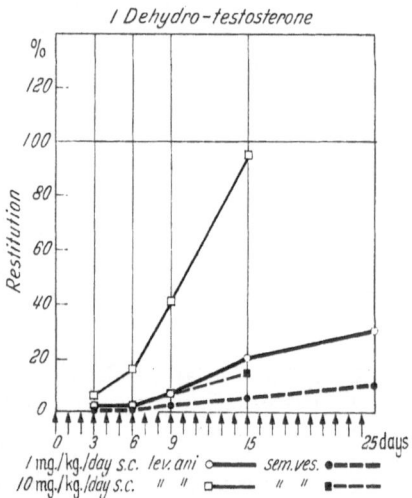

Fig. 3

Figs 2—5. Anabolic and androgenic effects of testosterone and various of its derivatives injected daily.

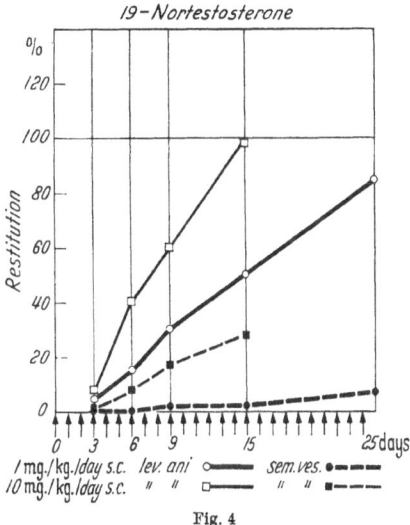

Fig. 4

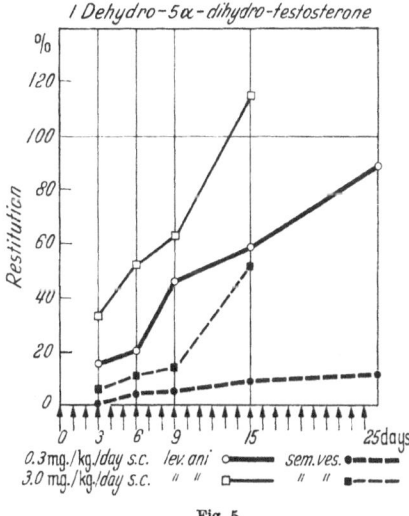

Fig. 5

Ordinate: percentage of restitution to normal organ
weights. Abscissa: duration of experiment in days

that dosage level. However, both 17α-methyl-1-dehydro -5α- dihydro-testosterone and 17α-methyl-19-nortestosterone are *more* active on a weight-for-weight basis, the former about 3 times more than the parent steroid. The degree to which the anabolic action of these derivatives is free from concomitant androgenic effects seems to be at best of about the same order of magnitude as for the parent substance.

It is evident from the experimental data presented that certain steroids are capable, depending upon dosage level and duration of administration, of completely restoring the weight of certain muscles in castrated rats to pre-castration levels, while at the same time having practically no effects on specific androgenic receptors. These results are in marked contrast to those obtained with testosterone, where, irrespective both of the doses and of the duration of treatment, such differentiation proves impossible.

These experiments also clearly demonstrate the differences in the rate of growth of the levator ani muscle and the seminal vesicles under the influence of anabolic steroids. It appears that a myotrophic effect,

Table 1. *Activity of testosterone derivatives on the restitution of the levator ani muscle and seminal vesicle weights to normal values, at different time intervals*

Substance	No. of animals	Dose mg./kg./day s.c.	3rd day Lev. ani %	3rd day Sem. ves. %	6th day Lev. ani %	6th day Sem. ves. %	9th day Lev. ani %	9th day Sem. ves. %	15th day Lev. ani %	15th day Sem. ves. %	25th day Lev. ani %	25th day Sem. ves. %
Testosterone	18	0.3							10	9		
	42	1	8	11	19	13	27	20	43	33	90	75
	18	3							65	50		
	30	10	17	20	40	55	80	63	105	109		
1-Dehydro-testosterone	12	0.3							12	4		
	36	1	2	∅	2	∅	7	2	20	5	30	10
	12	3							32	8		
	24	10	6	2	16	2	41	7	95	15		
1-Dehydro-5α-dihydro-testosterone	36	0.3							59	9		
	12	1	15	∅	20	4	46	5	73	33	87	20
	24	3							115	52		
	12	10	33	6	52	11	63	14	138	80		
19-Nortesto-sterone	12	0.3							20	3		
	36	1	5	∅	15	0	30	2	50	2	80	10
	12	3							72	9		
	24	10	8	1	40	8	60	17	98	28		

Table 2. *Activity of 17α-methyl-testosterone derivatives on the restitution of the levator ani muscle and seminal vesicle weights to normal values, at different time intervals*

	No. of animals	Dose mg./kg./day s.c.	3rd day Lev. ani %	3rd day Sem. ves. %	6th day Lev. ani %	6th day Sem. ves. %	9th day Lev. ani %	9th day Sem. ves. %	15th day Lev. ani %	15th day Sem. ves. %	25th day Lev. ani %	25th day Sem. ves. %
17α-Methyl-testosterone	18	0.3							25	5		
	42	1							50	40	74	65
	18	3	5	2	10	5	36	25	65	54		
	30	10	8	6	27	16	69	50	95	88		
17α-Methyl-1-dehydro-testosterone	18	0.3							10	1		
	18	1							24	2	119	20
	42	3	3	∅	10	∅	15	2	30	2		
	30	10	5	4	35	10	70	15	101	21		
	30	100	10	28	65	61	91	88	145	170		
17α-Methyl-1-dehydro-5α-dihydro-testosterone	36	0.1							53	10		
	12	0.3	3	3	17	5	30	6	70	18	85	14
	24	1	23	15	54	20	62	25	110	47		
17α-Methyl-19-nortestosterone	12	0.3							43	3		
	36	1							59	6	85	23
	12	3	7	2	35	2	49	2	93	30		
	24	10	20	11	50	15	64	21	115	65		

though producing only weak or partial restitution by the 9th or the 15th day of treatment, may completely restore the weight of the muscle by the 25th day in the absence of any significant androgenic effect.

Taking into account the slow rate of onset of the anabolic response compared to the rapid response of androgenic receptors, it appears that, irrespective of the intrinsic characteristics of the substances studied, the increased rapidity of onset of the anabolic response following an increase in the dosage administered is mainly due, in the case of the receptors used in this study, to concomitant androgenic effects. If administered in sufficiently high doses, all the substances would have effects which would be qualitatively similar to those of testosterone itself and difficult to distinguish from them (Fig. 7).

These comparisons show us the relation existing between the structure of these steroids and their type of activity, especially the influence of the introduction of the methyl group in C 17 position. They also underline the importance of both dosage level and duration of treatment for the assessment of anabolic activity.

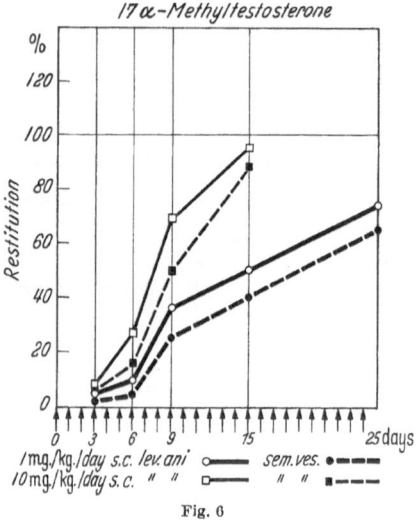

Fig. 6

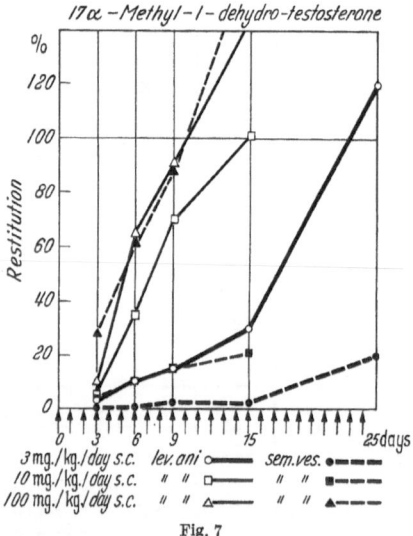

Fig. 7

Figs 6—9. Anabolic and androgenic effects of 17α-methyl-testosterone and various of its derivatives injected daily. Ordinate: percentage of restitution

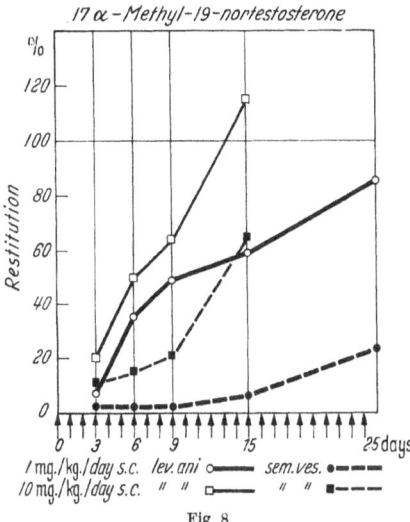

Fig. 8

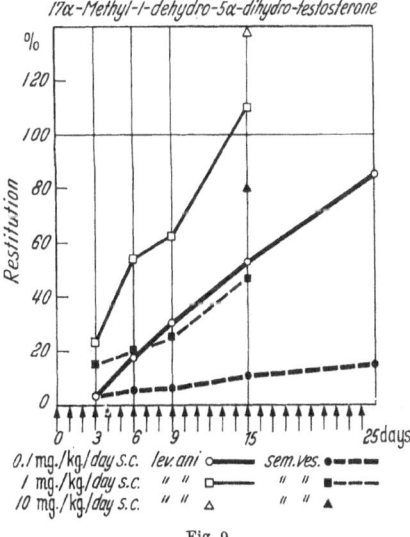

Fig. 9

to normal organ weights. Abscissa: duration of
experiment in days

In fact, various factors intervene in the anabolic response. It is well known that sex has marked effects on the electivity of anabolic responses to certain steroids. This phenomenon is very marked, for example, with 19-nortestosterone derivatives (*8, 9, 10*). Female animals are more sensitive than males. Age plays a certain role, young animals being, for instance, more sensitive than adults to the growth-promoting action of testosterone and its esters (*11, 12*). An explanation for these different actions is in all probability to be found in the multiple endocrine effects which, beside their anabolic action, are characteristic of these substances.

As anabolism can only be understood as the interplay of various hormonal and nutritional factors acting on protein synthesis, the anabolic properties of steroids may be further differentiated by studying the role played by other endocrine functions in influencing the intensity and electivity of their actions.

We therefore investigated the consequences of additional ablation of the thyroid, adrenals, or pituitary on the effects of an elective anabolic steroid by measuring simultaneously in the

12*

same animal the weight gain of the organs mentioned. In the present paper, only the effects of 17α-methyl-1-dehydro-testosterone will be reported (Fig. 10).

If we compare the effects of doses producing, respectively, an optimal anabolic response and an optimal androgenic response in castrated animals after 15 days of daily treatment, and if we give these doses the indexed value 100, we shall see that ablation of the

Index of relative activity of methandrostenolone in various endocrine conditions

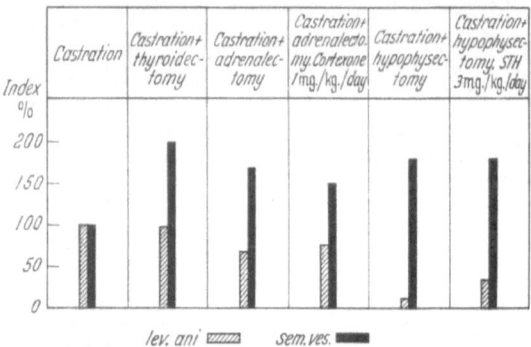

Fig. 10. Comparison of the effect of optimal anabolic and androgenic doses of 17α-methyl-1-dehydro-testosterone in castrated rats, as well as in castrated and thyroidectomised, castrated and adrenalectomised, and castrated and hypophysectomised rats

thyroid, adrenals, or pituitary markedly enhances the androgenic response to 17α-methyl-1-dehydro-testosterone, the intensity of the effect being increased from 50 to 100% above castrate values.

As regards the anabolic response, the results are different. Whereas ablation of the thyroid has practically no effect on the myotrophic action of 17α-methyl-1-dehydro-testosterone, this action is moderately reduced by adrenalectomy and very markedly reduced by hypophysectomy, the intensity of effect being, in the case of hypophysectomy, only about 10% of that produced in castrates.

Replacement therapy with cortisone or cortexone in adrenal-ectomised animals has only a slight effect on either action. Additional treatment with growth hormone in hypophysectomised animals restores the anabolic response to about 30% of its intensity in castrates without modifying the enhanced androgenic response.

The results illustrate the influence of the endocrine functions on steroid effects and compare well with similar observations made with testosterone. Ablation of the thyroid (*3, 13, 14*) or of the pituitary (*3, 14, 15, 16*) enhances the androgenic response and, judging from the available experimental evidence, reduces the intensity of anabolic effects (*16, 17, 3*) in hypophysectomised animals.

Discussion

From the experimental evidence available, it can be assumed that growth in different tissues follows essentially the same basic mechanisms. It has been demonstrated that the weight gain of muscle and of seminal vesicles following treatment with androgens is in all probability the consequence of increased cell hypertrophy and multiplication (*18*).

We have found that certain synthetic androgens show a high degree of dissociation in stimulating levator ani muscle growth without influencing the weight of seminal vesicles; an optimal dissociation is a function of dose and time, these two factors being inversely correlated.

On the basis of these results one could postulate that the higher rate of growth of the levator ani muscle is the result of a greater sensitivity of this muscle towards androgenic substances and that this myotrophic effect would thus be the most sensitive parameter for androgenic action. Such a hypothesis, however, is in contradiction with the fact that testosterone, the physiological androgenic hormone, stimulates growth both in seminal vesicles and in the levator ani muscle at about the same rate, irrespective of dose or time; there is no possibility of finding a dosage level at which one of these receptors shows stimulated growth and the other remains insensitive (*4*). We know furthermore that other naturally occurring androgens display a comparable lack of dissociation between myotrophic and androgenic action (*19*), or even produce — in the case of androsterone, for example — a much stronger stimulation of the androgen receptors than of the levator ani muscle (*19, 20*). This substance is without activity in other anabolic tests (*21, 22*).

From these considerations it appears that growth of the levator ani muscle is significant from a different point of view. On the basis of experimental evidence (*1, 3*), it can be considered for the purposes of screening procedures as a receptor which is highly sensitive to the overall trophic action (i.e. the anabolic effect) of testosterone derivatives, as long as its response is not associated with a reaction of androgenic receptors.

The lack of responsiveness of other muscles may be attributed to the fact that after castration or hypophysectomy these do not show marked involution; consequently, little, if any difference in weight can be observed. A further difficulty arises from the rate of restitution of these muscles, which differs perhaps from that of the muscles of the perineal complex.

Additional hormonal imbalance, or increasing the dose of the anabolic steroid above the optimum level, results in a rapid diminution in the degree to which muscle growth and growth of sexual organs can be dissociated; this dissociation may even become altogether impossible. The resulting pattern of activity tends to be qualitatively comparable to that of naturally occurring androgens, i.e. to that of testosterone itself.

These results tend to substantiate the hypothesis that the greater affinity of an organ for the growth-promoting action of a steroid is a consequence of both its synergism with and its antagonism towards other endocrine functions. We would emphasise here the role of thyroid function in limiting the growth response of sexual organs and the role of growth hormone in stimulating muscle growth only.

While the use of anabolic steroids affords a promising approach in all clinical situations where growth and/or enhanced protein synthesis is desirable, it must not be forgotten — and this is borne out by the experimental evidence presented — that the degree to which these steroids are free from marked hormonal activity depends largely on the hormonal balance of the organism.

Summary

An analysis has been made of the androgenic and anabolic properties of various steroids and of their 17α-methyl derivatives in terms of the dose employed, the duration of the test, and the role played by other endocrine functions in influencing the intensity and electivity of their actions.

The experiments demonstrate that certain steroids have the property of producing optimal anabolic effects, while at the same time being practically devoid of effects on specific androgenic receptors.

Apart from the stereochemical configuration of the steroids studied, these characteristics are a function of the specific rate of response of the different target organs, of the dose utilised, and of the duration of treatment, the two latter factors being inversely correlated. The results obtained are in marked contrast to those produced with testosterone, where, irrespective of both dosage and duration of treatment, such differentiation proves impossible.

The optimal anabolic effects of a given steroid are also dependent on sex, age, and other endocrine functions.

The role played by thyroidectomy, adrenalectomy, and hypophysectomy in modifying the anabolic and androgenic responses is described and discussed.

Zusammenfassung

Die androgenen und anabolen Eigenschaften verschiedener Steroide und ihrer 17α-Methylderivate werden analysiert im Hinblick auf die angewendeten

Dosen, die Versuchsdauer und die Rolle, die anderen endokrinen Faktoren auf Intensität und Spezifizität der Wirkung zukommen.

Die Versuche zeigen, daß bestimmte Steroide optimale anabole Effekte hervorrufen, ohne gleichzeitig eine Wirkung auf spezifische androgene Rezeptoren zu zeigen.

Neben der sterischen Konfiguration der untersuchten Steroide sind diese differenzierten Wirkungen eine Funktion der spezifischen Reaktion der einzelnen Erfolgsorgane, der angewendeten Dosis und der Dauer der Behandlung, wobei die beiden letztgenannten Faktoren sich umgekehrt proportional zueinander verhalten. Die Resultate stehen im Gegensatz zu den mit Testosteron erhobenen Befunden, mit dem unabhängig von Dosis und Behandlungsdauer eine derartige Differenzierung der Wirkung nicht nachweisbar ist.

Die optimalen anabolen Wirkungen eines gegebenen Steroids hängen in wesentlichem Maße vom Geschlecht, vom Alter und von anderen endokrinen Funktionen ab.

Die modifizierenden Einflüsse, welche die Entfernung von Schilddrüse, Nebennieren oder Hypophyse auf die anabolen und androgenen Eigenschaften ausübt, werden beschrieben und besprochen.

Résumé

On a procédé à l'analyse des propriétés anaboles et androgènes de différents stéroïdes et de leurs dérivés 17α méthylés. Cette analyse a été faite en fonction des doses, de la durée des essais ainsi que du rôle joué par d'autres facteurs endocriniens sur l'intensité et l'électivité de leurs actions.

Ces expériences démontrent que certains stéroïdes possèdent des effets anaboles optima tout en étant pratiquement dépourvus d'action sur les récepteurs spécifiques d'effets androgènes.

Ces caractéristiques sont fonction, outre de la configuration stéréochimique des stéroïdes étudiés, du degré de spécificité des réponses des différents organes, de la dose employée, et de la durée du traitement; ces deux derniers facteurs étant inversement proportionnels. Les résultats obtenus sont en contraste marqué avec ceux produits sous l'action de la testostérone, avec laquelle toute dissociation des effets est impossible, quelles que soient les doses ou la durée du traitement.

Les effets anaboles optima d'un stéroïde donné dépendent en outre du sexe, de l'âge et des autres fonctions endocrines.

Le rôle modificateur joué par l'ablation de la thyroïde, des surrénales et de l'hypophyse sur les réponses anabole et androgène est exposé et discuté.

References

1. DESAULLES, P. A., C. KRÄHENBÜHL, W. SCHULER, and H. J. BEIN: Schweiz. med. Wschr. 89, 1313 (1959). — 2. TSCHOPP, E.: Arch. internat. pharmacodyn. thérap. (Belg.) 52, 381 (1936).—3. DESAULLES, P. A.: Helvet. med. acta 27, 479 (1960).— 4. DESAULLES, P. A., and C. KRÄHENBÜHL: to be published shortly. — 5. BOUIN, P., and P. ANCEL: Compt. rend. Acad. sc. (Fr.) 142, 232 (1906). — 6. GORDAN, G. S.: Arch. Int. Med. (U.S.A.) 100, 744 (1957). — 7. SVEJKAR, G.: Praxis (Switz.) 45, 374 (1956). — 8. DUCOMMUN, P., S. DUCOMMUN, and M. BAQUICHE: Acta endocr. (Den.) 30, 78 (1959). — 9. NIMMI, M. E., and E. GEIGER: Proc. Soc. Exper. Biol. Med. (U.S.A.) 94, 606 (1957). — 10. ASCHKENASY, A.: Thérapie (Fr.) 14, 332 (1959). — 11. KOCHAKIAN, C. D.: Amer. J. Physiol. 160, 53 (1950). — 12. KOCHAKIAN, C. D., and

B. BEALL: Amer. J. Physiol. **160**, 62 (1950). — 13. SCOW, R. O., M. E. SIMPSON, C. W. ASTURG, C. H. LI, and H. M. EVANS: Anat. Rec. (U.S.A.) **104**, 445 (1949). — 14. Scow, R. O.: Endocrinology (U.S.A.) **60**, 273 (1957). — 15. KONEFF, A. A., R. O. Scow, M. E. SIMPSON, C. H. LI, and H. M. EVANS: Anat. Rec. (U.S.A.) **104**, 465 (1949). — 16. SIMPSON, M. E., W. MARX, H. BECKS, and H. M. EVANS: Endocrinology (U.S.A.) **35**, 309 (1944). — 17. FUJII, T.: Endocr. Japon. **6**, 46, 125 (1959). — 18. KASSENAAR, A., A. KOUWENHOVEN, and A. QUERIDO: Acta endocr. (Den.) **39**, 223 (1962). — 19. MEYER, C. J., C. KRÄHENBÜHL, and P. A. DESAULLES: to be published shortly. — 20. DESAULLES, P. A.: Seminar on various actions of Dianabol. London, 1960, p. 1.— 21. HOWARD, J. E., L. WILKINS, and W. FLEISCHMAN: Transact. Ass. Amer. Physicians **57**, 212 (1942). — 22. REIFENSTEIN, E. C.: Jos. Macy Found. N. Y. Sept. 1942. Metabol. Asp. Convalescence, No. 1.

Evaluation of long-acting anabolic steroids

G. A. Overbeek, J. van der Vies, and J. de Visser

"Long-acting" anabolic steroids would not require a separate discussion if they all had the same duration of action. Since, however, they may differ greatly in this respect, the questions arise, firstly, as to how they should be compared and, secondly, why their duration of action is different.

In the first part of this paper we shall explain what we consider to be the best manner of pharmacological comparison. In the second part some experiments will be described which may contribute to an elucidation of the second question.

1. Bioassay of long-acting anabolic steroids

When comparing two short-acting steroids which each have roughly the same duration of action, only the relation between dose and effect need be studied. This is done by drawing up dose-response curves for both substances. In order to enable one to calculate a potency ratio, valid for more than one dose level, these dose-response curves must be parallel. When two different pharmacological actions are involved, this same procedure should be followed separately for both actions. Hence, in the case of anabolic steroids, separate sets of dose-response curves have to be produced for the anabolic and for the androgenic effects. When both an anabolic and an androgenic potency ratio can be calculated, it is also possible to calculate an anabolic-androgenic quotient simply by dividing the anabolic by the androgenic potency ratio.

The differing duration of action of long-acting compounds adds a third dimension: the time factor. As might be expected, dose-response curves obtained at different times after injection usually have different slopes, and even when the curves for the two compounds are still parallel the potency ratios differ. Time-response curves should yield information on the duration of action, although the latter is also dependent on the dose. The higher dose almost always has the longer duration of action.

As in the case of short-acting compounds, the same sort of information has to be obtained for both the anabolic and the androgenic responses.

Thus, what is needed are two three-dimensional graphs — one for the anabolic and another for the androgenic effect — each

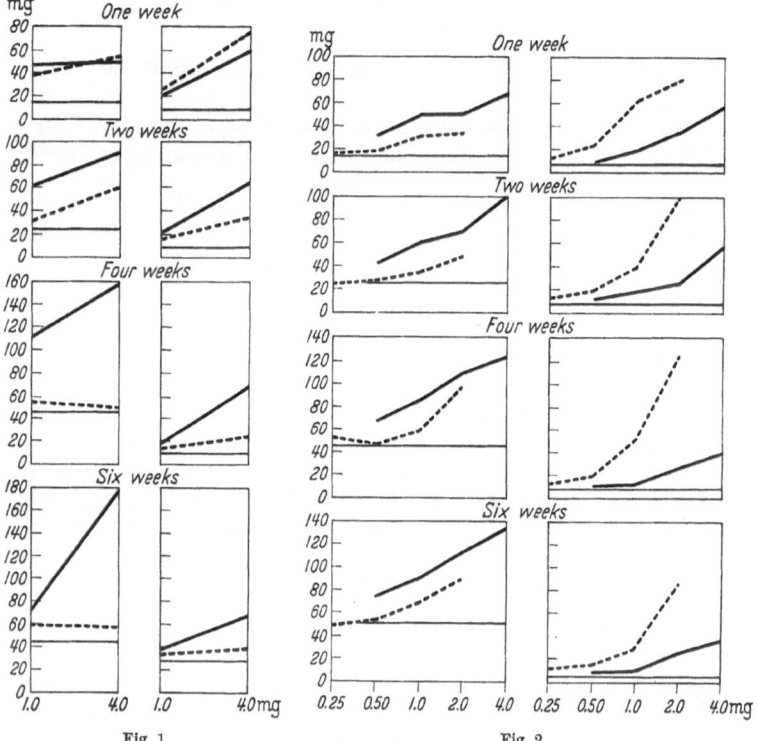

Fig. 1

Fig. 2

Fig. 1. Effect of two esters of nandrolone, at various time intervals after injection, on the weight of the levator ani muscle and seminal vesicles of the castrated rat. ———— nandrolone-decanoate; - - - - nandrolone-phenylpropionate. Left: levator ani muscle; right: seminal vesicles. Treatment: one single subcutaneous injection at zero time. Abscissa: dose; ordinate: response

Fig. 2. Same as in Fig. 1 for the decanoates of nandrolone and of testosterone. ———— nandrolone-decanoate; - - - - - testosterone-decanoate. Left: levator ani muscle; right: seminal vesicles. Treatment: one single subcutaneous injection at zero time. Abscissa: dose; ordinate: response

relating dose, time, and response. In practice several transverse sections are studied, i.e. either dose-response curves at various times or time-response curves at various dose levels. We undertake the former using the HERSHBERGER levator ani test, modified for the purpose as described by OVERBEEK and DE VISSER (1961). Fig. 1 demonstrates the situation for the two steroids nandrolone-phenylpropionate and nandrolone-decanoate. Vertical comparison shows that the duration of action of both esters is different, depending in each instance upon the dose. It will be seen that the dose-response curves are largely parallel for the seminal vesicles after one week and for the levator ani after two weeks, but in no instance is this the case simultaneously for both. Hence, it would be possible to calculate an androgenic potency ratio after one week and an anabolic potency ratio after two weeks, but in neither case could an anabolic-androgenic ratio be obtained.

Fig. 2 shows a comparison between the decanoates of nandrolone and testosterone. These two compounds are more closely comparable, as revealed at once by a glance at the curves. Statistical evaluation of the data showed that after one and two weeks the dose-response curves for the anabolic and for the androgenic effects did not deviate from parallel.

Table 1 lists the results of the calculation of the potency ratios and anabolic-androgenic ratios one or two weeks, respectively, after injection of the steroid. These calculations were impossible after four and six weeks, because there were not enough doses which elicited a significant effect.

Table 1. *Anabolic and androgenic ratios 1 and 2 weeks after injection of steroids*

		1 week	2 weeks
R anab.	nandrolone-decanoate / testosterone-decanoate	4.92	3.29
R andr.	nandrolone-decanoate / testosterone-decanoate	0.41	0.31
Q	nandrolone-decanoate / testosterone-decanoate	12.1	10.6

Treatment: one subcutaneous injection at zero time.
Autopsy: after 1 and 2 weeks.

The procedure described above is the only one by which to prove that the ratios obtained are *dose-independent*, this being a primary requirement for each bioassay. This is obviously not the case if one calculates a quotient from the mean weights of the levator ani and seminal vesicles, nor is the evidence conclusive if one determines the dose at which the atrophied levator ani and seminal vesicles (or prostate) of castrated rats are restored to their normal weight. The procedure outlined here could also be applied to criteria other than weight-increase of the levator ani muscle, etc.; for example, it could be applied to nitrogen retention, uptake of amino acids, etc. There are indeed several investigators who prefer these latter criteria, because they consider that from the qualitative point of view the levator ani assay is not a suitable test for anabolic activity. Although we agree that at first sight these other criteria may appear more attractive, their use is

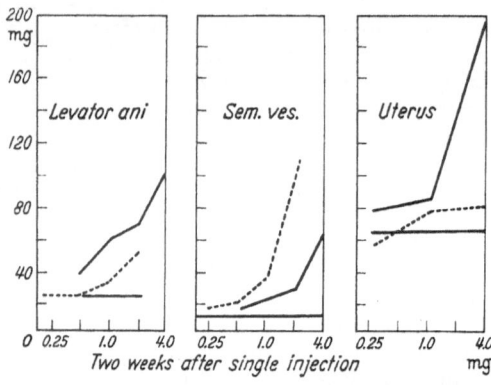

Fig. 3. Weight increase of levator ani muscle, seminal vesicles, and the uterus 2 weeks after a single injection of various doses of the decanoates of nandrolone and of testosterone in the castrated rat. ———— nandrolone-decanoate; - - - - testosterone-decanoate. Treatment: one single subcutaneous injection at zero time. Data two weeks after treatment

impracticable, since they are much less precise and more time-consuming. They could perhaps be used for short-acting compounds, but as the study of long-acting compounds requires many more rats (different time points!), it would be impossible to obtain a reasonably accurate figure with a reasonable expenditure of effort. Moreover, the arguments against the levator ani test as a criterion of anabolic activity (firstly, that the levator ani muscle is absent in the female rat and, secondly, that it is much more sensitive to stimulation by androgenic steroids than skeletal muscles) are in our opinion outweighed by the fact that the differential effects of certain steroids in these organs correlate with their clinical behaviour and also by the fact that the uterus (certainly not a "male" organ) reacts in the same way as the levator ani. The latter is a new argument and is demonstrated by Fig. 3. Whereas testosterone-decanoate is more active on the seminal vesicles than nandrolone-decanoate, the reverse applies not only to the levator ani muscle but also to the uterus. Histological inspection of these uteri revealed that the weight increase was mainly due to growth of the myometrium.

On the basis of the data presented here we consider the quantitative levator ani assay as the only practicable test by which to compare long-acting anabolic steroids. The available data indicate that there is no reason to be concerned about the qualitative aspects of this test.

2. Mode of action of long-acting anabolic steroids

Remarkably little is known as to why various esters of nandrolone differ with regard to their duration of action and their anabolic-androgenic ratio. We found that, when nandrolone is esterified with acids of differing chain length, the duration of action increases with the length of the chain. At the same time the anabolic-androgenic ratio increases, mainly because of a decrease in androgenic activity. Although both phenomena are probably interrelated, in this paper it is only the duration of action which will be discussed.

Two very obvious questions immediately come to mind:

1. Do the various esters disappear from the site of injection at a different rate ?

2. Are they hydrolysed, either at the site of injection or elsewhere, before reaching the end-organs or do they act as such ?

Additional questions concern the time the compounds remain in the circulation and their storage elsewhere in the body (liver, muscle, fat).

As radioactive material was not at our disposal, chemical determination of the steroids administered was necessary. This was done in the following way[1]:

An alcohol-ether extract of the organs or blood was prepared and evaporated to dryness. The residue was extracted with petroleum ether, the extract evaporated to dryness, and redissolved in absolute alcohol. The 3-ketosteroids present in this solution were converted into a dinitro-phenylhydrazone which was dissolved in acetone, after which the colour of the solution was measured at 515 mμ. In this manner 50 mcg. of 3-ketosteroids could still be detected. From 1 mg. injected into a muscle 95—98% could be recovered. Extracts of non-injected muscles, liver, or blood gave rise to no significant light absorption at 515 mμ.

---

[1] An extensive description of the procedure, together with standard curves, recovery, etc. will be published elsewhere.

The following compounds were studied and administered in such doses that all rats received the same amount of nandrolone:

nandrolone
nandrolone-phenylpropionate (C_8H_9COOR) (R = nandrolone)
nandrolone-decanoate $(C_9H_{19}COOR)$ (R = nandrolone)
nandrolone-oleate $(C_{17}H_{33}COOR)$ (R = nandrolone)
nandrolone-docosanoate $(C_{21}H_{43}COOR)$ (R = nandrolone)

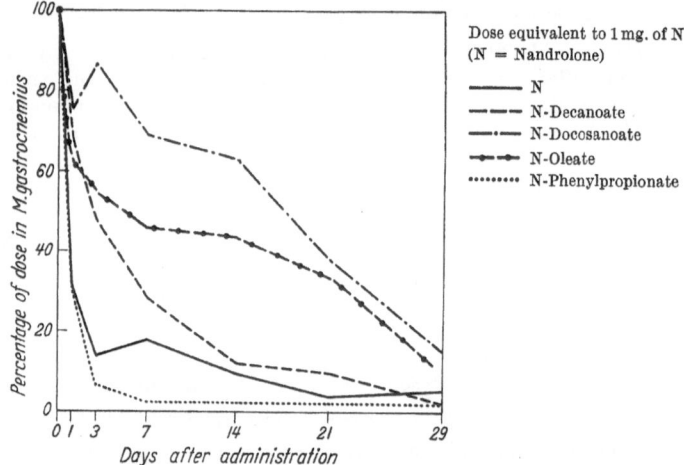

Dose equivalent to 1 mg. of N
(N = Nandrolone)

———— N
– – – – N-Decanoate
—·—·— N-Docosanoate
—•—•— N-Oleate
·········· N-Phenylpropionate

Fig. 4. Disappearance rate of various nandrolone esters from the site of injection. Treatment: one single injection into gastrocnemius muscle at zero time. Abscissa: days after injection. Ordinate: % recovered from the muscle at indicated time

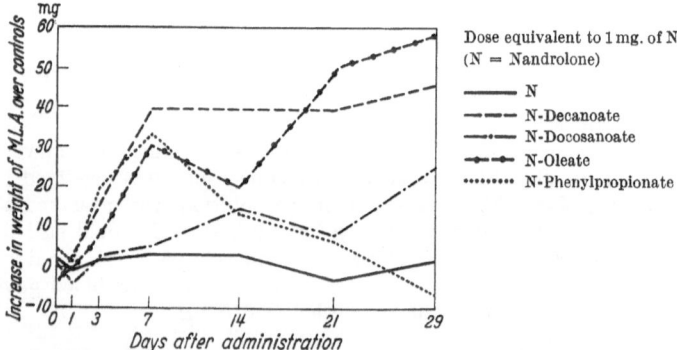

Dose equivalent to 1 mg. of N
(N = Nandrolone)

———— N
– – – – N-Decanoate
—·—·— N-Docosanoate
—•—•— N-Oleate
·········· N-Phenylpropionate

Fig. 5. Effect of a single dose of various nandrolone esters on the weight of the levator ani muscle in the castrated rat. Treatment: one single injection into gastrocnemius muscle at zero time. Abscissa: days after injection. Ordinate: increase in weight of levator ani muscle over controls at indicated time

For intramuscular injection the compounds were dissolved in peanut oil and for intravenous injection either in a 1% aqueous gelatin solution or in rat plasma.

In a first experiment 1 mg. nandrolone or the corresponding amount of the esters was injected into the gastrocnemius muscles of groups of 3 castrated rats (body-weight approx. 50 g.). Fig. 4 shows the rate of disappearance from the site of injection; at autopsy, which took place after 0, 1, 3, 7, 21, and 29 days, the gastrocnemius muscle was extracted according to the procedure described above. In Fig. 5 the weights of the levator ani at autopsy are shown.

Among the esters, there is thus a reasonable degree of correlation between the rates of disappearance and the duration and magnitude of the effect. However, the free nandrolone is not more rapidly absorbed than its phenylpropionate, whereas it is definitely less active and for a shorter period. Whatever the reasons for

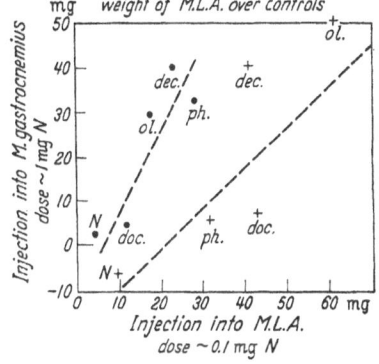

Fig. 6. Increase in weight of levator ani muscle over controls at indicated time after injection of a single dose of various nandrolone esters. Ordinate: treatment consisting of one single injection into gastrocnemius muscle at zero time. Abscissa: treatment consisting of one single injection into levator ani muscle at zero time.

● after 7 days   + after 21 days

this phenomenon may be, it makes it at any rate seem very improbable that the phenylpropionate is hydrolysed before taking effect, since otherwise there would be no reason why nandrolone and its phenylpropionate should have quite different effects.

Another experiment, in which 100 mcg. nandrolone, or corresponding amounts of the esters, were injected directly into the levator ani muscle, lends support for this supposition. This time 0.005 ml. was injected in groups of 6 immature castrated rats, and the animals were killed after 7 or 21 days. Fig. 6 shows that there is a remarkably good correlation between the effects after injection in either muscle. At first glance this looks surprising, since one might have expected that after local administration the existing differences in the rate of absorption would be eliminated and hence another factor, i.e. possible differences in effect because of differences in chemical structure, would prevail. There is no obvious reason why the influence of these two factors should be the same. However, as Fig. 7 shows, local administration into the levator ani muscle does not eliminate differences in the rate of absorption. It must be realised that probably the injected ester will be absorbed, and hence *disappear from the levator ani into the circulation*, just as well as it does from the gastrocnemius muscle. But in the former case the substance will not act, since the amount injected (100 mcg.) is far too small to cause any effect on the levator ani after previous circulation through the entire body. On the other hand, at least part of the much larger amount (1 mg.) injected into the gastrocnemius and absorbed into the circulation *is* effective on the levator ani. But, in both instances, effects will only occur as long as the substances are present at either site of injection. The various esters remain at these sites for different periods of time, and these periods will parallel the time of exposure of the receptors in the cells of the levator ani muscle to the anabolic agents. Hence,

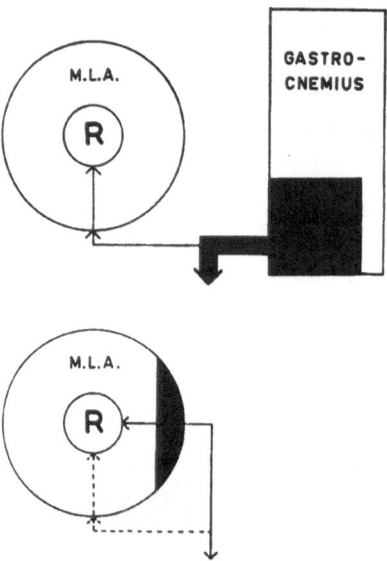

Fig. 7. Transportation to receptors (R) in levator ani muscle (M.L.A.) Top: from depot in gastrocnemius muscle. Bottom: from depot in levator ani muscle

although the absolute times of exposure for each single compound will certainly be different in the two experiments, the sequence of the duration of action of the various compounds can be expected to be the same. This is exactly what Fig. 6 shows to be the case.

This concept may also explain why the slowly absorbed compounds are more *active*. If the same number of receptors can bind the same amount of anabolic compounds, which does not seem to be unlikely, there remains a difference in the time during which they are exposed to the compounds. This time of exposure is determined by the time taken for the substances to become absorbed, either from the gastrocnemius or from the levator ani itself. When nothing is left in the depot, growth of the muscle continues for some time (cf. the curves for the phenylpropionate and the decanoate), but then it inevitably stops. This is in agreement with the results of experiments with testosterone-C^{14} (BUTENANDT et al., 1960) showing that growth of the seminal vesicle occurs after complete disappearance of the testosterone from this organ.

The possibility that the steroid might circulate for a considerable time in the blood-stream could at least be ruled out in the case of nandrolone-phenylpropionate. Even after intravenous injection this substance disappeared almost instantaneously from the blood-stream. The injected compound could not be detected in the liver, kidneys, or lungs of the rats.

The hypothesis that the receptors must be exposed for some time to the compound is also supported by the above experiments using intravenous injections. Here the effect of 1 mg. nandrolone-phenylpropionate was nil, a fact which at the same time excludes the possibility that the intravenously injected steroid is stored at some other site, from which it can still be released in an active form.

These experiments strongly suggest that the esters of nandrolone act as such and are not first hydrolysed to nandrolone. Their differing effects appear to be determined mainly by their differing rates of absorption. Nandrolone-phenylpropionate rapidly disappears from the blood-stream and is not stored and later released as such.

Summary

When assessing the properties of long-acting steroids, one must consider not only the type and intensity, but also the duration of their effect. Depending on the time that has elapsed after the administration of a preparation, dose-response curves show a variable pattern, in which the anabolic and androgenic activities may differ quantitatively. A comparison can be made either between dose-response curves at various times or between time-response curves at various dose levels.

The anabolic and androgenic effects of nandrolone-phenylpropionate, nandrolone-decanoate, and testosterone-decanoate were compared in castrated rats. Nandrolone esters varying in their duration of action did not display a uniform anabolic-androgenic ratio irrespective of the time factor, whereas at one and two weeks after administration the two decanoates did. The value of the levator ani test in castrated rats as a criterion by which to assess the activity of anabolic steroids is emphasised.

To investigate the mechanism involved in the prolongation of effect achieved by esterification with a fatty acid, the rate of disappearance of the esters from the site of injection was determined at various intervals and compared with the effect exerted at the same time on the levator ani muscle. A reasonable degree of correlation was observed between the rate of disappearance of the esters from the site of injection and the duration of effect. The findings suggest that the esters are not hydrolysed but are absorbed and take effect as such, regardless of whether they be injected directly into the levator ani or into another muscle.

Zusammenfassung

Für die Beurteilung lang wirkender Steroide ist neben der Qualität und Intensität der Wirkung deren Dauer zu berücksichtigen. Dosis-Wirkungs-kurven zeigen zu verschiedenen Zeiten nach Gabe des Präparates einen verschiedenen Verlauf, wobei anabole und androgene Wirkungsqualitäten quantitativ voneinander abweichen können. Es können entweder Dosis-Wirkungskurven zu bestimmten Zeiten oder Zeit-Wirkungskurven für einzelne Dosen zueinander in Beziehung gesetzt werden.

Nandrolon-phenylpropionat, Nandrolon-decanoat und Testosteron-decanoat wurden hinsichtlich anaboler und androgener Wirkung an der kastrierten Ratte untereinander verglichen. Für die verschieden lang wirkenden Nandrolonester ergab sich kein von der Zeit unabhängiges, konstantes anabol/androgenes Verhältnis, während dies für die beiden Decanoate nach ein und zwei Wochen der Fall war. Die Eignung des levator ani-Testes an der kastrierten Ratte als Kriterium für die Beurteilung anabol wirkender Steroide wird hervorgehoben.

Um den Mechanismus der Verlängerung der Wirkungsdauer abzuklären, die sich durch Veresterung mit einzelnen Fettsäuren erzielen läßt, wurde nach intramuskulärer Gabe das Verschwinden der Ester vom Ort der Injektion zu verschiedenen Zeiten bestimmt und mit dem gleichzeitigen Effekt auf den levator ani verglichen. Dabei ließ sich eine verhältnismäßig gute Überein-stimmung zwischen dem Verschwinden der Ester vom Ort der Injektion und der Dauer der Wirkung feststellen. Die Befunde sprechen dafür, daß die Ester nicht hydrolisiert, sondern als solche resorbiert werden und zur Wirkung gelangen, und zwar unabhängig davon, ob sie direkt in den levator ani oder in einen anderen Muskel injiziert werden.

Résumé

Dans l'étude des stéroïdes à effet prolongé, il faut tenir compte non seulement de la qualité et de l'intensité de leur action mais encore de sa durée. A différents intervalles après l'administration du produit, les courbes dose/effet ont un tracé différent; les effets anabolisant et androgénique peuvent être quantitativement différents. On peut comparer entre elles soit les courbes dose/effet à différents moments, soit les courbes temps/effet à différentes doses.

Les actions anabolisante et androgénique du phénylpropionate et du dé-canoate de nandrolone ainsi que du décanoate de testostérone sont comparées chez le rat castré. Pour les esters de nandrolone ayant des effets de différentes durées, on n'a pas constaté de rapport effet anabolisant/effet androgénique constant indépendant du temps, tandis que ce fut le cas pour les deux décanoates après 1 et 2 semaines. L'auteur souligne que le test du muscle releveur de l'anus chez le rat castré convient parfaitement pour apprécier l'activité des stéroïdes anabolisants.

Pour comprendre le mécanisme de la prolongation de la durée de l'effet après l'estérification avec un acide gras, on a enregistré la disparition de ces esters au point d'injection à différents intervalles après l'injection intramusculaire, et comparé les chiffres relevés avec l'effet obtenu à ce moment sur le muscle releveur de l'anus. On a constaté ainsi une assez bonne concordance entre la disparition de l'ester au point d'injection et la durée de l'effet. Ces observations montrent que les esters ne sont pas hydrolysés mais résorbés tels quels, et qu'ils agissent aussi bien qu'ils aient été injectés directement dans le muscle releveur de l'anus ou dans un autre muscle.

References

BUTENANDT, A., H. GÜNTHER, and F. TURBA: Zschr. physiol. Chem. (G.) **322**, 28 (1960). — HERSHBERGER, L. G., E. G. SHIPLEY, and R. K. MEYER: Proc. Soc. Exper. Biol. Med. (U.S.A.) **83**, 175 (1953). — OVERBEEK, G. A., and J. DE VISSER: Acta endocr. (Den.) **38**, 285 (1961).

Discussion

GROSS: Dr. DESAULLES showed us that it is possible to differentiate between the anabolic and androgenic activity of various steroids. I should like to ask him, as well as Dr. OVERBEEK, whether there are any steroids known which have an androgenic activity without any anabolic activity and, vice versa, whether there are any anabolic steroids completely lacking in androgenicity. In other words, are there any known steroids which in some way mimic somatotrophic hormone as regards protein metabolism? I also have a question for Dr. OVERBEEK concerning the effect of nandrolone-decanoate on uterine growth. Do you mean that this compound also has oestrogenic activities, and, if not, how do you explain this effect on the muscle of the uterus?

DESAULLES: The question asked by Dr. GROSS is very difficult to answer. The only point which we can answer with any degree of certitude is that, among the many androgenic steroid derivatives, we have so far not been

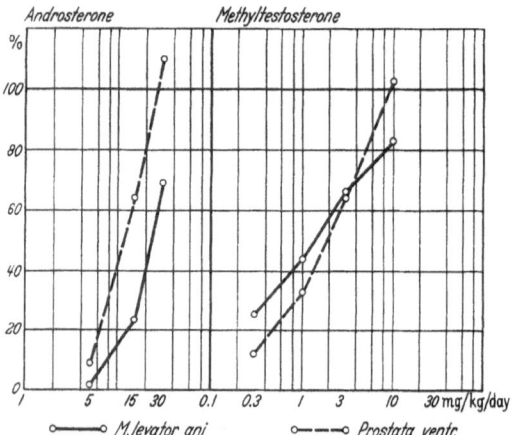

Fig. 1. Comparison of the effects of androsterone and methyl-testosterone on the restitution of the weight of castration-involuted organs to normal values. 15 days castrated adult male rats. Treatment 15 days. Abscissa: Doses in mg./kg. s. c. Ordinate: Relative restitution of organ weight in per cent of normal controls

able to detect any compounds which display only androgenic activity and no anabolic activity. At the moment, I know only of one group of steroids and, among these, only of one, androsterone, which shows a marked dissociation of the type referred to by Dr. GROSS. Androsterone is a more potent androgen than anabolic, although it must be stressed that its androgenic activity is somewhat lower than that of testosterone on a weight-for-weight basis. We have done some experiments using young castrated rats and taking the ventral prostate as the end-point of the androgenic effects, as it is known that in young animals the ventral prostate is more sensitive than the seminal vesicles, whereas in mature animals the reverse is true. The results are expressed in per cent of restitution to normal weights (Fig. 1). Compared with testosterone, it can be seen that androsterone over the whole dosage range has a more marked effect on the ventral prostate than on the levator ani muscle, but the difference is only slight.

Further evidence is afforded by the fact that, in balance studies on humans, androsterone has been shown to lack the property of producing nitrogen retention, although it does have definite androgenic effects. I am aware of the difficulties involved in making comparisons between animal and clinical observations, but these are the only results we know of.

OVERBEEK: Well, I don't know any substance which has only androgenic and no anabolic activity. Apparently androsterone still has some effect on the levator ani. I might add that, similarly, I know of no substance where the reverse is true, apart from the fact that growth hormone of course is an anabolic substance without androgenic activity. With the long-chain esters of nandrolone the androgenic activity practically disappears, but it is still not entirely eliminated (Fig. 2). These effects on the seminal vesicles belong to the same experiment I showed you with respect to the levator ani, and you see here that the oleate and the docosanoate have practically no

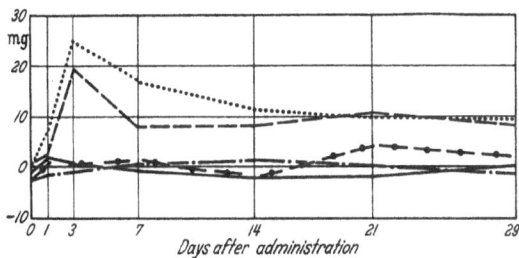

Fig. 2. Treatment: One single injection the into gastrocnemius muscle at zero time. Data: Increase in weight of seminal vesicles over controls at indicated time. Dose equivalent to 1 mg. of N. N = Nandrolone, ————— N, — — — N-decanoate, — · — · — N-docosanoate, -•- -•- -•- N-oleate, · · · · · · · · N-phenylpropionate

effect on the seminal vesicles. Still, I believe that if we had injected more, this higher dose would have had some effect. Thus, although in this instance we got perhaps the most marked dissociation in effect I have ever seen, I still don't think one can say that androgenic activity is totally absent as in the case of growth hormone.

Now, with regard to the other question Dr. GROSS asked concerning the effect on the uterus, I really think that what is involved here is a myotrophic effect comparable with the effect on other muscles. The nandrolone-decanoate is not oestrogenic; on the contrary, if anything, it is somewhat anti-oestrogenic. It may be slightly progestational, but progestational compounds do not cause growth of the uterine muscle, so I think that what we are measuring here is a real myotrophic action, which can also be seen in the histological slides, showing that it was precisely the myometrium that had grown.

FRASER: May I ask Dr. DESAULLES, in relation to his previous slide showing the different endocrine states responding to the anabolic steroids, whether the dietary intake was variable or constant between the different groups, and if he knows to what extent these differences arise from variations in appetite response?

DESAULLES: In all experiments the animals were kept on the same diet and were paired-fed, so that I think this source of error was reduced to a minimum, but I would like to stress that I am aware that the endocrine balance of the animals was so profoundly altered that these results also show how careful one has to be in assessing the activity of a steroid when the endocrine balance is disturbed.

QUERIDO: Am I correct in understanding from your answer that the marked decrease in effect which you noted in the hypophysectomised animals occurred while these hypophysectomised animals were receiving enough calories?

DESAULLES: On the basis of the food intake per body-weight, the quantity of food eaten by paired-fed animals was certainly reduced. We did not measure the caloric intake of the hypophysectomised rats.

FRASER: When you say paired, do you mean paired against animals that didn't get these steroids?

DESAULLES: They all got as much as the controls got.

FRASER: But what did you use as controls for the hypophysectomised animals?

DESAULLES: They were paired against castrated controls. The quantity of food was proportional to the animal's body-weight.

KASSENAAR: In connection with the effect of changes in the hormonal status of the animal on its response to androgenic steroids, insulin has come up in the literature several times. On the one hand, KOCHAKIAN[1] has presented evidence that the effect was largely independent of the secretion of insulin, whereas BEST's[2] group maintain that the opposite is true. Furthermore, it seems that in some tissue cultures testosterone only exerts an effect when insulin is present. This seems to me a very important point which needs clarification. Do you have any information on this aspect?

DESAULLES: I can give only a partial answer based on experiments of our own which are still unpublished. Together with Dr. HERZOG we have tried to repeat some of the experiments of SCOW[3] and to study in pancreatectomised rats the effects of anabolic steroids and of growth hormone. It appears that, 4 weeks after pancreatectomy, rats given no insulin show a marked involution of the levator ani muscle which compares fairly well with that observed following castration (Table 1). These experiments indicate that lack of insulin provokes marked involution of certain muscles. Under

---

[1] KOCHAKIAN, C. D., and G. COSTA: Endocrinology (U.S.A.) **65**, 298 (1959).

[2] SIREK, O. V., and C. H. BEST: Endocrinology (U.S.A.) **52**, 390 (1953).

[3] SCOW, R. O.: Rec. Progr. Horm. Res. (U.S.A.) **16**, 497 (1959).

Table 1. *Weight gain during 2 weeks and weight of various muscles in pancreatectomised rats*

| | Weight gain | Muscle weight (mg.) | | | | | | |
| | | Gastrocnemius | | Levator ani | | Heart | |
		Absolute	mg./100 g.	Absolute	mg./100 g.	Absolute	mg./100 g.
Sham-operated controls $n = 15$	34 ± 2.5 g.	1092 ± 40	510 ± 19	141.0 ± 6.7	67.1 ± 3.1	611.4 ± 21	291 ± 10.0
Pancreatectomised rats $n = 13$	16 ± 4.4 g.	730 ± 46	456 ± 28	51.0 ± 5.4	32.0 ± 3.0	520.0 ± 9	325 ± 5.6
Pancreatectomised rats $n = 10$ + STH 3 mg./kg./day s.c.	24 ± 7.0 g.	886 ± 103	520 ± 60	88.4 ± 14	52.0 ± 8.0	476.0 ± 14	280 ± 8.0

Table 2. *Plasma concentration of free fatty acids and lipids in pancreatectomised animals 4 weeks after operation and 2 weeks after the beginning of treatment*

| | n | Free fatty acids mg./100 g. | Phospholipids mg./100 g. | Cholesterol | | Blood glucose mg./100 g. |
				Total	Free	
Normal controls	15	234.0 ± 10.3	4.4 ± 0.1	58.4 ± 1.5	20.2 ± 1.5	96.4 ± 3.0
Sham-operated controls	16	238.5 ± 15.7	4.07 ± 0.16	56.8 ± 1.3	18.6 ± 0.9	105.8 ± 18.6
Sham-operated rats + STH 3 mg./kg./day s.c.	7	283.0 ± 6.5	4.1 ± 0.01	63.5 ± 0.7	23.5 ± 1.3	100.6 ± 12.3
Pancreatectomised rats	13	400.0 ± 34.0	5.8 ± 0.27	62.3 ± 1.6	25.0 ± 0.6	277.0 ± 8.7
Pancreatectomised rats + STH 3 mg./kg./day s.c.	11	224.0 ± 9.4	3.7 ± 0.1	65.0 ± 0.2	25.0 ± 1.0	275.0 ± 15.0

treatment with growth hormone for 2 weeks, the rats show a definite gain in muscle-weight, but no complete restitution. Treatment with growth hormone does not reduce the elevated blood sugar levels.

The free fatty acids, which in pancreatectomised animals rise to high values, are reduced to normal under treatment with growth hormone. These analyses were undertaken in our laboratories by Dr. W. SCHULER and Dr. W. ALBRECHT (Table 2). The fact that growth hormone partially restores muscle growth is, of course, only an incomplete answer to your question.

The action of insulin has been inconclusive in the experiments performed up to now. The rat is, as you know, not a very reliable animal for studying insulin, because in this animal very large amounts of insulin are required to normalise the blood sugar levels. In pancreatectomised rats it is necessary to give from 10 to 20 units per kg. daily in order to normalise the blood sugar levels.

KORNER: May I ask Dr. DESAULLES a supplementary question about his last answer ? First of all, are you sure that these were totally pancreatectomised animals ?

DESAULLES: The rats were as completely pancreatectomised as possible, the criterion of pancreatectomy being mainly the rapid rise in the blood sugar levels to values of over 250 mg.%.

KORNER: Can you tell me how long they would survive without insulin and how long after hypophysectomy you treated them with growth hormone and obtained the effects which you described ? SCOW's pancreatectomised rats, I believe, would only survive for a very short time without insulin.

DESAULLES: The rats survive for about 6 weeks following operation and can be subjected to treatment 2 weeks following operation. The survival rate depends, among other factors, on the way the rats are fed. We found that it was necessary to give them pancreatic enzyme orally in order to assure proper food utilisation.

The animals are subjected to experimentation from 1 to 2 weeks following operation.

One important problem is the measurement of the blood insulin levels in order to check the completeness of the operation and the possible regeneration of the pancreas.

ASCHKENASY: In correlation with the work of Dr. DESAULLES on the influence of adrenal function on the sensitivity of tissues to androgens, I should like to mention that I have carried out comparative studies on the weight changes of various organs and tissues in castrated and castrated-adrenalectomised male rats in response to injections of testosterone-propionate (TP) administered for 3 weeks[1]. I found that adrenalectomy has a distinct influence on certain tissular effects of this androgen and, in particular, that it amplifies the catabolic effects on the lymphoid organs.

In point of fact, the weights of the thymus and spleen in TP-treated animals are not very different in the presence or in the absence of the adrenals; however, as the same organs are much larger in adrenalectomised than in non-adrenalectomised rats, the gap between the treated and the non-treated animals appears to be more marked after the removal of the adrenals: in particular, one finds that TP induces a decrease in the weight of the spleen in

---

[1] ASCHKENASY, A.: Ann. endocr. (Fr.) **20**, 158 (1959).

adrenalectomised rats, whereas it does not significantly alter the weight of this organ in non-adrenalectomised rats. In any case, these various results suggest that the effects of androgens on the lymphoid organs are certainly not related to stimulation of the secretion of glucocorticoids.

DESAULLES: It seems to me that the enhancement of the androgenic or seemingly androgenic response as regards weight of the heart or kidney — a phenomenon which we have also observed in castrated-adrenalectomised rats — will be difficult to explain until we can express it functionally.

This weight gain is perhaps due simply to electrolyte and fluid shifts and not to cell multiplication — as QUERIDO and KASSENAAR[1] have found, for instance, in the case of the renotrophic effects of androgen in mice, where the RNA/DNA ratio rises markedly.

RABEN: Dr. DESAULLES, I was not quite clear as to whether you said that growth hormone has an effect on fatty acids in the completely pancreatectomised rats. Dr. Scow claimed, I believe, that when the rat was completely pancreatectomised the only substances he could find which affected ketosis, at least, were adrenal steroids. He did not measure the free fatty acids, so I am curious as to what your result was.

DESAULLES: We have not measured ketosis in our pancreatectomised rats.

RABEN: But what happened to the free fatty acids?

DESAULLES: They dropped to normal values as in sham-operated controls or normal controls.

RABEN: They were high when you gave growth hormone?

DESAULLES: They were very high in pancreatectomised controls.

RABEN: They dropped when you gave growth hormone?

DESAULLES: Yes.

---

[1] KASSENAAR, A. A. H., A. KOUWENHOVEN, and A. QUERIDO: Acta endocr. (Den.) **39**, 223 (1962).

The influence of nutrition and nutritional status on the effect of androgens and anabolic agents

By

J. H. Leathem

The concept of a dynamic state of protein metabolism has pictured a common metabolic pool of nitrogen (ALLISON, 1955). This metabolic pool is provided with nitrogen constituents not only from the changing state of tissue proteins but also from dietary amino acids. Since tissue-protein synthesis requires a proper orientation of amino acids, any anticipated action of an anabolic agent may, therefore, be varied by the amino-acid pool which exists at the time the agent is administered.

Feeding a low level of protein, despite adequate caloric intake, will deplete the body-protein stores. Tissue-protein depletion is not uniform, however, as some tissues contribute more readily to the metabolic pool than others. Thus, one tissue may be maintained at the expense of another, creating an imbalance in protein reserves. Hormone action may correct or augment the defect. Furthermore, protein quality, which reflects amino-acid patterns, can influence the nitrogen pool. A food protein which is deficient in one or more essential amino acids will restrict tissue-protein synthesis and can alter anticipated hormone action. By the same token, hormones can influence the utilization and nutritional value of proteins, thereby altering the potential for growth. Clearly, then, deficiencies in experimental results with hormones may be a problem of nutrition.

To evaluate the influence of hormones on anabolism, many of the same measures used to evaluate protein efficiency of foods can be employed, i.e. nitrogen balance, plasma proteins, repletion of depleted adults, and organ-protein content. Nitrogen balance permits a measure of body-protein stores, and relating nitrogen balance to nitrogen intake will indicate the rate at which protein stores are being filled during repletion. Nevertheless, nitrogen-balance data alone must not be considered as a complete evaluation of protein metabolism. Animals can be in positive nitrogen balance, have normal plasma proteins, and yet be losing organ nitrogen. Therefore, a study of organ proteins must be recommended as an important adjuvant to nitrogen balance.

The status of the body-protein reserves prior to an experiment will influence anticipated results, but is usually unknown. However, the tendency for the normal adult rat to continue in rather marked positive nitrogen balance when fed *ad libitum* must limit the possibility for observing an anabolic action of hormone supplements in this species.

It has been estimated that the intact rat requires four times more hormone than the castrated rat to cause nitrogen retention (KOCHAKIAN et al., 1950; GORDAN et al., 1947). Furthermore, a "wearing off" effect is noted even in castrated rats after 10—14 days. This decreasing effectiveness of testosterone-propionate with time is seen in hypophysectomized and adrenalectomized animals as well. The lack of a continued effect of androgen on body-weight may relate to an increased loss of body fat, or the "wearing off" effect may represent a filling of protein stores. Apparently, then, one should consider the action of an anabolic agent at a time when protein reserves are not filled. In this regard, tube feeding or reduced daily food allotments so as just to maintain body-weight (ARNOLD et al., 1956; RUPP and PASCHKIS, 1953) may well enhance the potential for response, despite the negative findings of PERLMAN and CASSIDY (1953) who considered that androgen action was not facilitated by a minimum nitrogen intake.

Protein nutrition and organ response to androgen

Androgens are well known to stimulate growth and protein metabolism in the accessory sex organs. Recent studies have shown that seminal-vesicle response increases the RNA-to-DNA ratio (ROBINOVITCH et al., 1951; KASSENAAR et al., 1962) and the specific activity of the amino-acid activating enzymes (KOCHAKIAN et al., 1961) as further evidence of protein synthesis. When food intake is inadequate, a subnormal response of the seminal vesicles to androgen may be observed in mice and rats. However, castration and 10 days on a protein-free diet may or may not prevent a normal weight response to testosterone-propionate. Furthermore, if an 18% casein, lactalbumin or gelatin diet was fed during the 3-day period in which the androgen was acting, no improvement in weight response was noted (LEATHEM, 1961).

Androgens have a general effect on body proteins, and it is assumed that nitrogen retention is largely due to a protein-anabolic action in the skeletal muscle. From this assumption, EISENBERG and GORDAN (1950) drew attention to the levator ani muscle as a test method for anabolism. The levator ani/seminal vesicle ratio to relate anabolism to androgenicity became widely used and is currently employed (OVERBEEK et al., 1962), although results must be interpreted with reservation (AHREN et al., 1962). Recently, SAUNDERS (1962) has found that the levator ani, but not the rectus femoris, is responsive to androgen as estimated by RNA:DNA ratios.

The kidney, especially of the mouse, is responsive to androgens (FRIEDEN et al., 1961; KOCHAKIAN and HARRISON, 1962) and has been used to distinguish steroid actions (ASCHKENASY-LELU and ASCHKENASY, 1959). Using RNA:DNA ratios to measure protein synthesis, KASSENAAR et al. (1962) characterized the androgen response as due to an increase in cell size. FISHMAN (1961) has called attention to renal β glucuronidase as sensitive to slight changes in the steroid molecule with the view toward androgenic to anabolic effects. Recently, OVERBEEK et al. (1962) reported on the anabolic properties of ethylestrenol; prior studies on renal β glucuronidase were thus confirmed.

A preliminary examination of the kidney enzymes has been made to assess two androgens, testosterone-propionate and Δ^1-17α-methyl-testosterone (Dianabol). The FISHMAN procedure was modified in that castrated Swiss strain mice were used and renal alkaline phosphatase was added. Otherwise, the procedure of injecting 1 mg. of steroid on alternate days for 7 injections was followed. Testosterone-propionate increased renal β glucuronidase significantly, but androgenicity was also expressed in the decrease of renal alkaline phosphatase. Dianabol increased renal β glucuronidase significantly, but was less active than testosterone-propionate and did not reduce alkaline phosphatase (Table 1).

Table 1. *Steroids and kidney enzymes in castrated mice*

	No. of mice	Body wt. g.	Kidney wt.		Alkaline phospha- tase mcg./mg./hr	Beta glucuroni- dase u./g./hr
			mg.	mg./100 g.		
Controls	9	33.3	104	1230	21.2	614
Testosterone-prop.						
1 mg. × 7 . . .	9	37.4	568	1524	7.4	2998
Dianabol 1 mg. × 7	10	33.9	447	1314	24.5	1381

Protein nutrition and aging

A reduction in protein reserves due to physiological adjustments could provide a responsive subject. In this regard, our attention was directed toward aging, where in man a reduction in lean body mass due to a loss in protein is not uncommon. Further-more, a reduction in urinary steroids is noted in aging, as well as an anabolic response to androgen (SHOCK, 1956). It became of interest to compare the 20—30-month-old rat with the 6-month-old rat for physiological differences. It was quite evident that the older animal excreted significantly less nitrogen when fed a protein-free diet for 4 days, suggesting reduced protein reserves (LEATHEM, 1956). Aging may also slow metabolism, and reduced calorie requirements would be met with less total food intake, and body-protein stores may decrease. Thirty-month-old Long Evans strain male rats that had been maintained on a com-mercial food (fox chow) were used. The animals were fed an agar based diet containing 18% ca-sein and 25% lard. Each animal exhibited a posi-tive nitrogen balance ini-tially and gained weight.

Table 2. *Steroids and nitrogen balance in old rats*

Age (months)	Nitrogen balance (4-day periods)		
	Control	Injection	Post-injection
Testosterone-propionate (0.25 mg.)			
22	+ 8	+129	+67
24	+ 9	+ 68	+74
Dianabol (1 mg.)			
24	+52	+114	—
27	+ 9	+112	+70

However, after 12 days body-weight gain stopped and nitrogen balance decreased toward equilibrium. In several animals, food in-take during the last 8 days of the 20-day balance study was roughly 50% of that recorded for the first 4 days and was sufficient to induce a state of nitrogen equilibrium. Old rats which were at nitrogen equilibrium were studied over a 4-day period for improve-ment in nitrogen balance in response to androgen. Testosterone-propionate (0.25 mg.) or Dianabol (1 mg.) improved nitrogen bal-ance significantly (Table 2). Liver and kidney protein content was improved.

Protein nutrition, calories, and androgen

The relationship between nitrogen retention and caloric intake indicates that a caloric deficit will cause the body to call upon protein reserves to meet the deficiency (MUNRO, 1951). In the rat, fed a constant nitrogen intake, a reduction in calories will increase urinary nitrogen and decrease nitrogen balance (ROSENTHAL and ALLISON, 1956). Furthermore, the loss in body-weight was pro-portional to the number of calories ingested, but a varied quality of whole proteins was without effect.

In the absence of a calorie restriction, there is no question that protein quality will influence nitrogen balance and protein stores (ALLISON, 1955). Thus, excretion of urinary nitrogen and rate of protein synthesis are influenced. The influence of androgen on nitrogen balance in animals fed proteins of different quality has had little consideration. However, NIMMI and GEIGER (1957) observed nitrogen retention following testosterone-propionate administration to castrated rats fed gelatin, zein, or a mixture of three non-essential amino acids, even though these dietary proteins do not normally support protein synthesis. It was suggested that androgen caused an increased tubular reabsorption of nitrogen in the kidney without enhanced protein synthesis.

The action of testosterone-propionate in rats subjected to caloric restriction as well as proteins of varied quality seems not to have been studied. To investigate this problem, adult male rats (300—350 grams in body-weight) were observed for an effect of the androgen on body-weight, nitrogen balance, and organ protein. The rats were initially fed 18% casein for four days. At the end of the control period, the animals were placed on an experimental diet for 20 days. The proteins incorporated in the diets were casein, wheat gluten, lactalbumin, and gelatin. Food intake was restricted to 100 calories per kilogram body-weight, estimated to provide 65% of the basal caloric requirements.

Body-weight loss was 15—17% in 20 days when casein, lactalbumin, and wheat gluten were fed, and a 25% body-weight loss occurred with gelatin in the diet fed at the 100 calorie/kg./day level. Testosterone-propionate. (0.5 mg. daily) failed to alter body-weight changes.

As anticipated, calorie restriction caused a decrease in the nitrogen retained when compared with an *ad libitum* fed rat. However, animals fed gelatin retained 4,000 mg. N_2 per kilogram over the 20-day period as compared with 3,200 mg., 1,800 mg., and 900 mg. for casein, wheat gluten, and lactalbumin respectively. Curiously, the gelatin-fed animals retained the most nitrogen while losing the most weight. Possibly, the retained nitrogen is not being utilized for protein synthesis, and loss of body fat is accentuated. On this premise, lactalbumin-fed animals may utilize retained nitrogen more efficiently. Administration of testosterone-propionate increased nitrogen retention to a total of 2,000 mg. when lactalbumin was fed and to 2,500 mg. when wheat gluten was fed, but failed to improve nitrogen retention when casein or gelatin served as the dietary source. Several possibilities are suggested to explain the influence of protein quality on hormone action: (1) The sulfur amino-acid content of the four proteins reveals that a hormone response was obtained with the two proteins that contain more than 4 g./100 g. of protein. Attention was drawn to sulfur amino-acids, as nitrogen-balance index can be improved by methionine supplementation in rats fed 12% casein; (2) Another factor may be the cystine-to-methionine ratio; and (3) The general amino-acid pattern of lactalbumin may be more suitable, particularly in comparison with gelatin, which lacks tryptophane. One may question the anabolic action of the androgen in rats fed wheat gluten, because this protein has a low nutritive value for growth. However, when wheat gluten is fed for maintenance, it has a nutritive value greater than casein (BLOCK and MITCHELL, 1947).

Calorie restriction will decrease total carcass protein. Over a 12-day period of 50% calorie restriction, total carcass nitrogen loss represented 9% of the total nitrogen (CALLOWAY and SPECTOR, 1955). In spite of the differences in nitrogen retention exhibited by calorically restricted rats fed lactalbumin, wheat gluten, or casein, gastrocnemius-muscle size and protein content did not differ. Gelatin feeding reduced muscle size and total protein. Androgen administration did not influence muscle weight, protein concentration or total protein. Previous investigations have revealed little or no effect of androgen on the size, weight or composition of the thigh muscles (KORNER and YOUNG, 1955; KOCHAKIAN, 1956; SCOW and HAGEN, 1957; ASCHKENASY-LELU and ASCHKENASY, 1959). SAUNDERS (1962) has shown that the RNA:DNA ratio decreases following castration and increases after testosterone-propionate in the levator ani, but not in the rectus femoris muscle.

Protein shifts induced by a stress may seriously impair body function, and a growing tumor may impose such a stress. The tumor can act as a nitrogen trap, thus releasing none of its protein for utilization by host tissues. The carcass of the host is depleted by a tumor, whereas the liver undergoes hypertrophy, and yet nitrogen balance may be unchanged. Testosterone propionate is of therapeutic value in breast cancer and may favor anabolism in the host. However, the protein nutritional base upon which a hormone must act in a tumor-bearing animal on restricted calories has had little consideration. To study the problem, adult male rats were fed 100 calories per kilogram body-weight, and the protein source was lactalbumin, wheat gluten, casein or gelatin. A mammary sarcoma was transplanted and followed for 20 days.

The presence of a transplantable tumor substantially aggravated body-weight loss in calorie-restricted rats fed 18% casein or wheat gluten. A degree of protection was obtained with testosterone-propionate (0.5 mg.) in casein and wheat gluten fed rats but not in animals fed other proteins. Nitrogen balance was not influenced by the tumor in rats fed lactalbumin or wheat gluten. Furthermore, nitrogen retention induced by testosterone-propionate in non-tumor animals on these diets was not negated by the presence of the tumor. When 18% casein served as the nutritional source, tumor-bearing animals retained less nitrogen than non-tumor animals. Testosterone-propionate in tumor animals fed casein exhibited a markedly adverse effect on nitrogen retention but had no effect in non-tumor animals. Tumor-bearing rats fed 18% gelatin exhibited a lower nitrogen balance than non-tumor rats; the androgen had no effect (Table 3).

Table 3. *Effect of dietary protein, testosterone-propionate, and a transplanted sarcoma on total retained nitrogen in caloric-restricted adult rats*

Diet 18%	Normal rat		Tumor rat	
	Control	Test.-prop.	Control	Test.-prop.
Casein	3200[1]	3200	2600	1900
Lactalbumin. .	900	2000	1000	2100
Wheat gluten .	1800	2500	2000	2300
Gelatin	4000	3500	2600	2600

[1] mg. nitrogen/kg. body-weight.

Of all the organs depleted by a tumor, the residual carcass is known to give up the most nitrogen. Therefore, it was not surprising that the transplanted sarcoma depressed gastrocnemius-muscle weight and reduced muscle protein. However, testosterone-propionate did not prevent the change in muscle weight or protein, although an increase in muscle-protein concentration was observed in rats fed lactalbumin or wheat gluten. BLOCH and his associates (1951) noted that incorporation of labeled amino acids by host tissues of tumor-bearing animals was least in muscle protein. However, other investigators related the muscle decrease to changes in water and lipid content (BOYD and CRANDELL, 1955). Current studies failed to reveal an effect of the tumor on muscle water concentration (Table 4).

Regardless of dietary-protein quality and calorie restriction, testosterone-propionate administration increased seminal-vesicle weight nearly 100% in each dietary group. A potential suggestion of anabolic-to-androgenic action of a steroid has resulted from computing a muscle-to-seminal vesicle ratio. This calculation has involved the responsive levator ani muscle, but it seemed worthy to check the gastrocnemius. If one computes the gastrocnemius muscle-to-seminal vesicle weight, this ratio is 3.7, 3.1, 2.3, and 2.1 for normal rats fed lactalbumin, gelatin, wheat gluten or casein, respectively. Testosterone-propionate reduced the muscle/seminal vesicle ratios but did not alter the dietary order; lactalbumin-fed rats were first in order. In tumor-bearing rats, the muscle/seminal vesicle ratios were increased in all dietary groups, except lactalbumin. Following androgen administration, the ratio declined and dietary influences were voided.

Table 4. *Effect of testosterone-propionate and a transplanted sarcoma on gastrocnemius muscle of caloric-restricted adult male rats fed 18% lactalbumin*

No. rats	Test.-prop. mg.	Muscle wt.		H_2O %	Protein		
		g.	g./100 g.		%	Total g.	Relative
			Normal rats				
6	0	1.50	0.44	75.4	87.5	0.32	0.094
		± .10	± .02	± 0.3	± 0.4	± .02	± .005
6	0.5	1.53	0.45	75.5	87.4	0.33	0.097
		± .06	± .01	± 0.4	± 0.7	± .02	± .002
			Tumor rats				
7	0	1.34	0.38	75.5	88.3	0.29	0.081
		± .04	± .01	± 0.3	± 1.4	± .01	± .003
1	0.5	1.27	0.36	75.8	91.0	0.28	0.079
		± .07	± .01	± 0.4	± 0.7	± .02	± .002

Protein nutrition and adrenal corticoids

The influence of adrenal steroids on protein metabolism has had considerable attention, but only modest effort has been given to the elucidation of the influence of protein nutrition on corticoid action. The physiological action of corticoids is influenced by available dietary protein (ASCHKENASY, 1961; LEATHEM, 1958), and mortality of adrenalectomized rats is inversely proportional to the level of protein in the diet (ASCHKENASHY, 1955). Furthermore, dietary proteins of different biological values each provide a different influence on adrenalectomized rats receiving a constant dosage of cortisone acetate. Biological value of dietary protein as estimated in adrenalectomized rats was frequently at variance with data obtained in normal rats (LEATHEM, 1958).

When immature male rats were adrenalectomized at 60 to 65 grams in body-weight, given 150 mcg. of cortisone acetate, and fed casein, wheat gluten or lactalbumin, the 20-day survival

percentage was 100, 70, and 20, respectively. Essentially 100%
survival of adrenalectomized rats was noted with 0.9% saline or
25 mcg. desoxycorticosterone acetate when casein or lactalbumin
were fed. Curiously, wheat gluten failed to support the adrenalectomized rat given saline. Organ analyses for nitrogen, lipid, and
glycogen have failed to reveal in what manner the nutrition is
modifying hormone action.

Table 5. *Serum proteins as influenced by diet in normal and
adrenalectomized rats*

Diet	Total protein	Albumin g./ 100 ml.	Globulin			
			Alpha-1	Alpha-2	Beta	Gamma
Casein-adx[1] . . .	5.87	2.09	.98	.83	1.41	.70
Casein-c[2]	5.92	2.20	.83	1.05	1.14	.92
Wheat gluten-adx .	5.29	2.17	.79	.57	1.07	.75
Wheat gluten-c . .	4.50	1.65	.58	.43	.86	.94
Lactalbumin-adx .	6.23	3.33	.68	.48	1.06	.66
Lactalbumin-c . .	5.81	3.12	.58	.37	.76	.85

[1] adx = adrenalectomized
[2] c = control, pair-fed.

Further studies have been attempted to relate dietary protein
to survival of cortisone-treated adrenalectomized rats. The measures included serum NPN and urea nitrogen, plasma proteins, and
S 35 dl methionine uptake into the protein of liver, muscle, and
plasma. Plasma proteins exhibited an increase in alpha-1 and beta
globulin and a decrease in gamma globulin in cortisone-maintained
adrenalectomized male rats fed casein, wheat gluten, and lactalbumin. Alpha-2 globulin decreased in adrenalectomized rats fed
casein, in comparison with pair-fed controls, whereas no change was
noted in other diet groups. Albumin concentration was greatest in
rats fed lactalbumin (Table 5). Previous studies revealed a subnormal albumin concentration in adrenalectomized rats maintained
on NaCl (Leathem, 1945; Aschkenasy, 1960).

Serum non-protein nitrogen and urea nitrogen are known to
increase in adrenal insufficiency. Following adrenalectomy, serum
NPN increased from 20 to 97 mg. % despite cortisone acetate
therapy. However, the dietary protein fed influenced the rate at
which the NPN increased. Maximal NPN levels were observed on
day 10 after adrenalectomy when wheat gluten was fed, on day 15
when casein was fed, and on day 20 when lactalbumin was fed.
Serum urea nitrogen levels duplicated the pattern presented by

the non-protein nitrogen. Accumulation of urea nitrogen could have been due in part to impaired renal function or to the catabolic action of cortisone. The data do not unravel the influence of diet on survival, however.

Since tagged amino acids can provide clues on shifts in body proteins, S 35 dl methionine was studied in adrenalectomized rats after 20 days' maintenance on 150 mcg. of cortisone acetate. Casein, lactalbumin, and wheat gluten served as the dietary proteins. The tissue-protein concentration of S 35 was unchanged in liver, reduced in gastrocnemius muscle, and significantly increased in serum protein, but no dietary effects were noted. The increased labeling of serum protein was noted previously (CLARK, 1953). An increased labeling was anticipated in the liver, as cortisone increases the availability of amino acids for protein synthesis and increases the ability of liver microsome to incorporate amino acids into protein (KORNER, 1960).

Hormones in recovery from protein depletion

Androgens induce nitrogen retention in hypogonadism, pan-hypopituitarism, breast cancer, thyrotoxicosis, and various types of cachexia. A reduction in body-protein reserves may be associated with these clinical conditions, permitting an enhanced potential for anabolism. For studies in the laboratory animal, therefore, one might consider reducing the protein stores prior to administration of a potentially anabolic agent.

As the labile proteins of the body are expended in protein depletion, organ function begins to change, the digestive tract is altered, and hormone production is reduced. The tissues soon begin to conserve nitrogen, and protein catabolism is reduced. The refeeding of protein reveals the enhanced tendency for tissue-protein anabolism. Data are needed to clarify whether hormones will aid adequate nutrition in the refilling of body-protein stores. If protein stores have been reduced by chronic starvation, androgen will aid repletion in castrated (KOCHAKIAN et al., 1950), but not in intact rats (GEIGER and EL RAWI, 1952). This difference in androgen response may be due to the increased sensitivity of the castrated animal. Nevertheless, if protein reserves are depleted by protein-free feeding, and castrated and non-castrated rats are compared during repletion, nitrogen balances during 20 days could not distinguish one from the other. Furthermore, liver and kidney protein returned to normal, indicating that protein repletion was taking place in the absence of the testis essentially as well as in the non-castrated rat (LEATHEM, 1958).

14*

Although castration did not retard protein repletion, the possibility remained that administration of steroids could enhance recovery rate of body-protein stores. Testosterone-propionate, 19-nortestosterone or 17-αethyl-19-nortestosterone were administered subcutaneously in daily amounts of 0.25 mg. during the refeeding period. The 19-nor compounds improved body-weight gain slightly over that of pair-fed controls. Nitrogen balances were strongly positive throughout the recovery period and slightly favored the androgen-treated group. Liver and kidney total protein were improved slightly, and androgenicity was slight, as previously reported (SAUNDERS and DRILL, 1956).

Hormonal influences on protein repletion were examined in normal adult male rats fed 18% casein for 20 days following 30 days of protein depletion. Slightly favorable effects on total nitrogen retention were observed for testosterone-propionate, 19-nortestosterone, 17α-ethyl-19-nortestosterone, and growth hormone in 0.25 mg. daily dosages, and with progesterone and cortisone

Table 6. *Hormonal influences on body-weight and nitrogen balance during protein repletion*

Treatment (10 days)	Body wt. gain g.	Nitrogen retained g./kg.
Testosterone-propionate (0.25 mg.)	104	8.1
Control (pair-fed)	97	5.6
19-nortestosterone (0.25 mg.)	110	10.6
Control.	108	8.7
17α-ethyl-19-nortestosterone (0.25 mg.) . .	134	10.7
Control.	131	8.5
Progesterone (1 mg.)	132	12.4
Control.	120	10.7
Cortisone acetate (1 mg.)	113	11.1
Control.	112	8.8
Estradiol benzoate (0.1 mg.)	—68	2.1
Control.	—25	2.5
Growth hormone (0.25 mg.)	155	13.9
Control.	151	11.2

acetate in 1.0 mg. daily dosages. Estradiol benzoate (0.1 mg.) enhanced body-weight loss (Table 6). Tissue analyses revealed that liver and kidney protein content was not significantly influenced by the androgens. Liver, kidney, and gastrocnemius muscle of control and progesterone-treated rats were comparable. It should be noted that progesterone is catabolic in man (LANDAU and LUGIBIHL, 1961). It was interesting to find that cortisone acetate, at a dosage which would maintain an adrenalectomized adult rat, was not catabolic in the normal rat. No adverse effect was noted on liver or kidney weight or protein concentration. Indeed, on a body-weight basis, the organs from the adrenal-steroid treated rats contained more protein (LEATHEM, 1961). Unfortunately, carcass protein was not determined to more fully evaluate internal shifts of proteins.

Protein-repletion studies with hormonal supplements were conducted in which caloric intake was restricted 50%. Adult male rats were fed a protein-free diet for 30 days, then refed either 18% casein or lactalbumin. Testosterone-propionate (0.25 mg.) did not influence body-weight loss or nitrogen balance on either diet. Similar results were obtained with insulin (one unit protamine-zinc insulin), but in this instance the lactalbumin-fed rats retained more nitrogen than those fed casein (Table 7). A lack of effect on nitrogen balance was not unanticipated (MANCHESTER and YOUNG, 1961).

Retention of nitrogen provides a good method for estimating anabolism, but the test is laborious for long-term studies. Adult male rats have been examined over a 4-day period following either 4 or 30 days of protein depletion. During the 4-day repletion period an 18% casein diet was fed at 50% of caloric needs. The best results were obtained with the 30-day depleted rats; both testosterone-propionate and Dianabol (2.5 mg.) induced a modest increase in nitrogen retention.

Table 7. *Influence of protamine-zinc insulin and diet on protein repletion in adult male rats*

| | 18% Casein[1] | | 18% Lactalbumin[1] | |
	Treated	Control	Treated	Control
Body weight change g.	—10	— 9	0	— 1
N$_2$ retained g./kg. . .	3.4	3.7	4.1	5.0
Gastrocnemius . . .				
weight g.	.59	.70	.65	.71
total protein g. . .	.12	.16	.15	.16
% protein	88.4	85.2	89.0	87.7
% glycogen	.65	.64	.82	.79

[1] Calorie intake reduced 50%.

Hypophysectomy and protein repletion

The precise nutritional requirements of the hypophysectomized rat are not known (MEITES and NELSON, 1960), but dietary influences on survival and on response to growth hormone have been noted (Gordan et al., 1947; Chow and Greep, 1948; Shaw and Greep, 1949).

Hypophysectomy is followed by a loss in body-weight, a decrease in nitrogen retained, a reduced uptake of amino acids into liver and muscle (Manchester and Young, 1959; Korner, 1961; Knobil, 1961), and alterations in liver and kidney cell fractions (Reid, 1956; Korner, 1961). Thus protein anabolism is reduced by hypophysectomy and restored by growth hormone (Russell, 1957). However, at least part of the physiological changes ascribed to pituitary ablation do not occur if the protein stores of the body are depleted prior to the operation. Hypophysectomy does not negate enhanced anabolism of repletion as determined by amino-acid incorporation into the diaphragm (Wool, 1960) or by nitrogen balance (Leathem, 1958). Hypophysectomy does slow the rate of protein repletion and does induce protein shifts internally which are not observed by modifications of the diet alone. Internal shifts of protein can be ascribed to hormonal influences, as the growth of a particular tissue depends upon the stimulus (Russell and Wilhelmi, 1958; Scow, 1959; Donovan and Jacobsohn, 1960).

Previous studies have shown that, if adult male rats are hypophysectomized at the end of one month of protein-free feeding and then refed an 18% casein diet, retention of nitrogen simulates that of a normal rat. However, body-weight gain is subnormal and liver-protein concentration improves, but the increase in total liver protein is only modest and kidney mass clearly decreases (Leathem, 1958).

Hypophysectomy is known to alter serum-protein concentrations by decreasing albumin and increasing globulin concentrations. If rats are fed a protein-free diet for 30 days, total serum protein decreases primarily because of a loss of serum albumin. Hypophy-sectomy and refeeding 18% casein resulted in a return of total serum protein to normal concentrations owing to an increase in globulin; serum albumin failed to improve. In contrast, the normal rat total serum protein levels increased because of an increase in serum albumin (Table 8).

Recent studies have been conducted with protein-depleted rats in which, following hypophysectomy, 18% lactalbumin was fed for 20 days to compare repletion with rats fed casein (Table 8). Body-weight gain, retention of nitrogen, loss of kidney protein, and changes in plasma proteins failed to distinguish the two diets. Total liver protein was improved by feeding lactalbumin.

Following protein depletion, the hypophysectomized rats have been fed *ad libitum* and, as is noted above, a gain in weight and retention of nitrogen was recorded. If, however, caloric intake was decreased to 50% of anticipated needs during refeeding of protein, sharply different effects were noted. Now the hypophysectomized rat lost weight in excess of the non-operated control and retained only $1/7$ as much nitrogen. Perhaps these circumstances will more clearly differentiate the effect of hypophysectomy on protein repletion.

Table 8. *Influence of hypophysectomy and protein repletion on kidney and serum protein*

	PFD × 30	18% Casein		18% Lactalbumin	
		Hypophx	Control	Hypophx	Control
Body weight change g. . .	—62	+30	+70	+31	+60
N_2 retained g./kg.	0	8.0	8.4	7.5	9.5
Kidney					
weight g.	2.1	1.6	2.2	1.4	2.1
protein g.	.30	.24	.33	.25	.37
Serum protein					
total g./100 ml.	4.50	6.04	5.85	6.14	5.87
albumin	1.45	1.70	2.79	1.87	2.63
a/g ratio	.48	.40	.91	.44	.82

The subnormal body-weight gain of rats hypophysectomized prior to *ad libitum* protein refeeding, and the decrease in kidney weight and protein which occurred despite a positive nitrogen balance, prompted a study of hormone supplements (Table 9). Test-

Table 9. *Hormonal influences on body-weight and nitrogen balance during protein repletion with casein*

Treatment	Body wt. gain g.	Total N_2 retained g./kg.
Hypophx	30	8.0
Normal, pair fed	70	8.4
Normal, ad lib.	153	10.0
Hypophx + test.-prop. (0.25 mg.)	30	5.2
Normal	53	7.1
Hypophx + insulin (0.5 u.)	44	7.6
Normal	103	9.0
Hypophx + growth hormone (0.25 mg.) . .	70	10.5
Normal.	82	9.7

osterone-propionate (0.25 mg.), and protamine-zinc insulin (0.5 unit) failed to improve body-weight gain or nitrogen balance in hypophysectomized rats being repleted on 18% casein. On the other hand, growth hormone (0.25 mg.) restored body-weight gain to that of a pair-fed normal rat, but did not enhance appetite.

Nitrogen retention does not assure uniform distribution of the nitrogen retained (Leathem, 1958; Scow, 1959). Furthermore, prior protein depletion reduces total protein content of muscle, kidney, and liver and, in fact, decreases the liver-protein concentration by 50%. Thus, the repletion process involves a considerable physiological correction. Nevertheless, the subnormal protein concentrations of the liver of protein-depleted rats do return to normal on refeeding protein in the absence of the hypophysis

Table 10. *Influence of hypophysectomy, diet, and hormones on organ weight*

	PFD × 30	Hypophx	Hypophx + STH	Hypophx + PZI	Hypophx + test.-prop.
Liver g.	9.30	8.4	11.2	9.0	10.3
Kidney	2.10	1.62	2.13	1.45	1.81
Heart	.86	.71	.88	.62	.84
Muscle	.90	1.01	1.43	.82	1.22

Hypophx fed 18% casein; STH — 0.25 mg.; PZI — 0.5 unit; Test.-prop. — 0.25 mg.

(Table 10). Administration of 0.25 mg. growth hormone to hypophysectomized rats increased liver and muscle size and total protein. A further decrease in heart and kidney weight which follows hypophysectomy was prevented by growth hormone, and some increase in kidney protein was associated with an increase in protein concentration. The dosage of growth hormone used did not permit the kidney and heart to keep pace with the normal controls. Hypophysectomized rats not prior protein-depleted respond to growth hormone with isometric growth of the kidney and heart (Korner and Young, 1955). Results to date with a daily dose of 0.5 unit of protamine-zinc insulin revealed that organ weights of treated and untreated hypophysectomized rats were similar although liver protein improved. Testosterone-propionate (0.25 mg.) in hypophysectomized rats increased the actual weight of the liver and muscle and partially prevented the further decrease in kidney, testis, and heart weight which occurs in untreated hypophysectomized rats. The improvement in liver and muscle weight was reflected in an increase in total protein. These data support prior publications on non-depleted hypophysectomized rats in showing the varied influences of hormones on internal shifts in protein. It is apparent, however, that hormones can play a role in protein repletion, but the best nutritional base upon which the hormones can act remains to be determined.

Summary

These studies emphasize that protein metabolism is influenced by hormones and nutrition. Although the endocrine system is clearly involved in protein metabolism, the state of the body-protein reserves at the time of hormone administration may determine whether an anabolic or catabolic response is obtained.

Dietary proteins are assigned a biological value which is related to anabolic tendencies as tested in normal animals. However, some reconsideration of the biological value of fed proteins may be necessary in endocrine deficiencies, and the need for amino acids may vary for full physiological expression of different hormones. Specific nutrition may enhance hormone usefulness therapeutically.

Zusammenfassung

Diese Untersuchungen zeigen deutlich, daß der Eiweißstoffwechsel durch Hormone und durch die Ernährung beeinflußt wird. Obwohl das endokrine System sicher an der Regulation des Eiweißstoffwechsels beteiligt ist, sind die im Organismus zum Zeitpunkt der Hormongabe vorhandenen Eiweißreserven maßgebend dafür, ob ein anaboler oder ein kataboler Effekt resultiert.

Am intakten Tier wird Nahrungseiweißen ein biologischer Wert zugeschrieben, der in Beziehung zur anabolen Tendenz steht. Bei Hormonmangelzuständen kann jedoch die biologische Wertigkeit von zugeführtem Eiweiß anders zu beurteilen sein, und um das vollständige physiologische Wirkungsbild eines Hormons zu erreichen, kann sich ein verschiedener Bedarf an Aminosäuren ergeben. Eine gezielte Ernährung vermag den therapeutischen Nutzen der Hormone zu verbessern.

Résumé

Ces recherches montrent que le métabolisme protéique est influencé par les hormones et par l'alimentation. Bien que le système endocrinien participe de façon évidente à la régulation de ce métabolisme, ce sont les réserves de protéines présentes dans l'organisme au moment de l'administration de l'hormone qui détermineront si l'effet sera anabolique ou catabolique.

Chez l'animal normal, on attribue aux protéines alimentaires une valeur biologique qui est en rapport avec la tendance anabolique. Dans les états de carence hormonale toutefois, cette valeur biologique de l'apport protéique peut devoir être considérée différemment, et les besoins en acides aminés peuvent être variables pour que telle ou telle hormone exerce sa pleine action physiologique. Une alimentation spécifique permet d'améliorer l'efficacité thérapeutique des hormones.

Acknowledgements

These investigations were supported by a research grant, A-462, from the Institute of Arthritis and Metabolic Diseases of the National Institutes of Health, United States Public Health Service, and by a grant from E. R. SQUIBB and Sons.

Grateful acknowledgement is made to the Endocrinology Study Section of NIH for the growth hormone and to CIBA Pharmaceutical Company and G. D. Searle & Co., who provided many of the steroids used.

References

AHREN, K., A. ARVILL, and A. HJALMARSON: Acta endocr. (Den.) **39**, 584 (1962). — ALLISON, J. B.: Physiol. Rev. (U.S.A.) **35**, 644 (1955). — ARNOLD,

218   J. H. Leathem: The influence of nutrition and nutritional status

A., J. S. Shad, and A. L. Beyler: Fed. Proc. (U.S.A.) 15, 543 (1956). —
Aschkenasy, A.: Ann. endocr. (Fr.) 16, 199 (1955); — Arch. internat.
pharmacodyn. thérap. (Belg.) 123, 406 (1960); — Ann. nutr. alim. (Fr.)
15, B 165 (1961). — Aschkenasy-Lelu, P., and A. Aschkenasy: World Rev.
Nutr. p. 33, 1959. — Bloch, H. S., C. R. Hitchcock, and A. S. Kremen:
Cancer Res. (U.S.A.) 11, 313 (1951). — Block, H. J., and H. H. Mitchell:
Nutr. Abstr. (G.B.) 16, 249 (1946—1947). — Boyd, E. M., and E. M. Cran-
dell: Cancer Res. (U.S.A.) 16, 198 (1955). — Calloway, D. E., and H.
Spector: J. Nutrit. (U.S.A.) 57, 73 (1955). — Chow, B. F., and R. O. Greep:
Proc. Soc. Exper. Biol. Med. (U.S.A.) 69, 191 (1948). — Clark, L.: J. Biol.
Chem. (U.S.A.) 200, 69 (1953). — Donovan, B. T., and D. Jacobsohn:
Acta endocr. (Den.) 33, 197 (1960). — Eisenberg, E., and G. S. Gordan:
J. Pharmacol. Exper. Therap. (U.S.A.) 99, 38 (1950). — Fishman, W. H.:
In: Mechanisms of Action of Steroid Hormones. Pergamon Press, 1960,
p. 157. — Frieden, E. H., E. H. Cohn, and A. A. Harper: Endocrinology
(U.S.A.) 68, 862 (1961). — Geiger, E., and I. El Rawi: Metabolism (U.S.A.)
1, 145 (1952). — Gordan, G. S., H. M. Evans, and M. E. Simpson: Endo-
crinology (U.S.A.) 40, 375 (1947). — Kassenaar, A., A. Kouwenhoven,
and A. Querido: Acta endocr. (Den.) 39, 223 (1962). — Knobil, E., and
P. F. Brande: Proc. 43rd Endocrine Soc. Meeting 1961. — Kochakian,
C. D., and D. G. Harrison: Endocrinology (U.S.A.) 70, 99 (1962). —
Kochakian, C. D., J. G. Moe, and J. Dolphin: Amer. J. Physiol. 162, 581
(1950). — Kochakian, C. D., R. Tanaka, and J. Hill: Amer. J. Physiol.
201, 1068 (1961). — Kochakian, C. D., and C. T. Tillotson: Endocrinology
(U.S.A.) 58, 226 (1956). — Korner, A.: J. Endocr. (G.B.) 21, 177 (1960); —
Mem. Soc. Endocr. 11, 60 (1961). — Korner, A., and F. G. Young: J.
Endocr. (G.B.) 13, 84 (1955). — Landau, R. L., and K. Lugibihl: Recent
Progr. Hormone Res. (U.S.A.) 17, 249 (1961). — Leathem, J. H.: Proc.
Soc. Exper. Biol. Med. (U.S.A.) 60, 260 (1945); — In: Hormones and the
Ageing Process (Ed. by Engle and Pincus), 1956, p. 79. — Recent Progr.
Hormone Res. (U.S.A.) 14, 141 (1958); — In: Inflammation and Diseases
of Connective Tissue (Ed. by Mills and Moyer). Saunders Press, 1961,
p. 367. — Lostroh, A. J.: Endocrinology (U.S.A.) 70, 747 (1962). —
Manchester, K. L., and F. G. Young: J. Endocr. (G.B.) 18, 381 (1959); —
Vitamins and Horm. (U.S.A.) 19, 95 (1961). — Meites, J., and M. M. Nelson:
Vitamins and Horm. (U.S.A.) 18, 205 (1960). — Munro, H. N.: Physiol
Rev. (U.S.A.) 31, 549 (1951). — Nimmi, M. C., and E. Geiger: Endocrinology
(U.S.A) 61, 753 (1957). — Overbeek, G. A., A. Delver, and J. de Visser:
Acta endocr. (Den.) 40, 133 (1962). — Perlman, P., and J. Cassidy: Proc.
Soc. Exper. Biol. Med. (U.S.A.) 83, 674 (1953). — Rabinovitch, M. L., C. U.
Junqereia, and H. A. Rothschild: Science (U.S.A.) 114, 551 (1951). —
Reid, E.: J. Endocr. (G.B.) 13, 319 (1956). — Rosenthal, H. O., and J. B.
Allison: J. Nutrit. (U.S.A.) 44, 423 (1951). — Rupp, J. J., and K. E.
Paschkis: Metabolism (U.S.A.) 2, 226 (1953). — Russell, J. A.: Amer. J.
Clin. Nutr. 5, 404 (1957). — Russell, J. A., and A. E. Wilhelmi: Amer.
Rev. Physiol. 20, 43 (1958). — Saunder, H. L.: personal communication,
1962. — Saunders, F. J., and V. A. Drill: Endocrinology (U.S.A.) 58, 567
(1956). — Scow, R. O.: Amer. J. Physiol. 169, 859 (1959). — Scow, R. O.,
and S. N. Hagen: Endocrinology (U.S.A.) 60, 273 (1957). — Shaw, J. H.,
and R. O. Greep: Endocrinology (U.S.A.) 44, 520 (1949). — Shock, N. W.:
In: Hormones and the Ageing Process (Ed. by Engle and Pincus), 1956,
p. 297. — Wool, L. G.: Amer. J. Physiol. 198, 357 (1960).

Discussion

QUERIDO: Dr. LEATHEM has discussed some questions which have a very pertinent bearing on this Symposium. May I ask for one clarification? Am I right in believing that in depleted animals you could not get a good increase in repletion by using an anabolic agent?

LEATHEM: If we use criteria such as the nitrogen balance and the tissue protein concentrations, which we have measured in the gastrocnemius, heart, liver, and kidney, I would say that in none of these instances have we seen anything more than a minor stimulus. The retention of nitrogen is increased by a total of roughly 2 g. over a 20-day period. We have seen this with almost every agent that has been used. I would have hoped for a greater difference, which might have enabled us to separate out these various agents.

QUERIDO: You mean to say that, if the body is at a phase of maximum repletion, you can't enhance this mechanism significantly?

LEATHEM: Correct.

YOUNG: I should like to ask Dr. LEATHEM if he has used muscles other than the gastrocnemius in the rat. The reason why I ask is that Dr. GREEN-BAUM and I[1] investigated the effect of treatment with growth hormone on different muscles in the adult normal ("plateaued") female rat. In these experiments we found that the gastrocnemius increased in size much less than did the quadriceps or sartorius. On the whole, in these experiments, the muscles near the middle of the body did better than the more peripheral ones, and I wondered why Dr. LEATHEM chose the gastrocnemius in particular.

LEATHEM: My heritage is Irish, and it is perhaps for this reason that I have a particular weakness for devilish things! We realised that the gastrocnemius was one muscle that never seemed to want to respond experimentally; nevertheless, we keep on trying to make it do something. We thought that, if we could get the gastrocnemius to respond, then certainly other muscles would be responsive. We have continued to see no response and we are about ready to give up on this and go on to some of the muscles you have shown to be truly responsive.

KORNER: Dr. LEATHEM, I was intrigued by the slide you have shown of the methionine uptake into liver fractions. In particular, you showed a higher uptake into the pH 5 fraction of normal rat liver than in any other fraction; and in the hypophysectomised rats you found a higher uptake in the nuclei. This finding is not in agreement with the results I have obtained, nor with those of other workers, and I wondered whether you had any explanation. How long after the injection of the radioactive amino acid did you remove the liver?

LEATHEM: The whole experimental design is a bit different, as you know, in that these animals have been subjected to protein depletion for 30 days,

---

[1] GREENBAUM, A. L., and F. G. YOUNG: J. Endocr. (G. B.) **9**, 127 (1953).

then hypophysectomised and refed for 20 days. The S^{35}-methionine uptake was a 2-hour uptake. All I can say at the moment is that I know these data are not in accord with yours and that we are gathering more data before I may take the plunge and say I'm sure that this is exactly what's happening.

WATERLOW: In all your experiments on repletion after a low-protein diet, you waited 20 days, did you not, and then killed the animals?

LEATHEM: That's correct.

WATERLOW: I think from the small amount of work we have done that the very early period of repletion may be interesting. At that time there may be lags in the restoration of some tissues and increased rates in others. I wonder whether this might not be a better situation in which to see the effects of some of these substances?

LEATHEM: Well, Dr. WATERLOW, we selected this particular time merely as an arbitrary one and, since we were at least getting continuous nitrogen balances at 4-day periods, we chose the 20-day period from some prior experiments. All I can say is that I can't disagree with you that earlier periods of response certainly would provide useful information, but I don't feel like going back and repeating it all.

WATERLOW: In my experiments I was using the gastrocnemius, not because I'm Irish, but because it is easy to dissect. In this muscle there is, for instance, a lag in new DNA formation compared with the liver. When you start to replete after depletion, almost immediately there is a rapid new formation of DNA in the liver, and then after about 4 days the same thing starts up in muscle. I would very much like to know whether these differences in pattern are under hormonal influences.

LEATHEM: I can't answer that at all. We have absolutely no information at an earlier period. We set out, you see, with a sort of preconceived notion that a 20-day period would be a maximum in a sense, particularly in view of the prior ideas that there was a so-called wearing-off effect of androgens in a 10 to 14-day period, which may simply be a matter of filling up the protein stores. So we proceeded a bit beyond that period in order to get what we hoped might be the maximum degree of potential response. As you say, of course, much more exciting results could be obtained in an earlier period. But then I think we might not perhaps have seen such drastic internal shifts as we are getting — with the liver changing, with the kidney decreasing just as if the animal were receiving no protein at all, and with the heart losing protein under these circumstances, although the animal is in positive balance. If anything, I think it's just to re-emphasise to us that, as we all know, single measures can lead us into horrible pitfalls of misunderstanding.

WATERLOW: Yes, I entirely agree.

SALA: We have used castrated male rats that had received a protein-free diet for 7 days; when we gave a normal diet and administered at the same time 1 mg./day of 4-hydroxymethyl-testosterone, we could show nitrogen retention. Under these circumstances, testosterone derivatives obviously exert a nitrogen-retaining effect.

LEATHEM: Yes, I think that our laboratory would also be able to demonstrate for you examples in castrated animals, on lower levels of protein, to show that you can get positive effects by nitrogen-balance measurements. My own personal reaction, though, is that these positive responses are still not very great. We have other studies, too. For example, if you subject the animal to caloric restriction and just measure a 4-day balance period, testosterone or Dianabol will show positive effects during this time. If your animal is protein-depleted for 30 days and then studied for a 4-day nitrogen period on a low casein intake, you can induce some enhancement of nitrogen retention, but it always seems to me relatively trivial by comparison with the effect of something like growth hormone.

OVERBEEK: We have always been struck by the fact that so many people, including ourselves, saw practically no effects on body-weight when anabolic steroids were administered to rats. This led us to wonder whether the animals are under sub-optimal conditions or whether, when already under optimal conditions, they grow so fast that it would be impossible to see anything. We have also considered the possibility that what they gain on protein they lose on fat. But we couldn't find any difference in body composition when we looked for the fat. So could it be that, on a diet with lactalbumin as a protein source, these anabolic steroids might produce better responses.

LEATHEM: We, too, have been hoping for results somewhat along these lines, and we have therefore made repeated attempts to subject animals to such nutritional variances and to determine the amino-acid patterns which would enable testosterone to yield a better response. I still feel that we can devise an experiment in which, offered the proper amino acid, the anabolic agents will produce more convincing results than if we simply use routine feeding. In this connection, I would cite our experiment with the tumour, for example. I might just add that we do have some preliminary data showing that the simple addition of methionine at 0.5% actually doubled the response to testosterone-propionate in terms of nitrogen retention in the refeeding process. This, to me, is most encouraging.

Effects of anabolic steroids on nucleic acid and protein metabolism

By

A. A. H. KASSENAAR, A. QUERIDO, and A. HAAK

In recent years, with the increased knowledge we now have of the fundamental biochemical processes which result in protein synthesis, much work has been devoted to the study of possible changes in these processes induced in target tissues after treatment with testosterone and related compounds.

The beautiful studies presented by Dr. WILSON at this Symposium illustrate this trend of investigation and show what great prospects and, at the same time, what difficulties this kind of work involves.

At the present moment it seems hard to reconcile this type of approach to the study of the action of these compounds with the older one, which might be called the pharmacological approach. In the latter the chief emphasis is placed on the study of differences in the reaction of various tissues and on the study of variations in the distribution pattern of these reactions which can be induced by the various agents in question. In such studies it is mainly the increase in the tissue mass of an organ as determined by its weight which is used as an indicator.

In the former type of approach, which might be described as the biochemical one, a much more unitarian point of view is taken, and it is more or less accepted that the pharmacological expression of the effects of these steroids can be explained by changes in one rate-limiting stage of cellular metabolism.

All three aspects of protein metabolism, which were mentioned by the Chairman in his Opening Address, have at some time been considered to be involved in this specific locus of action of these androgenic agents. Increased availability of amino acids for protein synthesis owing to increased cellular uptake was suggested as the mechanism responsible. Changes in energy metabolism, resulting in more TPNH being available to the cell, have also been considered in this respect.

Dr. WILSON's observations on the greater efficiency of hen oviduct ribosomes after testosterone stimulation point to the possibility that changes in the protein-synthesising mechanisms may be responsible for the effect observed.

There can be no doubt that at some phase of the reaction of a target tissue to testosterone these changes will be found. However, if we look at the tissue and at the cell as a functional unit in it, it seems to us that these changes, and the many others which have been described, fit into a general pattern of reactions which can be called tissue growth and which is not specific for the particular kind of tissue growth we are interested in. In view of these questions, we have investigated whether the way in which the tissue mass increases in various target tissues after testosterone treatment is constant for these different tissues, and furthermore whether these changes are specific for a particular growth-inducer.

In view of the fundamental role of the nucleic acids in cellular metabolism we have studied the concentration of RNA and DNA in target organs for these steroids in mice and rats. The seminal vesicles of both species were studied as being the most commonly used end-point for testing androgenic activity, while in the mouse the kidney and in the rat the levator ani was studied as being the most commonly used end-point for testing so-called anabolic activity.

RNA and DNA were determined in pooled tissues, which were stored frozen after dissection, the determinations being carried out by the orcinol (1) or the diphenylamine (2) reaction after isolation according to the method of SCHNEIDER (3). All the data refer to the mean values observed in 2—4 groups of 3—6 animals.

Fig. 1 shows the effect of castration on weight and on RNA and DNA content of the seminal vesicle in the mouse.

The values were calculated as percentages of those found in the control group. It can be seen that after castration a rapid decline in weight occurs, which reaches a plateau 5—6 weeks after castration. The total amount of RNA per seminal vesicle declines more rapidly than the weight of the organ, while the opposite is the case for DNA.

Obviously, castration results in important changes in the nucleic acid composition of this tissue. These results are in accordance with the observations made by KOCHAKIAN et al. (4) and published at the beginning of this year.

The changes can be reversed by treatment with testosterone, which induces a rise in the total amounts of RNA and DNA, so that the RNA concentration increases, while the DNA concentration decreases (Table 1).

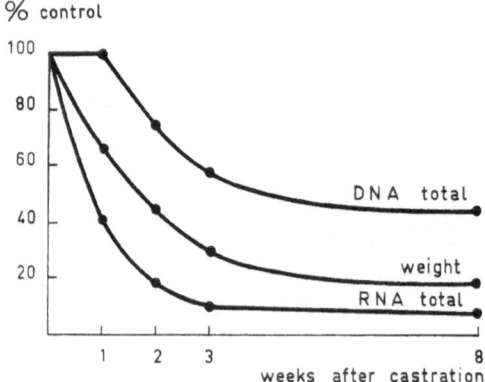

Fig. 1. Changes in weight and nucleic acid content of the seminal vesicles of mice after castration. The data are expressed as percentages of the values found in the seminal vesicles of normal animals

Assuming that the amount of DNA per cell does not change, these results indicate that the weight increase of the seminal vesicles after testosterone treatment is due to hyperplasia as well as to hypertrophy. These changes can already be observed a very short time after treatment with testosterone has been started.

Table 1. *Effect of testosterone treatment on nucleic acid composition in the seminal vesicles of castrated adult mice*

Group	Treatment	No. of ani-mals	Weight of sem. vesicles (mg.) ± s. d.	DNA total (mcg.)	RNA total (mcg.)	DNA (mcg./mg.)	RNA (mcg./mg.)	RNA/DNA
I	Oil	2×5	17.2 ± 3.9	183	103	10.65	5.96	0.56 (0.52—0.59)
II	Testosterone $(7 \times 12.5 \text{mcg./day})$	2×5	46.1 ± 7.4	350	374	7.59	8.11	1.07 (1.06—1.07)
III	Testosterone $(7 \times 25 \text{ mcg./day})$	2×5	51.7 ± 5.1	397	429	7.68	8.29	1.08 (1.03—1.13)
IV	Testosterone $(7 \times 50 \text{mcg./day})$	2×5	61.4 ± 6.4	456	502	7.43	8.17	1.10 (1.10—1.10)

24 hours after the administration of 50 mcg. testosterone the weight of the seminal vesicles is increased by about 20%, while the amount of RNA per organ is increased by about 35%. At this moment no changes in DNA content are found. After 48 hours the

increase in RNA content is much more pronounced and there is also an increase in DNA content (Table 2).

Table 2. *Early effect of testosterone (T) treatment on nucleic acid composition in the seminal vesicles of castrated adult mice*

Group	Time after 1st injection	Treatment	Weight (mg.) (6)	DNA total (mcg.)	RNA total (mcg.)	DNA (mcg./ mg.)	RNA (mcg./ mg.)	RNA/ DNA
I	—	Sesame oil	69	619	305	8.9	4.4	0.49
II	24 hrs	T (50 mcg.)	84	616	410	7.3	4.8	0.66
III	48 hrs	T(2×50 mcg.)	111	800	751	7.2	6.8	0.94

The data given in columns 4, 5, and 6 are the values found in the pooled tissue of 6 animals.

Following long-term treatment with testosterone (3 weeks with 50 mcg. per day), the same pattern is found. The results show that the cell size is increased and that the cells contain a large amount of RNA. Also, tissue hyperplasia is evident, as revealed by the increased amount of DNA.

In the mouse kidney, however, quite another pattern is found. In response to this long-term treatment, when the weight increase is about 60%, the total amount of DNA remains constant, while RNA increases roughly parallel to the weight increase (Table 3).

Table 3. *Effect of long-term treatment with testosterone (T) on nucleic acid composition in the kidneys and seminal vesicles of castrated adult mice*

Group	Treatment	Weight (mg.)	DNA total (mcg.)	RNA total (mcg.)	DNA (mcg./ mg.)	RNA (mcg./ mg.)	RNA/DNA
		(3)	*Kidneys*				
I	Sesame oil	360	2115	1924	5.83	5.30	0.91
II	T (21×50 mcg./day)	583	1920	3359	3.31	5.75	1.73
		(6)	*Sem. vesicles*				
I	Sesame oil	78	595	396	7.63	5.08	0.67
II	T (21×50 mcg./day)	453	1485	4097	3.28	9.04	2.76

The data given in columns 3, 4, and 5 are the values found in the pooled tissue of 3 animals (kidneys) or 6 animals (seminal vesicles).

In order to investigate whether this reaction of the kidney is specific for treatment with these steroids, we studied the effect of another growth stimulus for the kidney, namely unilateral nephrectomy, in which the remaining kidney displays compensatory hypertrophy. The results of one of our experiments are shown in Table 4. While the weight increase is largely the same as in response to testosterone treatment (about 50%), here, too, the results indicate that only tissue hypertrophy has occurred, whereas the RNA concentration in the organ has remained constant (5).

Table 4. *Effect of unilateral nephrectomy on nucleic acid composition in the remaining kidney 3 weeks after the operation. (Castrated adult male mice)*

Group	Treatment	Weight (mg.) (3)	DNA total (mcg.)	RNA total (mcg.)	DNA (mcg./mg.)	RNA (mcg./mg.)	RNA/ DNA
I	None	360	2115	1924	5.83	5.30	0.91
II	Unilateral nephr.	547	2170	2706	4.00	4.95	1.21

The data presented in columns 3, 4, and 5 are the values found in the pooled tissue of 3 animals.

In the light of these results, we wondered whether organs which are used as an end-point for the assay of so-called anabolic activity exhibit reactions comparable to those found in the mouse kidney and differing from those observed in the seminal vesicles. The results presented in Table 5 show that this is not the case. In the rat both the seminal vesicles and the levator ani muscle react to testosterone treatment with an increase in RNA content, as well as concentration, while the DNA content increases and its concentration decreases. This indicates that hyperplasia, as well as hypertrophy, also plays a part in the increase in weight found in the levator ani muscle following treatment with testosterone.

Table 5. *Effect of testosterone treatment (T) on nucleic acid composition in the seminal vesicles and levator ani muscle of castrated adult rats*

Group	Treatment	Weight (mg.) (4)	DNA total (mcg.)	RNA total (mcg.)	DNA (mcg./ mg.)	RNA (mcg./ mg.)	RNA/ DNA
			Seminal vesicles				
I	Sesame oil	35	520	112	14.9	3.2	0.22
II	T (7×250mcg./day)	180	1500	1550	7.9	8.6	1.09
III	T (7×375mcg./day)	310	1920	2600	6.1	8.5	1.38
			Levator ani muscle				
I	Sesame oil	81	160	139	1.95	1.7	0.86
II	T (7×250mcg./day)	234	251	790	1.10	3.4	3.08
III	T (7×375mcg./day)	275	285	957	1.05	3.5	3.36

The data presented in columns 3, 4, and 5 are the values found in the pooled tissue of 4 animals.

In a subsequent series of experiments we investigated whether the effects exerted on the nucleic acid composition of the seminal vesicles or levator ani muscle by testosterone, as an example of the androgenic steroids, and by 17α-ethyl-19-nortestosterone, as an example of the so-called anabolic steroids, are qualitatively or quantitatively different.

In these experiments we tried to select the doses of the two compounds in such a way as to produce identical effects on the weights of these organs, and at the same time to keep the doses within a range at which a change of dose would still have an effect on the relative concentration of RNA and DNA.

Table 6 shows some of our data on the nucleic acid composition of the seminal vesicles. These results indicate that 150 mcg. 17α-ethyl-19-nortestosterone daily for 7 days, or 20 mcg. testosterone daily for 7 days, produces the same effect on the weight of this organ. In these two groups identical effects on DNA and RNA content were found.

Table 6. *Effect of testosterone (T) and of 17α-ethyl-19-nortestosterone (N) on nucleic acid composition in the seminal vesicles of castrated adult rats*

Group	No. of animals	Treatment	Weight of sem. vesicles (mg. ± s.d.)	Total DNA (mcg.)	Total RNA (mcg.)	DNA (mcg./mg.)	RNA (mcg./mg.)	RNA/DNA
I	2 × 6	Sesame oil	10.2 ± 0.83	118	29	11.6	2.8	0.25
II	2 × 6	N (7 × 75 mcg./day)	17.3 ± 1.24	156	61	9.0	3.5	0.39
III	1 × 6	N (7 × 150 mcg./day)	21.8 ± 1.25	180	120	8.2	5.5	0.67
IV	2 × 6	T (7 × 20 mcg./day)	20.2 ± 1.18	179	115	8.8	5.7	0.65
V	2 × 6	T (7 × 40 mcg./day)	26.7 ± 1.68	203	203	7.6	7.6	1.00

In Table 7 the results of a similar experiment on the levator ani muscle are presented. In this organ identical weight increases were induced by 7 × 200 mcg. testosterone and 7 × 50 mcg. 17α-ethyl-19-nortestosterone. Here again, in these two groups, quantitatively the same changes in nucleic acid composition were found. These results show that in the seminal vesicles, as well as in the levator ani muscle, the effects of the two compounds are the same (6).

We should like to discuss these results in connection with the question formulated above, namely:

Is the pattern of reaction to treatment with these compounds constant among those organs which are sensitive, and is this pattern specific for a particular treatment?

15*

Of course, our information is too limited to enable us to answer this question definitely. More tissues must first be analysed and many more aspects of cellular activity taken into account. But with this restriction, we think that our present data on the effect of these steroids on nucleic acid composition allow us to draw some conclusions.

Table 7. *Effect of testosterone (T) and of 17α-ethyl-19-nortestosterone (N) on nucleic acid composition in the levator ani muscle of castrated adult rats*

Group	Treatment	Weight (mg.) (3)	DNA total (mcg.)	RNA total (mcg.)	DNA (mcg./mg.)	RNA (mcg./mg.)	RNA/DNA
I	Sesame oil	138	217	265	1.50	1.92	1.31
II	T (7 × 200 mcg./day)	285	318	895	1.12	3.14	2.80
III	T (7 × 400 mcg./day)	303	342	1035	1.13	3.42	3.03
IV	N (7 × 25 mcg./day)	243	380	609	1.15	2.85	2.50
V	N (7 × 50 mcg./day)	285	302	860	1.06	3.02	2.85
VI	N (7 × 100 mcg./day)	348	362	1205	1.04	3.46	3.35

The data presented in columns 3, 4, and 5 are the values found in the pooled tissue of 3 animals.

Comparison of the effect exerted by testosterone on the mouse kidney on the one hand, and on the seminal vesicles and levator ani muscle on the other, reveals an important difference. While the kidney reacts only by cell hypertrophy, without increased cell multiplication, the seminal vesicles and levator ani muscle increase in mass as a result of an increase in cell number as well as in cell size. Another difference is that in the mouse kidney no significant increase in RNA concentration is found, whereas this is very pronounced in the other organs.

These findings indicate that the pattern of reaction to treatment with these compounds in the various target organs is not constant. Furthermore, our results show that the changes found in an organ — which reacts to treatment with these steroids with an increase for the particular growth stimulus applied — seem to be dependent on the kind of tissue. This is inferred from the fact that the mouse kidney reacts to unilateral nephrectomy in the same way as to the administration of testosterone. Moreover, the changes which are found in the levator ani muscle or seminal vesicles are dependent not on the kind of steroid by which a certain growth is induced, but only on the magnitude of that growth.

These considerations lead us to the conclusion that both the questions raised have to be answered in the negative. It seems that the pattern of reactions to treatment with these compounds is *not constant* among the various target organs and that the metabolic changes which are induced in such organs are not specific for a particular treatment but are *organ specific*. Hitherto, in research into the mechanism of action of these steroids, most emphasis has been placed on the description of the metabolic changes which can be found in target tissues after stimulation with such steroids. Very important observations have been made, and it can be expected that much more information in this respect will become available in the near future.

However, if the statements we have made above are correct, it seems that, in order to explain the action of these steroids in chemical or physical terms, attention must be paid to the question: What makes a cell or a tissue willing to react to treatment with these steroids with a specific pattern of highly integrated reactions which might be described as growth ?

In another series of experiments we studied some aspects of the effect of testosterone on the *in vitro* incorporation of C^{14}-labelled glycine into the proteins and nucleic acids of the seminal vesicles of castrated mice.

As a rule the tissues were incubated in Krebs-Ringer phosphate buffer at pH 7.3 for three hours. Glucose, an amino-acid mixture, and penicillin were present in the medium. 2 μC. of 2-C^{14}- labelled glycine with a specific activity of 5.6 mC./mM. was added per vessel.

Table 8. *Effect of testosterone pre-treatment on the incorporation of 2-C^{14}-glycine into proteins of seminal vesicles in vitro. (Castrated mice, incubation in Krebs-Ringer phosphate buffer for 3 hours)*

Dose of testosterone (mcg.)	Time after testosterone (hours)	c.p.m./mg. protein
0	0	2000
1×100	5	2370
1×100	16	3550
1×100	24	5650
2×100	48	6100

After the incubation, a large excess of cold glycine was added and the samples were placed in ice. The tissues were washed in ice-cold Krebs-Ringer phosphate buffer with glycine. After this, the protein and nucleic acid fractions were isolated and measured. All the data represent the mean values of duplicate incubations with less than 15% difference.

Table 8 shows the results of an experiment in which the effect of short-term treatment was studied. A very significant increase in glycine incorporation in the protein fraction is found only 16 hours after the injection of 100 mcg. testosterone. This effect is much more pronounced 24 and 48 hours after the commencement of treatment.

Other experiments showed that the stimulation of amino-acid incorporation into protein, expressed per mg. protein, is much less pronounced after 14 days of treatment with testosterone than after 48—96 hours.

Table 9 illustrates the results of an experiment in which the incorporation of glycine into proteins and into nucleic acids was determined in the seminal vesicles of normal and castrated animals, as well as in a group of castrated animals which had received treatment for 48 hours with testosterone.

Table 9. *Incorporation of 2-C^{14}-glycine into protein and into RNA by the seminal vesicles of normal, castrated, and testosterone-pretreated castrated mice in vitro. (Incubation in Krebs-Ringer phosphate buffer for 3 hours)*

Group	Treatment	DNA (mcg./mg.)	RNA (mcg./mg.)	s. a. protein c.p.m./mg. C	s. a. RNA c.p.m./mg. tissue
I (castr.) . .	Sesame oil	9.8	2.9	1950	72
II (castr.) .	Testosterone (2 × 100mcg./day)	7.8	5.8	4800	166
III (normal)	Sesame oil	4.6	10.2	1960	120

We should like to discuss in particular the results obtained in the first and third groups, since it seems reasonable to assume that in these groups a steady state was present in the seminal vesicles, which facilitates interpretation of the results. The very pronounced differences observed in the nucleic acid composition of the seminal vesicles were similar to those described above.

If the data on protein synthesis are expressed in terms of the specific activity of the protein isolated after the incubation, no difference between normal and castrated animals is found in this experiment. In the normal animals, more nucleic acid is synthesised per mg. tissue than in the castrated animals. It seems advantageous for two reasons to express these data also per unit weight DNA. In the first place, the results will then be expressed in relation to the most stable cell constituent. This might be important in this particular case, where large differences exist between the tissue composition of the two groups being compared. In this way we obtain information on the amount of protein synthesised per cell.

Secondly, as discussed here by Dr. KORNER, the most advanced working hypothesis visualises that protein synthesis is regulated by the cell nucleus via regulation of RNA synthesis.

If we accept this, the data on the rate of protein and nucleic acid synthesis per unit weight DNA enable us to draw conclusions as to the efficiency of this function of the cell nucleus in the two groups. Such data, which are presented in Table 10, show that in the seminal vesicles of the non-castrated control animals more protein and more RNA is synthesised per cell nucleus and per unit time than in the same organs from castrated animals.

Table 10. *Incorporation of 2-C¹⁴-glycine into protein and into RNA by the seminal vesicles of the same groups of animals as shown in Table 9. Data expressed per mcg. DNA and per mcg. RNA*

Group	Treatment	spec. act. protein		spec. act. N. A.	
		c.p.m./mcg. DNA	c.p.m./mcg. RNA	c.p.m./mcg. DNA	c.p.m./mcg. RNA
I (castr.) . .	Sesame oil	19.8	67.2	7.4	24.8
II (castr.) .	Testosterone (2×100mcg./day)	61.5	83.0	21.5	28.7
III (normal)	Sesame oil	42.5	19.2	26.0	11.8

In this same table, the results are also calculated per unit weight RNA. From these data, which appear in columns 4 and 6, it can be concluded that the RNA present in the tissue, which has been under endogenous testosterone stimulation, is less efficient in promoting protein synthesis than the RNA in the seminal vesicles of the castrated animals. Parallel with this, the turnover rate of the RNA in the latter tissue is higher than that of the RNA in the tissue from the normal animals.

Summary

Studies on the nucleic acid composition in the seminal vesicles of mice and rats, in the kidneys of mice, and in the levator ani muscle of rats after castration and after treatment with testosterone are described. The results indicate that the seminal vesicles and the levator ani muscle increase in weight after testosterone treatment, both by cell hypertrophy as well as by cell hyperplasia, while the RNA concentration increases.

In the mouse kidney no change in DNA content was found, and it is therefore concluded that this organ increases in weight by cell hypertrophy. No change in RNA concentration in the mouse kidney was found either after treatment with testosterone or when the animals were subjected to unilateral nephrectomy.

Provided that the same weight increase had been induced, the changes in nucleic acid composition in seminal vesicles and in the levator ani muscle were qualitatively and quantitatively the same after treatment with 17α-ethyl-19-nortestosterone as after treatment with testosterone.

These observations lead us to the conclusion that the pattern of reactions to treatment with these compounds is not constant among the various target organs, and that the changes which are induced in such organs are not specific for a particular treatment, but are organ specific.

The results of studies on the *in vitro* incorporation of 2-C^{14}-glycine into proteins and nucleic acids of the seminal vesicles of normal, castrated, and castrated + testosterone-treated mice are discussed.

Zusammenfassung

Es wurden Untersuchungen über den Gehalt an Nucleinsäuren der Samenblasen von Mäusen und Ratten, der Mäuseniere und des m. levator ani der Ratte nach Kastration und Behandlung mit Testosteron vorgenommen. Die Resultate sprechen dafür, daß die Samenblasen und der m. levator ani unter Testosteronbehandlung an Gewicht zunehmen, und zwar sowohl infolge Hypertrophie als auch Hyperplasie der Zellen, während die Konzentration an Ribonucleinsäure ansteigt.

In der Mäuseniere fand sich dagegen keine Änderung im Gehalt an Desoxyribonucleinsäure. Daraus wird geschlossen, daß die Gewichtszunahme dieses Organs auf Zellhypertrophie beruht. Testosteron-Gabe oder unilaterale Nephrektomie waren ohne Einfluß auf den Ribonucleinsäuregehalt der Mäuseniere.

Nach Behandlung mit 17 α-äthyl-19-nortestosteron waren bei gleicher Gewichtszunahme die Änderungen in der Nucleinsäurezusammensetzung in den Samenblasen und im m. levator ani qualitativ und quantitativ gleich wie nach Testosteron-Gabe.

Diese Beobachtungen legen den Schluß nahe, daß die Reaktionen auf Behandlung mit diesen Verbindungen in den verschiedenen Erfolgsorganen nicht gleich sind, und daß die in diesen Organen hervorgerufenen Veränderungen nicht spezifisch für eine besondere Behandlung, sondern organspezifisch sind.

Die Befunde der Untersuchungen über den in vitro Einbau von 2-C^{14}-Glykokoll in Eiweiße und Nucleinsäuren der Samenblasen von normalen und kastrierten Mäusen mit oder ohne Testosteron-Behandlung werden besprochen.

Résumé

L'auteur a entrepris des recherches sur le taux des acides nucléiques, dans la vésicule séminale de la souris et du rat, dans le rein de la souris et dans le muscle releveur de l'anus chez le rat, après castration et traitement à la testostérone. Les résultats montrent que le poids des vésicules séminales et du releveur de l'anus augmente sous l'effet de la testostérone, tant par hypertrophie que par hyperplasie cellulaire, tandis que la concentration de l'acide ribonucléique augmente.

Dans le rein de la souris, on n'a pas trouvé de modification du taux de l'acide désoxyribonucléique. On en conclut que l'augmentation de poids de cet organe est due à une hypertrophie cellulaire. Ni la testostérone ni une néphrectomie unilatérale ne modifient le taux de l'acide ribonucléique dans le rein de la souris.

Pour une même hausse pondérale, les changements de composition de l'acide nucléique dans les vésicules séminales et le muscle releveur de l'anus furent qualitativement et quantitativement les mêmes après administration d'éthyl-17α-nor-19-testostérone qu'après administration de testostérone.

Ces observations sont un indice que les réactions à ces substances ne sont pas les mêmes dans les divers organes effecteurs, et que les modifications provoquées dans ceux-ci ne sont pas spécifiques à un traitement particulier, mais spécifiques à l'organe.

L'auteur expose enfin les résultats des recherches sur l'incorporation in vitro du glycocolle 2-C¹⁴ dans les protéines et dans les acides nucléiques des vésicules séminales de souris normales et castrées, traitées ou non traitées à la testostérone.

Acknowledgements

Our thanks are due to the Netherlands Organisation for Pure Research (Z.W.O.) for financial support. We gratefully acknowledge the collaboration of Miss A. KOUWENHOVEN, chemical engineer, and of Miss Y. HULSEBOSCH.

References

1. MEJBAUM, W.: Zschr. physiol. Chem. (G.) **258**, 117 (1939). — 2. DISCHE, Z.: Mikrochemie (Austria) **8**, 4 (1930). — 3. SCHNEIDER, W. C.: J. Biol. Chem. (U.S.A.) **161**, 293 (1945). — 4. KOCHAKIAN, C. D., and D. G. HARRISON: Endocrinology (U.S.A.) **70**, 99 (1962). — 5. KASSENAAR, A. A. H., A. KOUWENHOVEN, and A. QUERIDO: Acta endocr. (Den.) **39**, 223 (1962). — 6. KASSENAAR, A. A. H., A. KOUWENHOVEN, and A. QUERIDO: Acta endocr. (Den.) in press.

Discussion

DREYFUS: In regenerating liver, after resection of a part of the liver, there is an increase in the concentration of the RNA-synthesising enzyme, RNA-polymerase. I wonder whether its activity has been measured in the seminal vesicles, or in the levator ani, after castration and after treatment with growth hormone, anabolic agents, or other steroids. This could perhaps give a clue as to the mechanism of action of these drugs.

KASSENAAR: Thank you very much for this suggestion. I am afraid we have not done any work along those lines and I am not familiar with other investigations in which the activity of RNA-synthesising enzymes has been studied in the seminal vesicles under the influence of these steroids. Dr. WILSON, maybe you have information about that ?

WILSON: I just want to say that this is a very important and very interesting piece of work. The only comment I have to make is that WILLIAMS-ASHMAN did report earlier this year[1] that in the rat prostate there was a considerable fall in the level of the RNA-synthesising enzymes in the nucleus following castration.

KASSENAAR: But I think that, in very many of its reactions on this level, the prostate is different from the seminal vesicle. Would you agree with that ?

WILSON: Yes, at least quantitatively.

TREMOLIÈRES: I would like to say that we have been working along similar lines, studying the effect of cortisone on the liver of adrenalectomised rats. Here, we found exactly the same thing. Cortisone has an anabolic effect on liver proteins. RNA increases after 3 days, and later protein synthesis continues. So it seems that the effects you described with so-called anabolic agents on the seminal vesicle are exactly the same as with cortisone on the liver. We have not observed any influence on the liver in response to testosterone. So it seems that there is a general process of protein synthesis at the cellular level and that the difference in action is due to the site at which each hormone is efficiently metabolised[2].

KASSENAAR: If you study these anabolic agents in the regenerating liver, the effect — though slight — is, I believe, significant.

TREMOLIÈRES: In regenerating liver, cortisone (5 mg./24 hrs) inhibits regeneration. While stimulating the synthesis of cytoplasmic RNA, cortisone seems at the same time to inhibit cellular multiplication[3].

[1] HANCOCK, R. L., R. F. ZEUS, M. SHAW, and H. G. WILLIAMS-ASHMAN: Biochim. biophysica acta (U.S.A.) 55, 257 (1962).

[2] TREMOLIÈRES, J., R. DERACHE, and R. LOWY: Path. biol. (Fr.) No. 21/22, 1729 (1958).

[3] FRAYSSINET, C.: Etude chez le rat de l'hypertrophie compensatrice du foie après ablation partielle. Mise en évidence d'une caractéristique du tissu hépatique selon le type de croissance. — Thèse de Sciences, Paris, April 1962.

It would therefore seem that cortisone and testosterone derivatives have the same type of action. Under appropriate conditions they can increase microsome formation and cytoplasmic RNA and induce protein synthesis in the cells of the target organ. From this point of view they are all anabolic. But synthesis of this type is achieved either at the expense of other protein (this is the case with cortisone, and the nitrogen balance may become negative if the protein intake is too low) or, alternatively, at the expense of fat, in which case the nitrogen balance becomes positive.

Hence, the terms "anabolic" and "catabolic" appear very unsatisfactory. In practice the most obvious effect of the so-called anabolic hormones is an improvement in appetite, which after all is the best way to achieve true anabolism.

OVERBEEK: You certainly appear to have found very early changes in the RNA after administration of testosterone; I might add that, according to BUTENANDT[1], C^{14}-testosterone rapidly accumulates in the seminal vesicle and disappears again even before your changes have occurred after 15 hours. These facts suggest some kind of trigger mechanism. On the other hand, I have already mentioned that after intravenous injection nandrolone-phenyl-propionate disappeared extremely quickly from the blood-stream, but I forgot to say that the intravenous injection had no effect whatsoever. I also pointed out that quite a long time of exposure is required in order to get a good effect. These findings are not in agreement, and I cannot see how they could be made to agree. Have you any suggestions?

KASSENAAR: I am afraid we haven't. Using C^{14}-labelled testosterone given by intravenous injection, instead of by intra-abdominal administration, we found some increase in the level in the seminal vesicles over and above the plasma concentration — which indicates accumulation of these steroids in the seminal vesicles; but, when people are talking about concentrations of hormones in the tissue, I always wonder whether we can really draw any definite conclusions about the action of the hormone in question. Presumably, it is the concentration in some minor part of this whole cell machinery which we have to measure if we are to find the real target. If we look at the data on testosterone distribution, we know that, 15—30 minutes after administration of C^{14}-testosterone, the concentration in the liver is much higher than in the seminal vesicles, and yet it is only in cases where the liver is in process of regeneration that testosterone has some demonstrable metabolic effects. In the normal liver we don't see any effect on protein synthesis or on nucleic acid synthesis or any shifts in nucleic acid composition. For this reason, I always encounter problems when trying to use data on the concentration of a hormone in a tissue as an explanation for its action.

SZIRMAI: I would like to make a general comment with reference to Dr. KASSENAAR's paper, particularly concerning the general growth-promoting effect of androgens. The terms "growth", "hyperplasia", and "hypertrophy" all denote very complex processes. Several aspects have to be considered, including the life-span of the cells. In some tissues, like that of the capon comb, the tremendous increase in tissue mass (largely due to water imbibition) is not accompanied by cell division, the relative number of cells per unit volume remaining constant[2]. One would hesitate to describe

[1] BUTENANDT, A., H. GÜNTHER, and F. TURBA: Hoppe-Seyler's Zschr. physiol. Chem. (G.) **322**, 28 (1960).

[2] SZIRMAI, J. A.: Anat. Rec. (U.S.A.) **105**, 337 (1949).

such a process as growth: apparently this tissue is structurally capable of
reacting to androgens without the need for cell multiplication. In contrast,
in the chick comb, mitoses have been observed in the course of androgen
treatment[1], and obviously in this case the number of cells has to increase
prior to the androgen response proper. In the case of epithelial modulations
it is often difficult to decide whether a given cell itself undergoes a change
into, let's say, the glandular type, or whether the latter can proceed only
from the undifferentiated germinal cell. What is the extent of the atrophy
which occurs following castration in an epithelial cell like that of the seminal
vesicle ? Do the cells degenerate completely, and is there nothing to react
when the androgen is given — a situation requiring mitosis as the first but
not necessarily primary effect — or are the cells of the castrate themselves
capable of resuming their secretory function again ? This might differ from
tissue to tissue, but would not necessarily imply an essentially distinct
mode of action for a given hormone.

DE JONGH: An increase in the content of DNA indicates, as far as I
know, an increase in the number of nuclei. Is that right ?

KASSENAAR: We assume that it is, yes.

DE JONGH: But you conclude from this that an increase occurs in the number of cells, which is not quite the same. In general that may be true, but in
the case of striated muscle it is rather dangerous to draw the same conclusion,
because the muscular elements may have very many nuclei. Do you think
this may mean that the difference between your findings in the muscle and
those in the kidney are perhaps less fundamental than you have suggested ?

KASSENAAR: We don't know about the cell nuclei. It is very hard to
homogenise the seminal vesicles completely in order to do cell fractionation.
The cell nuclei of the seminal vesicles are terribly difficult to get at. The only
thing we have done is to determine in the microsomal fraction and in the
$105,000 \times G$ supernatant the RNA and protein content, both in the seminal
vesicles of castrated controls as well as in those of animals treated with
250 mcg. testosterone per day for 48 hours, in which a very considerable
increase in weight had occurred. Both in the microsomal fraction and in the
supernatant fraction the RNA over protein concentration is increased
considerably. So we think that in the supernatant as well as in the microsomal fraction there is an increase in RNA. We have no data on the nuclei,
for the reasons I have just mentioned.

KORNER: Dr. TREMOLIÈRES, you mentioned that one can regard the
adrenocortical steroids as being anabolic in the rat because of the changes
in the liver. Is this necessarily so ? We have results indicating that corticosteroids which have a catabolic effect on muscle cause the liver to synthesise
more protein. Nevertheless, the direct effect of the steroid on the liver might
well be catabolic, but — because the liver is flooded with amino acids from
the break-down of muscle protein — it is forced to synthesise more protein
despite the direct catabolic action exerted on it by the corticosteroids.

TREMOLIÈRES: As to the first question, FRAYSSINET[2] has determined the
RNA in the nucleus and in the cytoplasm. Cortisone increases cytoplasmic
but not nucleic RNA. — With regard to the second question, I think we are
faced with a circle here, and we don't know at which point it starts; but it

---

[1] LUDWIG, A. W., and N. F. BOAS: Endocrinology (U.S.A.) **46**, 291 (1950).
[2] FRAYSSINET, C.: loc. cit.

appeared that maximum nitrogen excretion generally occurs on the third day and that increases in RNA start sooner, i.e. as if protein synthesis in the liver were not due to the catabolism of lymphoid or muscle proteins.

We have advanced the hypothesis that hydrogenation of the A ring by TPNH was the starting point from which ribose phosphate and nucleotide synthesis are orientated [1, 2].

DIRSCHERL: I should like to say something to Dr. KASSENAAR regarding the labelled testosterone administered: 15 minutes after the administration of the testosterone, the liver shows a high level of radioactivity, but it no longer contains any testosterone, because the latter is rapidly metabolised into unknown compounds.

ARIAS: Have you or others had an opportunity to examine by radio-autography the incorporation of tritiated thymidine into kidney cells following testosterone administration ?

KASSENAAR: Of course, the reason why you are asking this is obvious: it looks very strange that the kidney doesn't respond with cell multiplication. We have not done any studies as suggested by you, although we agree that they might give very valuable information. I know there are studies — and we confirmed them — showing that the kidney can react with cell multiplication, but this is only true in very young animals. Dr. KENNEDY [3] from Cambridge has some very nice data on the effect of the animal's age on the capacity of the kidney for cell multiplication, but our findings are based on results in older animals. When subjected to histological examination following treatment with testosterone, the cells of mouse kidney look larger, so we think that the data presented here are in line with the morphological observations which have been made.

QUERIDO: Before closing the discussion I should like to comment on the remark made by Dr. DIRSCHERL on the radioactivity in the liver after injection of C^{14}-testosterone. The point I want to make is that in our discussions yesterday and today one important aspect has not yet been mentioned, and that is that the tissues might be differently equipped for metabolising an agent which we administer. We are generally inclined to think that it is the administered agent itself which acts, but it might be a metabolite. In this connection, I would draw your attention to the work recently published by HARDING [4], in which it was shown that in the seminal vesicles the C^{14}-testosterone is chiefly changed into androstenedione, whereas in the muscles such transformation is much less marked. I am not inferring that androstenedione is the active agent, but I do suggest that you may have cells which are equipped for making the active compound, whereas other cells are not so equipped. This could fit in with Dr. SZIRMAI's concept of graded responses of tissues to agents in the body.

[1] TREMOLIÈRES, J., R. DERACHE, and R. LOWY: Path. biol. (Fr.) No. 21/22, 1729 (1958).
[2] LOWY, R.: Effets de la cortisone sur les phosphorylations oxydatives. Thèse de Sciences, Paris, 1957.
[3] KENNEDY, G. C.: Ciba Foundation Colloquia on Ageing 4, 250 (1958).
[4] HARDING, B. W., and L. T. SAMUELS: Endocrinology (U.S.A.) 70, 109 (1962).

Effect of steroids on bone formation

By

K. KOWALEWSKI

It is now generally accepted that steroid hormones affect mesenchymal tissue and influence growth and development of cartilage and bone. It is also apparent that bone formation occurs even in the absence of these hormones, and one is inclined to consider steroids as merely accelerators or inhibitors of the spontaneous processes.

I. Spontaneous osteogenesis and osteolysis

Bone, a "growing tissue" *par excellence,* is one of the most plastic tissues in the body, with a large metabolic turnover and with ceaseless chemical and physiological activity throughout life. As it takes an active part in many metabolic processes, bone is constantly being destroyed and renewed. This process of remodelling is very sensitive to alterations in mechanical function, nutritional state, and hormonal balance.

Bone develops by a transformation of embryonic or adult connective tissue into a calcified connective tissue. Calcification of cartilage models of the bones occurs in embryonic life, and the growth in length of these bones continues after birth by a process of deposition of minerals in cartilage. Replacement of cartilage by bone is called endochondral ossification, in contrast to intramembranous ossification, which occurs directly by condensation and differentiation of mesenchymal tissue independently of cartilage formation. Bone cells play an essential role in the synthesis of matrix elements, particularly in the process of fibrogenesis (*16, 33*).

The matrix is formed of fibres set in an amorphous gel of ground substance. The fibres consist of a crystalline protein, collagen, rich in proline, hydroxyproline, and glycine. Ground substance contains mucopolysaccharides, chiefly hyaluronic acid, chondroitin sulphate, and glycoproteins, in varying degrees of polymerisation (*51*). Collagen represents 95% of the organic matrix of the bone and 25—35% of the total body protein (*27*).

Extracellular bone minerals represent about 65—70% of the dry weight of bone. Crystals are deposited in the matrix and appear to be oriented parallel to the axis of the collagen fibres. The main components of crystals are calcium, phosphate, carbonate, hydroxyl, and citrate with small amounts of other ions. The ultimate formation of bone, with its pattern of collagen and elastic fibres in relation to salt crystals and the appearance of the Haversian systems, is related to the development of the vascular pattern (26). Bone actively participates in body homoeostatic processes. In addition to the physical process of ionic exchange, the turnover of bone minerals is favoured by the constant remodelling of bone, involving a dissolution and a recrystallisation of bone salts.

Being a very active part of the animal's connective tissue, bone is affected by the hormones which influence the formation and metabolism of connective tissue.

Previous studies of the action of steroids on bone have been concerned with such problems as size, shape, weight, strength, chemical composition, and metabolism of various components of bone tissue. An adequate review of these studies has been published by other authors (58).

In the present report some effects of steroid hormones on various bone components will be reviewed in relation to bone formation. Morphological changes will then be discussed. Certain effects of steroids on bone will also be illustrated by a few experiments.

II. Effects of steroids on various bone components in relation to bone formation

1. Steroids and matrix. The organic part of bone, composed of cells, collagen fibres, and ground substance, is the active growing element. Proteins liberated from fibroblasts to the extracellular space are polymerised to form collagen fibres. Ground substance, rich in colloids, diffusible electrolytes, and water, may be considered as an ionic exchange resin reacting with diffusible blood electrolytes. The electrochemical properties of this system may be affected by those steroid hormones which influence the distribution of water and electrolytes (23). Hormones act on both fibrogenesis and the formation of ground substance (4), and bone matrix is particularly affected by steroids regulating protein and electrolyte metabolism.

Steroid hormones are known to influence the synthesis and break-down of proteins (45). Despite the fact that the terms "anabolic", "anti-anabolic", and "catabolic" are not clearly defined and may be misleading (57), it is generally considered that cortisone-like C-21 steroids are anti-anabolic and that the sex steroids are anabolic (18, 45, 46, 56).

(A) *Corticoids*. Corticoids suppress protein synthesis from amino acids, increase the rate of protein break-down, and produce a negative nitrogen balance (63). Corticoids prevent the formation of normal bone matrix and thus may shift the balance between growth and resorption of bone in favour of the latter. Connective-tissue cells, the synthesis and metabolism of mucopolysaccharides of ground substance, and collagen fibrinogenesis are affected by corticoids (4, 60).

Ground-substance mucopolysaccharides, detectable in cartilage and in the growing Haversian system, play an important role in chondrogenesis, osteogenesis, and repair. The sulphate exchange reaction in these mucopolysaccharides may be measured by the uptake of radiosulphur (20), which, when administered as inorganic sulphate, is utilised by the cells, particularly by the chondrocytes, in the synthesis of chondroitin sulphate (5, 10, 13).

Cortisone inhibits sulphate fixation and the formation of the chondroitin sulphate of connective tissue. It also blocks the synthesis of sulphated mucopolysaccharides *in vitro* and *in vivo* in embryonic and normal bone tissue (4, 15, 44). This steroid also inhibits the uptake of S^{35} in young growing bones and in regrowing injured adult bones (38, 39, 40). The mechanism of this action, however, is not clear. Decreased synthesis of radiosulphate compounds, or increased degradation of this material, or both processes together, may be responsible (17).

Hexosamine, a constituent of the mucopolysaccharides of ground substance, is also affected by hormones. Hexosamine content of bone may be reduced by cortisone treatment (4).

As bone grows, collagen is laid down at a greater rate than hexosamine. The observation that the ratio of hexosamine to collagen diminishes as bone grows is part of a general phenomenon of connective-tissue metabolism. This has been made the basis of certain theories of ageing (59, 62).

Formation and growth of bone crystal begin in apposition to collagen fibres of matrix. Synthesis and metabolism of collagen are, therefore, an essential part of the process of bone formation. As mentioned before, collagenous fibres, derived from fibroblasts, are rich in glycine, proline, and hydroxyproline. Hydroxyproline is found primarily in collagen, and the determination of hydroxyproline is considered an accurate measure of collagen (29). Deficient collagen synthesis or synthesis of an abnormal collagen may affect calcification, crystal aggregation, and growth, but many questions concerning the mechanism of this process are still unanswered (*32*).

Cortisone is known to inhibit the growth and activity of fibroblasts, the mother cells of collagen, both *in vitro* and *in vivo* (*4, 19, 60*). Cortisone may be responsible for the abnormal synthesis of hydroxyproline (*53*) and the defective formation of collagen fibres (*4*).

Disaggregation of connective-tissue matrix by anti-anabolic steroids represents a general effect of these hormones on matrix colloids. This effect is associated with changes in tissue electrolytes, particularly potassium and sodium, and with abnormal hydration of matrix (*35*).

To sum up, the steroids considered as anti-anabolic inhibit cellular elements of bone, synthesis of bone proteins, and formation and metabolism of bone matrix.

(B) *Anabolic steroids*. Anabolic androgens cause a retention of such elements essential for bone growth as nitrogen, potassium, phosphorus, and water (*45, 63*). They have also been found to promote synthesis of tissue nucleic acids (*37*). Progressive changes in the cells and matrix of ageing cartilage and bone, resulting in the loss of nitrogen and some tissue electrolytes (*30*), may be interpreted as being due to the reduced activity of anabolic sex steroids. Growth of matrix may be stimulated by anabolic androgens (*61*), and this was recently confirmed by bone biopsy studies (*34*). The effect of anabolic steroids on protein and electrolyte metabolism is generally opposite to the action of cortisone. Androgens antagonise, for example, the action of cortisone on sulphated mucopolysaccharides. The effect of anabolic androgens on S^{35} uptake simulates the action of growth hormone, a chondrotrophic hormone which has been known for a long time, and probably increases the rate of synthesis of radiosulphate compounds (*17*).

Anabolic androgens were found to stimulate the uptake of S^{35} by growing or healing bones *in vivo*, and to counteract the inhibitory action of cortisone on this uptake (*38, 39, 40*).

The action of steroids on various elements of bone matrix, discussed above, illustrates the importance of so-called anti-anabolic and anabolic hormones as regards bone formation.

It is apparent that faulty formation of bone matrix may result in abnormal deposition of minerals. Clinically, the amount of mineral deposited in the skeleton is of definite importance. Pathological signs of clinical and iatrogenic endocrine conditions associated with too much or not enough of some steroid hormones include bone demineralisation and a reduction in the amount of bone in the skeleton — a condition known as osteoporosis. Reduction of bone mass in osteoporosis may be due to impairment of new bone formation or to increased destruction. The protein-matrix hypothesis states that osteoporosis is due to the reduction of matrix formation. This hypothesis is based largely on the association between bone changes and endocrine disorders characterised by a negative nitrogen balance. It has been postulated that the reduction of matrix makes bones unable to contain or retain normal amounts of calcium (1). This hypothesis, however, has some weak points, because negative calcium balance is not always associated with negative nitrogen balance (50). Negative calcium balance affecting bone mineral can also be produced by low intake of calcium, or by malabsorption or excessive loss of this mineral. The modern view (50) distinguishes between abnormality of the serum chemistry (osteomalacia) and that which is due to negative calcium balance (osteoporosis).

2. **Action of steroid hormones on bone minerals.** Bone is actively involved in calcium homoeostasis. The major fractions of bone calcium — ionic, crystalline, and protein-bound — are normally in equilibrium, and are integrated with the organic components to form a colloidal aggregate. The process of skeletal calcium fixation in bone is a combination of ion exchange and crystallisation. In addition, there is the continuous process of recrystallisation. Bone calcium is removed to non-exchangeable positions at a very rapid rate (3). Any process altering one of the calcium fractions may change the chemical structure and morphology of bone.

An example of the chemical binding of calcium is the high affinity between calcium and chondroitin sulphate (43). The above-mentioned effects of cortisone on the synthesis of chondroitin sulphate in bones may indirectly produce the changes in bone calcium.

Changes in the colloidal state of ground substance, in the solubility of crystals, and in the milieu bathing the crystals affect the dissolution and redistribution of calcium. Altering the pH of the milieu which bathes the crystals by various metabolites (23), particularly by citrate and lactate, directly influences bone solubility (24), and the importance of oxidative enzymes in this process is apparent. Both C-21 steroids and sex hormones may influence citrate metabolism in animals. It remains, however, to be determined to what extent the enzymes of bone concerned in citrate formation or utilisation may be influenced by steroids (49, 54).

(A) *Corticoids.* Negative calcium balance appears to be the rule in animals and humans treated with cortisone (*47*). Part of the action of this steroid on bone calcium may be explained by its action on the kidney. Cortisone is known to reduce the tubular reabsorption of calcium (*42*). Increased calciuria is associated with abnormal loss of calcium by the intestine. Excessive biliary excretion and poor intestinal absorption of calcium, as confirmed also by *in vitro* studies with radiocalcium, occur after cortisone treatment (*28*). The uptake of radiocalcium and radiophosphorus by bones is increased after adrenalectomy (*54*), and this may be interpreted as being due to the elimination of the anti-anabolic action of adrenal steroids.

(B) *Anabolic steroids.* As may be expected, anabolic steroids counteract the action of cortisone on calcium and phosphorus balance (*11*), and they are used clinically to prevent or improve various forms of osteoporosis. Androgens such as Dianabol were extensively studied by balance and isotope methods and found to produce calcium retention in the bones (*2, 46*) and to counteract the action of cortisone on the bone (*56*). Oestrogens are also considered to be capable of promoting a positive calcium balance (*28*). Oestrogens may furthermore increase Ca^{45} uptake by bones (*52*).

The effect of steroid hormones on calcium metabolism is particularly evident in birds. Hypercalcaemia occurs at each ovulation period, and this may be duplicated by administration of oestrogen. The non-ultrafiltrable vitellin-bound calcium used for shell formation originates in bone, as shown by the changes in the amount of medullary bone formation occurring during ovulation or following oestrogen. The importance of oestrogens in maintaining calcium stores has been confirmed by balance studies. It has also been found that birds with cortisone-induced osteoporosis responded to oestrogen with phosphoproteinaemia, phospholipidaemia, and increase in the non-ultrafiltrable fraction of the serum calcium, and that oestrogen given with androgen or androgen alone afforded protection against osteoporosis (*28, 67*).

Briefly, then, osteoporosis may be produced by cortisone-like steroids, and some androgenic anabolisers may prevent or improve it. Also, osteoporosis due to sex hormone or nutritional deficiency may be improved by oestrogen and/or androgen treatment. The extensive literature on osteoporosis points out the great difficulties in finding uniform criteria of hormonal action (*16, 31, 50*).

Both experimental and clinical studies are rich in observations, but there is still no agreement about the mechanism of steroid action, nor is the term "osteoporosis" clearly defined.

III. Steroid hormones and bone morphology
1. Corticoids

Inhibition of cellular growth, defective matrix formation, and loss of bone minerals interfere with bone growth. Linear growth is inhibited by deficient proliferation of the epiphyseal cartilage and by reduced synthesis of cartilaginous matrix (*58*).

In vitro studies have demonstrated that corticoids inhibit the growth of explanted chick-embryo bone anlages, an effect already evident 24 hours after explantation. Hormone withdrawal is followed by recovery of active growth (*12*).

Hypercorticoidism of both endogenous and iatrogenic origin is associated with diminished osteoblastic activity and the thinning of bone trabeculae (*7*).

Rapid resorption of bone is observed in cortisone-treated rodents and is associated with increased vascularity and the promotion of osteoclastic activity. Spaces left by the removal of bony structures are filled at first with dilated blood vessels, later with bone marrow. Absence of a fibrous-tissue replacement of bone is consistent with the anti-anabolic action of cortisone on connective tissue. Intermittent administration of cortisone in the rabbit is associated with cycles of resorption and deposition of new bone in both growing and mature animals. Cartilaginous growth at the epiphysis of the young rabbit is also inhibited (*64*).

Histological changes in cortisone-treated animals, such as retardation of epiphyseal cartilage growth, diminished growth of bone in length, abnormal resorption of calcified matrix in metaphyseal spongiosa, increased osteolysis, and decreased osteoblastic activity have been observed (*36*). The results depend upon the dosage of steroid, endocrine balance in the animal, nutritional state, and species.

Retardation of longitudinal growth of bones in birds, rodents, and children treated with cortisone has been reported (*36, 67*). As the growth of bone in length results from endochondral ossification in relation to epiphyseal cartilage proliferation, the inhibitory action of cortisone on endochondral ossification (*14, 67, 68*) may well explain the retardation of bone elongation.

Associated with the post-cortisone delay in bone growth and repair is a reduced alkaline-phosphatase activity of bone (*66*).

In domestic birds, post-cortisone skeletal lesions consist of resorption of bone from the endosteal surface of the cortex, without formation of osteoclasts. Inhibited endochondral ossification is also found in birds. In general, birds develop post-cortisone bone lesions more easily than mammals (*67*).

2. Sex steroids

(A) *Androgens*. Sexual dimorphism of the skeleton is genetically determined, but sex hormones interact with genetic factors in the development of bones. Sex hormones not only influence the rate of bone formation but, by affecting epiphyseal union, they determine the duration of growth. Androgens influence bone growth particularly by hastening skeletal maturation and closure of epiphyses. They probably act also by inhibiting pituitary growth hormone (*63*). In prepuberal castration, retardation of epiphyseal closure and prolongation of the growth period permit some bones to overgrow in length, with a resulting disproportion in the skeleton. Susceptibility of bone to androgens depends upon the physiological age of the skeleton, the maximum response occurring shortly before puberty (*63*). Acceleration of bone growth was observed in children treated with Dianabol (*55*). Dianabol was also found, experimentally, to promote bone repair, vascularity and growth of cartilage, osteoblastic activity, and ossification (*14*).

(B) *Ovarian steroids*. A primary or experimental deficiency of oestrogen delays endochondral and intramembranous ossification of growing bones. Retardation in the appearance of centres of ossification and in the closure of the epiphyseal growth zones is also observed. Oestrogen accelerates bone development by hastening epiphyseal closure and the premature appearance of centres of ossification. Hyalinisation or calcification of the matrix is intensified and bone density increased. Acceleration of skeletal development may result in inhibition of linear growth. Stimulation of osteoblasts, increased hyalinisation of the ground substance, increased osteoclasis and osteolysis followed by excessive medullary bone formation, oestrogen-induced injury of the reticulum of bone marrow, and inhibition of bone resorption have been proposed as explanations for the effect of oestrogens on bone growth (*58, 65*). Local intramedullary application of oestrogens to bone induced local hyperostosis and, in distal metaphyses, endosteal proliferation (*65*).

IV. Summary of the review

Only certain aspects of the effects of steroids on bone formation have been discussed in this review, but it may be concluded that some steroid hormones affect bone formation. It is, however, necessary to remember that the anabolic or anti-anabolic processes in an organism are the result of the simultaneous interaction of various hormonal systems. The comparison of hormones only on the basis of their relative action on protein and mineral metabolism cannot be considered satisfactory, either from an experimental or from a clinical point of view. Steroids act on various body receptors, and the relative intensity of these effects may depend not only on sex and age, but also on endocrine balance.

Steroids considered anti-anabolic inhibit the growth of cellular elements of bone, inhibit or delay the synthesis of various elements of bone matrix, and produce a negative calcium balance.

Anabolic sex steroids promote the formation of bone matrix and produce a positive calcium balance.

Experimental information on the action of steroids on bone formation should be used, for clinical purposes, with great caution. It is apparent that many anabolic androgens are useful in the treatment of osteoporosis, irrespective of aetiology, and may counteract the destructive action of corticoids on bone.

Dianabol has been extensively investigated in recent years, and the action of this steroid on bone seems evident (18, 55, 56), but there are some difficulties in establishing a pattern of clinical treatment. Particularly delicate seems the problem of the use of this steroid in orthopaedic surgery and in paediatrics (8, 22, 46, 57). Careful attention should be paid to the possible dangers of promoting the linear growth of bone in children. Too much hormonal stimulation or too prolonged a treatment may result in premature closure of cartilage and may alter the final prognosis. The possible side effects of sex steroids on a young growing organism should also be considered.

Table 1. *Effect of cortisone (C), Dianabol (D), and cortisone plus Dianabol 3 weeks of treatment, compared with normal (N) controls. Determinations made mEq./kg. of bone.*

Components	Newborn	Stock birds	1st week			
			N	C	D	C + D
Number of birds .	30	24	20	20	20	20
Body-weight g. .	40	82	171	133	173	160
Femur wet/dry weight	1.40	1.96	2.30	1.60	2.21	1.93
Total N mg.. . .	5	12	31	7[1]	25	23
Total N mg./g. .	10	64	79	18[1]	63	62
Ca mg.	4	19	41	40	44	41
Ca mg./g. . . .	82	122	112	124	110	111
P mg.	5	9	24	16[1]	23	23
P mg./g.	54	56	64	51	58	62
Na mEq./kg. . .	415	279	273	238	309	301
K mEq./kg. . .	126	128	115	78[1]	102	101

[1] P less than 1% — cortisone-treated vs (N) and (D) groups, same period

V. Experimental

Some effects of steroid hormones on bone formation are illustrated in the following experiment. Young Sprague-Dawley female rats and young cockerels (*39*) were used. Rats were fed a special grain diet (*14*) and birds received a standard chicken starter. As the status of body-protein stores may influence the response of tissues to hormonal treatment (*45*), one group of rats was fed, prior to the administration of steroids, a protein-depletion diet (U.S.P. XV, Nutritional Biochemicals Co., U.S.A.). Two representatives of anti-anabolic and anabolic steroids, cortisone (Cortone) and methandrostenolone (Dianabol), were selected for this study. Young rats or birds of under 50 g. body-weight were injected with 1 mg. cortisone and/or 0.5 mg. Dianabol, four times per week. Rats and birds weighing more than 50 g. received 5 mg. cortisone and/or 1 mg. Dianabol, also four times per week. The animals from the large group, called "stock animals", were divided into various experimental groups (Table 1) and treated from one to three weeks. The untreated animals were considered as controls. The bone matrix and bone minerals of humeri (rats) and femora (birds) were analysed biochemically for hydroxyproline (*41, 48*), hexosamine (*9*), total nitrogen, calcium (*6*), phosphorus (*25*), sodium, and potassium (*21*). Histological studies were also performed. The results of this experiment are summarised in the accompanying tables and figures.

(C + D) on various components of femur in cockerels studied after 1, 2, and in fat-free dry bones. Results expressed in mg./bone, in mg./g. of bone, and in Mean values

	2nd week				3rd week		
N	C	D	C + D	N	C	D	C + D
20	20	20	20	20	20	12	18
268	219	316	248	416	340	416	375
2.10	1.26	1.82	1.88	1.94	1.82	1.79	1.62
45	22[1]	50	36	87	34[1]	59	47
61	45[1]	63	60	59	44[1]	55	49
101	59[1]	103	71	199	93[1]	126	122
134	121	129	118	137	119	117	125
47	30[1]	56	39	108	46[1]	66	68
62	61	72	64	74	59	62	70
283	246	249	306	263	225	291	282
94	63[1]	79	86	69	63	82	79

of treatment.

Results of experiment

(A) *Biochemistry*. Cortisone treatment inhibited body growth, altered the ratio of femur-weight to body-weight, and affected the water content of bone (Table 1). The anabolic androgen used in

Table 2. *Effect of cortisone (C) and/or Dianabol (D) and cortisone plus Dianabol (C + D) on hexosamine and hydroxyproline of humeri in rats and femora in cockerels, compared with normal controls (N). Treatment lasted 10 and 20 days. Determinations made in fat-free dry bones. Results expressed in mg./bone and in mg./100 g. of bone. Mean values*

Components	Stock ani- mals	After 10 days				After 20 days			
		N	C	D	C + D	N	C	D	C + D
Number of rats .	12	12	20	20	20	10	10	19	13
Body-weight g. .	49	93	62	85	78	121	63	91	73
Humerus wet/dry weight	2.45	2.32	2.33	2.39	2.36	2.30	2.33	2.35	2.42
Hexosamine mg..	0.06	0.12	0.05[1]	0.08	0.09	0.14	0.05[1]	0.10	0.08
Hexosamine mg./100 g. . .	140	129	76[1]	109	116	126	73[1]	118	104
Hydroxyproline mg.	0.58	0.81	0.54[1]	0.76	1.12	1.22	0.58[1]	0.97	1.20
Hydroxyproline mg./100 g. . .	1303	1267	874[1]	1085	1524	1340	852[1]	1042	1520
Number of cockerels . . .	20	24	24	—	—	26	24	—	—
Body-weight g. .	87	201	147	—	—	383	236	—	—
Femur wet/dry weight	2.37	2.41	2.43	—	—	2.27	2.23	—	—
Hexosamine mg..	0.12	0.44	0.15[1]	—	—	0.93	0.26[1]	—	—
Hexosamine mg./100 g. . .	87	95	51[1]	—	—	98	55[1]	—	—
Hydroxyproline mg.	1.20	3.49	2.63	—	—	7.25	5.55	—	—
Hydroxyproline mg./100 g. . .	811	757	940	—	—	718	1123	—	—

[1] P less than 1% — cortisone-treated vs controls, same period of treatment.

this study prevented the effects of cortisone on weight of body and of bone. The various elements of the matrix, and also the minerals of the bone, were significantly decreased in cortisone-treated birds (Tables 1 and 2). In nearly all instances, Dianabol prevented the effects of cortisone on the bone elements studied (Tables 1, 2, and 3). In the rats fed with a protein-depletion diet prior to steroid treatment, cortisone did not prevent the recovery of body-weight

and did not significantly alter bone hexosamine and hydroxyproline, as compared with other groups of rats (Table 3). It should
be noted, however, that both the hexosamine and the hydroxyproline of control rats, previously fed a protein-depletion diet,
were very low. It is apparent that some changes in bone matrix

Table 3. *Hexosamine and hydroxyproline in humeri of rats after cortisone (C)
and Dianabol (D) treatment, compared with normal controls (N). Rats were
fed with normal stock (n. s.) diet or protein-depletion (p. d.) diet, prior to
injection of steroids. Determinations made in fat-free dry bones. Results expressed in mg./humerus and in mg./100 g. of bone. Mean values*

Components	Normal-diet group (n.s.)			Depletion-diet group (p.d.)		
	N	C	D	N	C	D
Number of rats	10	10	10	24	21	29
Body-weight (b.w.) in g. stock rats	131.4	131.0	131.2	131.5	131.4	131.5
B.w. in g. 8 days after n.s. or p.d. diet	181.1	189.8	188.7	109.3	108.3	108.1
Change in b.w. in 8 days, in %	+42	+44	+43	—16	—17	—17
B.w. in g. after 2 weeks of steroid treatment on n.s. diet (repletion period). . .	219.0	181.0	220.0	188.0	192.0	214.0
Change in b.w. in 2 weeks, in %	+16	— 4	+16	+71	+77	+97
Humerus wet/dry weight . .	1.91	2.00	1.93	2.05	1.94	1.98
Hexosamine mg. humerus . .	0.26	0.11[1]	0.18	0.12	0.09	0.10
mg./100 g.	122	69[1]	95	90	63	60
Hydroxyproline mg. humerus	2.74	1.67[1]	2.10	1.53	1.38	1.49
mg./100 g.	1267	950[1]	1102	1002	858	1007

[1] P less than 1% — (C) vs (N) and (D), in n.s. group.

appearing in animals fed a normal diet and treated with steroid
hormones are absent when hormones are given to animals with
depleted protein stores. The rats which were fed only a normal diet
reacted to cortisone by a reduction in body-weight and by a
significant decrease in bone hexosamine and hydroxyproline
(Tables 2 and 3). It seems, therefore, that the state of the body-
protein stores may be an important factor influencing the reaction
of tissues to anti-anabolic or anabolic steroids. In view of the previous detailed discussion on the biochemical changes of bones in
steroid-treated animals, no further comment will follow the descrip-

tion of our experiment. A study of the tables shows that our results generally agree with previous findings quoted in the review.

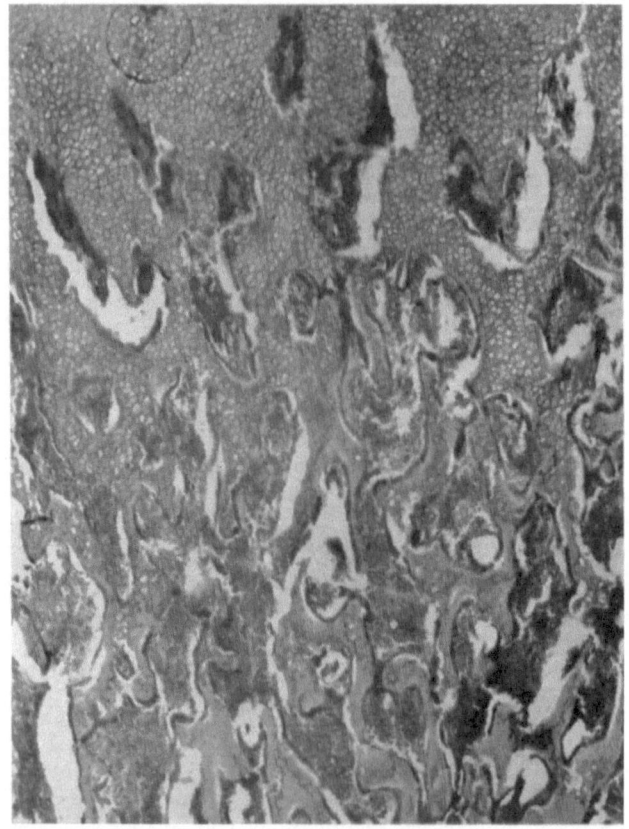

Fig. 1. Femoral epiphysis of normal cockerel, showing irregular zone of transition from hypertrophic cartilage (above) to trabeculae of primary spongiosa (below). Haematoxylin and eosin stain × 50

(B) *Histological findings.* Histological studies were done on birds and rats at the ages of 1, 2, 3, and 4 weeks. Treatment was started when the animals were 1 week old and lasted from 1 to 3 weeks. In addition to routine haematoxylin and eosin stain, alcian blue stain for acid mucopolysaccharides and GOMORI's method for alkaline phosphatase were used.

Cockerels. Histological differences between the control and the treated groups were first observed after 3 weeks. At this time, the epiphyseal end of the femur of the control animals revealed, with

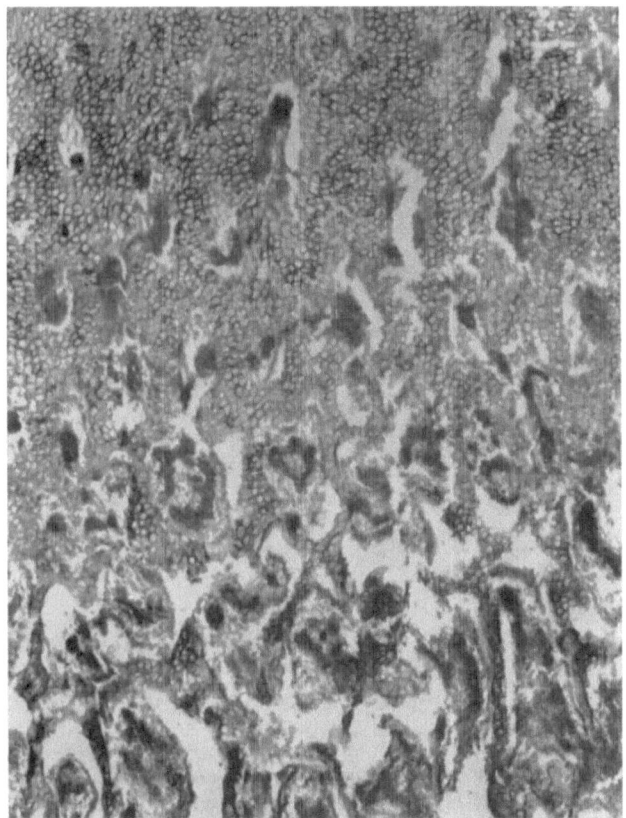

Fig. 2. Femoral epiphysis of cortisone-treated cockerel, showing formation of attenuated trabeculae (below). Haematoxylin and eosin stain × 50

the routine stain, a wide band of eosinophilic staining, proliferating cartilage and a thinner band of more basophilic, hypertrophic cartilage with larger lacunae. The junction between the hyper-trophic cartilage cells and the primary spongiosa formed an inter-digitating pattern (Fig. 1). The trabeculae in the primary spongiosa varied in thickness from 3 to 10 cartilage cells. The outer surface of these trabeculae was lined by one to several layers of spindle-shaped to cuboidal osteoblasts with vesicular nuclei and darkly basophilic cytoplasm. The intervening spaces were filled with the usual avian type of nucleated red cells.

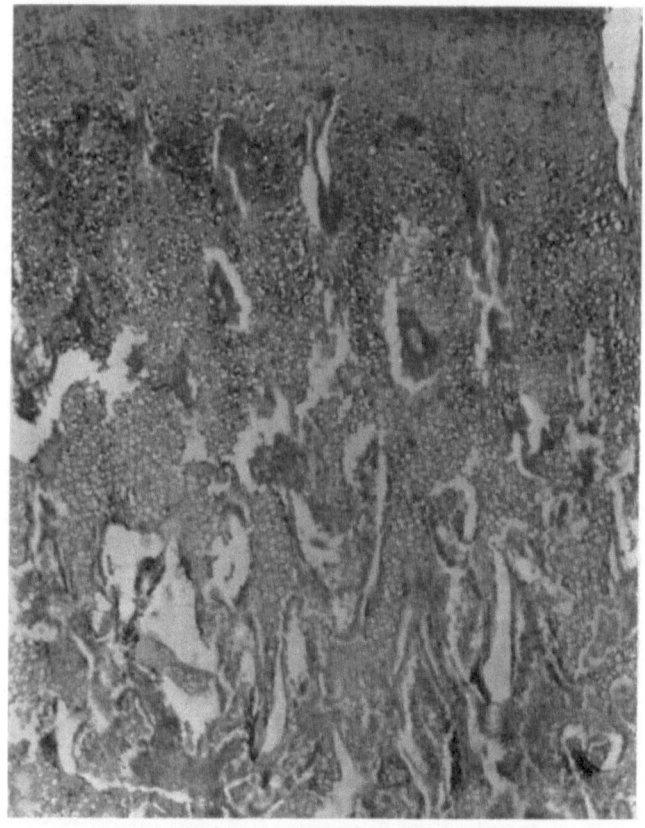

Fig. 3. Femoral epiphysis of normal cockerel, showing greatest concentration of acid muco-polysaccharides in hypertrophic zone of cartilage (upper part of picture). Alcian blue stain
× 50

In the cortisone-treated cockerels of the same age the widths of the various layers were not appreciably altered. However, the maturest part of the hypertrophic cartilage showed enlargement of cartilage cell lacunae (Fig. 2). The trabeculae of the primary spongiosa were thinner, being not more than 3 to 5 cartilage cells in width. The distribution of osteoblasts was more variable than in the control group, there being patchy areas of almost complete absence.

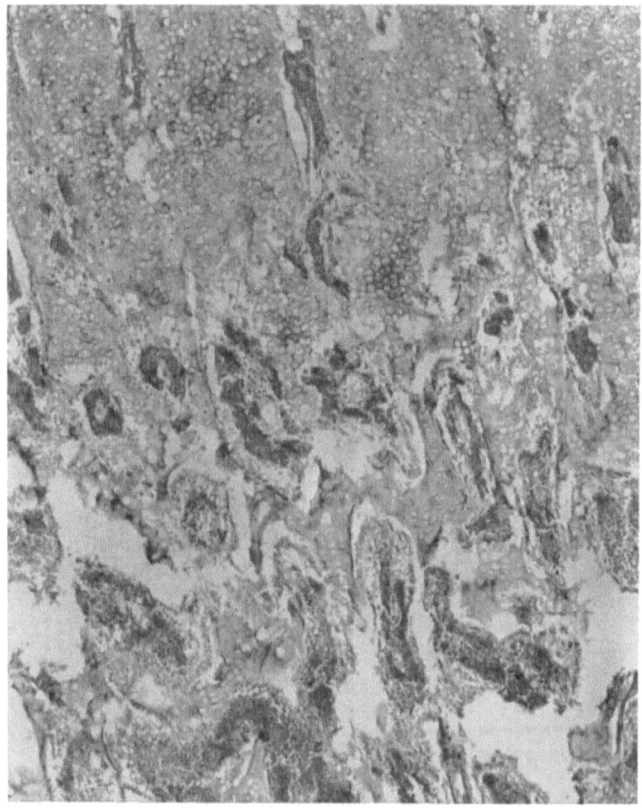

Fig. 4. Femoral epiphysis of cortisone-treated cockerel, showing overall decrease in staining for acid mucopolysaccharides in hypertrophic cartilage (above) and primary trabeculae (below). Alcian blue stain × 50

Alcian blue stain for mucopolysaccharides was most intense along the margins of the chondrocytic lacunae (Fig. 3). The staining was greatest in the hypertrophic zone, but gradually disappeared from the trabeculae of the primary and secondary spongiosa concomitantly with replacement of cartilage cells by osteoid. In the cortisone-treated group, a decrease in alcian blue staining was noted largely in the hypertrophic cartilage, partly due to enlargement of the component lacunae (Fig. 4). Staining of cartilage in trabeculae of the primary spongiosa showed a similar alteration.

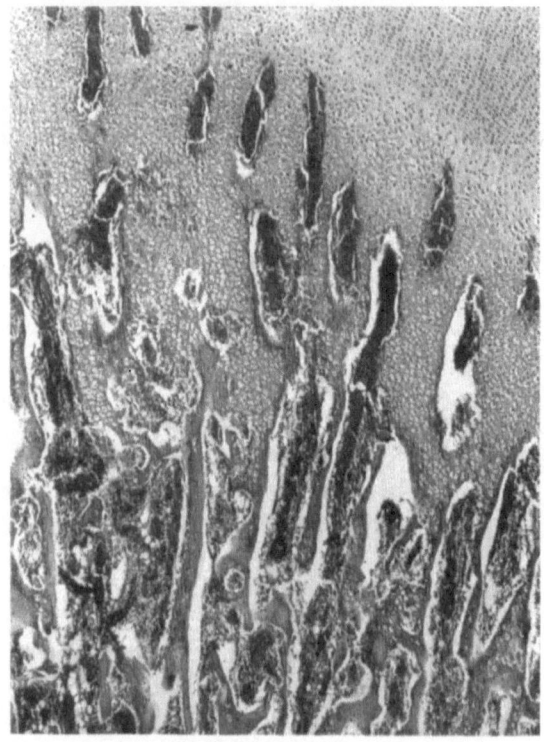

Fig. 5. Femoral epiphysis of cockerel treated with Dianabol and cortisone, showing normal pattern (see Fig. 1). Haematoxylin and eosin stain × 50

Alkaline phosphatase staining in the control animals resulted in black pigmentation of osteoblasts lining the trabeculae of the primary spongiosa. In the cortisone-treated cockerels, pigmentation was seen only in occasional areas where there was persistence of osteoblasts. No differences from the controls were noted with any stain in the case of the birds treated with Dianabol, or with Dianabol and cortisone (Fig. 5).

Rats. At 2 and 3 weeks the normal rat bone differs considerably from that of the normal cockerel of similar ages (Fig. 6): a) The rat develops a secondary area of ossification within the epiphysis

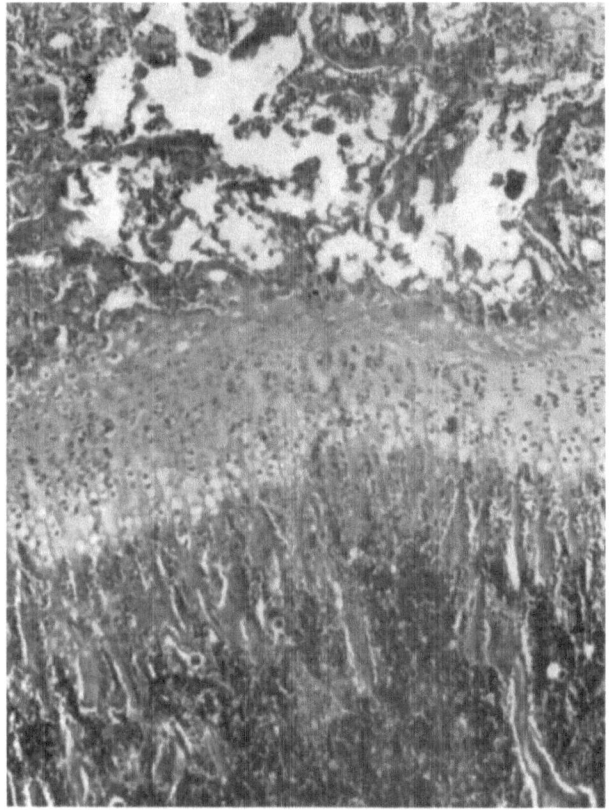

Fig. 6. Femoral epiphysis of normal young rat, showing (from above downward) epiphyseal ossification centre, hypertrophic cartilage zone, and trabecular primary spongiosa. Haematoxylin and eosin stain × 50

itself; b) The junction between the hypertrophic cartilage and the primary spongiosa is a relatively straight line; c) Chondrocytic lacunae in the hypertrophic cartilage are considerably larger; d) Cartilage cells are not so prominent in the trabeculae of the primary spongiosa; e) The bone marrow is more cellular and the red cells are not nucleated.

In the rats treated with cortisone for 2 weeks, the most striking difference from normal was marked suppression of the formation of

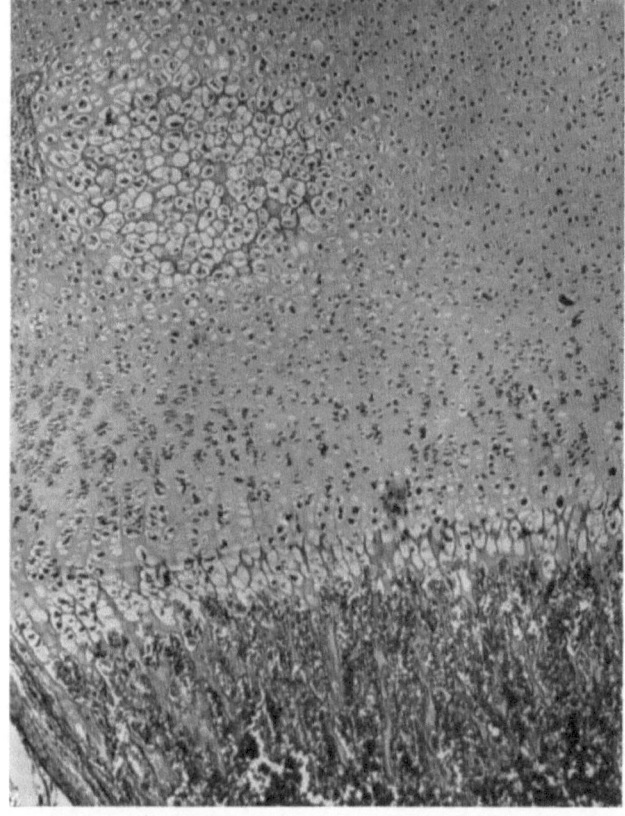

Fig. 7. Femoral epiphysis of cortisone-treated rat, showing (from above downward) proliferating cartilage with nidus of enlarged lacunae, hypertrophic cartilage zone, and mildly irregular zone of transition to trabecular primary spongiosa. Haematoxylin and eosin stain × 50

the epiphyseal ossification centre (Fig. 7). In a few of these animals a central nidus of markedly enlarged cartilage cell lacunae was seen, but in none was there actual formation of a trabecular spongiosa. In addition, the demarcation between hypertrophic cartilage and primary spongiosa was not so regular and well defined as in the control group, and the trabeculae of the primary spongiosa tended to be narrower and more irregular. There was little difference in the number of osteoblasts between the control and the cortisone groups.

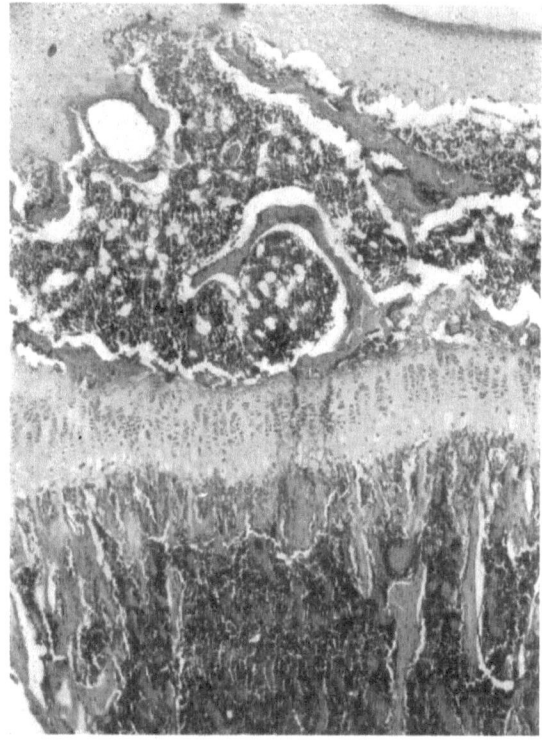

Fig. 8. Femoral epiphysis of rat treated with Dianabol and cortisone, showing normal pattern (see Fig. 6). Haematoxylin and eosin stain × 50

With alcian blue stains, both the normal rats and the cortisone rats showed the same pattern as in comparable groups of cockerels. Alkaline phosphatase staining revealed little difference between the control and the cortisone groups. No differences from the controls were noted with any stain in rats treated with Dianabol, or with Dianabol and cortisone (Fig. 8).

(C) *Comment on histology.* Some of the histological post-cortisone changes observed in this experiment have been reported previously (*7, 14, 67*). Acid mucopolysaccharides, as demonstrated by alcian blue staining, were decreased in the hypertrophic cartilage and the primary trabeculae of birds and rats treated with cortisone.

Although cortisone treatment has been reported as markedly decreasing alkaline phosphatase activity in bone healing (*66*), it is not clear as to whether the decrease in enzyme is due to inhibition of enzyme formation or to absence of osteoblasts themselves. Thus, in the cockerels treated with cortisone for 3 weeks, routine stains revealed large focal areas in which osteoblasts were absent. With alkaline phosphatase stains, no reaction was found in these areas, as might have been expected. However, in areas containing osteoblasts, phosphatase activity was present. Therefore, it seems likely that the decrease in phosphatase activity with cortisone therapy is merely a secondary result of decreased formation of osteoblasts.

To sum up, then, in the femora of cortisone-treated cockerels there was increased size of chondrocytic lacunae, thinning and irregularity of primary trabeculae, and focal absence of osteoblasts. These changes coincided with a decrease in acid mucopolysaccharides and a decrease in alkaline phosphatase. The alkaline phosphatase decrease was directly related to a decrease in the numbers of osteoblasts. Somewhat similar changes were seen in humeri of rats, except that changes in alkaline phosphatase activity were not so evident. No differences from the normal controls were observed histologically in birds and rats treated with Dianabol. However, Dianabol protected the bones of birds and rats from the deleterious effects of cortisone when both steroids were given simultaneously.

Summary

Being a part of the connective tissue, bone is affected by the hormones which influence the formation and development of this tissue. In the present report some effects of steroid hormones on various bone components are reviewed in relation to bone formation. It is apparent from a review of the literature that steroids considered anti-anabolic inhibit the formation of the organic matrix of growing bone. Fibrogenesis and the synthesis of mucopolysaccharides are inhibited or delayed by cortisone-like steroids. There is, on the other hand, considerable clinical and experimental evidence that some anabolic steroids have an opposite action to cortisone and may prevent the inhibitory effects of corticoids on bone. Corticoids also affect mineral metabolism in the bone. A negative calcium balance is found as a rule in animals and humans treated with cortisone. The action of this steroid on bone mineralisation is, in part, the result of a post-cortisone alteration in the absorption and elimination of minerals by the kidney and the gastro-intestinal tract. Anabolic steroids counteract this action of corticoids on bone minerals. There is also histological evidence that corticoids inhibit bone formation and that the anabolisers promote bone growth.

The problems discussed in this review are further illustrated by some experiments on young cockerels and rats. Experimental evidence is presented that cortisone has indeed an inhibitory action on various elements of bone matrix and on the mineralisation of the bones. Histological studies also confirm this general deleterious effect of cortisone on bone formation[1]. The anabolic steroid Dianabol, used in this experiment, protected the birds and rats against the effects of cortisone.

Briefly, those steroid hormones which are considered anti-anabolic inhibit the growth of cellular elements of bone, inhibit or delay the synthesis of various elements of bone matrix, and furthermore produce a negative calcium balance. On the other hand, anabolic steroids promote formation of bone matrix and produce a positive calcium balance; they may also counteract the destructive action of corticoids on bone.

Zusammenfassung

Als Teil des Bindegewebssystems wird der Knochen durch Hormone beeinflußt, die einen Effekt auf seine Bildung und Entwicklung besitzen. In der vorliegenden Arbeit wird über Wirkungen von Steroidhormonen auf verschiedene Knochenbestandteile in Beziehung zur Knochenbildung berichtet. Eine Übersicht über die Literatur ergibt, daß Steroide, denen eine antianabole Wirkung zugeschrieben wird, die Bildung der organischen Matrix des wachsenden Knochens beeinträchtigen. Faserbildung und Synthese von Mucopolysacchariden werden durch cortisonähnliche Steroide gehemmt oder verzögert. Andererseits sprechen klinische und experimentelle Daten dafür, daß einzelne anabole Steroide eine dem Cortison entgegengesetzte Wirkung haben und dessen hemmenden Einfluß auf den Knochen verhindern. Die Corticoide wirken auch auf den Mineralstoffwechsel im Knochen. Im allgemeinen findet sich unter der Behandlung mit Cortison bei Tieren und beim Menschen eine negative Calciumbilanz. Die Wirkung von Cortison auf die Knochenmineralisation ist zum Teil auf eine danach auftretende Änderung der Resorption und Ausscheidung von Mineralien durch den Magen-Darmtrakt und die Nieren zurückzuführen. Anabole Steroide wirken diesem Effekt der Corticoide auf den Knochen entgegen. Auch histologische Befunde sprechen dafür, daß Corticoide die Knochenbildung hemmen, und daß anabole Steroide das Knochenwachstum fördern.

Die in dieser Übersicht behandelten Fragen werden durch Untersuchungen an jungen Hähnen und Ratten ergänzt. Experimentelle Befunde sprechen dafür, daß Cortison tatsächlich einen hemmenden Einfluß auf verschiedene Elemente der Knochenmatrix und auf die Mineralisation besitzt. Histologische Studien bestätigen den schädigenden Einfluß von Cortison auf die Knochenbildung[2]. Das anabole Steroid Dianabol schützt bei beiden Tierarten gegenüber diesen Effekten von Cortison.

---

[1] The assistance rendered by A. E. RODIN, M. D., Department of Pathology, University of Alberta, in the interpretation of the histological slides is gratefully acknowledged.

[2] Für die Hilfe bei der Beurteilung der histologischen Schnitte bin ich Dr. A. E. RODIN, Department of Pathology, Univerity of Alberta, sehr zu Dank verpflichtet.

Zusammenfassend ist festzustellen, daß Steroide, denen antianabole Eigenschaften zugeschrieben werden, das Wachstum zellulärer Elemente des Knochens beeinträchtigen, die Synthese verschiedener Elemente der Knochenmatrix unterdrücken oder verzögern und außerdem eine negative Calciumbilanz hervorrufen. Anabole Steroide andererseits fördern die Bildung der Knochenmatrix, bedingen eine positive Calciumbilanz und können dem zerstörenden Einfluß von Corticoiden auf den Knochen entgegenwirken.

Résumé

Faisant partie du tissu conjonctif, l'os est influencé par les hormones agissant sur la formation et le développement de ce tissu. Ce travail expose quelques actions des hormones stéroïdes sur divers éléments constitutifs de l'os en rapport avec sa formation. La littérature montre que les stéroïdes considérés comme antianaboliques inhibent la formation de la matrice organique de l'os en cours de croissance. Les stéroïdes du genre cortisone inhibent ou retardent la fibrogénèse et la synthèse des mucopolysaccharides. Les observations cliniques et expérimentales indiquent d'autre part claire- ment que divers stéroïdes anabolisants ont un effet contraire à celui de la cortisone et peuvent s'opposer à son action inhibitrice sur l'os. Les corticoïdes agissent aussi sur le métabolisme minéral dans l'os. Pendant la médication cortisonique, le bilan calcique est, en règle générale, négatif, chez l'animal et chez l'homme. L'effet de la cortisone sur la minéralisation de l'os résulte en partie d'une perturbation de la résorption et de l'élimination des minéraux par le rein et par le tractus gastro-intestinal; les stéroïdes anabolisants s'oppo- sent à cet effet. Les recherches histologiques indiquent également que les corticoïdes inhibent la formation de l'os tandis que les stéroïdes anabolisants la favorisent.

Ces diverses questions sont illustrées par des expériences sur de jeunes coqs et rats. Leurs résultats prouvent que la cortisone inhibe effectivement divers éléments de la matrice osseuse, ainsi que la minéralisation des os. Les examens histologiques confirment l'influence délétère de la cortisone sur la formation des os[1]. Un stéroide anabolisant, le Dianabol, a par contre protégé les deux espèces d'animaux contre les effets de la cortisone.

En résumé, on constate que les stéroïdes considérés comme antianaboli- sants inhibent la croissance des éléments cellulaires de l'os, inhibent ou retardent la synthèse de divers éléments de la matrice osseuse, et déterminent en outre un bilan calcique négatif. Les stéroïdes anabolisants au contraire stimulent la formation de la matrice osseuse, entraînent un bilan calcique positif et peuvent s'opposer à l'action destructrice des corticoïdes sur l'os.

References

1. ALBRIGHT, F., and E. C. REIFENSTEIN, Jr.: In: Parathyroid Glands and Metabolic Bone Disease. Williams and Wilkins, Co., Baltimore, 1960. — 2. ALMQUIST, S., D. IKKOS, and R. LUFT: Acta endocr. (Den.) 38, 413 (1961). — 3. ARNOLD, J. S., and W. S. S. JEE: Proc. Exper. Biol. Med. (U.S.A.) 85, 658 (1954). — 4. ASBOE-HANSEN, G.: Physiol. Rev. (U.S.A.) 38, 446 (1958). — 5. AMPRINO, R.: In: Ciba Foundation Symposium on

---

[1] Je tiens à remercier ici le docteur A. E. RODIN du Département de pathologie de l'Université d'Alberta pour son aide dans l'interprétation des coupes histologiques.

Bone Structure and Metabolism. J. A. Churchill Ltd., London, 1956, p. 89. —
6. BARON, D. N., and J. L. BELL: J. Clin. Path. (G.B.) 12, 143 (1959). —
7. BAUER, G. C. H., A. CARLSSON, and B. LINDQUIST: In: Mineral Metabolism.
Ed. by COMAR, C. L., and F. BRONNER. Acad. Press, New York, London,
1961, p. 609. — 8. BERTOLOTTI, E.: Minerva med. (It.) 51, 987 (1960). —
9. BOAS, N. F.: J. Biol. Chem. (U.S.A.) 204, 553 (1953). — 10. BOSTRÖM, H.:
In: Connective Tissue in Health and Disease. Ed. by ASBOE-HANSEN, G., and
E. MUNKSGAARD. Denmark, 1954, p. 97. — 11. BROCHNER-MORTENSEN, K.,
S. GJØRUP, and J. H. THAYSEN: Acta med. Scand. 165, 197 (1959). —
12. BUNO, W., and H. GOYENA: Proc. Soc. Exper. Biol. Med. (U.S.A.)
89, 622 (1955). — 13. CAMPO, R. D., and D. D. DZIEWIATKOWSKI: J. Bio-
phys. and Biochem. Cytol. (U.S.A.) 9, 401 (1961). — 14. CLEIN, L. J., and
K. KOWALEWSKI: Canad. J. Surg. 5, 108 (1962). — 15. COLLINS, E. J.,
and V. BAKER: Metabolism (U.S.A.) 9, 550 (1960). — 16. COOKE, A. M.:
Lancet (G. B.) 268, 877 (1955). — 17. DENKO, C. W., and D. M. BERGEN-
STAL: Endocrinology (U.S.A.) 69, 769 (1961). — 18. DESAULLES, P. A.:
Helvet. med. acta 27, 479 (1960). — 19. DOUGHERTY, T. F., D. L. BERLINER,
and M. L. BERLINER: Metabolism (U.S.A.) 10, 966 (1961). — 20. DZIEWIAT-
KOWSKI, D. D.: J. Biol. Chem. (U.S.A.) 189, 187 (1951). — 21. EINBINDER,
J. M., C. T. NELSON, and C. L. FOX, Jr.: Amer. J. Physiol. 179, 347 (1954). —
22. ENDLER, F.: Wien. klin. Wschr. 73, 131 (1961). — 23. ENGEL, M. B.,
N. R. JOSEPH, and H. R. CATCHPOLE: A.M.A. Arch. Path. 58, 26 (1954). —
24. FIRSCHEIN, H. E., W. F. NEUMAN, G. R. MARTIN, and B. J. MULRYAN:
Recent Progr. Hormone Res. (U.S.A.) 15, 427 (1959). — 25. FISKE, C. H.,
and Y. SUBBAROW: J. Biol. Chem. (U.S.A.) 66, 375 (1925). — 26. FLOREY, H.:
In: Lectures on General Pathology. W. B. Saunders Co., Philadelphia and
London, 1954, p. 553. — 27. FOURMAN, P.: In: Calcium Metabolism and the
Bone. Blackwell, Oxford, 1960. — 28. GESCHWIND, I. I.: In: Mineral Meta-
bolism. Ed. by COMAR, C. L., and L. BRONNER. Acad. Press, New York,
London, 1961, p. 388. — 29. GROSS, J.: J. Exper. Med. (U.S.A.) 107,
247 (1958). — 30. HAGERTY, R. G., T. B. CALHOON, W. H. LEE, and J. T.
CUTTINO: Surg. Gyn. Obstetr. (U.S.A.) 110, 3 (1960). — 31. INNACCONE,
A., J. L. GABRILOVE, S. A. BRAHMS, and L. J. SOFFER: Ann. Int. Med.
(U.S.A.) 52, 570 (1960). — 32. JACKSON, D. S.: In: Connective Tissue. —
Symposium: Council for Intern. Org. Med. Sc. C. C. Thomas, Springfield,
1957, p. 62. — 33. JACKSON, S. F., and J. T. RANDALL: In: Ciba Foundation
Symposium on Bone Structure and Metabolism. J. A. Churchill Ltd.,
London, 1956, p. 47. — 34. JESSERER, H.: CIBA Symposium 8, 217 (1960).—
35. JOSEPH, N. R., M. B. ENGEL, and H. R. CATCHPOLE: A.M.A. Arch.
Path. 58, 40 (1954). — 36. KAHN, D. S., and S. C. SKORYNA: Laborat.
Invest. (U.S.A.) 8, 703 (1959). — 37. KOCHAKIAN, C. D., and D. G. HARRI-
SON: Endocrinology (U.S.A.) 70, 99 (1962). — 38. KOWALEWSKI, K.:
Endocrinology (U.S.A.) 62, 493 (1958). — 39. KOWALEWSKI, K.: Endo-
crinology (U.S.A.) 63, 759 (1958). — 40. KOWALEWSKI, K.: Proc. Soc.
Exper. Biol. Med. (U.S.A.) 97, 432 (1958). — 41. KOWALEWSKI, K.: Acta
endocr. (Den.) 38, 427 (1961). — 42. LAAKE, H.: Acta endocr. (Den.) 34,
60 (1960). — 43. LACROIX, P.: In: Ciba Foundation Symposium on Bone
Structure and Metabolism. J. A. Churchill Ltd., London, 1956, p. 36. —
44. LAYTON, L. L.: Proc. Soc. Exper. Biol. Med. (U.S.A.) 76, 596 (1951).—
45. LEATHEM, J. H.: Recent Progr. Hormone Res. (U.S.A.) 14, 141 (1958). —
46. LEDERER, J.: Schweiz. med. Wschr. 90, 1379 (1960). — 47. MOLINATTI,
G. M., F. CAMANI, and M. OLIVETTI: Acta endocr. (Den.) 34, 323 (1960). —
48. NEUMAN, R. E., and M. A. LOGAN: J. Biol. Chem. (U.S.A.) 180, 549
(1950). — 49. NEUMAN, W. F., and M. W. NEUMAN: In: The Chemical

Dynamics of Bone Mineral. University of Chicago Press, 1958. — 50. NOR-
DIN, B. E. C.: In: Clinical Endocrinology. Ed. by E. B. ASTWOOD. Grune
and Stratton, New York, London, 1960, p. 233. — 51. PARTRIDGE, S. M.,
and H. F. DAVIS: In: Ciba Foundation Symposium on Chemistry and
Biology of Mucopolysaccharides. Little, Brown, and Co., Boston, 1958,
p. 93. — 52. RANNEY, R. E.: Endocrinology (U.S.A.) **64**, 783 (1959). —
53. ROBERTS, E., A. A. KARNOFSKY, and S. FRANKEL: Proc. Soc. Exper.
Biol. Med. (U.S.A.) **76**, 289 (1951). — 54. RUFFO, A.: A. M. A. Arch. Surg.
80, 172 (1960). — 55. SCHÄRER, K., H. HABICH, and A. PRADER: Helvet.
med. acta **27**, 530 (1960). — 56. SCHWARTING, G., and R. NETH: Schweiz.
med. Wschr. **90**, 1092 (1960). — 57. SCOW, R. O.: Amer. J. Physiol. **196**,
859 (1959). — 58. SILBERBERG, M., and R. SILBERBERG: In: The Biochemistry and Physiology of Bone. Ed. by G. H. BOURNE. Acad. Press, New
York, 1956, p. 624. — 59. SINEX, F. M.: Science (U.S.A.) **134**, 1402 (1961). —
60. SIUKO, H., J. SAVELA, and E. KULONEN: Acta endocr. (Den.) **31**, 43
(1959). — 61. SOBEL, E. H., C. S. RAYMOND, K. V. QUINN, and N. B. TALBOT:
J. Clin. Endocr. (U.S.A.) **16**, 241 (1956). — 62. SOBEL, H.: J. Bone Surg.,
Am. Ed. **40**, 1111 (1958). — 63. SOFFER, L. J.: In: Diseases of the Endocrine
Glands. Lea and Febiger, Philadelphia, 1956. — 64. STOREY, E.: Endo-
crinology (U.S.A.) **68**, 533 (1961). — 65. SUZUKI, H. K.: J. Bone Surg.
40, 435 (1958). — 66. TONNA, E. A., and J. A. NICHOLAS: J. Bone Surg.,
Am. Ed. **41**, 1149 (1959). — 67. URIST, M. R., and N. M. DEUTSCH: Endo-
crinology (U.S.A.) **66**, 805 (1960). — 68. WIANCKO, K. B., and K. KOWALEW-
SKI: Acta endocr. (Den.) **36**, 310 (1961).

Discussion

LARON: Mr. Chairman, if you will permit me, 1 should like to describe very briefly some experiments which we did on the same lines as Dr. KOWA-LEWSKI has reported. We, too, were very much interested in the potential value of anabolic steroids to protect animals, and possibly man, against the effects of corticosteroids. In earlier experiments on rats, we had found that testosterone could not afford protection against the effect of cortisone on

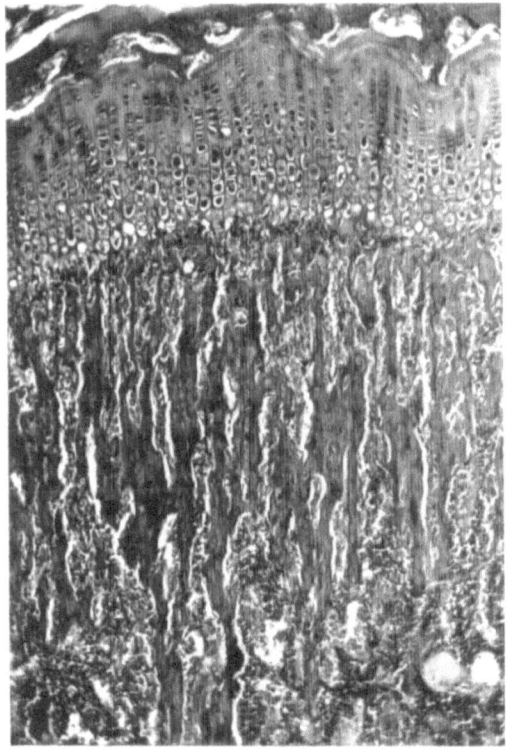

Fig. 1 A

Fig. 1 A—C. Histological appearance of the proximal epiphysis and metaphysis of the rat tibia. A Medrol (6-methylprednisolone) 1 mg. daily; B Medrol 1 mg. + BGH 0.5 mg.; C Medrol 1 mg. + SGH 1 mg. (BGH = bovine growth hormone; SGH = sheep growth hormone; both are identical in effect)

bone. In later experiments, using Durabolin on normal rats[1] we again found that with the doses used — I would like to stress that we were always dealing with pharmacological doses — we could not achieve protection against the ill effects of cortisone. Cortisone thins the epiphyseal cartilage and pro-

Fig. 1 *B*

duces marked osteoporosis. In response to Durabolin we see that there is no marked difference between cortisone and cortisone plus Durabolin.

We now went on to study rats which had been given a low-calcium diet[2], and we found that in these rats there was some effect when the same amount of Durabolin was given, but these animals already had disturbed growth from the beginning. We also found in these rats that, although the calcium balance was positive as compared with the calcium balance in rats treated with cortisone alone, the bone structure remained essentially pathological.

---

[1] LARON, Z., and J. H. BOSS: Endocrinology (U.S.A.) **69**, 608 (1961).
[2] LARON, Z., A. KOWADLO, J. KALISH, and S. KENDE: in preparation.

And now I come to the last experiment, in which we used growth hormone instead of the anabolic steroid[1]. Here, to our great surprise, we found that bovine growth hormone completely reversed the deleterious effect of the cortisone-like steroid Medrol (Fig. 1). The latter (6-methylprednisolone)

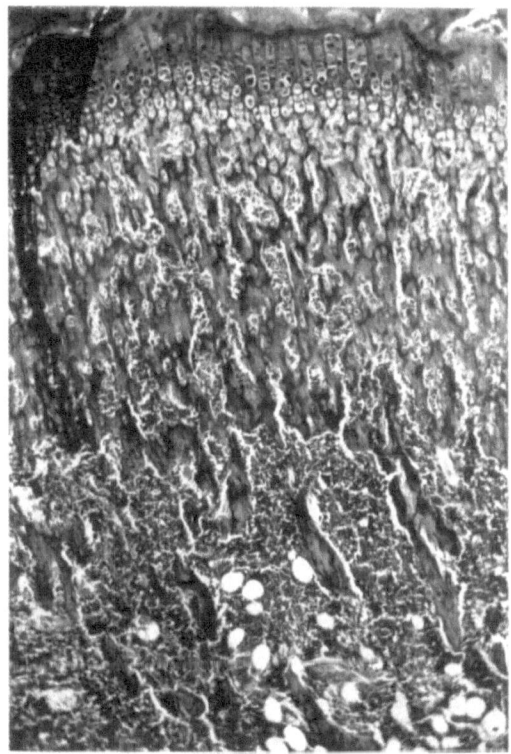

Fig. 1 C

causes marked thinning of the epiphyseal cartilage, while at the same time the structure of the trabeculae and encased cartilage becomes pathological. If the same amount of Medrol is given and if 0.5 mg. of bovine growth hormone is added, we can see quite a widening of the epiphyseal cartilage, although the structure of the bone still resembles that of the rat treated with Medrol alone. Increasing the dose of BGH leads to an even more marked widening of the epiphyseal cartilage and a bone structure which closely resembles that found in our normal controls. I would like to mention another effect. Although the bone was well protected, we could not offset the diminished rate of weight

[1] LARON, Z., B. Z. ARIE, and S. KENDE: Endocrinology (U.S.A.) in press.

gain. In none of the animals studied did the anabolic substance, whether Durabolin or growth hormone, raise the weight of the treated animal to that of the controls. The weight of all the animals was roughly the same as that of those treated with cortisone or Medrol alone.

GROSS: I would like to ask Dr. KOWALEWSKI about the significance of his data obtained with Dianabol treatment and with the combination of cortisone plus Dianabol. In your first slide the difference between normal controls and cortisone-treated animals was significant. I wonder if the same difference exists between cortisone-treated animals and those treated with the anabolic agent or with the anabolic agent plus cortisone. From your figures and from the columns you had in your slides I gained the impression that these data did not differ very much from those obtained with cortisone alone. As far as I remember, control values were never reached, and that would correspond with what Dr. LARON has just said, i.e. that he, too, couldn't completely offset the effect on weight or on the nitrogen content of bone by giving Durabolin in addition.

KOWALEWSKI: We considered differences as significant when the P values were less than 1%. We found that only in rats and birds treated with cortisone alone were the values obtained significantly lower than in any other groups. There was also a great overlap between the values observed with Dianabol alone and with Dianabol plus cortisone, and these values were within the range for normal untreated controls. We did not say in our report that Dianabol alone had a significant promoting action on the parameters studied in our experiment. When given simultaneously with cortisone, Dianabol did, however, prevent some changes in the bone which were considered to be due to the cortisone.

DYMLING: I believe that you should interpret these results on animals with the utmost caution when you compare them with those obtained in human beings. Some of our results in human adults are in definite disagreement with those reported by Dr. KOWALEWSKI. However, I don't think the data are comparable. On the other hand, Dr. KOWALEWSKI's paper gave the impression that his results were directly comparable to those in human adults, which I don't feel is the case.

KOWALEWSKI: I did not discuss the comparability of the experimental results in animals with the studies done on human adults. Clinical studies of anabolic steroids done on humans are admittedly mentioned in my review of the literature, but in my own report only experiments on rats and birds are described. Studies on osteoporosis in man are difficult, as there are no precise criteria for evaluating changes following hormonal therapy.

OVERBEEK: We tried to produce calcium losses in rats, first with hydrocortisone and then with dexamethasone. We didn't do histological studies like you did, but we couldn't find any losses by excretion in the urine or in the faeces. I don't understand why not. Can you offer any explanation?

KOWALEWSKI: We did calcium determinations only in bones, and we just mentioned that cortisone is known to affect secretion and excretion of calcium by the kidneys and the intestinal tract. We did not study calcium loss in hormone-treated animals.

OVERBEEK: My second remark is to Dr. LARON. We also tried to offset the decrease in body-weight which occurs in rats in response to corticosteroids. With nandrolone esters and ethyloestrenol we always succeeded in compensating for the loss of body-weight. Other groups have found the same. It is certainly not a common finding that you were unable to counteract the effect of the corticoids on body-weight.

LARON: Well, I guess it's not only difficult — if not impossible — to compare human with animal experiments, but that it is actually wrong to try to do so in this particular instance. It is also difficult to compare even one animal experiment with another, because the dose of any drug used and the relationship between the catabolic and the anabolic agent are, as such, rarely comparable. This poses the question as to what we really gain from these experiments as regards clinical purposes. We have to remember that the patients in whom we are interested are receiving long-term treatment with large amounts of cortisone or cortisone-like substances, e.g. lupus patients or cases of rheumatoid arthritis, nephrosis, etc. I am not sure about this, but perhaps you use smaller doses of the drugs, and this may explain the difference in our results. It would be interesting to hear what dosages you employed, and maybe Dr. KOWALEWSKI's dosages as well.

OVERBEEK: I don't remember the exact dosages off-hand, but I doubt whether the doses used in animals will give any indication as to what doses are likely to be required in humans.

LARON: I can only repeat that in our experiments we failed to observe any beneficial effect on the animal's body-weight, and I was glad to hear that Dr. DESAULLES also agrees with our results. With no anabolic substance were we able to obtain results comparable with the control animals: even growth hormone, which completely reversed the bone changes, did not offset the deleterious effect of cortisone on the rats' body-weight.

DESAULLES: At this point, I should just like to emphasise certain aspects of Dr. LARON's report. If my memory is good, it was shown histologically some years ago that, in the epiphyseal cartilage of the long bones of young, immature rats, the site of action of the steroids (androgens and oestrogens) and of growth hormone were different. The site of action of the steroids is located at the level of the cartilage culumns, excluding the hypertrophic cartilage. It has also been shown that full activity was only obtained when the presence of growth hormone was supplemented by the action of a steroid. In the absence of growth hormone following hypophysectomy, the steroids were virtually ineffective.

These experiments provide strong evidence to suggest that stimulation of growth of the epiphyseal cartilage of long bones under the influence of steroids is dependent, among other things, upon the presence of growth hormone.

It seems to me that these observations confirm the relationship shown in hypophysectomised animals treated with anabolic steroids, in which the myotrophic effects, which are regarded as evidence of an anabolic response, are greatly impaired and cannot be completely restored by the use of steroids alone.

VERMEULEN: I was very impressed by Dr. KOWALEWSKI's results. I think the ultimate idea behind these experiments is to protect our patients from the side effects of corticoids. Now, by giving anabolic steroids to our corticoid-treated patients, we can counteract the catabolic effect on the nitrogen balance, we can counteract the osteopenia, and we can probably counteract the diabetogenic effect. The therapeutic effect of corticoids has, I think, something to do with inhibition of fibroblast proliferation, with the decrease in cellular permeability, and perhaps with their lympholytic effect. Now my question is this: is there any evidence that association of anabolic steroids with corticoids also inhibits these therapeutic effects of corticoids ?

KOWALEWSKI: I'm afraid I cannot answer that question at the moment.

Introduction to the General Discussion:

Effects of protein deficiency on the endocrine system

By

A. ASCHKENASY

During this session we have had some very valuable and stimulating papers, both methodological and biochemical, on the tissular action of androgens, as well as on the role played by nutritional status in this action, and finally we have listened to a convincing study of the anabolic effect of androgens on bone tissue.

In opening the general discussion, I should like, with your permission, to submit to you a few reflections concerning the interactions between protein starvation and the secretion of androgens, glucocorticoids, and thyroid hormones in the rat.

The classic approach, which was introduced 20 years ago by MULINOS and POMERANTZ (1940) and which has since been adopted in almost all reviews devoted to this question, likens every type of malnutrition — as far as its endocrinological effects are concerned — to a functional pseudohypophysectomy.

In point of fact, such a comparison seems to be correct only with regard to gonadal function, and not with regard to thyroid or adrenocortical function.

Indeed, in protein-deprived rats only the gonads, particularly the male gonads, exhibit a weight decrease which is disproportionate to the reduction in body-weight. Histological signs of involution are observed at a much earlier stage in the interstitial tissue than in the seminiferous canaliculi (ASCHKENASY, 1953).

It is worth noting that atrophy of the seminal vesicles and prostate represents, together with involution of the thymus, one of the first visceral changes induced in the rat by protein deprivation (ASCHKENASY, 1953).

Contrary to the testicles, the adrenals seem to exhibit — at least according to histological criteria — a functional activity which is normal or perhaps even proportionately higher than that of normally nourished rats of the same initial weight (ASCHKENASY, 1953).

In any case, the maintenance of a certain level of glucocorticoid secretion is absolutely indispensable for the survival of protein-deprived rats, whereas the same hormonal secretion can be suppressed without any noticeable damage when the diet is rich in proteins, on condition that the animals receive physiological saline as drinking water (ASCHKENASY, 1954).

As Table 1 shows, young female rats adrenalectomised before the beginning of protein starvation do not survive for long on this diet despite the administration of salt. However, survival can be prolonged by giving cortisone in adequate doses: with a dosage level of ≥ 150 mcg. per day all the adrenalectomised rats are still alive at the end of the experimental period (28 days of protein deprivation). The absence of accessory adrenal tissue is demonstrated by the fact that all these animals die after a few days when they are put on a diet rich in proteins but containing no salt.

Table 1. *Protection of adrenalectomised female rats by cortisone against the lethal effect of protein starvation* [From A. ASCHKENASY: Compt. rend. Soc. biol. (Fr.) **154**, 1783 (1960)]

Dose per day (mcg.)	Number of rats	Mean duration of survival (days $\pm$ S.E.M.)	Number of surviving rats after 28 days of protein starvation	Number of days of survival ($\pm$ S.E.M.) on a protein-rich, saltless diet following the protein starvation period
—	10	6.8 $\pm$ 0.6	—	—
5	8	8.5 $\pm$ 1.0	—	—
25	12	12.0 $\pm$ 1.2	1	3
50	18	14.6 $\pm$ 1.7	3	2; 3; 3
100	10	13.4 $\pm$ 1.9	5	3.4 $\pm$ 0.7
150	10	—	10	6.0 $\pm$ 1.1
250	10	—	10	3.5 $\pm$ 0.4
500	9	—	9	8.6 $\pm$ 1.1

Thus, protein deprivation not only does not decrease the need for glucocorticoid hormones, as it does for the sexual steroids; on the contrary, it even increases the need for these hormones.

This fact provides a very convincing demonstration of the relationship between nutritional status and the need for certain hormones.

As regards the thyroid, it must be said that, contrary to current opinion and to our own initial conclusions based exclusively on histological data (ASCHKENASY et al., 1952), the secretory activity of the gland is not at all inferior to that of a normal thyroid, taking into account the reduced body-weight of the undernourished animals.

Indeed, experiments performed with radioactive iodine do not permit of any other conclusion (ASCHKENASY et al., 1962).

All these data indicate the extent to which the inhibitory influence of protein deprivation on the secretion of androgens is peculiar to these hormones, at least in the rat.

One may wonder whether it is useful or not to administer sexual steroids in protein malnutrition states.

There is some reason to suppose that androgens such as testosterone are noxious rather than beneficial, because in protein-starved animals their predominant action consists in the transfer of proteins from certain organs, such as the thymus and other lymphoid organs, to accessory sex organs (ASCHKENASY, 1953). The aggravation by androgens of the lymphoid atrophy induced by protein starvation might decrease still further the immunological resistance of the undernourished animals against microbial aggression.

On the other hand, testosterone-propionate does not restore the body-weight to any appreciable extent if given together with a 7% casein diet to protein-depleted adult male rats (ASCHKENASY and DRAY, 1954), although it produces noticeable effects (some stimulating, some inhibitory) on the restoration of specific organs and tissues.

In the case of steroids with a weaker androgenic activity, such as methandrostenolone or 19-nortestosterone derivatives, the internal shift in tissular proteins is directed more towards certain muscles than towards the accessory sex organs. Nevertheless, some reservations should be made concerning the beneficial effects of these steroids.

These hormones do not exert their protein-anabolic effect on all the muscles: some are refractory to this effect — for instance, the gastrocnemius and the heart muscle in the castrated rat treated with 19-nor-17α-ethyltestosterone (ASCHKENASY, 1959) (Table 2); in the castrated guinea-pig, androgens, though stimulating the growth of the majority of skeletal muscles, produce widely differing effects in each of them (KOCHAKIAN and TILLOTSON, 1957).

In any case, the most sensitive muscles are the perineal ones (especially the levator ani and the bulbocavernosus in the rat), the physiological weight changes of which seem to be related primarily to their role in the sexual activity of the animal.

For this reason it is perhaps somewhat hazardous to consider the action of an androgen on the latter muscles as an absolute criterion of the general anabolic potency of the hormone.

Table 2. *Relative weights (mg. per 100 g. of terminal body-weight ± S.E.M.) of some muscles in castrated male rats either untreated or treated with 19-nor-17 α-ethyltestosterone (norethandrolone) (5 mg. orally per day for one month). Dry weights in brackets* [From A. ASCHKENASY: Thérapie (Fr.) 14, 332 (1959)]

Group (number of rats)	Bulbo-cavernosus	Levator ani	Gastro-cnemius	Masseter	Heart
Untreated castrates (7)	80 ± 12 (22 ± 4)	17.5 ± 2 (7 ± 1)	615 ± 14 (159 ± 4)	303 ± 5 (74 ± 1)	315 ± 9
Castrates treated with nor-ethandro-lone (8)	195 ± 13 (49 ± 3)	66 ± 6 (17 ± 1)	601 ± 14 (156 ± 3)	358 ± 8 (89 ± 2)	305 ± 5

As a matter of fact, it may be that in human beings the number of extra-perineal muscles stimulated by androgens is much larger than in the rat. However, it seems that hitherto no systematic clinical investigations have been performed concerning the comparative action of different androgens on the functional and, especially, electromyographic reactions of various muscles. Nor has it been established that there is any relationship between the myotrophic action of these hormones in man and their effect on the levator ani in the rat.

The beneficial action of certain androgens on various muscles and on some other extra-genital tissues and organs — such as the kidney [in the rat and the mouse, but not in the guinea-pig (KOCHAKIAN et al., 1956)], bone tissue, erythropoietic bone marrow or salivary glands — contrasts, in laboratory animals, with their harmful effect on the thymus and also, but only if the diet is poor in proteins, on the other lymphoid organs (ASCHKENASY, 1953, 1959a, b) and (at least as far as testosterone is concerned) on the production of eosinophilic granulocytes (ASCHKENASY and DRAY, 1954).

At all events, according to the experimental data obtained from laboratory animals, administration of anabolic androgens seems to be justified in clinical pathology, especially where it is desired to protect certain tissues or organs which are particularly sensitive to the beneficial action of these hormones, and provided that these tissues or organs are in a deficient functional state: for instance, bone tissue altered by senile or cortisone-induced osteoporosis, bone marrow in cases of hypoplastic anaemia, kidneys in cases of nephritis, and also, perhaps, certain skeletal muscles in amyotrophia induced by undernutrition or prolonged immobilisation.

272 General Discussion

References

ASCHKENASY, A.: Ann. endocr. (Fr.) **14**, 353 (1953); — Compt. rend.
Acad. sc. (Fr.) **239**, 1000 (1954); — Ann. endocr. (Fr.) **16**, 199 (1955); —
Thérapie (Fr.) **14**, 332 (1959a); — Ann. endocr. (Fr.) **20**, 158 (1959b). —
ASCHKENASY, A., J. P. BENHAMOU, and G. J. ROLLAND: Compt. rend. Soc.
biol. (Fr.) **146**, 44 (1952). — ASCHKENASY, A., and F. DRAY: Ann. endocr.
(Fr.) **15**, 441 (1954); — Sang (Fr.) **25**, 461 (1954). — ASCHKENASY, A., B.
NATAF, and M. SFEZ: Compt. rend. Soc. biol. (Fr.) **155**, 986 (1961). — ASCH-
KENASY, A., B. NATAF, C. PIETTE, and M. SFEZ: Ann. endocr. (Fr.) **23**, 311
(1962) — KOCHAKIAN, C. D., and C. TILLOTSON: Endocrinology (U.S.A.)
60, 607 (1957). — KOCHAKIAN, C. D., C. TILLOTSON, J. AUSTIN, E. DOUGHERTY,
V. HOAG, and R. COALSEN: Endocrinology (U.S.A.) **58**, 315 (1956). —
MULINOS, M. G., and L. POMERANTZ: J. Nutrit. (U.S.A.) **19**, 493 (1940).

General Discussion

QUERIDO: Thank you, Dr. ASCHKENASY. Of course, one would like to know which muscles react in the human.

VERMEULEN: Mr. Chairman, you raised the question of the role played by the metabolism of steroids in relation to their effects. Now Dr. DESAULLES has shown (in his Table 3) that after thyroidectomy there is a shift in the ratio of anabolic to androgenic activity of testosterone derivatives. It has been observed in hypothyroid patients that testosterone is mainly metabolised to aetiocholanolone, whereas in hyperthyroid patients the main metabolite of testosterone is androsterone. In thyroid dysfunction, alterations have also been noted in cortisol metabolism, oestrogen metabolism, and 11-hydroxy-androstenedione metabolism. Now I would like to ask the audience — and it is perhaps a rather naive question — whether the shift in the ratio of the effects of the testosterone derivatives might be related to an alteration in the metabolism of these testosterone derivatives in rats. I would suggest as a working hypothesis that, if a steroid has a different effect on the levator ani and the seminal vesicle, this may be due to a difference in tissue sensitivity, which in turn is probably related to a difference in the enzymatic equipment of the two targets; this might cause the metabolism of the steroid to differ in the two targets. Thyroid dysfunction might alter the enzymatic equipment of the cells, which might in turn be reflected in alterations in the metabolism of the steroid.

DESAULLES: I should be very happy if I could answer this question. Androsterone, which certainly is a poor anabolic steroid, also has only weak androgenic properties. But in general it has to be admitted that the pharmacology of the metabolites of natural androgens in various pathophysiological situations is still only in its beginnings.

DREYFUS: Together with Mrs. GERDAY-LAURENT[1] we have done an experiment on the incorporation of radioactive leucine into the diaphragms of rats which had received Dianabol, and we have also made *in vitro* studies by adding various doses of Dianabol to the rat diaphragms. No definite conclusions could be drawn, but there was no significant difference in the

---

[1] Unpublished results.

incorporation of radioactive amino acid into the diaphragms of treated as compared with non-treated animals.

TANNER: I have been a little worried all through the discussion this morning at the choice of the levator ani as the test muscle for anabolic effects. It is after all a very peculiar muscle in the rat and is perhaps more comparable to the bulbocavernosus muscle than to an ordinary skeletal muscle, at least in function. I should have thought that the actions on the levator ani in the rat had little significance so far as muscles in the human go. The fact that the uterus also grew really makes the criticism more, rather than less, pertinent. There is one occasion in the human when half the population receives a dose of testosterone, and one can follow the events at that time, known as adolescence, quite easily. And it is a fact that at adolescence there is an increased growth rate of voluntary muscles. The details of this are not really statistically well enough known for us to be able to answer the question as to which muscles grow most in man, although it is obvious in some other creatures — like the gorilla — that there is more growth in certain muscles than in others.

Some years ago, my colleagues and I[1] showed that there is a relation in healthy young men between the excretion of 17-ketosteroids and the muscle mass. It is not an exceedingly close relation, but it is statistically very significant. I have a feeling that this reflects the manner in which I suppose testosterone or other androgens enter into the metabolism of skeletal muscle.

One last remark: what is the substance which causes the increase in bone and muscle growth in the female at adolescence ? We rather suppose that the female doesn't secrete a great deal of testosterone or, if she does, that it must be metabolised in a different fashion from testosterone in the male. Administered testosterone has a virilising effect in young women, whereas the naturally secreted product, whatever it is, is not virilising.

ASCHKENASY: I should like only to suggest the possibility that androgens of adrenal origin may be more myotrophic in human subjects than the testicular ones. Muscular hypertrophy has in fact been observed in certain cases of adrenogenital syndrome.

DESAULLES: With regard to Dr. ASCHKENASY's very pertinent suggestion I should only like to add that we, too, have been thinking along these lines. My colleagues Dr. MEYER and Dr. KRÄHENBÜHL[2] have studied the properties of all the known androgens of possible adrenal origin that were available. It was found that those of them which exerted a potent androgenic effect on certain targets all displayed only a poor degree of dissociation between their androgenic and anabolic activity.

The only steroid showing a somewhat more pronounced dissociation between these two types of effect was 11β-hydroxy-4-androstene-3,17-dione, and even here the degree of freedom from myotrophic effects as compared with androgenic activity is still small.

OVERBEEK: Dr. TANNER still feels uneasy about the levator ani, and I can understand his feelings, but I think he has given the answer himself. All muscles, both in human males and females, react. We know very well that in males it is the anabolic action of the testosterone — of the male

---

[1] TANNER, J. M., M. J. R. HEALY, R. H. WHITEHOUSE, and A. C. EDGSON: J. Endocr. (G.B.) **19**, 87 (1959).

[2] MEYER, C., C. KRÄHENBÜHL, and P. A. DESAULLES: Acta endocr. (Den.): in press.

hormone — which causes us to have bigger muscles than women. I would also remind you that very marked results have been obtained in progressive muscular dystrophy after treatment with anabolic steroids. As long as there is some muscle present, you can actually see how the skeletal muscles, which were quite atrophied, grow and recover. I think the main problem is not really that the levator ani muscle of the rat is not comparable to human skeletal muscles. The main problem is why, in the rat, skeletal muscles do not react. Everybody would be happy if they did react, and the question which you put, and which many people have asked here and elsewhere, is: why is this levator ani such a special muscle and why is it only special in the rat? Once again, why is there no reaction of the *skeletal muscle* in the rat? You can see a reaction in humans, you can see one in monkeys, and you can see that in any other animal the skeletal muscles react.

ASCHKENASY: If I may revert to Dr. DESAULLES' remarks, I should say that the negative results you have encountered in your research depend, like all experimental results, on the method that has been employed. If your assessment of the anabolic action is based only on the levator ani test, then it is very possible that your adrenal androgens which have no notable action on the levator ani may nevertheless have an anabolic effect on skeletal muscles in humans.

DESAULLES: Yes, that is certainly possible, but it is always difficult to assess a shift or a change in the spectrum of properties of a steroid between the rat and man. It is quite conceivable that certain anabolic steroids may exist which are very active in humans and which are less active, or not active at all, in rats. But so far, to my limited knowledge, no anabolic steroid of this type, fulfilling these conditions, has ever been found. Assessing the relative activity of compounds in different species is undoubtedly one of the most challenging problems. Dr. LEATHEM said yesterday that between mouse, rat, and guinea-pig — all of them animals belonging more or less to the same group — there is some slight change in sensitivity. Maybe this provides a possible lead, but I think we still have a very long way to go before we have completely understood this problem.

QUERIDO: Is there anybody in the audience who knows whether or not a normal male of the human species will react to an anabolic steroid with an increase in muscle mass or a positive nitrogen balance?

GROSS: Certainly not.

VOICE: What do you consider a normal male?

QUERIDO: I mean people like the members of this Symposium!

ASCHKENASY: In normal, non-castrated, and non-undernourished male animals, androgens do not seem to exert any significant general anabolic action, i.e. on the body-weight. KOCHAKIAN and WEBSTER[1] have shown that testosterone-propionate actually decreases the body-weight of normal adult rats in proportion to the dose administered — this effect being essentially due to the mobilisation of body fat, but also to an inhibition of appetite.

Even in castrated rats, testosterone-propionate inhibits growth if it is given in a dose of 1 mg. daily[2]. This inhibitory effect is not observed in castrated, adrenalectomised rats.

[1] KOCHAKIAN, C. D., and J. A. WEBSTER: Endocrinology (U.S.A.) **63**, 737 (1958).

[2] ASCHKENASY, A.: Ann. endocr. (Fr.) **20**, 158 (1959).

Protein metabolism in human pathological states

The effect of injury upon human protein metabolism

By

J. M. KINNEY

The metabolism of the adult human body is characterized by a complicated series of reactions which provide a dynamic but relatively steady state during health. Injury modifies these reactions in many ways, but particularly with respect to nitrogen metabolism (1—14). The importance of protein nutrition has usually been emphasized to the surgeon in relation to pre-operative weight loss, hypoproteinemia, or the deficient healing of wounds. These factors are of obvious importance, but the material in this discussion is directed toward the influence of injury on protein metabolism as a part of total energy exchange in the human body. During the past five years, we have studied over seventy surgical patients by a modification of indirect calorimetry which provides a means for calculating the amounts of fat, carbohydrate, and protein being oxidized per day. When this information is combined with that provided by a research dietitian on the intake of food, it is possible to calculate a daily balance for each foodstuff (15). Thirty-five of these patients have been studied for ten to sixteen days each, while undergoing a variety of elective operative procedures. In addition, we have had the opportunity to study the early and late convalescence of patients after major fractures and a few cases of major sepsis and burns. Fig. 1 represents the course of an average male patient undergoing elective subtotal gastrectomy for chronic duodenal ulcer disease. His intake during early convalescence was approximately 400 calories of intravenous carbohydrate per day. The total metabolic expenditure of this individual showed no significant increase as a result of injury of this magnitude. Our studies are consistent with estimates for the specific dynamic action of an average mixed diet (16) and the physical activity of a hospitalized patient (17) accounting for 15 to 25 per cent of the total pre-operative caloric expenditure. These requirements are essentially abolished for several days after operation. Therefore, it appears

J. M. KINNEY:

that the post-operative basal metabolic expenditure has been increased by an amount which roughly offsets the few hundred calories that had been expended for SDA and activity. The custom-

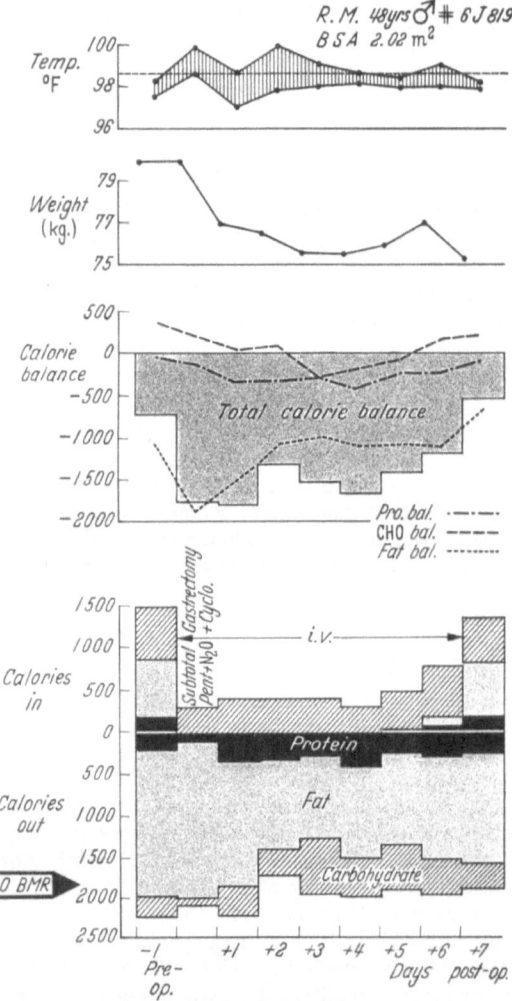

Fig. 1. The daily energy exchange during early convalescence of a patient following elective subtotal gastrectomy

ary increase in post-operative nitrogen excretion is shown here, not in grams of nitrogen per day, but in protein calories per day represented by the excreted nitrogen. It is evident that the extra energy available from this increase in protein oxidation from 220 calories per day to 400 calories per day in Fig. 1 meets only a small fraction of the caloric demands per day, most of which are being provided by the body fat stores. Post-operative patients, such as the patient in Fig. 1, commonly receive a small but important part of their daily caloric demands as intravenous carbohydrate. Dietary studies (without the complicating factors of injury) suggest that a few hundred calories of carbohydrate are sufficient to prevent the ketosis of starvation (18) and to provide the maximum protein-sparing that is unique to carbohydrate and not fat (19). The time of the post-operative increase in the oxidation of protein appears to approximate to the interval when the basal metabolic expenditure is increased.

Influence of trauma on nitrogen excretion and metabolic expenditure

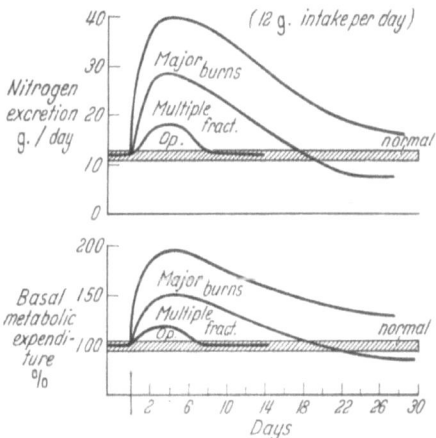

Fig. 2. The approximate ranges and duration of increase in human nitrogen excretion and basal metabolic expenditure after uncomplicated elective operation (indicated as "Op."), multiple fractures, and severe burns

The surgical patients we have studied have nearly always shown an increase in nitrogen excretion at times when their basal metabolic expenditure was elevated, and likewise a decrease in nitrogen excretion when their basal metabolic expenditure was less than normal. Fig. 2 indicates the ranges of alteration in nitrogen excretion and basal metabolic expenditure (BME) which we have come to expect when adult males undergo a) elective operations (20), b) major injury such as multiple fractures (21), or c) extensive third-degree burns (22, 23). The correspondence of these two factors is not as precise or regular from day to day as shown in Fig. 2, but the overall extent and duration of each is remarkably similar. The influence of age, sex, body build, and previous nutrition all appear to modify both factors in the same direction and to approximately the same extent. Studies of certain disease processes have also suggested that these alterations tend to occur in parallel (24). The exact relationship of these two factors is unknown, but is probably an important key to the understanding of metabolism after injury.

CUTHBERTSON (1) first emphasized the increase in nitrogen loss which occurs following injury. This has since been confirmed in many other laboratories. The increase may include various nitrogenous products in the urine, but the increase in urinary urea is quantitatively of greatest importance. After a major surgical operation, the nitrogen excretion is usually increased for three to seven days, but may continue elevated for three weeks or more with more extensive injury. During this time, the patient is often in negative nitrogen balance and, therefore, suffers a gradual depletion of body protein. During the later phases of convalescence, the urinary nitrogen excretion falls, and as the nutritional intake is increased, the patient reaches a positive nitrogen balance. The extent of the nitrogen loss after injury is roughly proportional to the magnitude of the injury and is most marked in the previously healthy young adult male. The reaction is less in females, children, the elderly, and the malnourished. The surgical literature reflects major differences of opinion regarding the importance of this loss of nitrogen and the effectiveness with which it can be treated. There is inadequate information on which to establish optimum nutritional support for the injured patient, but a consideration of four questions is relevant to this objective:

Is the increased nitrogen loss following injury

a) An obligatory part of the metabolic response, or the result of starvation and immobilization ?

b) Of any survival value to the patient ?

c) An effect of adrenal (or other hormonal) secretion ?

d) A decrease in anabolism or an increase in catabolism ?

Is the nitrogen loss obligatory ?

In early convalescence, many influences may contribute to the net response to injury, including particularly immobilization and partial starvation. The metabolic effects of immobilization have been studied in volunteer subjects immobilized for extended periods (25, 26). There was a minimal increase in nitrogen excretion rates on a constant intake, although significant demineralization of the skeleton was noted.

Many investigators have wondered whether the loss of nitrogen and potassium which follows injury or operation might not be related to the partial starvation which commonly occurs under such circumstances. It is important to examine uncomplicated starvation to obtain perspective for the role of starvation in the response to injury. During a total fast of 31 days, BENEDICT (27)

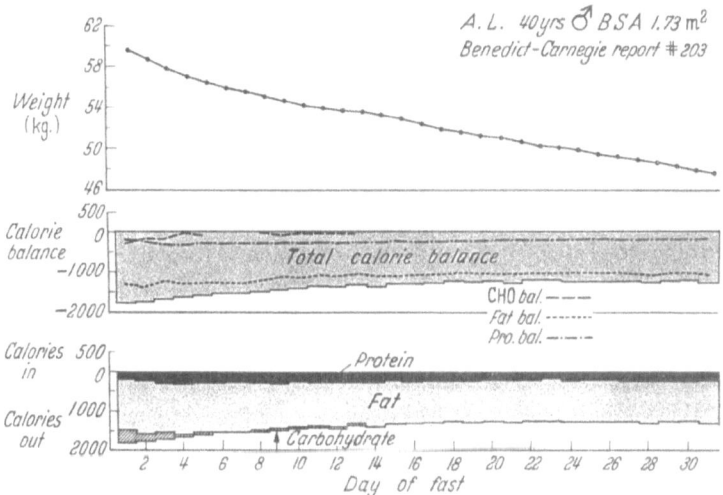

Fig. 3. The daily expenditure of fat, carbohydrate, and protein during human starvation. Note the parallel changes in basal metabolic expenditure and protein oxidation as indicated by nitrogen excretion. Data from BENEDICT (27)

noted in an adult male a weight loss of from 60.6 to 47.4 kg. associated with a total loss of 277 grams of nitrogen (Fig. 3). The level of nitrogen excretion steadily decreased from the fourth day (12.0 g.) to the eighteenth day (8.0 g.), while the heat production fell from 1750 calories on the second day to 1250 calories on the eighteenth day. Thereafter both nitrogen excretion and heat production remained at approximately the same levels. In the Minnesota experiment on semi-starvation (28), thirty-two men received an average of 1570 calories daily for six months following a control period that required roughly 3200 calories for nitrogen equilibrium. This regimen resulted in a 24 per cent weight loss (average starting weight was 69 kg.) and a total nitrogen loss which varied between 450 and 900 grams. During this time, the average BMR values per man had dropped by 39 per cent. Therefore, it appears that the body is able to conserve tissue during any significant period of starvation, by reducing both the basal level of nitrogen excretion and the basal level of energy expenditure.

The nutritional state of the surgical patient prior to studying nitrogen balance is an important factor in determining the extent of nitrogen loss (*29, 30*). But reports vary as to the effectiveness of large nutritional intakes in reducing the increased nitrogen loss of early convalescence in the previously well-nourished patient. Some investigators, after a study of patients undergoing elective cholecystectomy and subtotal gastrectomy, feel that post-operative weight loss and the wasting of nitrogen and potassium are primarily the result of the inadequate nutritional intake which is usually provided (*31*). Other workers have felt that the increased excretion during this phase is largely an obligatory loss representing a fundamental part of the body's response to injury (*32*).

In an effort to make a sharp distinction between the influence of total starvation in normal subjects and in patients undergoing operation, WILKINSON (*33*) performed balance studies on four volunteers and compared them with studies on fifteen patients undergoing various stomach operations. Both volunteers and surgical patients had their water intake stopped for two days and their food intake stopped for four days. It was clearly evident that the uninjured volunteers had smaller rates of weight loss and nitrogen excretion than the patients immediately after operation. Starvation alone was noted to cause only a small increase in potassium excretion and a slight reduction in sodium excretion. Under similar conditions of environmental temperature, humidity, and bodily activity, the combination of operation and starvation was shown to cause a large increase in nitrogen excretion within forty-eight hours, an increase in potassium excretion within twenty-four hours, and a sharp reduction in sodium excretion.

WILKINSON comments: "It seems unlikely that this response to injury can be readily modified, except by the prevention of starvation for food and water, and it is uncertain how far this can be profitably achieved in the biological sense by parenteral, as opposed to oral feeding. Modification of the portion of the response which can be attributed to the infliction of injury would require interference with the normal whole body response to injury, the nature of which is still imperfectly understood and the disturbance of which might do more to hinder healing than good in the prevention of tissue catabolism." (*33*)

Probably both points of view are to some extent correct, the differences having arisen, in part, from the study of different clinical situations. Starvation plays a definite role in the loss of body tissues after injury, but the dynamics of this loss are quite different from those of starvation alone. Minor trauma induces negligible losses of body nitrogen readily prevented by suitable feeding. As one progresses up the scale to more severe injury, larger and larger amounts of nitrogen and calories are required to maintain neutral balance in the post-traumatic patient. Large caloric intakes in the early convalescent phase can be expected to prevent some of the customary weight loss. But the provision of a high-calorie, high-nitrogen intake does not "cure" the fundamental metabolic alteration;

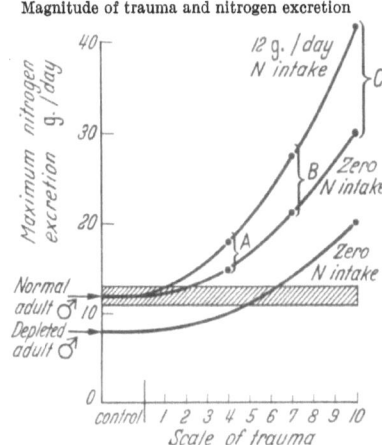

Magnitude of trauma and nitrogen excretion

Fig. 4. Approximate values for the maximum nitrogen excretion of an adult male after sustaining injury of increasing magnitude on a scale of one to ten

rather, it appears to provide enough amino acids to offset the increased nitrogen excretion and, therefore, achieve a neutral or sometimes positive balance. In Fig. 4 approximate values are shown for maximum nitrogen excretion of an adult male sustaining injury of increasing magnitude on a scale of one to ten. Moderate injury, such as an uncomplicated subtotal gastrectomy, is indicated at Point A, where the nitrogen excretion is 15 grams per day, with zero nitrogen intake. This excretion may be expected to increase to perhaps 18 grams per day with a 2000 calorie intake including 12 grams of nitrogen (which would have provided neutral balances of calories and nitrogen in the hospital before operation). It is evident that this magnitude of injury is associated with the additional loss of only a small to moderate proportion of the nitrogen intake (*34*), and the retained nitrogen will minimize the postoperative loss of lean tissue. The situation becomes more critical with more severe forms of injury. At Point B, as after multiple fractures, one might expect the peak nitrogen excretion on zero intake to reach 20 grams per day. In our experience, the provision of high-nitrogen intakes at the height of the nitrogen excretion may increase excretion by 50—70 per cent of the ingested nitrogen. The most extreme situation might be represented by a third-degree burn of over 50 per cent surface area, indicated in this figure at Point C. The nitrogen excretion here, with no intake, can be expected to reach 30 grams or more, and intravenous nitrogen therapy will increase this excretion so that extra nitrogen is excreted in amounts which approach that which has been infused. This extreme injury has so altered the "set" of the patient's tissues that he now utilizes one and a half to two times the calories needed for basal demands under normal conditions. Metabolic studies on burns of 35 per cent surface area or less (*35*) indicate that, as time goes by, the extremely high nitrogen loss will diminish, and nitrogen equilibrium can be achieved with progressively lower nitrogen intakes. The injury-induced change in nitrogen metabolism may therefore be described as "increasing the urinary loss when there is no intake and increasing the intake required to avoid loss." (*36*).

Does nitrogen loss after injury have any survival value?

It is important to consider the significance of nitrogen loss after injury in the light of the major pathways of intermediary metabolism. Such pathways can be considered most effectively in terms of the body composition of the average human adult male. Approximate values for adult body composition are shown in Fig. 5. About 40 per cent of the body consists of protein and depot fat, with a few hundred grams of glycogen representing total carbohydrate stores. Fig. 6 presents

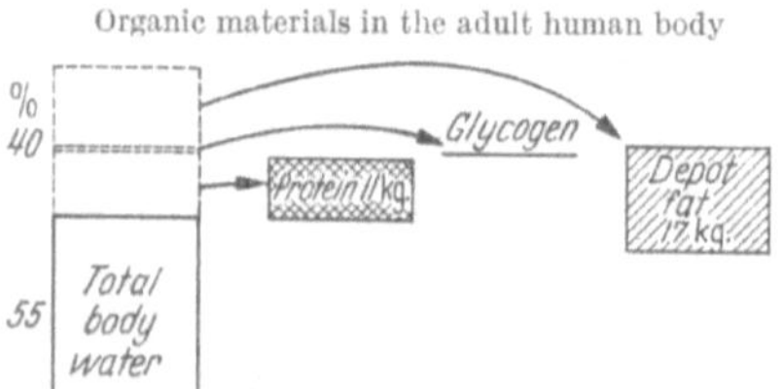

Fig. 5. Approximate values for the fat, carbohydrate, and protein content of an average adult 70 kg. male in relation to the water and mineral content of the body

some of the major pathways that relate fat, carbohydrate, and protein. Many details of intermediary metabolism have been deliberately omitted in order to emphasize certain clinical aspects of the pathways which are shown. It is recognized that the tissue amino acids include molecules which are ketogenic, or neutral, as well as the glucogenic ones indicated in the diagram, but the contribution to carbohydrate intermediates far outweighs the direct contribution of "two-carbon" fragments when amino acids are de-aminated. The non-essential amino acids, whose carbon chains can be synthesized in the body, are almost synonymous with the glucogenic amino acids. For the purposes of this discussion, triose phosphates, pyruvate, and the intervening members of the Embden-Meyerhoff pathway, as well as certain derivatives of the tricarboxylic acid cycle, have been indicated together in the figure

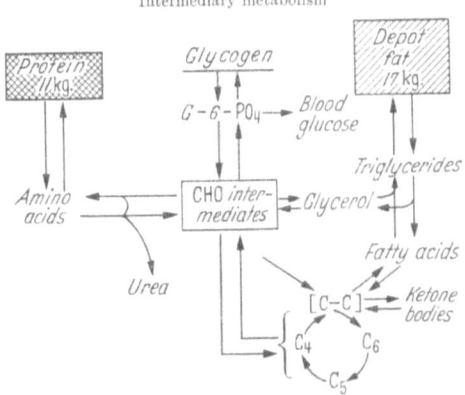

Fig. 6. Certain intermediary metabolic pathways which relate fat, carbohydrate, and protein in the body. See text for details

as "carbohydrate intermediates." It is important to recall that the pathway from pyruvate to acetylcoenzyme A is a one-way reaction, which is of both biochemical and clinical significance. Note that all three foodstuffs can supply "two-carbon" fragments which represent the basic unit of fuel, but that carbon chains from fatty acids have not been shown to provide a net gain of carbohydrate or protein (37, 38). Nutritional studies in man, and isotope work in animals and isolated systems, are consistent with the concept that the human body handles its energy sources on a priority basis, which normal nutrition tends to mask. These priorities are revealed only with a significant degree of undernutrition. The body's needs for foodstuffs may be listed in decreasing order of importance as follows:

1. Obligatory for immediate survival:

a) "Two-carbon fuel" for the tricarboxylic acid cycle to provide usable energy;

b) A continuous supply of intermediates for this cycle; and

c) Maintenance of blood glucose levels.

2. Obligatory for ultimate survival:

a) Synthesis of body protein, which seems to have its own spectrum of importance, ranging from hemoglobin and certain of the enzymes and hormones down to a wide variety of cellular proteins.

3. Energy storage in the form of:

a) Glycogen in liver and muscle, and

b) Fat depots.

Current evidence suggests that trauma probably does not introduce new pathways but inserts the wound high on the priority list for protein synthesis.

Clinicians have considered nitrogen loss after injury as one of the following: 1) the result of starvation; 2) a detoxification mechanism; 3) a means of providing extra materials for healing the wound; or 4) the destruction of protein to meet the sudden demand for extra fuel. These suggestions seem unlikely for various reasons. Urea is the major form of increased nitrogen excretion and is relatively non-toxic. The amount of protein represented by the extra nitrogen loss is often far greater than the calculated requirements for new wound protein, and surgical wounds frequently heal satisfactorily in the absence of any extra nitrogen loss when there has been pre-operative depletion. The most common suggestion, that protein is being destroyed to supply tissue fuel, is usually based upon the concept that there are extremely large caloric demands after injury. Studies from this laboratory have shown that these caloric increases seldom exceed one and a half times the normal basal metabolic expenditure, in contrast to the usual estimates of two or three times normal. The extra calorie needs appear to be supplied by the oxidation of fat without evidence of ketosis, and the extra nitrogen loss provides relatively few calories. Therefore, it seems more reasonable to consider the nitrogen loss after trauma as an outward reflection of a mechanism which allows an increased supply of carbohydrate intermediates at a time of actual or threatened loss of exogenous carbohydrate, and the possibility of increased tissue requirements. Body protein represents the only large tissue source, since fatty acids cannot supply carbohydrate intermediates.

There is also the possibility that injury is followed by increased protein break-down, in tissues such as muscle, in order to supply increased amounts of essential amino acids to the liver which is charged by nature with the responsibility for the synthesis of large amounts of extracellular protein. Since the advent of microbiological assay and chromatography, information regarding the concentration of individual amino acids in the plasma has begun to appear in the literature. Certain workers have wondered if there was a specific need for extra amounts of some particular amino acid after acute injury or disease. Experimental injury in animals has suggested that additional methionine might improve the healing of skin wounds (39), but this finding remains controversial, and no comparable situation has been shown to exist in man. The partition of nitrogen by the liver after injury does not suggest that any one amino acid is the limiting factor in protein metabolism (6). Additional nitrogen intake after injury is associated with an increase in urinary nitrogen as urea plus ammonia. Other nitrogenous materials are not consistently affected. Therefore, if the response is related to the provision of essential amino acids, it would appear to be a need for many rather than a single molecular configuration.

Is the nitrogen loss due to endocrine secretion?

Recent information has been summarized regarding the endocrine activity which has been observed in association with human injury (40). An increase in the secretion of one or more of the following hormones:— adrenal glucocorticoids, catecholamines or thyroid hormones — has been suggested as the mechanism producing the characteristic alterations in energy exchange and protein metabolism following human injury.

The observations of BROWN (41), SELYE (42), and ALBRIGHT (43) and their associates, during the decade of World War II, centered interest on the adrenocortical hormones following injury. Subsequent studies have confirmed the rise in blood and urinary 17-hydroxycorticoid levels after ether anesthesia (44), elective operations (45), and unanesthetized injury (46). Since one of the principal actions of glucocorticoids is to increase urinary urea excretion in association with increased gluconeogenesis (47), and since the urinary nitrogen loss was increased in a similar manner (mainly as urea) when rats were subjected either to fracture of a femur or to subcutaneous implantation of cortisone pellets (48), the entire metabolic response to injury has been interpreted by some as the result of a simple increase in adrenocortical secretion. But the demonstration of the "permissive" role of glucocorticoids (49, 50), and the observation that the metabolic changes, including nitrogen loss, occurred after injury in adrenalectomized animals (51) as well as adrenalectomized man (52) receiving a constant dose of hormone, suggested that the response to injury involved other factors than adrenal secretion. The interpretation of the role of the adrenal is made particularly difficult, since injury may not only stimulate glucocorticoid secretion, but also modify the intermediary metabolism and excretion of the steroid molecule (45, 51, 53).

There is much clinical evidence, such as tachycardia, vasoconstriction, and narrowed pulse pressure, to suggest sympathomedullary secretion in numerous surgical situations. Studies of post-operative patients have often failed to show increased secretion of epinephrine and norepinephrine (54, 55) unless complications were present. Increased circulating levels of both materials have been demonstrated following major skeletal and thermal trauma (56, 57). It seems probable that smaller elevations in circulating catecholamines will be demonstrated with lesser degrees of injury as more sensitive assay methods are utilized. However, the significance of catecholamine secretion upon protein metabolism after injury remains to be established.

The occurrence of an increased oxygen consumption after injury in animals and man has naturally directed attention toward the thyroid. COPE and associates (23) found normal radioactive iodine uptake and protein-bound iodine levels during the convalescence of surgical patients, a finding which was associated with increased oxygen consumption. GOLDENBERG, HAYES, and associates (58, 59) have studied thyroid and adrenal function after operation and suggested that thyroid activity may increase or decrease after injury as a primary event before alterations in adrenal function. However, other laboratories have not been able to demonstrate a primary role for the thyroid in the metabolic response to injury.

Decreased anabolism versus increased catabolism

The excess urinary nitrogen after injury could reflect an increased protein break-down or a decreased protein synthesis. REISS (60), studying infections in rats, MADDEN (61), studying turpentine abscesses in dogs, LEVENSON and associates (62), studying burns in rats, and BLOCKER and associates (63), studying human burns, were all able to demonstrate that amino-acid incorporation into certain proteins was normal or increased at the height of the reaction to infection or injury. Therefore, the increased urinary nitrogen excretion under these circumstances could not be due to a simple decrease in anabolism. Both anabolism and catabolism might be accelerated, with catabolism predominating, to account for the net loss of nitrogen. But all tissues do not participate equally in this response. The amino-acid incorporation in the above studies dealt with the proteins of rat liver and kidney (*60*), dog plasma protein (*61*), proteins of rat liver, gastro-intestinal tract, and plasma protein (*62*), and human plasma protein (*63*). In burned rats (*62*), the protein content of organs such as the liver had a markedly increased turnover rate, but changed very little in absolute amount. In contrast, the protein content of the carcass, despite a slower turnover rate, fell and accounted, in fact, for most of the extra urinary nitrogen loss. Starved rats with infection (*60*) showed no decrease in the protein content of the liver and kidneys, whereas starved control animals lost protein in these organs. This is consistent with the differences in the influence of glucocorticoids on protein metabolism, which depends upon the particular tissue or organ being studied. It appears that glucocorticoids may lead to increased protein synthesis in the liver, despite the induction of a negative nitrogen balance elsewhere in the body, especially in muscle.

Internal versus external nitrogen balance

A consideration of alterations in body protein on the basis of changes in conventional nitrogen balance measurements overlooks the central role of free amino acids in nitrogen dynamics. Fig. 7 represents an effort to correlate "external" or conventional nitrogen balance with a tentative framework for considering the "internal" nitrogen balance which exists between free amino acids and certain major groups of body proteins. The massive amount of literature regarding the "external" nitrogen balance of man under varying circumstances is in sharp contrast to the limited studies which have explored the "internal" nitrogen balance of either animals or man.

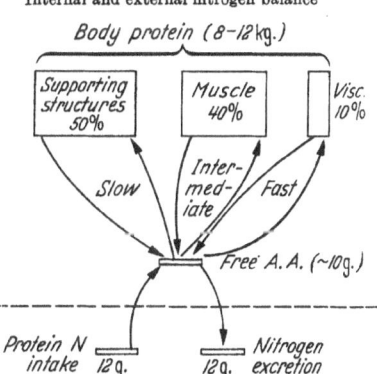

Internal and external nitrogen balance

Fig. 7. A tentative view of the "internal" nitrogen balance of the human body, in contrast to the "external" or conventional balance

A normal 70 kg. adult male has 8 to 12 kg. of protein (*64*), which can be divided for convenience into three groups of protein: 1. Approximately half existing in collagen, elastin, dermis, bone matrix, and other supporting structures; 2. Forty per cent as muscle protein; and 3. The remaining ten per cent as visceral protein (*65*).

Only limited evidence is available on overall rates of protein synthesis in man (*66, 67*), but it appears to be consistent with the idea
that visceral protein has a more rapid turnover than muscle
protein. This is of particular interest in view of the estimate of
GRANDE (*68*) that fifty per cent of the basal oxygen consumption
of the human body is required for liver, brain, and kidneys, while an
additional thirty-five per cent is needed for heart and skeletal
muscle. Thus, the tissues with a rapid protein turnover appear to
have a high resting oxygen consumption.

Reported figures for free amino acids in blood and tissues suggest
that the total amount in the body is in the range of 10 or 15 grams
(*69*), and this is in agreement with isotope studies in man (*66*).
The free amino acids of various tissues have been found to be five
to ten times greater than the circulating level in plasma (*70*). The
normal plasma level of four to eight mg. per cent of alpha amino
nitrogen is remarkably resistant to change. An increase of plasma
amino acids up to 20 per cent may occur after meals for brief
periods. The relative capacity of the various tissues to take up
circulating amino acids after intravenous injection (*69*) reveals
that the liver will acquire the greatest amount but that the kidney
also participates significantly. Other tissues, such as muscle, remove
much smaller quantities of injected amino acids. However, the
uptake in liver may be relatively transient, since this tissue rapidly
deals with cytoplasmic excesses of amino acid largely by a) incorporation into protein (*71*) or b) de-amination with the production
of urea (*72*).

The specific dynamic action of protein has been shown to be
associated with the de-amination of amino acids, either in the
metabolism of the carbon residues or in the synthesis of urea.
DOCK (*73*) studied the rate of oxygen consumption in rats receiving
casein, immediately before and after ligating the blood supply to
various regions of the body. There was a prompt decrease in oxygen
consumption of the abdominal viscera after feeding protein, and it
was concluded that at least 80 per cent of the specific dynamic
action following protein ingestion is due to the increased energy
liberated by hepatic cells. MYERS (*74*) has demonstrated the prompt
increase in oxygen consumption and urea production across the
human splanchnic circulation following intravenous injections of
amino acids. Since the nitrogen loss after injury is associated with
an increased production and excretion of urea, there seems little
doubt that the liver is of central importance in understanding
altered protein metabolism after injury. CAIRNIE and associates (*75*)
have shown that a fracture of the femur in rats is followed by an

increase in the minimum metabolic rate corresponding to the specific dynamic action which might have been expected from the increased oxidation of protein (Fig. 8).

Regardless of the significance of the increased nitrogen loss after injury, the increase in urinary urea must have arisen from an

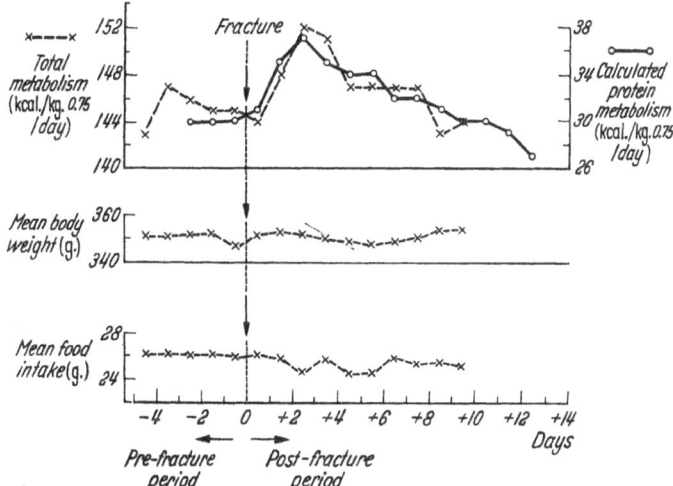

Fig. 8. The parallel response of caloric expenditure and nitrogen excretion in a rat sustaining a fracture of a long bone (75). (This is reproduced with the kind permission of Dr. D. P. CUTH-BERTSON and the Editor of the Brit. J. Exper. Path.)

increase in amino-acid concentration in the cytoplasm of liver cells. This could have occurred by one or more of the following five mechanisms:

The cytoplasmic protein of liver cells could have undergone break-down to yield a local excess of amino acids. It has been shown by many workers that the protein content of the liver is dependent upon the quantity of the dietary protein intake. ADDIS (76) and co-workers demonstrated that on fasting the liver lost protein more rapidly than any other tissue. KOSTERLITZ (77) studied the influence of diet on rat liver protein and confirmed the responsiveness of liver protein to dietary protein intake; he proposed that the easily lost fraction of liver cytoplasm be referred to as "labile" liver cytoplasm. Further work from the same laboratory (78) indicated that at least 60 per cent of the extra nitrogen excreted in the urine during the first five days of a protein-free diet for rats could be accounted for by the nitrogen loss from the liver. Information of this type has been quoted in support of a significant proportion of nitrogen loss arising from liver cytoplasm after human injury (*79*). However, the protein content of liver cytoplasm may not behave in the same way after injury as after simple protein deprivation. As noted previously, rats with infections and burns did not show a decrease in liver protein content, despite increases in the rate of protein turnover (*60, 62*).

Amino-acid concentration in liver cytoplasm might be increased after injury because of an absolute increase in the rate of hepatic blood flow, thereby presenting more amino acids to the liver per unit time. No direct information is available on hepatic blood flow after operation in man, but, if there were a major increase, one might expect some increase in the resting cardiac output. Studies on cardiac output after elective abdominal operations (*83, 84, 85*) indicate a mild to moderate increase during the first 24 hours, followed by a mild decrease below pre-operative levels for the following five to seven days.

An increased amino-acid concentration in the liver cytoplasm might also occur if the extraction ratio of amino acids from the hepatic circulation were increased after trauma. CHRISTENSEN and co-workers (*86*) have suggested from studies with alpha-aminoisobutyric acid that glucocorticoids may increase the hepatic uptake of amino acids. Clear-cut evidence on this point has not been obtained with naturally occurring amino acids.

Another mechanism for increased amino acids in the cytoplasm of liver cells would be an increase in the concentration of amino acids in the portal blood stream. CUTHBERTSON (*80*) has recently presented data showing the rapid increase in the amount of nitrogen in the contents of the duodenum and upper part of the small intestine following injury in the sheep and in the rabbit. Whether this would provide an increase in amino acids presented to the liver remains to be determined.

Extra amino acids might reach the liver cytoplasm owing to a rise in the concentration of amino acids in the systemic circulation. This could occur as a result of the rate of protein break-down exceeding the rate of synthesis in tissues such as muscle. INGLE and co-workers (*81*) and BONDY and co-workers (*82*) have shown that glucocorticoids produce a negative nitrogen balance in muscle tissue. As mentioned above, human injury is often associated with increased levels of 17-hydroxycorticoids in the blood and in the urine. However, circulating corticoid levels are often elevated for 36 hours or less after operation and may be normal during the period of maximum increase in nitrogen loss. Studies in man under circumstances of increased nitrogen excretion, but without circulatory failure, have shown decreases after infection (91) as well as after surgical operations (92). The absence of increased levels of alpha amino nitrogen in the blood following operations, where it is probable that there is increased glucocorticoid secretion with aminoacid output from muscle, suggests that some tissue or organ has an increased amino-acid uptake. Levels of amino nitrogen or amino acids were normal or decreased in battle casualties (93, 94) and were within normal ranges in severe burns after hemoconcentration was no longer present (95), but were reported to be elevated during the first 24 hours (96). This lack of consistent elevation of circulating amino acids is of particular interest, since a significant aminoaciduria was observed in both battle casualties and burns, although the major urinary nitrogen loss was in the form of urea. These findings are in contrast to hemorrhagic shock, which is characterized by rising levels of amino acid nitrogen in the rat (87, 88) and in the dog (89), with decreased urea formation. Studies on rat liver slices after hemorrhagic shock (90) have shown decreases in oxygen uptake, de-amination, and urea synthesis.

EVERSON and FRITSCHEL (97) found that the essential amino acids in plasma were decreased by surgical operation and by ether anesthesia alone, but returned to normal levels in one to three days after operation, while MAN (92) reported a decrease in total amino nitrogen of plasma for a longer period, which corresponded to the interval of increased nitrogen excretion. LEVENSON and associates (94) found that the first two weeks after injury in soldiers were associated with a pattern of plasma amino acids which did not correspond to whether they were essential or non-essential amino acids. They were grouped as follows: glycine, histidine, threonine, proline, and glutamic acid remained near normal, whereas leucine, isoleucine, lysine, valine, tyrosine, and alanine rose moderately during the first week and then fell, while phenylalanine and methionine also rose during the first week but to a greater degree. Taurine bore little or no relationship to the other amino acids, showing marked elevations at unpredictable times. Two patients in this study who became markedly malnourished showed decreased total plasma amino nitrogen levels, in agreement with the findings of others (92, 97). NARDI (95) observed that the amino-aciduria accompanying severe burns consisted of an increase in the excretion of non-essential amino acids of lower molecular weight normally found in the urine, together with the essential amino acids of higher molecular weight which were usually not present in the urine.

19*

There was some suggestion that both the qualitative and quantitative aspects of the amino-aciduria might be in proportion to the severity of the trauma. Patients who were not burned severely enough to endanger their lives, and patients undergoing minor surgical procedures, showed a minimal amino-acid excretion with a preponderance of non-essential amino acids. Those dying of their burns had a massive excretion of both essential and non-essential amino acids. EMERSON and BINKLEY (98) administered a synthetic mixture of ten essential amino acids to young male soldiers one to two months after they had received severe penetrating wounds. For a period of one to two weeks following the administration to subjects who were already receiving a high-protein, high-calorie diet, they were able to demonstrate a marked increase in nitrogen retention, which exceeded by two to three times the total increase in nitrogen intake.

LEVENSON and associates (94) were able to demonstrate a heterogenous amino-acid peptide or conjugate in plasma samples of battle casualties. The amino-acid composition of this material varied from patient to patient and in the same patient from day to day. In general, the patients with the highest non-protein nitrogen in the blood had this substance in greatest concentration. EADES and associates (99) also reported the urinary excretion of peptides following human burns. The significance of these materials remains uncertain.

Summary

The metabolic response to injury is discussed in terms of both alterations in protein metabolism and energy exchange. Injury appears to produce parallel increases in nitrogen excretion and basal metabolic expenditure.

The nitrogen loss after injury is discussed in terms of: a) the role of starvation and immobilization; b) the possible survival value; c) the relation to endocrine stimuli; and d) decreased anabolism versus increased catabolism.

The central role of circulating amino acids and hepatic de-amination is emphasized in comparing "internal" nitrogen dynamics with "external" or conventional nitrogen balance information.

Five mechanisms are considered which could produce an increased level of amino acids in the cytoplasm of liver cells and thereby account for the characteristic increase in urea production and excretion which follows injury.

Zusammenfassung

Die Reaktionen des Stoffwechsels auf schwere Verletzungen werden im Hinblick auf Änderungen des Eiweißumsatzes und auf den Energiehaushalt besprochen. Als Folge eines Traumas steigen gleichzeitig die Stickstoffausscheidung und der Grundumsatz an.

Der Stickstoffverlust nach Verletzungen wird unter folgenden Gesichtspunkten diskutiert: a) die Rolle von Nahrungskarenz und Ruhigstellung,

b) die mögliche Bedeutung für das Überleben, c) die Beziehungen zum endokrinen System und d) verminderter Anabolismus gegenüber gesteigertem
Katabolismus.

Beim Vergleich der Dynamik des intermediären Stickstoffumsatzes mit
den Schlußfolgerungen, die sich aus den üblichen Bilanzuntersuchungen
ergeben, wird auf die Bedeutung zirkulierender Aminosäuren und der
Desaminierung in der Leber hingewiesen.

Es werden fünf verschiedene Mechanismen in Betracht gezogen, die eine
erhöhte Konzentration von Aminosäuren im Cytoplasma der Leberzellen
und dadurch die charakteristische Zunahme der Harnstoffbildung und -aus-
scheidung hervorrufen können, wie sie als Folge einer schweren Verletzung
auftritt.

Résumé

L'auteur discute des modifications du métabolisme sous l'effet de traumatismes graves en ce qui concerne le métabolisme des protéines et le bilan
énergétique. Après tout traumatisme, il y a augmentation simultanée de
l'excrétion d'azote et du métabolisme de base.

La déperdition azotée consécutive aux traumatismes est étudiée sous les
angles suivants: a) rôle de la sous-alimentation et de la mise au repos, b) importance pronostique éventuelle, c) rapports avec le système endocrinien
et d) diminution de l'anabolisme face à l'augmentation du catabolisme.

En comparant la dynamique du métabolisme intermédiaire de l'azote
et les conclusions fournies par les bilans azotés usuels, on attire l'attention
sur le rôle des acides aminés circulants et de la désamination dans le foie.

On envisage cinq mécanismes différents qui pourraient déterminer un
accroissement de la concentration des acides aminés dans le cytoplasme des
cellules hépatiques et par là l'augmentation caractéristique de la production
et de l'excrétion d'urée après un traumatisme grave.

Acknowledgement

This work was supported in part by grants from the National Institute
of Arthritis and Metabolic Diseases (A-815), the John A. Hartford Foundation,
and the U.S. Army Medical Research and Development Command.

References

1. CUTHBERTSON, D. P.: Quart. J. Med. (G.B.) 1, 233 (1932). — 2. TAY-
LOR, F. H. L., S. M. LEVENSON, C. S. DAVIDSON, N. C. BROWDER, and C. C.
LUND: Ann. Surg. (U.S.A.) 118, 215 (1943). — 3. CoTUI, M. D., A. M.
WRIGHT, J. H. MULHOLLAND, B. CARABBA, I. BARCHAM, and V. J. VINCI:
Ann. Surg. (U.S.A.) 120, 99 (1944). — 4. HOWARD, J. E., W. PARSON, K. E.
STEIN, H. EISENBERG, and V. REIDT: Bull. Johns Hopkins Hosp. (U.S.A.)
75, 156 (1944). — 5. SPENCE, H. Y., E. I. EVANS, and J. C. FORBES: Ann.
Surg. (U.S.A.) 124, 131 (1946). — 6. PETERS, J. P.: Amer. J. Med. 5, 100
(1948). — 7. WERNER, C. S.: Ann. Surg. (U.S.A.) 126, 175 (1947). — 8. POL-
LACK, H., and S. L. HALPERN: Therapeutic Nutrition. Washington, D. C.
National Academy of Sciences, National Research Council, 1952. Publ. 234,
p. 47. — 9. RHOADS, J. E.: Surg. Gyn. Obstetr. (U.S.A.) 94, 417 (1952). —
10. MOORE, F. D., and M. BALL: The Metabolic Response to Surgery. C. C.
Thomas, Springfield, 1952. — 11. ABBOTT, W. E., S. LEVEY, and H. KREIGER:
Metabolism (U.S.A.) 8, 847 (1959). — 12. BEAL, J. M., G. M. CORNELL, and

H. GILDER: Surgery (U.S.A.) **36**, 468 (1954). — 13. DENOTTER, G.: Arch. chir. Neerl. **7**, 227 (1955). — 14. CANNON, P. R.: J. Amer. Med. Ass. **135**, 1043 (1947). — 15. KINNEY, J. M.: Bull. N. Y. Acad. Med. **36**, 617 (1960). — 16. BENEDICT, F. G., and T. M. CARPENTER: Food ingestion and energy transformation. Washington, D. C. Carnegie Institution of Washington, 1918. Pub. 261. — 17. LUSK, G.: Science of Nutrition, 4th Edition. W. B. Saunders, Philadelphia, 1928. — 18. SOSKIN, S., and R. LEVINE: The role of carbohydrates in the diet. In: Modern nutrition in health and disease. 2nd Edition. Ed. by WOHL, M. G., and R. S. GOODHART. Lea and Febiger, Philadelphia, 1960, pp. 152—171. — 19. MUNRO, H. N.: Physiol. Rev. (U.S.A.) **31**, 449 (1951). — 20. CAHILL, J. L., J. M. KINNEY, S. NICHOLS, and L. S. BRANN: Surg. Forum. (U.S.A.) **11**, 93 (1960). — 21. ZAREM, H. A., J. M. KINNEY, and S. NICHOLS: Surg. Forum. (U.S.A.) **11**, 450 (1960). — 22. RABELO, A., R. G. CLARK, and J. M. KINNEY: Surg. Forum (U.S.A.) **12**, 462 (1961). — 23. COPE, O., G. L. NARDI, M. OUIJANO, R. L. ROVIT, J. B. STANBURY, and A. WIGHT: Ann. Surg. (U.S.A.) **137**, 165 (1953). — 24. DUBOIS, E. F.: Basal metabolism in health and disease. 3rd Edition. Lea and Febiger, Philadelphia, 1936. — 25. DEITRICK, J. E., G. D. WHEDON, and E. SHORR: Amer. J. Med. **4**, 3 (1948). — 26. CUTHBERTSON, D. P.: Biochem. J. (G.B.) **23**, 1328 (1929). — 27. BENEDICT, F. G. L.: A study of prolonged fasting. Washington, D. C. Carnegie Institution of Washington, 1915. — 28. KEYS, A., J. BROZEK, A. HENSCHEL, O. MICKELSON, and H. L. TAYLOR: The biology of human starvation. Vols 1 & 2. Univ. Minn. Press, Minneapolis, 1950. — 29. MUNRO, H. N., and M. I. CHALMERS: Brit. J. Exper. Path. **26**, 396 (1945). — 30. FORSYTH, B. T., M. E. SHIPMAN, and I. C. PLOUGH: J. Clin. Invest. (U.S.A.) **34**, 1653 (1955). — 31. ABBOTT, W. E., H. KREIGER, and S. LEVEY: Lancet (G.B.) **1958/I**, 704. — 32. VANITALLIE, T. B., F. D. MOORE, R. P. GEYER, and F. J. STARE: Surgery (U.S.A.) **36**, 720 (1954). — 33. WILKINSON, A. W.: Lancet (G.B.) **1961/II**, 783. — 34. STRICKLER, J. H., C. O. RICE, and A. E. TRELOAR: Surgery (U.S.A.) **39**, 152 (1956). — 35. SOROFF, H. S., E. PEARSON, and C. P. ARTZ: Surg. Gyn. Obstetr. (U.S.A.) **112**, 159 (1962). — 36. MOORE, F. D.: Canad. Med. Ass. J. 788, 85 (1958). — 37. PETERS, J. P., and D. D. VAN SLYKE: Quantitative clinical chemistry. Vol. I. Interpretations. Williams & Wilkins, Baltimore, 1946, p. 640. — 38. WHITE, A., P. HANDLER, E. L. SMITH, and S. DE WITT: Principles of biochemistry. McGraw-Hill, New York, 1959, p. 469. — 39. CROFT, P. B., and R. A. PETERS: Lancet (G.B.) **1945/I**, 266. — 40. MOORE, F. D.: Recent Progr. Hormone Res. (U.S.A.) **13**, 511 (1957). — 41. VENNING, E. H., M. M. HOFFMAN, and J. S. L. BROWNE: Endocrinology (U.S.A.) **35**, 49 (1944). — 42. SELYE, H.: J. Clin. Endocr. (U.S.A.) **6**, 117 (1946). — 43. ALBRIGHT, F.: Cushing's Syndrome: its pathological physiology and its connection with the problem of the reaction of the body to injurious agents. Harvey Lecture Series **38**, 123 (1943). — 44. HAMMOND, W. G., L. D. VANDAM, J. M. DAVIS, R. D. CARTER, M. R. BALL, and F. D. MOORE: Ann. Surg. (U.S.A.) **48**, 199 (1958). — 45. STEENBURG, R. W., R. LENNIHAN, and F. D. MOORE: Ann. Surg. (U.S.A.) **143**, 180 (1956). — 46. HUME, D. M., D. H. NELSON, and D. W. MILLER: Ann. Surg. (U.S.A.) **143**, 316 (1956). — 47. THORN, G. W., G. F. CAHILL, Jr., and A. E. REYOLD: Bull. Acad. méd. Belgique **25**, 74 (1960). — 48. CAMPBELL, R. M., G. SHARP, A. W. BOYNE, and D. P. CUTHBERTSON: Brit. J. Exper. Path. **35**, 566 (1954). — 49. INGLE, D. J., E. O. WARD, and M. H. KUIZENGA: Amer. J. Physiol. **149**, 510 (1947). — 50. ENGEL, F. L.: Recent Progr. Hormone Res. (U.S.A.) **6**, 277 (1951). — 51. STEENBURG, R. W., and W. F. GANONG:

Surgery (U.S.A.) **38**, 92 (1955). — 52. MASON, A. S.: Lancet (G.B.) **1955/I**, 632. — 53. GOLD, N. I., D. A. MACFARLAND, and F. D. MOORE: J. Clin. Endocr. (U.S.A.) **16**, 282 (1956). — 54. FRANKSSON, C., C. A. GEMZELL, and U. S. VON EULER: J. Clin. Endocr. (U.S.A.) **14**, 608 (1954). — 55. HAMMOND, W. G., L. ARONOW, and F. D. MOORE: Ann. Surg. (U.S.A.) **144**, 715 (1956). — 56. GOODALL, M., C. STONE, and B. W. HAYNES, Jr.: Ann. Surg. (U.S.A.) **145**, 479 (1957). — 57. PEKKARINEN, A.: The effect of operations and physical injury on the adrenal glands in the vegetative nervous system in man. In: The Biochemical Response to Injury. Ed. by STONER, H. B., and C. J. THRELFALL. Blackwell, Oxford, 1960. — 58. GOLDENBERG, I. S., P. J. ROSENBAUM, and M. A. HAYES: Ann. Surg. (U.S.A.) **142**, 786 (1955). — 59. GOLDENBERG, I. S., P. J. ROSENBAUM, C. WHITE, and M. A. HAYES: Surg. Gyn. Obstetr. (U.S.A.) **104**, 295 (1957). — 60. REISS, E.: Metabolism (U.S.A.) **8**, 151 (1959). — 61. MADDEN, S. C.: Plasma protein formation in diseased states. In: Plasma Proteins. Ed. by YOUMAN, J. B. C. Thomas, Springfield, 1950, p. 62. — 62. LEVENSON, S. M., J. W. BRAASCH, H.MUELLER, and L. CROWLEY: Nitrogen metabolism following thermal injury. Quoted in: LEVENSON, S. M., E. J. PULASKI, and H. L. UPJOHN: Metabolic changes associated with injury. In: Physiologic principles of surgery. Ed. by ZIMMER-MAN, L., and R. LEVINE. W. B. Saunders, Philadelphia, 1957. — 63. BLOCKER, T. G., Jr., W. C. LEVIN, W. W. NORWINSKI, S. R. LEWIS, and V. BLOCKER: Ann. Surg. (U.S.A.) **141**, 589 (1955). — 64. FORBES, R. M., A. R. COOPER, and H. H. MITCHELL: J. Biol. Chem. (U.S.A.) **203**, 359 (1953). — 65. MAGNUS-LEVY, quoted in: WEST, E. R., and W. R. TODD: Textbook of Biochemistry, 2nd Edition. Chapter 8. MacMillan, New York, 1957. — 66. SAN PIETRO, A., and D. RITTENBERG: J. Biol. Chem. (U.S.A.) **201**, 457 (1953). — 67. TSCHUDY, D. P., H. BACCHUS, S. WEISSMAN, D. M. WATKIN, M. EUBANKS, and J. WHITE: J. Clin. Invest. (U.S.A.) **38**, 892 (1959). — 68. GRANDE, F.: Nutrition and energy balance in body composition studies. In: Techniques for Measuring Body Composition. Ed. by BROZEK, J., and A. HENSCHEL. Washington National Academy of Sciences, National Research Council, 1961. — 69. VAN SLYKE, D. D.: Science (U.S.A.) **95**, 259 (1942). — 70. WHITE, A., P. HANDLER, E. L. SMITH, and S. STETTEN: Principles of Biochemistry. McGraw-Hill, New York, 1959, p. 498. — 71. MCMENAMY, R. H., W. C. SHOEMAKER, J. E. RICHMOND, and D. ELWYN: Amer. J. Physiol. **202**, 407 (1962). — 72. VAN SLYKE, D. D., G. E. COLLEN, and F. C. MCLEAN: Proc. Soc. Expcr. Biol. Med. (U.S.A.) **12**, 93 (1915). — 73. DOCK, W.: Amer. J. Physiol. **97**, 117 (1931). — 74. MYERS, J. D.: The circulation in the splanchnic area. Fourth Conference on Shock. Josiah Macy, Jr. Foundation, N. Y., 1955. — 75. CAIRNIE, A. B., R. M. CAMPBELL, J. D. PULLAR, and D. P. CUTHBERTSON: Brit. J. Exper. Path. **38**, 504 (1957).   76. ADDIS, T., L. J. Poo, and W. LEW: J. Biol. Chem. (U.S.A.) **116**, 343 (1936). — 77. KOS-TERLITZ, H. W.: J. Physiol. (G.B.) **106**, 194 (1947). — 78. CAMPBELL, R. M., and H. W. KOSTERLITZ: Biochem. J. (G.B.) **43**, 416 (1948). — 79. TAYLOR, H. L., and A. KEYS: Ann. N. Y. Acad. Sc. **73**, 465 (1958). — 80. CUTHBERT-SON, D. P.: The disturbance of protein in metabolism following physical injury. In: The Biochemical Response to Injury. Ed. by STONER, H. B., and C. J. THRELFALL. Blackwell, Oxford, 1960, p. 193. — 81. INGLE, D. J., M. C. PRESTRUDE, and J. F. NEGAMUS: Proc. Soc. Exper. Biol. Med. (U.S.A.) **67**, 321 (1948). — 82. BONDY, D. K., D. J. INGLE, and R. E. MEEKS: Endo-crinology (U.S.A.) **55**, 354 (1954). — 83. CARLSTEN, A., O. NORLANDER, and L. TROELL: Surg. Gyn. Obstetr. (U.S.A.) **99**, 227 (1954). — 84. CLOWES, G. H. A., Jr., L. R. DELGUERCIO, and J. BARWINSKY: Arch. Surg. (U.S.A.)

81, 212 (1960). — 85. HEILBRUNN, A., and F. F. ALBRITTEN: Ann. Surg. (U.S.A.) **152**, 197 (1960). — 86. NOALL, M. W., T. R. RIGGS, L. W. WALKER, and H. N. CHRISTENSEN: Science (U.S.A.) **126**, 1002 (1957). — 87. ENGLE, F. L., M. G. WINTON, and C. N. H. LONG: J. Exper. Med. (U.S.A.) **77**, 397 (1943). — 88. RUSSELL, J. A., C. N. H. LONG, and F. L. ENGLE: J. Exper. Med. (U.S.A.) **79**, 1 (1944). — 89. VANSLYKE, D. D., R. A. PHILLIPS, P. B. HAMILTON, R. M. ARCHIBALD, V. P. DOLE, and K. EMERSON, Jr.: Transact. Ass. Amer. Physicians **58**, 119 (1944). — 90. WILHELMI, A. E.: Ann. Rev. Physiol. (U.S.A.) **10**, 259 (1948). — 91. FARR, L. E., W. C. MCCARTHY, and T. FRANCIS, Jr.: Amer. J. Med. Sc. **203**, 668 (1942). — 92. MAN, E. B., P. G. BETTCHER, C. M. CAMERON, and J. P. PETERS: J. Clin. Invest. (U.S.A.) **25**, 701 (1946). — 93. GREEN, H. N., H. B. STONER, H. J. WHITELY, and D. EGLIN: Clin. Sc. (G.B.) 8, 65 (1949). — 94. LEVENSON, S. M., J. M. HOWARD, and H. ROSEN: Surg. Gyn. Obstetr. (U.S.A.) **101**, 35 (1935). — 95. NARDI, G. L.: J. Clin. Invest. (U.S.A.) **33**, 847 (1954). — 96. LEVENSON, S. M., M. A. ADAMS, R. W. GREEN, C. G. LUND, and F. H. L. TAYLOR: N. England J. Med. **235**, 467 (1946). — 97. EVERSON, T. C., and M. J. FRITSCHEL: Surgery (U.S.A.) **30**, 931 (1951). — 98. EMERSON, K., and O. F. L. BINKLEY: J. Clin. Invest. (U.S.A.) **25**, 184 (1946). — 99. EADES, H., Jr., R. L. POLLACK, and J. D. HARDY: J. Clin. Invest. (U.S.A.) **34**, 1756 (1953).

Discussion

GEMZELL: I wonder whether you have been testing growth hormone in connection with surgical operations. The reason why I ask is that we examined the content of growth hormone in the pituitaries of patients of various kinds and found that it was extremely low in those who had been operated upon and had died a day or two after the operation. We also found low values in patients who had died following severe burns. The values were less than 1 mg. per pituitary, whereas patients who had died following an accident had levels of about 3—4 mg. per pituitary. In view of this finding we employed a biological technique in an attempt to determine growth hormone in the serum of patients who had been operated upon. We treated the serum with TCA, injected the precipitate into hypophysectomised rats, and examined the epiphyseal lines. We found that, within 5 or 6 days after the operation, the growth-hormone content was elevated, whereas later we were unable to detect any growth hormone. We also recorded elevated levels of growth hormone in patients 4 to 5 days after sustaining severe burns. This finding led us to give injections of growth hormone in patients with major burns, starting at about the 5th or 6th day after the injury and continuing for about 10 days. The results show that the nitrogen excretion decreased markedly and that the status of the patients improved. We also gained the impression that the patients were easier to handle and that their wounds healed better. It is of course very difficult to make any definite statements in this connection, because, after the short treatment with growth hormone, it was a long time before these patients completely recovered.

KINNEY: I think this is fascinating. We have studied growth hormone, using the preparation Dr. RABEN has been kind enough to let us have, but we have been interested not so much in attempting to produce anabolism as to find out whether or not the mobilisation of fat appears to be similar with growth hormone and with testosterone. We do not have a growth-hormone assay. Perhaps you have come across the thing we have been looking for over a long time, namely, good evidence of some circulating endocrine agent which might have something to do with the stages in convalescence — in other words, to determine the point at which the patient goes into anabolism.

FRASER: I would like to ask Dr. KINNEY if he made any similar studies on diabetic subjects undergoing an operation. One might anticipate that there would be differences in their response. We have been doing pituitary implants on various subjects including diabetics. The diabetics have differed from all the others in their response to this rather unusual operative insult in that they have had a considerable amount of post-operative vomiting. It has been a somewhat striking and surprising experience that this could be alleviated by large doses of cortisone. I wonder, therefore, whether you have noticed any correlation in your studies between the post-operative vomiting and the nitrogen loss. I know you cannot make such a correlation in some cases, such as gastrectomy, but perhaps this is possible in the case of extra-abdominal operations?

KINNEY: No, we have not. We have studied only a few diabetic patients, and I have not noticed any correlation with that.

YOUNG: I should like to comment on the relation of growth hormone to nitrogen loss under these conditions. At the outbreak of the war in 1939, Dr. CUTHBERTSON and I investigated the influence of treatment with pituitary extracts, which contained growth hormone, on the healing of wounds and on the wasting of muscles induced by the fracture of a limb in rats. We found[1] that, although the pituitary treatment induced a strongly positive nitrogen balance, it exerted no favourable effect either on the healing of a wound, consisting of removal of a disc of skin, or on the wasting of the muscles around the fractured bone. I question whether growth hormone can reasonably be expected to assist the movement of amino acids into specific sites and whether there is any value in preventing the loss of nitrogen in your experiments.

KINNEY: We are wondering exactly the same thing. We feel that there is no good evidence for the routine use of anabolic agents in treating surgical patients.

TUCHMANN-DUPLESSIS: We have made observations very similar to those of Dr. YOUNG in rats. We removed the pituitary in rats and observed that wound healing was just the same as in the control animals. We also performed hypophysectomy in newts and then cut a part of, or the entire leg and looked for signs of regeneration. In these cases we found that the rate of regeneration was just the same in the hypophysectomised animals as in the controls. Like many others, we have the feeling that the local growth, wound healing, and regeneration do not depend on the pituitary.

IKKOS: I should like to make a short remark on the relation between starvation and basal metabolism. Among the data of KEYS and BROZEK which you showed, there was a decrease of approximately 30 to 50% in the basal metabolic rate regardless of the way in which the results were expressed. Now, the situation in chronic starvation might be different. If you examine cases of anorexia nervosa, for example, and if you express the values for the basal metabolic rate per unit of cell mass, instead of body-weight or body-surface area, you get quite normal values.

KINNEY: In the slide which I showed from the Minnesota starvation study, the curve of oxygen consumption that most closely approximated the drop in body-weight was one in which they had attempted to calculate, from body-composition studies, the oxygen consumption per kg. of "active cells". I think that, once again, we are faced with the problem of trying not only to relate nitrogen excretion and energy expenditure, but also to make some sort of "dissection" of the body protein.

McCANCE: I have five small points to raise concerning this paper. First of all you made no mention of losses of faecal nitrogen. I take it you neglected them, but perhaps you can give us some estimate of their size. Secondly, there are great individual variations in the basal metabolic rate, and it seems a little dangerous to assess it without measurement in a particular patient. You could perhaps obtain a figure for it in some patients by getting them back into hospital for one night, when they are better, and making the determination then. Thirdly, have you considered the possibility of the

---

[1] CUTHBERTSON, D. P., G. B. SHAW, and F. G. YOUNG: J. Endocr. (G. B.) **2**, 468, 475 (1940/41).

thyroid gland being responsible for raising the metabolic rate and perhaps for causing some of the protein break-down ? Fourthly, I take it you are satisfied that you were always giving enough carbohydrate, intravenously or otherwise, to minimise gluconeogenesis from protein, because that of course was not done in the starvation experiments of BENEDICT. Fifthly, have you considered the suprarenal gland coming in ? We know that following operations and severe injuries there are major changes in the excretion of water and sodium — and also, I think, of potassium.

KINNEY: We have considered the faeces. In surgical patients who have been under study for longer than approximately ten days, we have done faecal analyses both for nitrogen and for lipids by organic solvent extraction. We have not usually done this during the early days after operation, because as a rule the percentage of faecal loss in the total amount is extremely small. We are troubled by the fact that we do not know the total caloric losses, and are now setting up a bomb-calorimetric method to check up on this point.

As far as variations in the conventional isolated BMR measurements are concerned, I could not agree with you more. One of our problems was the fact that, in trying to start this study, we felt that the individual variations in BMR would make it very difficult to draw any conclusions if we simply did one measurement a day by conventional techniques. We feel that doing numbers of these measurements (12—15 or more a day in the same individual previously trained in the technique) represents something with considerably less variation than a conventional BMR measurement.

As for thyroid function after injury, COPE and co-workers[1] made an effort to study thyroid function in surgical patients and, despite the increase in oxygen consumption, radioactive-iodine uptake and protein-bound iodine were uniformly normal.

Regarding carbohydrate administration to prevent gluconeogenesis, all we can say is that roughly 100 g. appears to provide a reasonable protein-sparing effect in dietary studies without trauma. Whether this amount is enough in injury, we have no idea. I think there may well be an increased need under circumstances involving injury and sepsis.

As for the suprarenal gland, I am sure that there are very important changes. In our department, and in others, an increase in circulating 17-hydroxycorticosteroids has been demonstrated during and immediately after the operation. However, we feel that this is not the whole story. During the past decade it has been fashionable to attribute all of the metabolic changes to increased adrenocortical secretion. There are a number of lines of evidence to indicate that injury produces more than this, although it may be important in initiating the sequence of convalescence. The major losses of nitrogen occur in a time sequence that extends well beyond the point at which the peak circulating 17-hydroxycorticosteroids are evident, either in the blood or in the urine. There are a number of other pieces of evidence which suggest that adrenocortical secretion may be important at the time of injury, but a lot else enters the picture.

---

[1] COPE, O., G. L. NARDI, M. OUIJANO, R. L. ROVIT, J. B. STANBURY, and A. WIGHT: Ann. Surg. (U.S.A.) 137, 165 (1953).

Human liver cirrhosis secondary to nutritional deficiency

By

R. SUBRAMANIAM

Liver cirrhosis as seen in western countries differs considerably from the type commonly encountered at the General Hospital in Madras. Whereas alcohol is traditionally accepted as playing the principal role in this connection, it is not such an important factor in cases seen in South India. The incidence of liver cirrhosis in our hospital for the last ten years has been as follows (Tables 1 and 2).

Table 1. *Number of patients with liver cirrhosis at the Madras General Hospital*

	Total number of cases for the Madras General Hospital	Number of cases in Dr. R. S.'s unit
1952	284	79
1953	396	60
1954	375	44
1955	444	60
1956	446	85
1957	238	75
1958	424	86
1959	398	95
1960	358	72
1961	334	65

Table 2. *Incidence of alcoholism among patients with liver cirrhosis*

	Number of cases	
	1960	1961
Alcoholics	12	23
Non-alcoholics . . .	29	34
Not known	15	9
Total	56	66

Of the 3,187 autopsies performed at our hospital between the years 1928 to 1961, 215 (0.67%) proved to be cases of liver cirrhosis.

In this paper I shall be confining my data to cases seen in the years 1960 and 1961. Total prohibition was introduced in this State of the Indian Union in 1949. Since then, alcohol in any form has been available only on health grounds or for foreign visitors, and the class of patient treated in our hospital is not able to procure alcoholic drinks. Stray cases of illicit drinking have been recorded, but the incidence is very low.

The patients' ages average between 20 and 40 years, the numbers declining sharply at either end (Table 3). "Children" are not admitted to my wards. Hence, the youngest patient must be at least 10 years old. The minimum age among cases of cirrhosis was 10 years and the maximum 72 years.

The *main clinical feature* of these cases of nutritional cirrhosis is the *lack of toxic manifestations*. Even in cases with pronounced ascites, the patients are still able to look after themselves. They seem to be mechanically embarrassed by the collection of fluid

Table 3. *Incidence of liver cirrhosis according to age and sex*

Age groups (in years)	Age and sex distribution					
	1960 Number of cases			1961 Number of cases		
	Male	Female	Total	Male	Female	Total
0—10	—	—	—	—	—	—
10—15	—	—	—	2	—	2
15—20	2	1	3	1	—	1
20—25	2	3	5	5	1	6
25—30	3	4	7	5	2	7
30—35	4	—	4	7	—	7
35—40	9	2	11	9	—	9
40—45	7	1	8	8	—	8
45—50	8	1	9	7	1	8
50—55	3	—	3	8	—	8
55—60	1	2	3	4	—	4
60—65	2	—	2	1	—	1
65—70	1	—	1	4	—	4
70—75	—	—	—	1	—	1
over 75	—	—	—	—	—	—
Total	42	14	56	62	4	66

and obtain considerable relief from abdominal paracentesis and from removal of fluid with a mercurial diuretic. Unlike patients suffering from toxic liver cirrhosis, these patients tolerate paracentesis very well, and it is not unusual for them to undergo paracentesis on 10—12 occasions without any supportive therapy until death finally ensues. With special care, the fluid becomes absorbed, and the patients are sometimes able to return to active work. Unlike patients with alcoholic or post-infective liver cirrhosis, they maintain their appetite until the terminal stage or until coma sets in. In such cases, some patients have survived for as much as 10 years after the ascites had first occurred. In all these instances, the *diagnosis was confirmed by needle biopsy*.

In our series of cases male patients far outnumbered those of female sex (Table 3). Their diet consisted principally of boiled rice with buttermilk and a little green chilli. Even among the so-called non-vegetarians the intake of high-quality animal protein, such as meat or fish, was low. The dietary habits of the patients are analysed in Table 4. The financial circumstances of the patient often prevent him from enjoying a more generous non-vegetarian diet. The economic status of the patients treated during the years 1960 and 1961 is indicated in Table 5. It has been noted by HIMSWORTH (1) that, when infective hepatitis occurs in nutritionally defective people, the mortality rate is low but the morbidity high. This was well borne out in all our cases.

Table 4. *Dietary habits of patients with liver cirrhosis*

	Number of cases	
	1960	1961
Vegetarian.	9	5
Non-vegetarian. . .	30	55
Not known	17	6
Total	56	66

Table 5. *Distribution of patients according to economic status*

Income groups (Rs/-per month)	Number of cases	
	1960	1961
Below 50	33	39
50—100	6	9
100—150	4	4
150—200	3	3
200—300	2	1
300—500	1	1
500 and above . . .	4	—
Not known	3	9
	56	66

In a fairly high percentage of patients the liver was enlarged and biopsy revealed fatty infiltration with incipient fibrosis before the onset of ascites. The fibrosis in such patients is periportal in character. Very few cases are diagnosed at this stage. Unless one is "cirrhosis conscious", the extremely vague symptoms associated with the early stages of cirrhosis fail to suggest the correct diagnosis, especially in a busy out-patient clinic. It was also noticed that some of the patients suffer from xerophthalmia or night-blindness and develop ascites months later. Once ascites has set in, the diagnosis is seldom missed. As I have already said, clinically the most prominent negative finding is the lack of toxic manifestations. Clinical examination reveals gross ascites with a liver which is not palpable. Though nutritional cirrhosis is common all over India, it seems to be especially common in those areas of the country where rice is the staple diet.

The other aetiological factor which has been observed either to precipitate liver cirrhosis or to aggravate existing cirrhosis is the occurrence of amoebiasis. In my wards it is a routine practice in cases of liver cirrhosis to enquire for a history of dysentery and to examine the stools three or four times for amoebae or amoebic cysts. On finding amoebae or cysts, a full course of anti-amoebic treatment is instituted, from which the patient usually derives considerable benefit. Such anti-amoebic treatment often causes the ascites to clear, the patient's appetite improves, and flatulence is relieved.

A history of infective hepatitis was obtained in 18 cases: 6 in 1960 and 12 in 1961. In many of these cases, infective hepatitis seems to be an added factor, and it appears to precipitate cirrhosis in a liver that is already damaged. The course of the illness is not particularly worsened by the fact of the cirrhosis having been precipitated by infective hepatitis. Other less common aetiological factors responsible for liver cirrhosis among patients seen in our hospital are syphilis, anaemia, malaria, kala-azar, tuberculosis, and ankylostomiasis. Malaria has been almost completely eradicated. The number of cases of malaria admitted to the hospital has been considerably reduced, and now only one or two are recorded annually. Among the tropical diseases, kala-azar is fairly common, and the antimony employed in its treatment may also precipitate liver cirrhosis. But in cases of liver cirrhosis subsequent to treatment for kala-azar, since the precipitating factor will have acted only for a short period and since such treatment is not likely to be required thereafter, the rate of spontaneous cure is high and, with a little supportive treatment, the patient may well effect a rapid and complete recovery.

The presence of helminthic infestation, such as ankylostomiasis of the intestinal tract, aggravates the existing malnutrition, but parasites of this type cannot be considered as direct aetiological factors.

Signs and symptoms. I propose to confine my remarks here to the signs and symptoms encountered in the ascitic or decompensated stage. This is a stage commonly seen and recognised in nutritional cirrhosis. Portal hypertension is never a marked feature, and haematemesis is very rare. No haematemesis has occurred in this series during the last ten years. The patient's main complaint at this stage is of distension of the abdomen, with or without ankle oedema. The swelling first starts in the abdomen, and it is only in those cases where the serum protein content is low that oedema of the ankles is seen. It is very rare for ankle oedema to precede ascites. Once abdominal paracentesis has been performed and the ascites relieved, the ankle oedema clears spontaneously.

Anorexia may be met with on rare occasions as an early symptom of cirrhosis; if the anorexia is pronounced, weight losses may also occur. Nausea, vomiting, and flatulence are further common complaints. Patients may also complain of gaseous distension. Weight loss may not be evident in the early stages owing to the ascites, which tends to mask loss of weight, but the patient himself is likely to complain that his appetite is deteriorating.

Pain in the abdomen may occur, but does not correspond to any other known cause of intra-abdominal pain. It varies in its quality from dull and mild to sharp and agonising, and it may recur at periodic intervals. The site of the pain is usually anterior, near the right hypochondrium. If pain occurs at the back of the thigh, the additional possibility of carcinoma of the liver must be borne in mind. In nutritional liver cirrhosis it can be taken for granted that haematemesis is never present. Where haematemesis occurs, one should consider the possibility that the cirrhosis may be due to causes other than nutritional deficiency.

Splenomegaly is also relatively rare. In the early stages, the liver may be palpable about a finger or a finger and half below the costal margin. Once ascites is well established, the liver shrinks and is no longer palpable.

These patients, as already mentioned, are able to tolerate repeated abdominal paracentesis, and it is nothing unusual for such patients to undergo paracentesis on 12 or 13 or even more occasions before death occurs. The usual text-book statement that a patient with liver cirrhosis will not tolerate paracentesis on more than 2 or 3 occasions does not apply to these cases. The need for paracentesis gradually ceases in patients who are carefully treated on a high-protein and low-salt diet, together with glucose intravenously, as in such instances the ascites eventually disappears. Though the ascites disappears, the pathological process in the liver is not reversed in all cases. A certain amount of liver-cell hypertrophy continues and may eventually result in primary cancer of the liver. The type of primary liver cancer encountered in these cases is the hepatoma or large liver-cell carcinoma. Here again, a marked difference is apparent as compared with the usual text-book descriptions of hepatoma. After we have made a biopsy diagnosis of hepatoma, the patient may live for as long as six months to one year. In other words, death does not ensue rapidly after a clinical and laboratory diagnosis of hepatic carcinoma has been made. In most cases of hepatoma we have not observed any metastases in other organs. We have had about 70 cases of hepatoma in the last ten years. In none of them were any metastases seen in other organs, the malignancy being confined to the liver.

Another noteworthy feature is that anorexia is not an important symptom in cases of nutritional cirrhosis in which hepatoma supervenes. In spite of the large, hard, nodular liver and the obvious wasting and emaciation, the patient retains his appetite. This clinical observation makes it easy to determine whether one is dealing with primary cancer of the liver or a cancer of the stomach with secondaries in the liver; in the latter instance the patient loses his appetite, whereas in primary cancer of the liver he does not lose his appetite until the very end. It should also be mentioned that, in cases of hepatoma, distinct glandular metastases are not to be found. The presence of VIRCHOW's glands points towards a diagnosis of cancer of the stomach and against a diagnosis of hepatoma.

Another change which may be expected is absorption of the fibrous tissue that has been laid down in cirrhosis. The first change observed is the disappearance of lipoid infiltration; later, the fibrous tissue clears — a fact which indicates that the formation of fibrous tissue in liver cirrhosis, and particularly in nutritional liver cirrhosis, is not an irreversible process.

Owing to the nutritional deficiency there may also be clinical manifestations of avitaminosis (vitamins A and B). Accordingly, one may find cases of pellagra and night-blindness in the presence of cirrhosis, but frank scurvy has not been seen in the cirrhotic patients in my wards.

The anaemia occurring in these cases is often of the hypochromic, microcytic type and, as in other cases of hypochromic anaemia, koilonychia may develop. In none of our own nutritional cirrhosis cases, however, was the macrocytic type of anaemia observed. The leucocyte count is generally within the normal range. Liver biopsy with a VIM-SILVERMANN

Table 6

	1960	1961
No. of cases confirmed by liver biopsy	19	32

needle was performed in all cases. The number of cases in which biopsy confirmed the diagnosis is shown in Table 6. In the late stages it may prove impossible to carry out a liver biopsy in this way, because the liver is too hard for the needle to penetrate it. Except in these late cases it is always possible to obtain a specimen of liver tissue by needle biopsy. Examination of the material with the naked eye usually reveals that the tissue is pale yellow in colour and not dark brown as in the case of a healthy liver. The microscopic appearance is also highly characteristic. In the early stages gross fatty infiltration is evident and fibrous-tissue deposition minimal. Even when the disease is advanced, the amount of fibrous tissue is always far less than in post-infective and toxic varieties of cirrhosis. One hardly ever encounters the dense band of fibrous tissue, dividing the liver into islands, which is characteristic of post-infective liver cirrhosis.

Serum electrophoresis by means of paper chromatography and estimation of serum albumin was carried out in these cases. It was found that the total protein may be as low as 3.25 g.%; the albumin was much more affected than the globulin, and there was reversal of the albumin-globulin ratio. Sometimes the globulin was as much as 3 times higher than the albumin. Cases that improved rapidly were those in which the albumin-globulin ratio was not grossly affected. Where the albumin-globulin ratio was 1:1 or 1:2, rapid improvement occurred, especially if the gamma-globulin fraction was not particularly elevated.

In persons over 40 the response to treatment was poor. Although treatment had no effect on the globulin moiety, the albumin moiety increased, and when the serum albumin content rose above 3.5 g. % the ascites cleared. Clearance of the ascites was accompanied by a steady improvement in the patient's general health, and, provided he did not revert to a low-protein diet, he was able to resume his normal activities. Unfortunately, in quite a high percentage of cases, the patients were obliged to revert to a low-protein diet upon discharge from hospital, and such patients sooner or later had to be re-admitted suffering from ascites.

Two of the cases under observation were first treated in 1952, i.e. 10 years ago. In one case, liver biopsy showed complete reversal of fibrous-tissue deposition, and the latest biopsy findings indicated that the structure of the hepatic tissue had reverted to normal. Clinically also, the patient appeared to be normal in all respects. Incidentally, it should be mentioned in this connection that LEVY (2), too, has reported regression of fibrosis in response to treatment for liver cirrhosis.

The second patient also started treatment in 1952 and is at present in hospital with a recurrence of ascites. On the last few occasions when we performed abdominal paracentesis the fluid was found to be blood-stained. The liver was palpable as a large nodular mass, indicating malignancy. The patient also exhibits myopathic changes and has developed claw hands. For a time he suffered from hemiparesis, from which, however, he has meanwhile fully recovered. In spite of the blood-stained ascitic fluid, he has a good appetite, and his usual complaint is that he is not being given enough to eat. The changes to which I have just referred were first observed over a year ago. We performed a liver biopsy on this patient, but it failed to reveal any hepatoma; the chances are that we missed the site of malignancy. Not much clinical deterioration has taken place during the past 12 months. Though survival for 12 months may normally be regarded as being about the limit for malignancy, a longer survival period can be expected in cases of primary liver cancer. Liver function tests in these cases showed a certain degree of impairment of liver function, but nothing like as severe as in toxic cirrhosis or post-infective cirrhosis.

Biochemical tests undertaken in these cases include — besides measurement of protein — the thymol turbidity test, the VAN DEN BERGH reaction, and the alkaline phosphatase test.

Patients of this kind die on account of protein deficiency, which gives rise to gross ascites and encourages secondary infections such as tuberculosis. In cases under treatment in hospital, death normally seems to occur as the result of a supervening hepatoma. A small percentage of patients die in hepatic coma, although in such cases hepatic coma does not run a rapidly fatal course as it does when it occurs in post-infective cases. By administering chlortetracycline (500 mg. i.v. in glucose) we were able to save our cases on four or five occasions before they eventually died.

Table 7. *Results of treatment for liver cirrhosis*

Result	Number of cases	
	1960	1961
Cured	3	—
Relieved.	25	21
Expired	13	7
Others[1]	15	38
	56	66

[1] This includes cases discharged against medical advice or at the patient's own request, as well as cases where the result was not specified or where the patient absconded.

Table 8

	1960	1961
Average duration of treatment given[1]	28[2]	28

[1] In number of days rounded off to the nearest integer
[2] In calculating this average, three very extreme cases were omitted.

The coma may persist for as long as 72 hours before recovery sets in.

Treatment. The results of treatment obtained during the years 1960 and 1961 are listed in Table 7. The average duration of

20*

treatment is shown in Table 8. Every case should be investigated for the presence of associated intestinal infections, which if found must be corrected. If the serological findings are positive, this should also be corrected. Patients with ascites tend to progress favourably in response to a little bed rest at the start. In the early stages, the response to proteins and essential amino acids, such as methionine, and lipotrophic factors, such as choline, vitamin B factors, and whole-liver extract, is extremely gratifying. The patients tolerate a high-protein diet very well. When treating cases of this kind, I have used a leaf extract from the plant *Wodalia calendulae*, which is prepared by drying the leaves; the extract is produced in the form of tablets of 0.5 g., which are administered in a dose of 3 tablets daily before meals. The leaves have been analysed and found to contain beta-sitosterol, flavone, and a small amount of sodium chloride and resin. It is evidently the combination of flavone with beta-sitosterol which exerts a beneficial effect on the liver. Treatment with this extract seems to improve the appetite, in addition to which it has a mild laxative and a mild diuretic action. No adverse effects have been noted, even in cases where the extract was taken for a number of years. Finally, it should be added that blood transfusion hastens recovery in these patients.

Summary

Cases of liver cirrhosis due to nutritional deficiency are characterised by a lack of toxic manifestations and a relatively benign course, the patient normally surviving for about one year, even where virtually no treatment is given. With a little medical attention, patients may survive for 10 to 12 years or longer. If the condition is diligently treated at an early stage, complete recovery may even be achieved. Haematemesis or other complications due to portal hypertension are rare. The patients are able to tolerate repeated abdominal paracentesis. There may be associated nutritional disorders, such as pellagra and avitaminosis, as evidenced by night-blindness or xerophthalmia. The condition may also be accompanied by intestinal infections, e.g. amoebiasis or helminthic infestation, correction of which normally results in a considerable improvement in the patient's condition. Two cases are described in detail in order to illustrate differences in the course of the disease.

Zusammenfassung

Lebercirrhose infolge Mangelernährung ist charakterisiert durch das Fehlen toxischer Schädigungen und einen verhältnismäßig gutartigen Verlauf. Ohne jede Behandlung überleben die Patienten für etwa ein Jahr, bei einem Minimum an Pflege und ärztlicher Kontrolle 10—12 Jahre oder sogar länger. Bei sorgfältiger Behandlung im Frühstadium kann vollständige Heilung erzielt werden. Haematemesis oder andere Komplikationen infolge Pfortaderhochdruck sind selten. Die Patienten vertragen wiederholte Ascitespunktionen. Häufig liegen gleichzeitig andere Ernährungsstörungen vor, z. B. Pellagra und Avitaminose, die sich in Nachtblindheit und Xerophthalmie äußern kann. Auch Amoebiasis oder Verwurmung können als Begleitkrankheiten auftreten; ihre Beseitigung führt im allgemeinen zu einer erheblichen Besserung im Allgemeinbefinden. Zwei Fälle werden näher beschrieben, um die verschiedenartigen Verlaufsformen der Erkrankung zu illustrieren.

Résumé

La cirrhose du foie due à une insuffisance alimentaire se caractérise par l'absence de manifestations toxiques et par une évolution relativement bénigne; non traités, les malades survivent environ 1 an; avec un minimum de soins et sous contrôle médical, ils peuvent encore vivre 10 à 12 ans ou même plus. Un traitement judicieux au stade initial permet d'obtenir la guérison complète. L'hématémèse ou les autres complications dues à l'hypertension portale sont rares. Les malades supportent des ponctions d'ascite répétées. Souvent ils présentent simultanément d'autres troubles nutritionnels, par exemple pellagre, avitaminose, qui se traduisent par de l'héméralopie et de la xérophtalmie. Peuvent être également associées des infections intestinales telles que l'amibiase, ou des infestations helminthiques; leur guérison améliore en général considérablement l'état du sujet. Deux cas font l'objet d'une description détaillée, illustrant les différentes formes évolutives de la maladie.

References

1. HIMSWORTH, H. P.: Lectures on the Liver and its Disorders. Oxford, 1947. — 2. LEEVY, C. M.: In: Practical Diagnosis and Treatment of Liver Disease. P. C. Hoeber Inc., New York, 1957, p. 210. — 3. SUBRAMANIAM, R.: Antiseptic (Ind.) 56, 355 (1959).

Discussion

ARIAS: Dr. SUBRAMANIAM's observation that fibrosis of the human liver may be reversed by nutritional therapy is very interesting. Dr. PATEK and his associates[1] in New York have described a similar sequence of events in rats in which cirrhosis was produced by dietary means. Quantitation of hepatic fibrosis is, of course, not possible; however, it is generally believed that hepatic fibrosis, with either alcoholic or post-necrotic cirrhosis, is not reversed by dietary therapy.

SUBRAMANIAM: Reversibility of fibrosis in humans has been reported from the United States by LEEVY[2] in his book on liver diseases.

WATERLOW: I was very much interested in what Dr. SUBRAMANIAM said, because some years ago I worked on a similar problem in West Africa, i. e. cirrhosis in children. We tried to study this by biopsies in all age groups. In this part of Africa, the Gambia, the nutritional state, although not good, is not as bad as in some parts of the world, and yet fibrosis and cirrhosis of the liver are common in quite young people. We came to the conclusion that to produce this liver change you need two damaging stimuli. One, affecting the parenchyma, is marginal malnutrition, including perhaps particularly cyclical malnutrition due to alternating hungry and good seasons. I think there is experimental evidence that this has a worse effect on the liver than a steady deficiency state. The second stimulus is something which affects the connective and vascular elements of the liver. In the Gambia this seemed to us almost certainly to be malaria, which is hyperendemic there. I wonder whether Dr. SUBRAMANIAM would agree with this dual-aetiological theory of cirrhosis. The rice diet of Madras should not be too bad inasmuch as rice is a good staple food as compared, for instance, with the cassava of equatorial Africa or equatorial America. And, although malaria has been eradicated, presumably it wasn't eradicated at the time when the patients who are now 30 or 40 years old were developing their disease.

The second question I would like to ask is this: in some tropical countries you very commonly see symptomless hepatomegaly in children; this has been described in particular from Indonesia, New Guinea, the West Indies, and some parts of Africa. Nobody knows what this is or what it develops into. I wonder whether Dr. SUBRAMANIAM or his colleagues have any observations on this and whether it occurs in his population.

SUBRAMANIAM: The question as to whether malaria was involved in these cases is very easy to answer, because, once malaria has occurred, it always leaves behind its imprint in the shape of an enlarged spleen. As I pointed out in my paper, splenic enlargement is surprisingly absent in these

---

[1] PATEK, A. J., Jr., D. E. OKEN, A. SAKAMOTO, N. FRITSCH, and M. BEVANS: A. M. A. Arch. Path. **69**, 168 (1960).

[2] LEEVY, C. M.: In: Practical Diagnosis and Treatment of Liver Disease. P. C. Hoeber Inc., New York, 1957, p. 210.

cases, so malaria cannot have been an operating factor. Even if the patients had had malaria, it was nothing worth mentioning, since in most of them the splenic enlargement was absent.

Now with regard to the second question, if you consume rice, and rice only, with nothing else, the protein content, though better than that of cassava, is still very poor. The food used to supplement the rice is buttermilk, which again is not a very rich source of protein. Now if, in addition to these two dietary handicaps, the patient is exposed to some other factor, e.g. hepatitis, then cirrhosis is likely to be precipitated. Though my cases are all adults, in most of our patients, and even in infants, the precipitating factor has been shown to be infective hepatitis. When infective hepatitis occurs without characteristic icterus, it is very difficult to diagnose it except by complement fixation tests. There are also other infections which may precipitate cirrhosis of the liver, and in these patients even a slight imbalance caused by intestinal infections tends to worsen the condition.

HOFFENBERG: With regard to the reversibility of cirrhosis, I know of two patients who were admitted to hospital more than 10 years ago suffering from alcoholic cirrhosis with liver failure and haematemesis from oesophageal varices. They are both quite well today with no evidence — clinical or bio-chemical — of liver disease. Liver biopsies had not been performed initially, so that I am unable to say whether liver fibrosis was reversed or not, but the clinical features of profound liver disease certainly were.

Dr. SUBRAMANIAM's patients are interesting, since they could tolerate high-protein diets and multiple abdominal paracenteses and were given intravenous, rather than oral, antibiotics, and yet recovered. I wonder whether their behaviour is not best explained by a lesser degree of severity, as suggested by the absence of marked fibrosis ?

SUBRAMANIAM: The type of disease that I have described certainly runs a very much milder course and, since nutritional deficiency is the main aetiological factor, the patients tolerate a high-protein diet – though of course only in the early and not in the very late stages. In the terminal stages you can induce hepatic coma by giving a high-protein diet. Intravenous antibiotics were only used in the treatment of hepatic coma; otherwise we don't use antibiotics at all.

KOWALEWSKI: I would like to ask Dr. SUBRAMANIAM whether he fre-quently observed neurological signs of coma in his cases of cirrhosis. Did he study blood ammonia in these patients ? In our experience a high-protein diet is not indicated in some cases of cirrhosis and may lead to coma, just as in the terminal stages to which Dr. SUBRAMANIAM has referred.

I wonder what could be the significance of the absorption of fibrous tissue observed in the case of cirrhosis responding to the treatment. Is it really a true absorption of connective tissue ? As your histological studies were done on biopsy specimens, I wonder whether regeneration of hepatic tissue did not mask the fibrous changes. Biopsy specimens might well contain the new tissue and be poor in fibrous tissue. Is it really an absorption which you observed ?

SUBRAMANIAM: With regard to the question of neurological complications in connection with a high-protein diet, it depends on the stage at which you are giving the protein. In the early cases, the patients are suffering essentially from a nutritional deficiency without any toxic features, and they do very well on a protein diet. In these early stages, the high-protein diet does not give rise to any neurological complications. It is only in the advanced stages that such a diet may precipitate coma and also neurological complications.

As to the question whether the fibrosis is reversible, we must realise that what is happening in the liver is after all a diffuse process, and that what obtains in one area is therefore likely to be a good sampling of the whole organ. In spite of the regeneration of hepatic tissue, you simply cannot find any fibrous tissue in the liver at all, and I take this as meaning that the process has indeed been reversed. Moreover, I have been able to confirm this observation not only in patients who were still alive, but also in those who for some reason or other had died; here, too, we find that the amount of fibrosis is reduced.

TREMOLIÈRES: I would like to report some facts which I think might perhaps explain how alcohol may affect protein metabolism in cirrhosis of the liver. In France, as you know, about 80% of cases of cirrhosis are related to alcohol. When we studied our alcoholic cirrhotics we found that they were more severely protein-depleted than any other type of patient, i.e. even more than patients we had studied upon their return from concentration camps. On a high-protein diet (120—150 g./day), they are able to store nitrogen at the rate of 5—8 g. per day for more than one month[1]. This was our starting point. It was our impression that, in alcoholics, alcohol — instead of having its classic nitrogen-sparing effect — actually has a wasting effect. We have therefore undertaken a prolonged study on this. We have seen that alcohol in alcoholics, given at a level exceeding 0.6 g./kg., produces a situation similar to that described by Dr. KINNEY in the catabolic phase of surgery, i.e. an increase in caloric expenditure with an increase in nitrogen mobilisation as reflected not in the nitrogen balance, because this cannot be done in a short-term study, but in a rise of non-protein nitrogen in the plasma. Later on, we were able to demonstrate that, in alcoholics receiving alcohol in excess of 0.6 g./kg., alcohol was oxidised through various per-oxidasic systems (xanthine-oxidase-catalase and amino acid oxidase) which appeared in the plasma[2, 3]. We then turned to experiments on rats. When you give rats more than 2 g. of alcohol per kg., which is a toxic dose, you induce the system which was demonstrated *in vitro* by KEILIN and HARTREE[4] long ago.

In the rat, you have in the plasma a mobilisation of the xanthine-oxidase-catalase and amino-acid oxidase systems, which can oxidise alcohol. In the course of further studies, we were able to demonstrate that, in the liver and pancreas of rats receiving a toxic dose of alcohol, you get a reduction in RNA and soluble nucleotides, and that, finally, in response to repeated toxic doses of alcohol you get acute haemorrhagic and necrotic pancreatitis very similar to that observed in man.

---

[1] TREMOLIÈRES, J., A. MOSSE, L. LYON, J. PASCHOUD, and C. SAUTIER: Presse méd. (Fr.) 87, 1862 (1954).

[2] TREMOLIÈRES, J., and L. CARRÉ: Compt. rend. Acad. sc. (Fr.) 251, 2785 (1960).

[3] TREMOLIÈRES, J., and L. CARRÉ: Compt. rend. Soc. biol. (Fr.) Vol. CLV, No. 5 (1961); — TREMOLIÈRES, J., and L. CARRÉ: Rev. alcoolisme (Fr.) 5, 199 (1959); — TREMOLIÈRES, J., and L. CARRÉ: Rev. alcoolisme (Fr.) 7, 202 (1961).

[4] KEILIN, D., and E. HARTREE: Proc. Roy. Soc. (G. B.) 119, 141 (1936) and Biochem. J. (G. B.) 39, 293 (1945).

To sum up, it seems that, in alcoholics and in certain nutritional conditions, alcohol — when given in excess of a certain dosage — is oxidised not only by the physiological ADH system, but by various peroxidasic systems which couple the oxidation of alcohol to nucleotides or amino acids. This pathway of oxidation is highly dangerous and may explain both the nitrogen-wasting effect of alcohol and also some of the necrosis observed in tissues such as the pancreas.

PRADER: From your tables, Dr. SUBRAMANIAM, it seems that you have seen more male than female patients. Is the disease really more common in males or are males more frequently sent to hospital? If the male sex is really more frequently affected, how can this be explained?

SUBRAMANIAM: The incidence may be somewhat higher in males than in females. On the other hand, in South India there are still rather strong prejudices among women against the idea of going to hospital — which means that fewer female patients get admitted to hospital. If possible, women avoid hospital. But it is also true that the economic status of the woman is slightly better in the sense that the men always try to look after their womenfolk better than they look after themselves.

I^{131}-albumin metabolism in human adults after experimental protein depletion and repletion

By

R. HOFFENBERG, S. SAUNDERS, G. C. LINDER, E. BLACK,
and J. F. BROCK

The recognition of protein deficiency in the human being is easy only when the deficiency is gross, as in kwashiorkor in infants or in severe hypoalbuminaemic states in adults. In 1961 BROCK (*1*) stressed the lack of adequate clinical and biochemical tests for protein deficiency, "at least where the deficiency is mild". While acknowledging that a low serum albumin level was a reasonably good index of marked deficiency, he recognised that a marginal range of hypoalbuminaemia might be evidence of impending or early protein deficiency, and pointed out that minor degrees of protein deficiency or "protein subnutrition" could conceivably exist without lowering of the serum albumin level.

In the present study, mild protein deficiency was induced in human subjects by the use of low-protein diets. In an effort to explore the earliest consequences of this protein deprivation, I^{131}-labelled albumin was used to measure intra- and extra-vascular pool-sizes before and after depletion and subsequent repletion. The use of albumin labelled *in vitro* with I^{131} has become widely accepted as a means of studying albumin distribution and turnover. Despite some early doubts about the validity of the technique, it is generally regarded as a reliable tracer method, provided that conditions of albumin fractionation and labelling are kept fairly constant (*2, 3*).

Materials and methods

Subjects. Eight adult male volunteers were studied. All were ambulant, and in no case was there a manifest underlying abnormality in protein metabolism. Subjects were admitted to a metabolic ward where initial I^{131}-albumin studies were made while they were on a "normal" balanced diet. Two further studies with I^{131}-albumin were performed — the first after 3 to 6 weeks of low-protein diet (not more than 10 g. protein daily) and the final study after a similar period of protein repletion (average intake about 150 g. protein daily).

Methods. Pure albumin was prepared by fractionation of human plasma through a carboxymethyl-cellulose column using an acetate buffer (4). Iodination was achieved by use of a potassium iodide/iodate mixture (5). This method was preferred to the iodine monochloride method which requires I¹³¹ free of reducing agent. The iodide/iodate method allows the use of I¹³¹ in a thiosulphate solution, which was more readily available. Passage through a Dowex resin column after labelling removed almost all free iodine, trichloroacetic acid precipitation of the eluate showing not less than 99% of the eluate radioactivity to be protein-bound. Carrier albumin was added to reduce radiation damage, and the whole was Seitz-filtered to render it sterile. Preparations were used within a few days of iodination, and regular electrophoretic and ultraviolet absorption determinations showed no deviation from expected *rf* and peak absorption before or after labelling. As further evidence of the integrity of the labelled compound, it may be stated that urine excretion showed no early rise in radioactivity, such as might indicate components capable of rapid degeneration.

LUGOL's iodine or sodium iodide was administered to subjects for 48 hours prior to intravenous injection of 10—20 μc. of I¹³¹-albumin. The exact amount administered was determined by weighing the syringe, and a weighed aliquot of the injected dose was used for preparation of counting standards. Heparinised venous blood samples were taken at 10 minutes, 3 hours, and daily for 7—10 days. Twenty-four-hour urine samples were collected for the same period[1]. Radioactivity was assayed in a well-type scintillation counter to give about 2% statistical accuracy.

Plasma volume was determined from the 10-minute blood sample and daily plasma radioactivity and urinary excretions were plotted as described by VEALL and VETTER (6). Relevant pool-sizes, turnover rates, etc. were calculated by the "equilibrium time" method, based on the assumption that labelled albumin behaves in the body in terms of an open two-compartment mamillary system. For the period of study the subject was regarded as being in a steady state, i. e. the rate of albumin production was equal to the rate of loss or degradation. This assumption was necessary in order to allow conclusions to be drawn regarding rate constants.

---

[1] In the early experiments, stool radioactivity was measured, but consistently low results were obtained, suggesting that faecal excretion did not materially alter the results based on urinary excretion alone. Faecal radioactivity was not measured in later experiments.

The fact that no consistent change in plasma albumin concentration was found during the course of any single test-period lends support to this assumption.

During the periods of depletion and repletion the subjects were on continuous nitrogen balances (Table 1). These confirmed the estimates of the protein content of the diets. The cumulative balances were measured from the first day of the depletion diet to the day of the injection of the labelled albumin for the albumin measurements in depletion and repletion. The high and low nitrogen diets were continued through the period of the measurements. At the time of the first depletion measurements the average balance was —58 g., at the time of the second (where two "depletion" studies were made) —92 g., and at that of the repletion measurements +42 g. Stable albumin was measured in quadruplicate by the biuret method (7) several times during the course of each test. Very slight fluctuations (within the limits of error of the method) were observed, and mean plasma albumin figures for each test were used in calculations.

Results

(a) **Pool-sizes and plasma albumin levels.** Table 1 shows the plasma albumin concentration and intra- and extra-vascular albumin pool-sizes, as derived from each study. Fig. 1 illustrates these findings graphically, total and intravascular albumin pools being

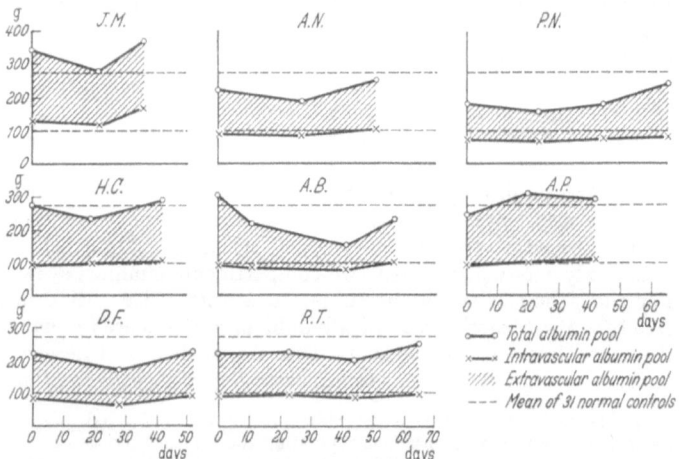

Fig. 1. Albumin pool-sizes measured after protein depletion and repletion

Table 1. *Nitrogen balance data and albumin pool-sizes during depletion and repletion of protein*

	Nitrogen intake	Depletion or repletion, days	Cumulative nitrogen balance	Intravascular albumin		Extravascular albumin		Plasma albumin conc.
	g./day		g.	g.	g./kg.	g.	g./kg.	g. %
Normal controls (mean of 31 cases) .				98.8	1.61	173.8	2.82	3.92
I. J. M. Control . . .	19.0			129.3	1.89	210.6	3.08	4.18
Protein depletion . .	0.57	22	—64	116.5	1.84	162.4	2.56	3.69
Protein repletion . .	22.5	12	+15	169.4	2.57	200.4	3.05	4.66
II. H. C. Control. . .	14.8			96.6	1.56	181.5	2.93	3.78
Protein depletion . .	0.76	20	—59	97.4	1.61	137.1	2.27	3.24
Protein repletion . .	23.3	24	+115	108.1	1.78	184.9	3.03	3.92
III. D.F. Control . .				82.0	1.32	140.0	2.26	3.70
Protein depletion . .	(0.21	10)	—65	66.6	1.16	103.2	1.80	3.40
	(2.1	17)						
Protein repletion . .	17.8	26		96.8	1.66	133.4	2.29	4.04
IV. A. N. Control . .				90.0	1.45	136.0	2.19	3.69
Protein depletion . .	0.84	28	—82	85.1	1.46	106.2	1.81	3.28
Protein repletion . .	19.2	25	+39	106.7	1.73	149.3	2.42	3.98
V. A. B. Control. . .				96.6	1.62	212.6	3.58	3.71
Protein depletion . .	0.34	12	—44	87.2	1.51	135.6	2.35	3.74
Protein depletion . .	0.34	42	—112	80.4	1.46	74.0	1.35	3.41
Protein repletion . .	23.1	15	—31	100.9	1.83	131.9	2.40	3.95
VI. R. T. Control . .				92.6	1.77	131.4	2.51	4.08
Protein depletion . .	0.57	24	—48	92.5	1.86	133.8	2.68	4.08
Protein depletion . .	0.57	45	—101	88.7	1.83	111.8	2.31	4.06
Protein repletion . .	22.5	22	—1	97.0	1.91	154.5	3.04	4.12
VII. P. N. Control .				75.3	1.60	106.9	2.26	4.34
Protein depletion . .	0.57	24	—38	69.1	1.51	89.8	1.95	3.83
Protein depletion . .	0.57	45	—64	74.7	1.67	105.3	2.36	4.15
Protein repletion . .	22.5	21	+79	82.7	1.74	160.1	3.36	4.35
VIII. A. P. Control .				91.3	1.36	157.5	2.35	4.04
Protein depletion . .	0.54	21	—68	101.4	1.60	212.7	3.35	3.63
Protein repletion . .	23.3	23	+82	111.4	1.76	184.6	2.92	4.34

plotted, so that the difference (hatched area) indicates the extra-vascular component. Our mean normal figures, based on 31 studies, are represented by the dotted lines. Convergence of the boundaries of the hatched area indicates a disproportionate decrease in extra-vascular pool-size, and divergence a disproportionate increase.

It is clear that 7 of the 8 patients showed essentially similar results. Case VIII. — A. P. differed from the others in that his extravascular albumin pool appeared to increase during protein depletion and decrease during repletion. No obvious explanation was found, but the consistency of the remaining findings suggests the possibility of a technical fault in this particular case.

In all other subjects a consistent drop in extravascular albumin pool-size followed protein depletion. This drop was considerable in 5 cases and exceeded by far any change in the intravascular pool-size. In 2 cases (VI. — R. T. and VII. — P. N.) the fall in extravascular pool-size was small, but protein repletion was followed by a

Table 2. *Albumin turnover rates during depletion and repletion of protein*

	Turnover rate		Plasma rate constant
	g./day	g./day/kg.	%/day
Normal controls (mean of 31 cases)	8.7	0.143	8.9
I. J. M. Control	12.5	0.183	9.7
Protein depletion	6.3	0.099	5.4
Protein repletion	10.7	0.162	6.3
II. H. C. Control	6.4	0.103	6.6
Protein depletion	3.3	0.054	3.4
Protein repletion	6.7	0.110	6.2
III. D. F. Control	7.4	0.119	9.0
Protein depletion	4.5	0.078	6.7
Protein repletion	7.1	0.123	7.4
IV. A. N. Control	8.2	0.132	9.1
Protein depletion	5.1	0.087	6.0
Protein repletion	9.2	0.149	8.6
V. A. B.[1] Control	?	?	?
Protein depletion (2 weeks)	5.7	0.099	6.5
Protein depletion (6 weeks)	5.6	0.102	7.0
Protein repletion	8.2	0.148	8.1
VI. R. T. Control	8.5	0.163	9.2
Protein depletion (3 weeks)	6.7	0.135	7.3
Protein depletion (6 weeks)	7.3	0.150	8.2
Protein repletion	9.0	0.177	9.3
VII. P. N. Control	9.3	0.196	12.3
Protein depletion (3 weeks)	4.7	0.102	6.8
Protein depletion (6 weeks)	5.0	0.112	6.7
Protein repletion	7.7	0.162	9.3
VIII. A. P. Control	9.9	0.147	10.8
Protein depletion	5.0	0.080	5.0
Protein repletion	10.5	0.166	9.4

[1] Control figures were considered invalid for technical reasons.

marked expansion. In these 2 cases the initial pre-test readings were the lowest of the series: the failure to drop further during protein depletion might indicate that these subjects were marginally depleted of protein when the study started. Analysis of the results in all 8 cases shows the mean fall in the extravascular pool to be about 20% after protein depletion. (If one excludes subject A. P., who gave contrary results, the mean fall is $\pm 28\%$.) At the same time plasma albumin concentration fell by only 9%. These results indicate that the extravascular albumin pool is capable of a disproportionate fall during protein depletion. Whether this precedes the fall in plasma albumin concentration cannot be said with certainty from our experiments, since the plasma albumin levels also fell during protein depletion in most cases. But subjects A. B. and R. T. show falls in extravascular pool without change in plasma albumin concentration. It is tempting to think that lowering of the plasma albumin concentration is a relatively late reflection of protein depletion and that marginal protein deficiency might exist before it is shown by plasma albumin levels.

(b) **Turnover studies.** In all 8 cases protein depletion was attended by a slower turnover of labelled albumin to figures usually well below our mean normal value. This is reflected by a drop in degradation rate and in plasma rate constant (Table 2). Since degradation rate is assumed to equal rate of synthesis of albumin, it would appear that production of albumin is much reduced during protein depletion. Following protein repletion this trend was reversed in the direction of, or even beyond, the mean normal value.

Conclusions

Protein depletion under the conditions of this experiment seems to be associated with two effects. In the first place there is a primary fall in the extravascular albumin pool. This may precede the fall in the plasma albumin pool or plasma albumin concentration, although this has not been shown convincingly. Certainly the fall in the extravascular pool is out of proportion to the fall in plasma albumin, which suggests that lowering of the plasma albumin concentration may be a relatively late manifestation of protein depletion. In two patients, protein refeeding was associated with a very marked increase in extravascular pool-size above the initial experimental reading. In neither of these cases did the plasma albumin concentration alter significantly.

A second constant finding in the course of protein depletion was the decreased rate of albumin turnover. This was found in all cases, whether measured in absolute terms or in terms of the smaller pool-size. Protein refeeding reversed this change.

Discussion

Cohen et al. (3) have drawn attention to the marked variation that has been reported in different studies regarding the absolute rate of albumin turnover in adult human subjects. In their analysis of 11 subjects a mean figure of 185 mg./kg. body-weight per day was obtained. This appeared to be in accord with other published figures. Cohen and Schamroth (8) found lower figures in apparently healthy Africans (113—150 mg./kg. body-weight; mean 127 mg./kg.). In our study a mean figure of 143 mg./kg. body-weight per day was found in 31 apparently normal individuals. These included members of all racial groups who were considered to be well-nourished at the time of testing. Our finding of a lowered turnover of albumin in protein-depleted individuals may explain why both Cohen and Schamroth's values for Africans and our own normal values for South Africans of all races are lower than other reported values, i.e. we may both be dealing with individuals, some of whom, at least, are marginally depleted of protein, although

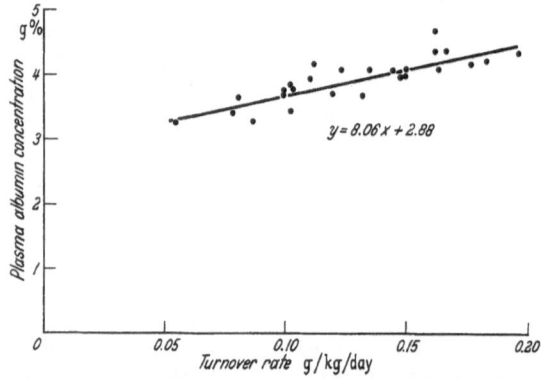

Fig. 2. Albumin turnover rate plotted against plasma albumin concentration with calculated regression line

not sufficiently to be reflected in lower plasma protein levels. This possibility is supported by a perusal of our findings, since slow turnover rates were found in some of our subjects undergoing depletion at a time when lowered extravascular albumin pools suggested protein depletion, despite normal or almost normal serum albumin levels.

Fig. 2 shows plasma albumin levels of all 8 subjects plotted against albumin turnover rates, as estimated throughout the course of the experiment. These data show a good correlation, indicating a distinct fall in albumin turnover in relation to a falling serum albumin level. A linear relationship seems to exist at levels of serum albumin between 3.24 g. % and 4.66 g. %. At its lowest

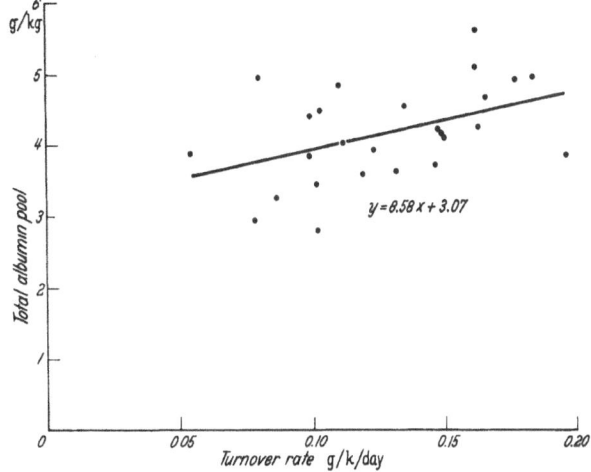

Fig. 3. Albumin turnover rate plotted against total albumin pool with calculated regression line

point albumin turnover would be approximately 50 mg. daily per kg. body-weight. It is interesting to speculate what might happen when the serum albumin is below 3.0 g. %, and studies are in progress to explore the possibilities in individuals whose serum albumin levels are low.

Fig. 3 shows a similar attempt to correlate turnover rate of albumin with the total body albumin pool. Although the scatter is greater around the calculated regression line, a rough correlation of the same order is again seen to exist.

Summary

Eight human subjects have been studied by means of I¹³¹-labelled albumin before and after experimental partial protein depletion and repletion. Analysis of pool-sizes and turnover rates following depletion shows a marked lowering of the extravascular albumin compartment, out of proportion to the fall in plasma albumin concentration, and, possibly, preceding it. At the same time, albumin synthesis is greatly diminished. This pattern was constant in 7 of 8 subjects studied. Protein repletion reversed these changes.

Zusammenfassung

An 8 gesunden männlichen Probanden wurden Untersuchungen mit I^{131}-markiertem menschlichen Plasmaalbumin vor und nach mäßiger experimenteller Eiweißverarmung (10 g Eiweiß pro Tag) und Wiederauffüllung von Eiweiß (150 g pro Tag) durchgeführt. Die Analyse der extra- und intravasculären Eiweißmenge sowie der Umsatzgröße ergab im Zustand der Eiweißverarmung eine deutliche Verminderung des extravasculären Albuminraumes, welche die Abnahme der Plasmaalbumin-Konzentration übertrifft und ihr möglicherweise vorausgeht. Gleichzeitig sind der Umsatz von markiertem Albumin und die Albuminsynthese herabgesetzt. Dieses Verhalten fand sich bei 7 von 8 Versuchspersonen. Durch Wiederzufuhr von Eiweiß ließen sich diese Veränderungen aufheben.

Résumé

Chez 8 sujets de sexe masculin en bonne santé, on a procédé à divers contrôles en utilisant de l'albumine plasmatique humaine marquée à l'I^{131} avant et après une réduction modérée de l'apport de protéines (10 g par jour), puis après restitution d'un apport protéique convenable (150 g par jour). L'étude de la quantité de protéines extra- et intravasculaires et de l'importance des transformations métaboliques a montré au cours de la restriction protéique une nette diminution des protéines extravasculaires hors de proportion avec la baisse de la concentration de l'albumine plasmatique qu'elle peut précéder. En même temps, le métabolisme de l'albumine marquée et la synthèse de l'albumine diminuent. Cette constatation a été faite régulièrement chez 7 sujets sur 8. La reprise de l'administration de protéines a fait disparaître ces changements.

Acknowledgements

We would like to express our sincere thanks to the following for their interest and support: Sister N. L. Daniels and the staff of the Metabolic Ward, Dr. J. G. Burger, the Superintendent of Groote Schuur Hospital, and Miss C. Liadsky of the Protein Research Laboratory; and to the following funds for financial assistance: The Council for Scientific and Industrial Research of South Africa, the International Atomic Energy Agency, Vienna, and the Staff Research Fund of the University of Cape Town.

References

1. Brock, J. F.: Recent Advances in Human Nutrition. Ed. by J. F. Brock. J. & A. Churchill Ltd., London, 1961, p. 50. — 2. McFarlane, A. S.: Ann. N. Y. Acad. Sc. **70**, 19 (1957). — 3. Cohen, S., T. Freeman, and A. S. McFarlane: Clin. Sc. (G.B.) **20**, 161 (1961). — 4. Peterson, E. A., and H. A. Sober: J. Amer. Chem. Soc. **78**, 751 (1956). — 5. Matthews, C. M. E.: Phys. in Med. Biol. (G.B.) **2**, 36 (1957). — 6. Veall, N., and H. Vetter: Radioisotope Techniques in Clinical Research and Diagnosis. Butterworth and Co. Ltd., London, 1958, p. 318. — 7. Wolfson, W. Q., C. Cohn, E. Calvary, and F. Ichiba: Amer. J. Clin. Path. **18**, 723 (1948). — 8. Cohen, S., and L. Schamroth: Brit. Med. J. **1958/I**, 1391.

Discussion

QUERIDO: Studies on albumin turnover rates are so nice, because one is studying a function of one cell, the liver cell. It would be very interesting to know whether this cell system as such is under the influence of, or could be influenced by, hormonal substances. So I should like to ask whether you have any data available concerning the action of growth hormone, testosterone, or so-called anabolic agents on the albumin turnover rate.

HOFFENBERG: I can offer some very limited information about the action of anabolic steroids, which I would prefer to give after the papers of Dr. BAUER and Dr. DYMLING, since it refers to the action of anabolic steroids in osteoporosis. But I have no information regarding growth hormone. One study in an acromegalic was normal, and studies in hypothyroid subjects showed a slower turnover rate than normal.

SCHREIER: I am very much interested in the disproportional behaviour of albumin in plasma and in the pool. From injection of labelled albumin we know that equilibration takes place very quickly. In the normal individual a decrease in the peripheral albumin pool would mean that a decrease in extracellular fluid takes place — or don't you agree ?
Just one other brief remark about factors which have an influence on albumin break-down. The turnover rate of albumin is influenced by growth. In young animals the half-life of albumin is only about 50% — or even less — of the half-life in older animals. I feel it is necessary to study breakdown in respect to growth hormone, too.

HOFFENBERG: We found no consistent changes in plasma volume in the course of our experiment, but we did not measure extracellular fluid volume as such. We have not studied normal children with this technique, because we were reluctant to expose them to possible radiation hazards and to subject them to the inconvenience of a full metabolic study.

ASCHKENASY: Among the hormones which control the level of plasma proteins, glucocorticoids seem to play a particularly important role as activators of the production of serum albumin[1, 2]. This effect may be related to the anabolic action which these hormones have long been known to exert on the liver[3].

HOFFENBERG: I have made studies in three patients with Cushing's syndrome before and after operation. In two, adrenal adenomas were removed; in the third, bilateral adrenalectomy was performed. I have also studied two other patients before and after the administration of adrenal steroids.

---

[1] WINTER, C. A., R. H. SILBER, and H. S. STOERK: Endocrinology (U.S.A.) **47**, 60 (1950).
[2] ASCHKENASY, A.: J. physiol. (Fr.) **49**, 16 (1957).
[3] SILBER, R. H., and C. C. PORTER: Endocrinology (U.S.A.) **52**, 518 (1953).

I could not demonstrate any consistent change in pool sizes or in turnover rates, although STERLING[1] reported such changes a few years ago.

WATERLOW: We would confirm from the results I gave you the other day that in normal children the albumin turnover rate is higher than in adults, and the half-life is about half, i.e. some 12 days. As to the question of the dosage of radioactivity, we give the child a total of about 5 μc. and make measurements on urine by precipitating the iodide with silver nitrate. This can be done quantitatively, and in this way the dose can be kept low.

FRASER: In connection with the inferred contraction of extracellular fluid, may I comment that we have made some rather similar studies in another field — on sodium depletion over the course of a week. A similar pattern is seen here, too, i.e. the serum sodium concentration remains unaltered, but there are quite striking contractions in the total exchangeable and presumably extracellular sodium.

IKKOS: Dr. HOFFENBERG, you mentioned that the extravascular pool decreased during protein depletion. I wonder if there was any correlation between the decrease in the extravascular pool and the decrease in the degradation or turnover rate during protein depletion.

My second question concerns the findings in the two cases in which you did not observe any significant changes during the phase of protein depletion. You suggested that these two patients might have had a protein depletion already before the experimental period. I wonder if the protein turnover rate before the experimental period was changed as in suspected protein depletion.

HOFFENBERG: The same order of correlation exists between extravascular pool size and albumin turnover rate, but the scatter is greater around the calculated regression line. The two patients you mention, who showed diminished albumin pools at the start of the study, had normal serum albumin levels, but the turnover rates were not especially slow.

QUERIDO: I wonder whether you or anybody else would like to speculate on the degradation site of albumin. If you see these regression lines, it appears that the size of the albumin pool determines the amount of albumin broken down. This — as has been suggested, but I think never proven — is not in contradiction with the fact that leakage in the gut could be a normal process for the degradation of albumin. On the other hand, if you say that hypothyroidism diminishes the disappearance rate, you would rather be thinking in terms of the basal metabolic rate as the determining factor.

HOFFENBERG: I would not like to speculate about the site of degradation. As I said earlier, we looked for intestinal leakage in our early cases, but abandoned the search, as the faecal excretion showed virtually no radioactivity at any stage of the study. I should perhaps stress that the turnover rate dropped, whether expressed in absolute terms or as a percentage of the pool.

SCHREIER: Perhaps I may be able to shed a little light on this point. From studies in which we injected S^{35}-labelled albumin into young and adult rabbits, it appeared that albumin is mainly stored in the skin, muscle, and kidneys. In these organs albumin had the same half-life as in blood. This means that albumin is merely stored there without very much break-down. But in the liver a more or less steady labelling is detectable — presumably owing to the break-down of albumin molecules.

[1] STERLING, K.: J. Clin. Invest. (U.S.A.) **39**, 1900 (1960).

Discussion 325

ARIAS: I think that a word of clarification is necessary regarding the question of protein leakage into the intestinal tract. Protein in the intestinal lumen is hydrolysed and the amino acids are largely reabsorbed. For this reason, quantitation of protein loss into the intestine has proven to be a difficult problem. Dr. JEEJEEBHOY, in Dr. FRASER's laboratory, has developed a method which may permit quantitation of normal protein leakage into the gut. Perhaps Dr. FRASER will comment on this.

FRASER: Dr. JEEJEEBHOY[1] has been doing studies mostly concerned with patients with intestinal disorders, but by a technique which involves giving anionic resins orally, so that the radioactive label which leaks into the gut

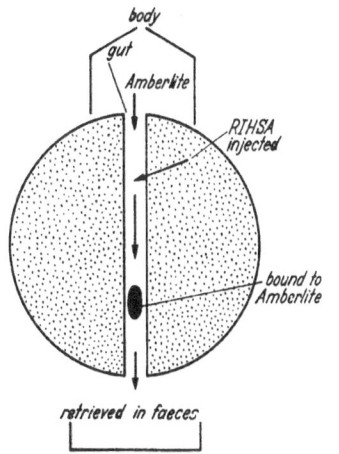

Fig. 1. Use of Amberlite in measuring break-down of injected I¹³¹-albumin

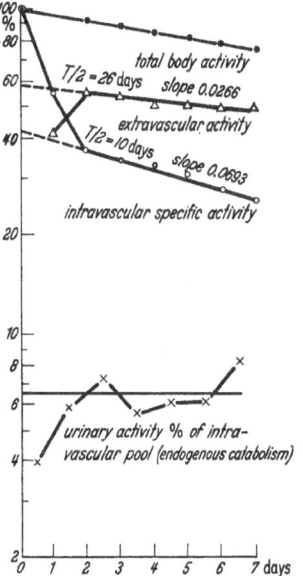

Fig. 2. Measurements on a patient after intravenous injection of I¹³¹-albumin

will stay there; in this way, of course, you can measure the leak of radioactive iodide plus radioactive albumin into the gut. After establishing a normal rate, it is possible to derive what is probably a fair index of any increased protein leakage into the gut (Fig. 1). The urinary leakage of radioactivity gives an index of the degradation going on by other procedures in the body, and these two together give you a total degradation. Fig. 2 illustrates one of Dr. JEEJEEBHOY's studies, just as an example, showing the measurements made. In the normal subject the loss into the gut is perhaps half of the total loss, the rest being indicated by the urinary loss. Hence, presumably, this leakage into the gut does not explain all the metabolism of albumin.

---

[1] JEEJEEBHOY, K. N., and N. F. COGHILL: J. Brit. Soc. Gastroenterol. **2,** 123 (1961).

Muscular protein metabolism in normal and diseased states

By

J.-C. Dreyfus, J. Kruh, and G. Schapira

PART I

General considerations

The study of protein metabolism nearly always calls for the use of C^{14}- or H^3-labelled amino acids. Such methods were first employed in *in vivo* experiments, and later in *in vitro* experiments, using tissue slices and — more recently — cell-free systems. Work of this kind has provided a good knowledge of the mechanism of protein biosynthesis. The specificity of protein biosynthesis is determined by desoxyribonucleic acid, which carries the requisite genetic information and transmits it to a ribonucleic acid known as "messenger RNA"; the latter is a copy of the desoxyribonucleic acid molecule and acts in the ribosomes as a template for the proteins to be synthesised. The proteins are produced inside the ribosomes, using certain non-particulate components, enzymes, and so-called transfer ribonucleic acid. The overall mechanism of protein synthesis appears so far to be universal, although several metabolic pathways have been described. Most of the work on cell-free systems has been carried out on bacteria, liver, and reticulocytes. Muscle has rarely been used in these studies for several reasons:

a) It is very difficult to obtain pure nuclei, which form a sediment with the bulk of the myofibrils.

b) Muscle is poor in mitochondria and in ribosomes.

c) It is very difficult to homogenise muscle without damaging the subcellular particles.

Muscle is poor in ribosomes and in ribonucleic acid, a finding which is bound up with the fact that protein synthesis is less active in muscle than in many other tissues.

In addition, half of the ribonucleic acid is located neither in the ribosomes nor in the nucleole, but in the myofibrils (HAMOIR, 1951; PERRY and ZYDOWO, 1959), a fact which raises interesting questions as to the site of the biosynthesis of myofibrillar proteins.

It is these difficulties which account for the scarcity of work on muscle cell-free systems; our own attempts in this connection have met with little success.

Little use has been made of muscle slices, as they are difficult to obtain without damaging the cells. Nevertheless, some incorporation studies have been performed on rat diaphragm, although the diaphragm cannot be considered as a typical skeletal muscle; similarly, heart muscle or pigeon breast muscle, which show greater metabolic activity, cannot really be compared with mammalian skeletal muscle.

The work to be described below was carried out on living animals treated with radioactive amino acids in order to study their fate in the muscle proteins. Before describing the experiments we should like to emphasise several points concerning the interpretation of isotopic data.

1. Evaluation of the overall activity of an organ or tissue

The rate of incorporation of an amino acid into the proteins of a given tissue provides a fairly reliable clue to the activity of the tissue with regard to protein metabolism. However, we must be clear as to the precise significance of such data. Fast incorporation within a tissue in a steady state could mean:

a) Rapid turnover of the proteins.

b) Replacement of removed cells (small intestine).

c) Synthesis of proteins which will be secreted into external ducts (pancreas) or into the blood plasma (liver).

Muscle and kidney are both stable tissues and they do not secrete their cell proteins. Nevertheless, kidney incorporates radioactive amino acids into its proteins several times faster than does muscle, and the rate of turnover is also much faster: the overall half-life of rat kidney protein is about four days, as opposed to more than 20 days in the case of muscle protein.

2. Evaluation of anabolism

A change in the rate of incorporation into muscle protein has sometimes been advocated as a criterion by which to evaluate a change in anabolism, either in pathological states or in response to treatment. It seems to us, however, that such results call for very careful interpretation.

An increase in the rate of incorporation might be due to several factors: accelerated penetration into the cell, which in the case of muscle, as opposed to other tissues, may be a limiting factor;

inhibition of other routes of disposal, e.g. oxidation of the amino
acids, which would leave more of the material available for incorporation; a diminution in the pool of intracellular amino acids:
a diet poor in proteins may have various repercussions as regards
the incorporation of amino acids, but it will always cause a slowing
down, apparent or real, of the turnover rate owing to re-utilisation
of the amino acids, which is much greater on a low protein diet.

On the other hand, an increase in the rate of incorporation may
be accompanied by a decrease, or an increase, in the loss of radioactivity from the tissue proteins. In the first instance, metabolism
is probably predominantly anabolic, although this could be determined just as well, or perhaps even better, by means of balance studies; in the second instance, there will be excessive catabolism,
as will be apparent in the case of muscular dystrophy. Thus, as far
as the investigator working with isotopes is concerned, it is by no
means easy to define anabolism.

The above considerations emphasise, and possibly overempha-
sise, the difficulties encountered in studying the metabolism of
tissues in general and of muscle in particular.

However, in many respects muscle is a privileged tissue as regards such studies, and this for the following reasons:

a) Muscle is the tissue whose proteins have so far been best
studied, and from muscle it is possible to isolate many proteins in a
satisfactory state of purity and in a good yield.

b) From muscle one can isolate proteins varying in origin and
function, such as soluble enzymes and fibrous-structure proteins.

PART II

Metabolism of muscular proteins in the normal state

1. Metabolism of aldolase

2. Metabolism of myosin

3. Metabolically stable proteins

*4. The interconversion of amino acids after their incorporation
into myosin.*

The fate of muscle proteins may present several metabolic
patterns:

1. Some are subjected to a molecular turnover: we shall describe
aldolase as an example. The same probably applies to all proteins
of the sarcoplasm.

2. Some apparently stay unchanged, or relatively unchanged,
throughout the life-span of the cell, or at least of their subcellular
unit: this is true, for example, of the myofibril.

3. Some proteins possibly remain completely inert during the whole life of the animal. Collagen, and perhaps other proteins as well, may belong to this category.

Methods

Our methods have been described in detail elsewhere (DREYFUS et al., 1960).

White male rats of pure Wistar strain, weighing 200—250 g. and fed on an identical diet, were used in each series of experiments. The animals were given three types of diet:

a) A *standard diet* containing approximately 25% of protein.

b) A *protein-rich diet* containing 45% protein.

c) A *diet poor in protein* containing only 8% of animal and vegetable protein.

The animals were injected intraperitoneally with C^{14}-glycine in a dosage of 5 μc. (100 mcg.) per 100 mg. body-weight.

Myosin was prepared according to the method of SZENT-GYÖRGYI (1951); its purity was ascertained by moving-boundary electrophoresis. The absence of actomyosin was always confirmed by the absence of a decrease in viscosity under the action of adenosine triphosphate.

Aldolase was crystallised three times by the method of WARBURG and CHRISTIAN (1943) modified by TAYLOR et al. (1948). The purity of the protein was verified in preliminary experiments by electrophoresis and by the absence of several enzymatic activities. Special experiments were performed to check the absence of serum proteins (which would be much more heavily labelled) and of small adsorbed molecules, such as glutathione (SCHAFIRA et al., 1960).

Water-soluble proteins were prepared by extraction of crushed muscle with 4 vol. of water and, after filtration, by precipitation of the protein with 1 vol. of 20% trichloro-acetic acid.

Myofibrils were prepared according to the method of PERRY (1953), and *globin* was prepared according to that of ANSON and MIRSKY (1930).

Isolation of glycine and serine and of phenylalanine and tyrosine from proteins: Proteins were hydrolysed by heating under reflux with 3,000 vol. of 6N-HCl for 48 hrs at 112°. Glycine and serine were isolated as dinitrophenyl derivatives.

Free glycine was isolated as its dinitrophenyl derivative from a trichloro-acetic acid extract.

Estimation of the DNP derivatives, of phenylalanine and tyrosine, and measurement of radioactivity have been previously described (Dreyfus et al., 1960). The results were expressed as counts/min./μMole.

1. Metabolism of aldolase (Schapira et al., 1960)

The decrease in radioactivity is exponential-like. The duration of the half-life calculated from the first part of the curve is 20 days (Fig. 1).

Other muscle proteins: The water-soluble muscle proteins display an exponential-like decrease in radioactivity. The half-life is only an average value, since we are dealing with a mixture of proteins. The faster rate of decrease may be due to the presence of plasma protein. The duration of the half-life calculated from the first part of the curve is 7 days.

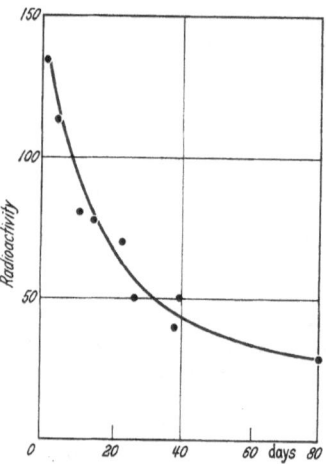

Fig. 1. Specific activity of rat muscle aldolase plotted against time. The radioactivities are expressed in c.p.m. of 30 mg. aldolase spread on 1.52 sq. cm. planchets

In order to demonstrate that the exponential decrease in aldolase radioactivity affords evidence of molecular turnover in proteins, it must be established that there is no secretion of aldolase outside the muscle and that the life of the muscle cell is long.

The role of aldolase secretion calls for further discussion, because there is some aldolase in plasma. The concentration of this enzyme is 3 Meyerhof units per g. of muscle and 0.1 to 0.2 units per 100 ml. of plasma. As muscle represents 40% and plasma 5% of the body-weight, it can be calculated that there is about 12,000 times more aldolase in muscle than in plasma. If the half-life of muscle aldolase is 14 days, and if this value were merely the result of secretion of muscle aldolase into the plasma, then the half-life of plasma aldolase would work out at only 1 to 2 minutes. Such a short half-life would seem to be unlikely. It was also found in our laboratory that there was no difference between the levels for arterial and venous femoral blood in the rabbit and the human; this would not be the case if there were a marked secretion of protein (Dreyfus et al., 1958).

In order to try and solve the problem of the life-span of aldolase in plasma we used the technique of labelling radioactive iodine (SCHAPIRA et al., 1962).

Aldolase was prepared according to the method of TAYLOR et al. (1948) and crystallised 4 times. Iodation was performed according to the procedure described by BERSON (1953); 1 mg. of labelled aldolase was injected in four successive experiments into the marginal vein of the rabbit's ear, and bleeding was carried out at different times after the injection.

If the results are to be valid, the aldolase must remain in a native state; otherwise its life-span will be shortened. The enzymatic activity is the same for labelled as for unlabelled aldolase, i.e. allowing for the duration of the necessary dialysis. Electrophoretic control has demonstrated that the rate of migration of labelled and unlabelled aldolase is also identical and that the radioactivity and the enzyme itself are situated in the same place.

The decay of the radioactivity follows an exponential law: the time of the half-life is 198 min. $\pm$ 108.

We can compare these results with the 12 hours which we found it took for a return to the initial value in a dystrophic patient following an exchange transfusion.

In view of the lack of mitotic activity, it can be regarded as certain that muscle cells have a long life. It can therefore be stated that muscle aldolase undergoes a phenomenon of molecular turnover according to the law of dynamic equilibrium; there are exchanges between bound and free amino acids which may or may not involve complete hydrolysis and resynthesis of the protein molecule.

2. Metabolism of myosin

The incorporation of glycine into myosin is maximal by the 36th hour; the radioactivity of the myosin *remains constant* until about the 30th day and then decreases to a lower plateau in animals fed with a diet containing 24% of protein.

Fig. 2 shows the results obtained with all the rats. The curve for the specific activity of glycine isolated from myosin is similar, and the specific activity of free glycine in muscle decreases exponentially.

However, the spread of the values indicated in Fig. 2 is such that other evidence must be adduced if we are to interpret the shape of the radioactivity curve for myosin. The plateau is obtained from the constant ratio of the radioactivity of myosin to globin between the 5th and the 30th day (Fig. 3).

This indicates that myosin and globin behave identically during the period in question (later there seems to be a temporary decrease owing to a diminution in the radioactivity of myosin). The initial decrease in the ratio is due to the time-lag in the appearance of

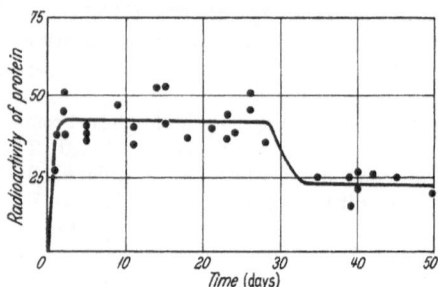

Fig. 2. Radioactivity of myosin after injection of C¹⁴-glycine into rats fed on the standard diet

circulating radioactive red cells: the specific activity of haemoglobin increases for one week, whereas that of myosin reaches its maximum within 2 days after the injection. The decrease in the radioactivity of myosin is also suggested by the shape of the curve

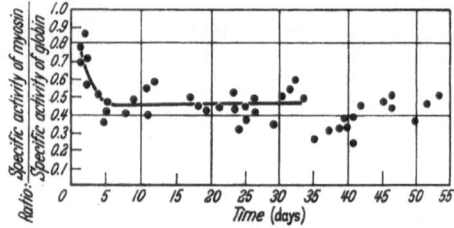

Fig. 3. Ratio of the radioactivity of myosin to that of globin

for the ratio of the radioactivity of aldolase or soluble muscle protein to that of myosin. Fig. 4 shows that the ratio curve descends at first, owing to the decrease in the radioactivity of aldolase and of the soluble muscle proteins; after 32 days the slope of the curve is reversed — a fact which can be explained by the sudden drop in the radioactivity of myosin. After 40 days the curve declines again, because the myosin has now reached a new plateau. A similar result was obtained in one series in which C¹⁴-valine was used instead of C¹⁴-glycine.

LEBLOND and WALKER (1956) have discussed the concept of cellular renewal in connection with a large number of cell populations; the types that have been most thoroughly investigated are the cells of the haemopoietic tissues and the red cells. In the case of fixed tissues, the epithelial cells are continuously being renewed; on the other hand, the duration of life of hepatic cells has been estimated to be at least 4—5 months (SWICK, KOCH, and HANDA, 1956); that of the muscle cells has not so far been studied. Their life-span seems to be very long in view of the absence of mitotic activity; the rate of turnover of the various cellular subunits varies widely, and we can interpret our findings in the light of the life-span of a subcellular unit likely to be the myofibril. This aspect has been considered by several histologists. Whereas division of the myofibrils has been observed only very rarely, by ADAMS, DENNY-BROWN, and PEARSON (1954), it has been noted that the phenomena of muscle hypertrophy or atrophy were accompanied by a variation in the number of myofibrils.

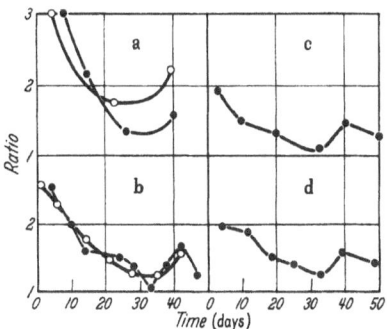

Fig. 4. Ratio of radioactivities of muscle proteins. (a) Aldolase/myosin, after injection of C^{14}- glycine into rats fed on a diet containing 24% of protein (two different series of animals were used). (b) Muscle soluble protein/myosin under the same conditions (two different series of animals were used). (c) Muscle soluble protein/myosin after injection of C^{14}-valine into rats fed on a diet containing 24% of protein. (d) Muscle soluble protein/myosin after injection of C^{14}-glycine into rats fed on a diet containing 45% of protein

In our experiments the plateau was reached after two days and the decline in the radioactivity of the myosin took place between the 28th and 35th day. The molecules of myosin are therefore destroyed about 30 days after having been synthesised. Since the radioactivity of myosin is very similar to that of the total myofibril proteins, it would seem that the curve for the radioactivity of myosin reflects the life-span of the myofibrils. The decline in the overall radioactivity of the myofibril protein, compared with that of myosin, can be explained by the fact that the myofibril contains a small amount of protein with a fairly rapid turnover, but the bulk of its protein is composed of myosin and is perhaps metabolically inert and endowed with the same radioactivity as myosin. These results permit us to ascribe to the myofibrils of the rat a life-span of 30 days (experiments have shown that mice myofibrils have a life-span of 20 days). The values given are only average values for the whole animals, since the incorporation of the amino acid may vary depending on the site of the muscle, as has previously been shown in the case of N^{15}-glycine by Schapira, Dreyfus, Coursaget, and Schapira (1953).

However, our results do demonstrate that it is possible to determine the life-span of a subcellular unit by biochemical methods.

Myosin and dynamic equilibrium: The experiments performed with rats fed on a diet containing 24 % of protein led us to conclude that myosin is metabolically inert or almost inert. But this inertness may perhaps only be apparent, since free muscle glycine remains significantly radioactive for several weeks, as we have already shown (Schapira et al., 1958). This radioactive precursor should tend to increase the specific activity of myosin, even after the first few days; if myosin were completely inert for 30 days, the specific activity of free glycine would be sufficient to increase the overall activity of myosin, because at that time only non-radioactive myosin would be destroyed. Since we observed a plateau and no increase, it is probable either that some molecules of radioactive myosin are destroyed during the life of the myofibril (compensating for the new molecules of radioactive myosin which are being synthesised) or that the myosin molecule is subject to some turnover which progressively decreases its radioactivity; the plateau would, in this case, represent the net result of the two mechanisms. To demonstrate this, we decided to feed rats on a protein-enriched diet, in order to dilute the free radioactive glycine as fast as possible.

High-protein diet: In rats given a protein-rich diet, the curve has a different shape and does not show a plateau. The values are so scattered that it is difficult to plot a curve; however, they would appear to indicate a slow decrease without any marked drop. Nevertheless, the ratio of the radioactivity of the soluble muscle protein to that of myosin presents the same reversal of the slope as seen in rats fed on the standard diet.

Steinbock and Tarver (1954) have shown that a protein-rich diet apparently accelerates the renewal of proteins — possibly owing to a decrease in the re-utilisation of the radioactive amino acid. In our experiments with a protein-rich diet, we observed a disappearance of the plateau and a progressive decrease in radioactivity with a wide dispersion of the individual values. It appears that under these conditions molecular renewal of myosin can be demonstrated. It is impossible to deduce from the curves the life-span of myofibrils: our conclusion, however, is supported by the fact that by about the 30th day the slope of the curve of radio-activities (water-soluble protein/myosin) is reversed.

We can therefore conclude as follows: the myofibrils display a well-defined life-span of about 30 days; the myosin is in a state of relative inertness and its metabolic behaviour can be compared with that of haemoglobin, although the inertness is not complete; if the radioactive free glycine is diluted by the use of a high-protein diet, the myosin molecule appears to have a slow but definite molecular turnover.

3. Metabolically stable proteins

Since the first observation by SPRINSON and RITTENBERG (1946) that the turnover of proteins varies according to the tissue in question, the hypothesis has been advanced that some muscle and nerve proteins may be completely inert. THOMSON and BALLOU (1953), using H^3-labelled amino acids, divided muscle into two kinds of component, one active and the other very stable. DAVISON (1961), after injecting radioactive glycine into 6-day-old rats, observed that the radioactivity still persisted after 250 days, and he concluded that some proteins of muscle and nerve are metabolically quite inert.

No such completely stable constituents were found, however, when radioactive amino acids were injected into adult rats. In view of the small number of animals studied (6 young and 6 adult rats) it is not possible to draw any definite conclusions; nevertheless, these experiments deserve to be noted, repeated, and extended, since they may enable us to identify this class of stable proteins, which would then take its place beside collagen, whose extreme stability was first demonstrated by NEUBERGER.

4. The interconversion of amino acids after their incorporation into myosin (SCHAPIRA et al., 1962)

KRUH, DREYFUS, SCHAPIRA, and PADIEU (1957) have shown that haemoglobin is not uniformly labelled when either radioactive glycine or phenylalanine is injected (KRUH, DREYFUS, and SCHA-PIRA, 1960), but that the labelling does become uniform as the red blood cells increase in age. These results suggested several hypotheses; one of them was that haemoglobin does not remain inert but that it undergoes intramolecular changes during the life of the red cell.

Comparison of the specific activities of glycine and serine, after the injection of radioactive glycine, and of the specific activities

of phenylalanine and tyrosine suggested to us that intertrans-
formations of amino acids could have occurred after the incorporation of the precursor amino acids and during the life of the red
blood cells and of the myofibrils. Our study of this problem was
facilitated by the fact that haemoglobin and myosin are not renewed, or only very slowly renewed, which enables one to exclude the
possibility of rapid exchange between bound and free amino acids.

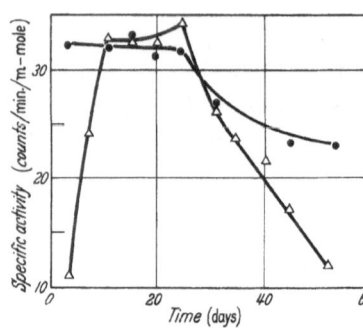

Fig. 5. Variation with time of the specific activities
of phenylalanine (●) and tyrosine (△) incorporated
into myosin

Preliminary experiments
have shown that when the
animals are fed on a normal
diet the ratios of the specific
activities of glycine to serine
in myosin rapidly reach unity after the injection of radioactive glycine. On the
other hand, these ratios approach unity much more
gradually when the rats are
fed on a low-protein diet;
the interconversion of amino acids after their incorporation into protein can thus
be more readily followed.

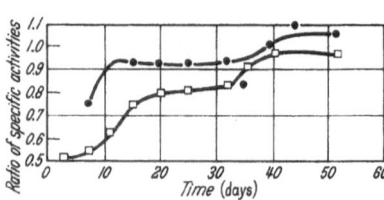

Fig. 6. Variation with time of the ratios of specific
activities of serine/glycine (□) and tyrosine/phe-
nylalanine (●) from myosin

Myosin: The results ob-
tained with myosin were
essentially the same as those
obtained with globin. The
radioactivity of the myosin
is constant for 25 days and
then decreases. The curve
for the total activity of
myosin, glycine, and serine
remains constant during the
first 20 days and then ra-
pidly declines. On the other hand, the specific activity of myosin-
glycine decreases and that of myosin-serine increases, the specific
activity of the phenylalanine of myosin having the same characteristics as that of globin. The plateau descends slightly until the
25th day, after which the curve falls away sharply. The shape of
the curve for the specific activity of phenylalanine gives us the best
evidence confirming the life-span of myosin (Fig. 5). The specific
activity of tyrosine is very low on the third day, after which it
increases rapidly, reaching 90% of that of phenylalanine within

11 days; it then remains unchanged until the 25th day and later decreases like that of phenylalanine. The ratio of the specific activities of tyrosine/phenylalanine increases much more rapidly than that of the specific activities of serine/glycine (Fig. 6) or than the ratio in the case of globin.

The ratio of the sum of the activities of the four amino acids to the radioactivity of myosin is plateau-shaped between the 3rd and 40th day. It then decreases, probably because of progressive transformations into other amino acids. The sum of the radioactivities of phenyl-alanine plus tyrosine gives a plateau between the 10th and 20th day (Fig. 7).

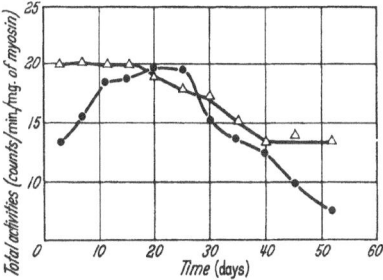

Fig. 7. Variation with time of the total activities of glycine + serine (△) and of phenylalanine + tyrosine (●)/mg. of myosin. The total activity of each amino acid was obtained by multiplying its specific activity by its relative amount in myosin

Interpretation of the results is the same for globin and for myosin: the evidence favours the view that glycine and phenylalanine are transformed into serine and tyrosine after having been incorporated into the proteins, and not that serine and tyrosine arise from a common amino-acid pool, which would become heavily labelled at a later period after injection of radioactive glycine or phenyl-alanine.

This phenomenon can be interpreted in terms of a three-step mechanism involving liberation of the amino acid from the protein, passage through a pool not mixed with the general pool, and re-incorporation of the amino acid into the protein chain. Temporary liberation of an amino acid and its re-incorporation is one of the mechanisms assumed to be involved in the intramolecular turn-over. But this turnover would presumably be purely internal, i.e. within the myofibrils, which would take no part in the general metabolic turnover of the rest of the body, since there is no overall decay in the radioactivity of myosin.

In the present work, one more step is involved, i.e. the transformation of glycine into serine and of phenylalanine into tyrosine. It should take place in the local pool before re-incorporation.

PART III

Metabolism of muscular proteins in diseased states

1. *Muscular dystrophy in mice*
2. *Atrophy due to nerve section*
3. *Dystrophy and atrophy in man*

1. Abnormalities of muscle protein metabolism in mice with muscular dystrophy

The discovery of a myopathic mutation in a colony of strain 129 mice (Michelson et al., 1955) provided an opportunity for studying possible abnormalities in the metabolism of muscle proteins with the use of labelled amino acids.

We compared the metabolism of myosin and water-extractable muscle proteins in dystrophic mice, in their normal littermates, and in another normal strain of mice. Although we did not prepare pure aldolase, which would have required too many animals, previous evidence had confirmed that the total water-extractable proteins have the same gross behaviour as aldolase.

Results

Normal mice (Strain RAP)

Myosin: Two series of experiments involving 40 mice showed a constant level of radioactivity between the 2nd and 20th day with a mean value of 30 c/min. $\pm$ SD $= 2$. The radioactivity then decreased and remained constant until the 38th day at a level of 19 ± 2.5.

Water-extractable proteins. The radioactivity of the water-extractable proteins was much higher than that of myosin on the 2nd day and then decreased exponentially.

Dystrophic mice and normal littermates (Strain 129)

The mice were killed at the 6th hour and on the 1st, 2nd, 4th, 6th, and 15th day after the injection of C^{14}-glycine.

Myosin (Fig. 8): In the normal littermates the radioactivity was constant between the 1st and the 15th day. In the dystrophic mice the radioactivity increased until the 4th day, by which time it was twice as high as in the normal mice. After the 6th day it decreased, and on the 15th day the radioactivity value was close to that of the normal mice.

Water-extractable proteins (Fig. 9): In normal as well as in dystrophic mice the radioactivity increased until the 24th hour and then decreased, but the variation as a function of time was much larger in the dystrophic than in the normal mice. The ratio of the radioactivities was 2 in the 24th hour and 1.3 on the 15th day.

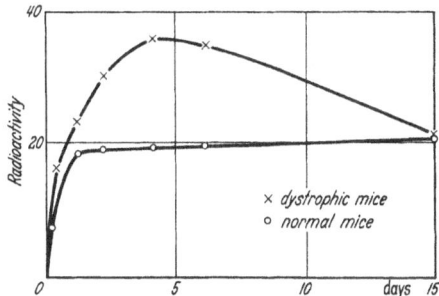

Fig. 8. Radioactivity of myosin. The radioactivities are measured on 40 mg. of protein on 1.52 sq. cm. planchets and expressed in counts per minute

Free muscle glycine: The specific activity of free glycine decreased exponentially until the 1st or 2nd day, after which the slope of the semi-logarithmic curve changed. The decrease, however, remained exponential until the 15th day, confirming the results previously obtained. The specific-activity curves were approximately the same in the normal and in the dystrophic mice.

Acceleration of protein metabolism in dystrophic mice has been demonstrated by SIMON et al. (1958, 1962) and by COLEMAN et al. (1959). These authors injected C^{14}-leucine into mice and observed a faster incorporation of the amino acids into the proteins and a faster rate of disappearance from the proteins in the dystrophic animals. The difference in behaviour cannot be ascribed to a modification in the permeability of the muscle cell to glycine, since we have not found any

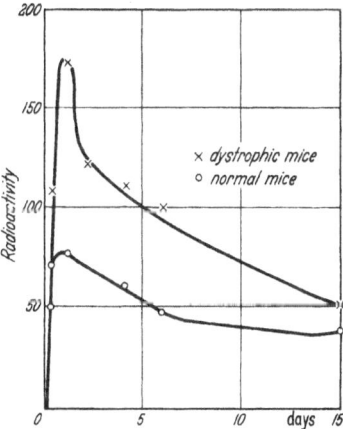

Fig. 9. Radioactivity of water-extractable proteins. The radioactivities are measured on 40 mg. of protein on 1.52 sq. cm. planchets and expressed in counts per minute

22*

significant difference between the specific activity of free glycine in normal and in dystrophic mice. Nevertheless, a difference in the degree of penetration of radioactive glycine into the cells during the first hours following the injection is a possibility which cannot be excluded. Such a difference would explain the more rapid increase in radioactivity but not the more rapid subsequent decrease. Recent knowledge concerning the life-span of myofibrils and the isolation of pure myosin has provided some information on the localisation of the biochemical lesion. The lesion may be localised at a molecular, a subcellular, or a cellular level.

At a molecular level, the results could be partly explained by an acceleration of muscle protein turnover according to the hypothesis advanced by Simon and co-workers. We have previously shown that in rats submitted to a high-protein diet, myosin shows a slow turnover, which, in rats on a standard diet, was disguised owing to the re-utilisation of radioactive glycine. In rats on such a diet, the radioactivity of myosin declined progressively. Under pathological conditions, with an accelerated turnover of myosin, the turnover can become apparent even where the animals are receiving the standard diet. The finding of an acceleration in the turnover of water-extractable muscle proteins is in keeping with a general acceleration in the turnover of all the muscle proteins. However, an acceleration in the turnover should lead to a faster rate of incorporation of labelled glycine into myosin.

Fig. 8 shows that in dystrophic mice there is a rise in the radioactivity of myosin during the first four days, whereas in normal mice the radioactivity reaches a maximum after only one day. It is possible that this discrepancy might be due to the release of amino acids derived from the degradation of radioactive water-extractable proteins and to their utilisation inside the cell for the synthesis of new myosin molecules.

At the subcellular level, these results could be interpreted as being due to a shorter life-span or to the absence of a myofibril life-span, as in the case of the red cells in haemolytic and sickle-cell anaemia. This hypothesis, however, would only explain the acceleration of the myosin turnover. Owing to the relatively small number of experiments performed, it is impossible to determine whether the myofibrils have no life-span or simply a shorter life-span, which would involve a short plateau.

At the cellular level, the results could be accounted for by a shorter life-span of the muscle cells, coupled with an accelerated formation of new cells. In dystrophic mice, Michelson and co-workers observed division of muscle fibres into two or three parts.

A more detailed study is still needed, however, since the life-span of the muscle cell has not yet been established. It seems that the results presented here could be explained in terms of an acceleration in the turnover of muscle proteins or, alternatively, a shorter life-span of the myofibril. It is not possible at present to determine which of these two hypotheses may be correct, but the first seems more likely, i.e. an acceleration in the turnover of the muscle proteins.

2. Atrophy of nervous origin

Experimental denervation brings about very rapid muscular atrophy. This poses the question as to whether the atrophy results from suppression of anabolism or from accelerated catabolism.

Long-term experiments are impracticable because the tissue is not in a steady state and because the muscle weight diminishes very considerably within a few weeks. Studies therefore have to be confined chiefly to the rate of incorporation.

Most workers have investigated atrophy of relatively long duration and have administered the labelled amino acid one to several weeks after denervation. SCHAPIRA et al. (1953), using N^{15}-glycine, have shown that section of the sciatic nerve induces after ten days a decrease in the incorporation of glycine into the gastrocnemius muscle. Similar results have been obtained with S^{35}-methionine (SEVES, 1955; FERDMAN, 1955), although FERDMAN observed that after 30 days the rate of incorporation increased. The only work pertaining to the incorporation of amino acids into muscle proteins at short intervals following denervation was undertaken by PADIEU (1959) in our laboratory: 10 to 24 hours after denervation, rabbits were perfused with C^{14}-glycine for 4 hours; 3 days later they were killed and their muscle proteins prepared. The specific activity of the muscle soluble proteins, myosin, and free glycine in both gastrocnemius muscles was determined.

The results revealed a decrease in the rate of incorporation into both myosin and soluble proteins: for a mean of 9 rabbits, the specific activity of myosin was 55%, and that of the soluble proteins 77% of the control value (contralateral, non-denervated limb).

It can be tentatively accepted that, a few hours after denervation, the incorporation of glycine into muscle proteins decreases, and that this decrease may play a part in the subsequent atrophy. However, one must be cautious in interpreting the findings, because the specific activity of free glycine is also lower in the denervated muscle at the time of the determination (3 days after the injection) — a fact which might be due either to a decrease in the penetration of radioactive glycine into the cells of denervated muscle or, on the other hand, to a faster release of radioactive glycine from the denervated muscle.

3. Dystrophy and atrophy in man

Neither in human progressive muscular dystrophy nor in human neurogenic atrophy were we able to study the metabolism of muscle proteins with labelled amino acids.

We shall summarise the results yielded by biochemical analysis of muscle after biopsy and by a study of serum enzymes, which afforded evidence of an escape of muscular enzymes into the blood stream owing to an increase in muscle permeability.

Muscle

Methods of study. Biochemical analysis of muscle tissue poses a whole series of problems which have already been discussed elsewhere (SCHAPIRA and DREYFUS, 1959).

Results

In progressive muscular dystrophy (PMD) the muscular tissue undergoes numerous changes. The main change is a replacement of the muscle tissue itself by fibrous and fatty tissue. Many biochemical changes have also been ascertained. Here, however, we shall comment only on the changes affecting the proteins.

The loss of muscular tissue in PMD involves mainly a loss of specific proteins. We must remember that the components of a muscle cell include:

a) Sarcoplasm, in which are found many enzymes (particularly those of glycogenolysis) and myoglobin.

b) Sarcosomes, which may be described as the mitochondria of muscle with their oxidative and respiratory enzymes.

c) Myofibrils with the specific and contractile proteins: myosin and actin.

There is a serious loss of proteins from the sarcoplasm. We have shown that glycogenolytic enzymes and total glycogenolysis fall to 25% of the control value. This loss is not merely an apparent one; it is not simply the result of replacement of muscle tissue by collagenous tissue, but also represents a true loss of sarcoplasm. Our own results were recently confirmed by RONZONI et al. (1958) and by VIGNOS and LEFKOWITZ (1959), except in the case of dystrophy of late onset.

A diminution in myoglobin also occurs. Although PERKOFF et al. (1958) failed to demonstrate a molecular disease by means of electrophoresis, quite recently a modification was described which would suggest that this protein (WHORTON et al., 1961) or part of it (PERKOFF et al., 1962) retains its foetal structure. The existence of a special foetal myoglobin, however, is not yet accepted by all authors (SCHNEIDERMAN, 1962). In the rat, by way of contrast, the electrophoretic pattern of myoglobin changes following denervation (LAURENT et al., 1961).

Enzymes of the mitochondria remain unchanged, a diminution in activity being observed only if the fibrous and fatty degeneration are not taken into account.

The myofibril proteins myosin and actin show a diminution. A change occurs in the double refraction of flow (DREYFUS et al., 1953; SCHAPIRA et al., 1954) and in the ultracentrifugation pattern (HORVATH, 1958) of myosin such as can also be observed in other pathological states.

Significance of changes in muscle biochemistry. The main diagnostic problem encountered in man is that of distinguishing between two types of muscular atrophy, i.e. myogenic and neurogenic atrophy, which appear to differ widely from the biochemical standpoint (FISCHER, 1952, 1955; GUTMAN, 1959; SCHAPIRA and DREYFUS, 1959). Most of the signs and symptoms, however, are similar, e.g. creatinuria, a diminution in glycolytic enzymes, in muscle potassium, in adenosine triphosphate, and in phosphocreatine, and maintenance of oxidative enzymes (KNOWLTON and HINES, 1934; SCHMIDT and SCHLIEF, 1956). The only difference as regards muscle composition is in the non-haeminic iron, which is increased in nerve atrophy and not in muscle dystrophy (DREYFUS, 1951). The local response of muscle is thus very similar in both diseases.

Serum enzymes

In the juvenile forms of PMD, an increase in serum aldolase (SCHAPIRA et al., 1953), glutamic pyruvic transaminase (DREYFUS and SCHAPIRA, 1955), phosphohexoisomerase (SCHAPIRA et al., 1955) and lactic acid dehydrogenase (SCHAPIRA and DREYFUS, 1957) is often observed.

Serum creatine kinase (EBASHI et al., 1959; DREYFUS et al., 1960), which occurs almost exclusively in muscular tissue, may be even more markedly elevated. In neurogenic atrophy few abnormal serum enzyme activities are encountered.

It seems most likely that the serum enzymes in PMD emanate from muscle. But, obvious though this might seem, we have had great difficulty in proving that these serum enzymes originate from muscle.

a) As shown above, many enzymes in the muscle display a considerable decrease in activity, and this decrease, which progresses with the evolution of the disease, may account for the diminution in aldolasaemia at a later period. If a constant fraction of the tissue enzymes escapes into the plasma within a given unit of time, it is clear that the amount of enzyme released will decrease in the course of the disease.

b) A comparison has been made between the aldolase level of arterial and venous blood in order to demonstrate directly the flow of enzyme from muscle to blood. No difference was found in normal rabbits. However, when the same determinations were performed in the femoral blood of 10 patients suffering from PMD, we observed in 5 cases a slightly greater activity in the vein than in the artery — which seems to suggest that the flow of aldolase from muscle to plasma can be directly demonstrated.

c) The main body of evidence for the theory that serum enzymes come from muscle is based on the concept of specificity. In this connection, two sets of data can be utilised. The first is a negative finding: "hepatic" aldolase (1-phospho-fructo-aldolase) is absent from the serum of PMD patients. The second is a positive one: the presence of high levels of creatine kinase, which affords proof of the muscular origin of serum enzymes in muscular dystrophy.

Mechanism of the liberation of tissue enzymes. Tissue necrosis is the most widely accepted cause of an increase in plasma enzymes in pathology. Though this may certainly apply to such acute conditions as myocardial infarction, toxic hepatitis, or crush injury of muscle, it cannot be accepted as such in the case of PMD. The essential feature of this chronic disease is the very slow and progressive destruction of the muscle tissue. We have seen that in this disease the mechanism of muscle wastage is probably based on the escape of muscle protein into the plasma. The problem is to find out why enzymes pass into the circulation in PMD, whereas they do not in muscle disease of nervous origin, a disease in which the atrophy is often more rapid. It may be assumed that there are differences in permeability or in metabolism in the two forms of disease. It may well be that, in myopathies, the muscle secretes whole proteins into the blood stream, whereas in neurogenic atrophy local catabolism may go one step further, the proteins themselves being destroyed *in situ*.

It is quite probable that PMD is associated with a change in permeability. But present concepts of permeability are still too confused to provide us with more than a mere theoretical explanation. In the case of PMD there are certain other factors which might be considered:

a) The role of anoxia may be worth investigating further. Some degree of anoxia is likely to occur in PMD. Vascular disorders, such as RAYNAUD's disease, are quite frequent. Our collaborators J. DEMOS and J. ECOIFFIER have demonstrated circulatory disturbances in PMD patients, as shown by arteriography and by measurement of the speed of circulation (DEMOS and ECOIFFIER, 1957). In addition, together with DEMOS we have recorded a diminution in the oxygen consumption of the dystrophic muscle. The most probable explanation, although it is by no means proven, is that transient anoxia damages the cell and allows proteins to escape through its membrane. The fact that creatine kinase increases obviously means that the disorder is located in the muscle itself.

b) That the release of enzymes is subject to hormonal regulation can be assumed on the basis of experiments performed by F. SCHAPIRA: in the rabbit, ACTH and cortisone give rise to marked hyperaldolasaemia (F. SCHAPIRA, 1954, 1959). It is thus possible that the adrenals play a role in regulating the level of serum enzymes (COPE and POLIS, 1959). However, 5 cases of myopathy due to triamcinolone showed no increase in serum enzymes (unpublished observations).

c) There is one last factor which may have some influence in increasing the permeability of dystrophic muscle, i.e. an hereditary factor. Such a factor has been demonstrated both in the human disease as well as in hereditary dystrophy in mice. Investigations into the parents of patients suffering from PMD have further emphasised the importance of this factor (F. SCHAPIRA et al., 1960).

Summary

With the aid of radioactive amino acids (C^{14}-glycine, C^{14}-valine, and C^{14}-phenylalanine), attempts were made to obtain information on the turnover of various muscle proteins in rats. The experiments were chiefly concerned with aldolase, myosin, and water-soluble muscle proteins under normal and under certain pathological conditions. Aldolase, as an example of a protein with a rapid turnover, shows an exponential-like decrease in radioactivity with a half-life of 20 days, while for other water-soluble muscle proteins the calculated half-life is 7 days. By contrast, the half-life of aldolase in the plasma works out at 198 minutes.

The incorporation of glycine into myosin is maximal by the 36th hour, the radioactivity remaining constant until about the 30th day, when it then decreases to a lower plateau in animals fed on a diet of average protein content. Between the 5th and 30th day, globin behaves identically following administration of glycine. It would seem that the radioactivity of myosin, a protein which may be regarded as largely inert, reflects the life-span of the myofibrils, which probably amounts to 30 days. The stability of myosin does not preclude the possibility that radioactive amino acids are transformed after their incorporation into the molecule. Glycine is transformed into serine, and phenylalanine into tyrosine, the total radioactivity of the myosin molecule remaining unchanged. Thus, myosin behaves in the same way as globin.

The metabolism of the above-mentioned muscle proteins undergoes characteristic changes in various diseases of the muscles. In mice with hereditary muscular dystrophy, there was a stronger initial increase in radioactivity, but also a more rapid decrease — which may be interpreted as evidence of an accelerated protein turnover or a shortened myofibril life-span.

In rabbits with atrophy due to nerve section, there is a decrease in the rate of incorporation of glycine into myosin and water-soluble proteins within only 24 hours. In progressive muscular dystrophy (PMD) in man, analysis of muscle tissue obtained by biopsy revealed a loss of sarcoplasm and a diminution in myoglobin and myofibril proteins. An increase in aldolase, creatine kinase, lactic acid dehydrogenase, and other enzymes in the serum is particularly marked in juvenile forms of PMD, whereas in neurogenic atrophy few abnormal serum enzyme activities are encountered. It seems probable that the enzymes present in the serum in increased quantities in PMD emanate from muscle. Various mechanisms which might be responsible for the rise in serum enzymes are discussed.

Zusammenfassung

Mit Hilfe von radioaktiv markierten Aminosäuren (C^{14}-Glykokoll, C^{14}-Valin, C^{14}-Phenylalanin) wurde versucht, an Ratten Aufschluß über den Umsatz verschiedener Muskeleiweiße zu erhalten. Untersucht wurden vor allem Aldolase, Myosin sowie wasserlösliche Muskeleiweiße unter normalen und einzelnen pathologischen Bedingungen. Aldolase als Beispiel für ein rasch umgesetztes Eiweiß zeigt einen exponentiellen Abfall der Radioaktivität mit einer Halbwertszeit von 20 Tagen, für andere wasserlösliche Proteine berechnet sich eine Halbwertszeit von 7 Tagen. Die Halbwertszeit von Aldolase im Plasma beträgt dagegen 198 Minuten.

Der Einbau von Glykokoll in Myosin ist nach 36 Stunden maximal; die Radioaktivität bleibt dann bis zum 30. Tag konstant, um anschließend bei Tieren mit durchschnittlichem Eiweißgehalt der Nahrung abzunehmen. Globin verhält sich zwischen dem 5. und dem 30.Tag nach Gabe von markiertem Glykokoll gleich. Es wird angenommen, daß die Radioaktivität des Myosin, das als weitgehend inertes Eiweiß anzusehen ist, die Lebensdauer der Myofibrillen widerspiegelt, die 30 Tage betragen dürfte. Die Stabilität des Myosin schließt die Umwandlung von radioaktiv markierten Aminosäuren nach ihrem Einbau in das Molekül nicht aus. Glykokoll wird in Serin, Phenylalanin in Tyrosin umgewandelt, wobei die gesamte Radioaktivität des Myosinmoleküls unverändert bleibt. Myosin verhält sich damit gleich wie Globin.

Der Stoffwechsel der genannten Muskeleiweiße ist bei verschiedenen muskulären Erkrankungen in charakteristischer Weise verändert. Bei der hereditären Muskeldystrophie der Maus fand sich für Myosin ein initialer stärkerer Anstieg der Radioaktivität, aber auch ein rascherer Abfall als Ausdruck eines gesteigerten Eiweißumsatzes oder einer verkürzten Lebensdauer der Myofibrillen.

Bei der Atrophie nach Denervierung ist am Kaninchen bereits innerhalb von 24 Stunden ein verminderter Einbau von Glykokoll in Myosin und wasserlösliche Proteine zu beobachten. Bei der progressiven Muskeldystrophie (PMD) des Menschen ergaben analytische Untersuchungen von bioptisch gewonnenem Muskelgewebe eine Abnahme von Sarkoplasma sowie Verminderung des Myoglobin und der myofibrillären Proteine. Aldolase, Kreatinkinase, Milchsäuredehydrogenase und andere Enzyme nehmen im Plasma besonders bei der juvenilen Form der PMD zu, während sie bei neurogenen Atrophien kaum verändert sind. Die im Plasma vermehrt vorliegenden Fermente sind wahrscheinlich muskulären Ursprungs. Verschiedene Mechanismen, die für den Anstieg der Plasmafermente in Betracht kommen, werden diskutiert.

Résumé

A l'aide d'acides aminés marqués (glycocolle-C^{14}, valine-C^{14}, phénylalanine-C^{14}), on a essayé d'obtenir des renseignements sur les transformations de diverses protéines du muscle chez le rat. On a étudié en particulier l'aldolase, la myosine, ainsi que les protéines hydrosolubles du muscle, à l'état normal et dans des conditions pathologiques. L'aldolase, exemple typique de protéine rapidement renouvelée, présente une décroissance exponentielle de la radio-activité, avec une demi-vie de 20 jours; pour d'autres protéines hydrosolubles, on calcule une demi-vie de 7 jours. La demi-vie de l'aldolase dans le plasma est en revanche de 198 minutes.

L'incorporation de glycocolle dans la myosine atteint son maximum au bout de 36 heures; la radio-activité reste ensuite constante jusqu'au 30e jour, puis elle diminue chez les animaux dont la nourriture a une teneur moyenne en protéines. La globine se comporte de même entre le 5e et le 30e jour après l'administration de glycocolle. La radio-activité de la myosine, substance qui doit être considérée comme une protéine à peu près inerte, reflèterait la durée de vie des myofibrilles, laquelle devrait être de 30 jours. La stabilité de la myosine n'exclut pas la transformation des acides aminés marqués, après leur incorporation dans la molécule. Le glycocolle est transformé en sérine et la phénylalanine en tyrosine, la radio-activité totale de la molécule de myosine restant inchangée. La myosine se comporte donc de la même façon que la globine.

Le métabolisme de ces protéines musculaires subit des modifications caractéristiques dans diverses affections du muscle. Dans la myopathie héréditaire de la souris, on a constaté une forte augmentation initiale de la radio-activité de la myosine, mais aussi une diminution rapide, ce qui peut être interprété comme la preuve d'une transformation accélérée des protéines ou d'une durée de vie plus brève des myofibrilles.

Dans l'atrophie provoquée par la section des nerfs chez le lapin, on observe au bout de 24 heures déjà une diminution de l'incorporation du glycocolle dans la myosine et dans les protéines hydrosolubles. Dans la myopathie de l'homme, l'analyse des tissus musculaires obtenus par biopsie a révélé une diminution des enzymes du sarcoplasme, ainsi qu'une réduction de la myoglobine et des protéines myofibrillaires. L'aldolase, la créatine kinase, la déshydrogénase de l'acide lactique et d'autres enzymes augmentent dans le plasma, en particulier dans la forme juvénile de la myopathine, tandis qu'ils restent à peu près inchangés dans les atrophies neurogènes. Les ferments qui se trouvent en quantité accrue dans le sérum sont probablement d'origine musculaire. Les auteurs discutent des divers mécanismes qui entrent en ligne de compte pour l'augmentation des enzymes dans le plasma.

Acknowledgements

This work was supported by grants from the Institut National d'Hygiène, the Caisse Nationale de Sécurité Sociale, the Commissariat à l'Énergie Atomique, the Délégation Générale à la Recherche Scientifique et ses Comités scientifiques (Fonds de Développement) (France), the Muscular Dystrophy Associations of America, Inc., New York (U.S.A.), and the Division of General Medical Sciences, Public Health Service (Research Grant No. RG-6016) (U.S.A.).

References

ADAMS, R. D., D. DENNY-BROWN, and C. M. PEARSON: Diseases of Muscle. A Study in Pathology. Paul B. Hoeber Publ., New York, 1954. — ANSON, M. L., and A. E. MIRSKY: J. Gen. Physiol. (U.S.A.) 13, 369 (1930). — BERSON, S. A., R. S. YALOW, S. S. SCHREIBER, and J. POST: J. Clin. Invest. (U.S.A.) 32, 746 (1953). — COLEMAN, D. L., and M. E. ASHWORTH: Amer. J. Physiol. 197, 839 (1959). — COPE, F. W., and B. D. POLIS: J. Aviat. Med. (U.S.A.) 30, 90 (1959). — DAVISON, A. N.: Biochem. J. (G.B.) 78, 272 (1961). — DEMOS, J., and J. ECOIFFIER: Rev. franç. étud. clin. biol. 2, 489 (1957). — DREYFUS, J. C.: Proc. 1st Med. Conf. Muscular Dystrophy Ass. America, New York, April 1951, p. 16. — DREYFUS, J. C., M. JOLY, G. SCHAPIRA, and L. RAEBER: Compt. rend. Acad. sc. (Fr.) 236, 2351 (1953). — DREYFUS, J. C., J. KRUH, and G. SCHAPIRA: In: Confer. intern. Radioisotopes dans la Recherche Scientifique, Paris, U.N.E.S.C.O. (1957), No. 89; — Biochem.†J. (G.B.) 75, 574 (1960). — DREYFUS, J. C., and G. SCHAPIRA: Compt. rend. Soc. biol. (Fr.) 149, 1934 (1955). — DREYFUS, J. C., G. SCHAPIRA, and J. DEMOS: Clin. chim. acta (Neths) 3, 571 (1958); — Rev. franç. étud. clin. biol. 5, 384 (1960). — DREYFUS, J. C., G. SCHAPIRA, and F. SCHAPIRA: J. Clin. Invest. (U.S.A.) 33, 794 (1954). — EBASHI, S., Y. TOYOKURA, H. MOMOI, and H. SUGITA: J. Biochem. (Jap.) 46, 103 (1959). — FERDMAN, X.: Internat. Confer. Peaceful Uses of Atomic Energy, Geneva, June 1955. — FISCHER, E.: In: Le Muscle. Etudes de Biologie et de Pathologie, p. 275 (Expansion Scientifique Française, Paris, 1952); — Amer. J. Physic. Med. 34, 212 (1955). — GUTMAN, E.: Amer. J. Physic. Med. 38, 104 (1959). — HAMOIR, G.: Biochem. J. (G.B.) 48, 146 (1951). — HORVATH, B.: Neurology (U.S.A.) 8, suppl. 1, 52 (1958). — KNOWLTON, G. C., and H. M. HINES: Amer. J. Physiol. 109, 200 (1934). — KRUH, J., J. C. DREYFUS, G. SCHAPIRA, and G. O. GEY, Jr.: J. Clin. Invest. (U.S.A.) 39, 1180 (1960). — KRUH, J., J. C. DREYFUS, G. SCHAPIRA, and P. PADIEU: J. Biol. Chem. (U.S.A.) 228, 113 (1957). — LAURENT, R., J. C. DREYFUS, and G. SCHAPIRA: Bull. Soc. chim. biol. (Fr.) 43, 416 (1961). — LEBLOND, C. P., and B. E. WALKER: Physiol. Rev. (U.S.A.) 36, 255 (1956). — MICHELSON, A. M., E. S. RUSSELL, and P. T. HARMAN: Proc. Nat. Acad. Sc. (U.S.A.) 41, 1074 (1955). — NEUBERGER, A.: Symp. Soc. Exper. Biol. (G.B.) 9, 72 (1955). — PADIEU, P.: Bull. Soc. chim. biol. (Fr.) 41, 57 (1959). — PERKOFF, G. T., D. M. BROWN, and F. H. TYLER: J. Clin. Endocr. (U.S.A.) 17, 1489 (1958). — PERKOFF, G. T., R. L. HILL, and F. H. TYLER: J. Clin. Invest. (U.S.A.) 41, 1391 (1962). — PERRY, S. V., and M. ZYDOWO: Biochem. J. (G.B.) 72, 682 (1959). — RONZONI, E., L. BERG, and W. LANDAU: 38th Ann. Meet. ARNMD, New York, 1958. — SCHAPIRA, F.: Compt. rend. Soc. biol. (Fr.)

148, 1997 (1954); — Thèse Doct. Sci. (Paris 1959). — SCHAPIRA, F., J. C. DREYFUS, G. SCHAPIRA, and J. DEMOS: Rev. franç. étud. clin. biol. 5, 990 (1960). — SCHAPIRA, F., J. C. DREYFUS, and G. SCHAPIRA: Rev. franç. étud. clin. biol. 7 (1962). — SCHAPIRA, G., and J. C. DREYFUS: Compt. rend. Soc. biol. (Fr.) 151, 22 (1957); — Amer. J. Physic. Med. 38, 207 (1959). — SCHAPIRA, G., J. C. DREYFUS, J. COURSAGET, and F. SCHAPIRA: Bull. Soc. chim. biol. (Fr.) 35, 1309 (1953). — SCHAPIRA, G., J. C. DREYFUS, and J. KRUH: Biochem. J. (G.B.) 82, 290 (1962). — SCHAPIRA, G., J. C. DREYFUS, J. KRUH, D. LABIE, and P. PADIEU: Bull. Soc. chim. biol. (Fr.) 41, 469 (1959). — SCHAPIRA, G., J. C. DREYFUS, and F. SCHAPIRA: Sem. hôp. Paris 29, 1917 (1953). — SCHAPIRA, G., J. C. DREYFUS, F. SCHAPIRA, and J. KRUH: Amer. J. Physic. Med. 34, 313 (1955). — SCHAPIRA, G., M. JOLY, and J. C. DREYFUS: Compt. rend. Soc. biol. (Fr.) 148, 1056 (1954). — SCHAPIRA, G., J. KRUH, J. C. DREYFUS, and F. SCHAPIRA: J. Biol. Chem. (U.S.A.) 235, 1738 (1960). — SCHMIDT, C. G., and H. SCHLIEF: Zschr. inn. Med. (G.) 127, 53 (1956). — SCHNEIDERMAN, L. J.: Nature (G.B.) 194, 191 (1962). — SEVES, G. S.: Biokhimya (U.S.S.R.) 20, 152 (1955). — SIMON, E. J., C. S. GROSS, and I. M. LESSEL: Arch. Biochem. Biophysics (U.S.A.) 96, 41 (1962). — SIMON, E. J., C. S. GROSS, I. M. LESSEL, and A. T. MILHORAT: Fed. Proc. (U.S.A.) 13, 311 (1958). — SPRINSON, D. B., and D. RITTENBERG: J. Biol. Chem. (U.S.A.) 209, 127 (1954). — STEINBOCK, H. J., and H. TARVER: J. Biol. Chem. (U.S.A.) 209, 127 (1954). — SWICK, R. W., A. L. KOCH, and D. T. HANDA: Arch. Biochem. Biophysics (U.S.A.) 63, 226 (1956). — SZENT-GYÖRGYI, A. G.: J. Biol. Chem. (U.S.A.) 192, 361 (1951). — TAYLOR, J. F., A. A. GREEN, and G. T. CORI: J. Biol. Chem. (U.S.A) 173, 591 (1948). — THOMSON, R. C., and J. E. BALLOU: J. Biol. Chem. (U.S.A.) 208, 883 (1954). — VIGNOS, P. J., Jr., and M. LEFKOWITZ: J. Clin. Invest. (U.S.A.) 38, 873 (1959). — WHORTON, C. M., P. C. HUDGINS, and J. J. CONNERS: N. England J. Med. 265, 1242 (1961).

Discussion

KORNER: Dr. DREYFUS, in the slide you showed of the labelling of myosin the total counts were rather low, and a considerable amount of protein was present in each sample counted (30 or 40 mg.). Glycine is an important precursor for purine synthesis and also for haem synthesis. Do you know how much nucleic acid or haem is present in the myosin as a contaminant which might account for the small amount of radioactivity?

DREYFUS: No. Certainly there is no protoporphyrin and very little nucleic acid. Some electrophoretic experiments have been done: they show only one peak, whereas, when there is nucleic acid, they generally go far ahead of the proteins.

KORNER: When preparing protein for radioactive counting, do you use procedures which will remove the nucleic acid?

DREYFUS: Not in all experiments, but in some of them.

KASSENAAR: Do you think that the second plateau which you observed in the studies on the turnover of myosin is due to recirculation of the label?

DREYFUS: I must say first that the slope of the curve subsequently declines very slowly, and after 3 months you can still very easily count the myosin, so I am not quite sure that there is a new plateau. But there certainly is much recirculation in muscle, because we have done some experiments on the specific activity of free glycine over periods of several weeks or months after giving one injection of glycine, and it can be seen that glycine in muscular tissue retains a very high specific activity for much longer periods (amounting to months) than in any of the other tissues, including even the brain. And liver, too, shows a lower specific activity than even plasma or red-cell free glycine, probably because of dilution by food amino acids. But the activity of muscle glycine is twice as high as that of plasma glycine, and it persists for as long as 6 months.

KORNER: Dr. DREYFUS, I wasn't clear whether you thought that the interconversion of glycine to serine takes place while the amino acid is *in* the protein, or that the protein breaks down first so that the interconversion is of the free amino acids.

DREYFUS: This is a debatable point, but our final conclusion is that probably it takes place inside the molecule, after the molecule has been made. Our main reason for reaching this conclusion is that the decrease in the two amino acids is simultaneous. The increase is not simultaneous. And so we believe that, if serine had been incorporated later than glycine in younger red blood cells, then it should have survived and also have been released later. That is our main argument.

KORNER: Do you think that the mechanism of serine formation from glycine inside the protein is the same as with the free amino acids? The interconversion of glycine and serine is a complex reaction, requiring pyridoxal phosphate, tetrahydrofolic acid, active one-carbon transfer, and so on.

DREYFUS: Yes, I can make only guesses which are not very interesting. The general mechanism is the same, but it is made in a kind of bound state so that the amino acid has no communication with the external pool; but there is an internal pool in the red cell, let's suppose, and no free access to the external pool. But this is only hypothesis.

STAEHELIN: In your last sentence you developed the idea that the permeability might be facilitated in the case of muscular dystrophy. Do you think this might also account for the higher labelling of the muscles if it were true not only for the proteins but also for the amino acids? I mean that the proteins would not have a higher turnover, but the amino acids would enter the cells more easily, and therefore an intracellular amino acid might be more strongly labelled.

DREYFUS: I think that this is probably not the case, because when we look for free glycine it has approximately the same specific activity at any time — as was shown in one slide — both under normal conditions as well as in the case of dystrophy. Besides, it cannot simply be a question of permeability, because the descending part of the curve is much faster. There could be less re-utilisation, of course, but I do not think it can be merely a problem of permeability, since we work mainly on the descending part of the curve.

ARIAS: Was the temporal relationship the same between the inter-conversion of radioactivity in amino-acid groups in haemoglobin and myosin?

DREYFUS: Myoglobin has not been investigated. Haemoglobin and myosin have been, and the temporal relationship is not exactly the same, either for the two amino acids of for myosin and globin. The conversion of glycine to serine is slower, and the conversion of phenylalanine to tyrosine is faster in myosin than in globin.

PRADER: Dr. DREYFUS, in your last sentence you speculated that in muscular dystrophy anoxia of the muscle cells might explain the increase of various muscular enzymes in the plasma. But would that explain why one enzyme, the creatine kinase, is so much more increased than all the other enzymes, as you have shown?

DREYFUS: That does not explain everything, certainly not. In the case of creatine kinase, the explanation may be that the basal line of creatine kinase is much lower than that of aldolase, for example, or of phosphohexo-isomerase, since muscle is the only tissue contributing. So if the basal line is much lower, then the relative increase will be much greater. I think that could be an explanation. I must add one point for the endocrinologists. For some reason which I do not understand, triamcinolone myopathy, of which we have had 6 cases, does not give rise to any increase in serum enzymes.

QUERIDO: Since Dr. LEATHEM has told us that hypophysectomised animals are able to replenish their protein stores, apparently there must be large protein stores that are able to synthesise protein independently of growth hormone. I'd just like to know whether you have any studies on, for instance, aldolase turnover in hypophysectomised animals?

DREYFUS: We have not made any studies of this kind so far, but it would certainly be interesting to do so.

Introduction to the General Discussion:

Some aspects of protein metabolism in surgical trauma and injury

By

K. SCHREIER

Since I have had no experience with undernourished humans and understand little about muscle metabolism, I shall confine my remarks to the more general aspects of the influence of surgical trauma on protein metabolism in animals and humans, based on our own studies.

The complex of reactions encountered in connection with an operation can be divided into several phases. In the *pre-operative period* the *psychic stress* — due to the impending operation and reflected in anorexia, nausea, and so on — dominates the metabolic processes. The exact mechanism by which the psychic state of "fear" or "anxiety" is transmitted to the different metabolic cycles is not known.

HOFMEISTER, in 1890 (*1*), was aware that carbohydrate homoeostasis is altered, and he coined the term *"Fesselungsdiabetes"*. THOMAS and MURPLY (*2*) showed that psychic stress leads to hyperlipaemia. We have some evidence that the amino acids also tend to show an increase under these conditions. Polyuria and pollakiuria are further well-known signs, and even fibrinolysis is changed. The higher incidence of complications met with in cases where the patient operated upon happens to be a doctor himself, or a nurse, for example, is undoubtedly due to the different mental attitude of these professional people.

The preparatory period is terminated by the anaesthesia. The latter has a significant influence not only on brain metabolism, but also — depending on its duration — on the turnover of many metabolites throughout the organism. Some modern anaesthetics may influence protein synthesis in the liver and so on, as we have shown, for instance, with regard to chlorpromazine, etc.

The events occurring in the *second phase* have been brilliantly elucidated by Dr. KINNEY. The extent of the metabolic trauma is again determined not only by the size of the wound, the volume of blood loss, the age of the patient, and so on, but also by the *organs* that are involved; for instance, in bone fractures and liver operations, high nitrogen losses are a typical feature. The degree of sensitivity to pain, the will to recover, as well as hormonal and other regulatory factors, are likewise of importance. Furthermore, one must not forget that immobilisation of large parts of the body also has an influence, which affects not only the nitrogen turnover, but also that of calcium and phosphorus, as demonstrated by SCHONHEYDER (3). There is no doubt that changes in protein metabolism may have a vitally important bearing on the success or failure of an operation.

To gain some insight into these complex problems, we studied the nitrogen balance of 85 patients with various operations until their discharge from the ward, and simultaneously we determined the amino acids in their blood and urine [using paper chromatography, microbiological techniques, the STEIN and MOORE procedure, etc. (4)]. In the immediate post-operative period there was a decrease in the amino acids in the blood and later a rise — a finding which is in general agreement with the data of EVERSON and FRITSCHEL (5) and others.

To obtain better information on the basic processes, we did a few animal experiments, using labelled amino acids (6). Young rats were given S^{35}-methionine by mouth and subjected to a standardised operation under ether anaesthesia. The incorporation rate and the behaviour of the specific activity of the proteins in several organs was followed up to the 6th post-operative day. We found that regulation of the amino-acid turnover apparently proceeds in two opposite phases.

In the immediate post-operative period, the rate of incorporation of S^{35}-methionine into the proteins of the liver, plasma, muscles, etc. showed a statistically significant increase in the traumatised animal. The free specific activity of plasma was lower and the excretion in the urine was diminished. (The latter was not due to a decrease in the volume of urine.) As was to be expected, the specific activity in the area of the wound was particulary high.

However, after 48 hours, and especially after 6 days, the excretion of S^{35} showed a marked increase in the operated animals as compared with the uninjured rats. Correspondingly, the specific activity of the proteins in the kidney, in the muscles, and even in the brain was as much as 6 σ lower in the first group of animals. This tendency was not so clearly evident in the liver (Table 1). As to which factors are possibly responsible for these so divergent metabolic trends, I presume that the trauma syndrome is initiated by substances proper to the tissue, which are formed in the wound.

Hormones do not presumably play a primary role, since Ingle (7) has shown in animals, and Mason (8) in human beings, that adrenalectomised individuals display roughly the same type of reaction as normal ones. Of course, the trauma syndrome affects the adrenal glands, as well as the pituitary and the rest of the endocrine system. Some of the hormones modify the intensity of the

Table 1. *Specific activity of protein in different organs 24 hours, 48 hours, and 6 days after intubation with S^{35}-methionine*

(impulses/min./10 mg.)

Organ	Operated animals			Normal controls		
	24 hours	48 hours	6 days	24 hours	48 hours	6 days
Liver	1,275	977	586	790	799	617
Intestine	1,131	571	199	1,107	544	197
Kidney	412	367	233	351	541	241
Muscles	248	134	128	192	158	154
Wound	331	276	137	—	—	—
Brain	157	261	112	106	282	131
Serum	3,292	1,380	586	2,050	1,207	617
Urine	2,200	18,816	41,100	7,231	13,224	21,700

organism's reaction. I should like to remind you in this connection of so-called "trauma-resistant" animals and of the evidence that certain human beings injured in war-time show hardly any metabolic deviation after quite a severe trauma.

The question as to the liberation from wounds of a metabolically active principle immediately reminds one of Menkin's polypeptides. However, since I have no desire to introduce an element of mystery into our discussions, I think it would be more important to stress the fact that the periphery regulates the extent of protein break-down in terms of a homoeostatic process. Which factors it is that interfere with the steering gear, as it were, and turn a primarily anabolic tendency into an anti-anabolic type of reaction, I do not know. Certainly glucocorticoids, aldosterone, adrenaline, and other hormones play a role in the manifestation of single symptoms.

I have tried to divide the entire trauma syndrome into several phases. So far, I have referred to the first two of these phases. To conclude, I should like to give you a short outline of the remaining ones. The *third phase* is marked by the start of convalescence. After a medium operation, this begins around the 4th day. After having been hitherto concerned only with his own tormented ego, the patient now slowly regains interest in his surroundings.

In the plasma the ions begin to normalise. The nitrogen balance is still negative, though this is usually less marked than during the preceding days. It is at this stage, and *not* earlier, that digestible proteins or protein hydrolysates should be supplied. (In this connection, I would beg you to remember the severely burned patient described by Dr. KINNEY.)

The *fourth period*, starting from the first day on which the nitrogen balance becomes positive, is characterised by strong anabolic impulses, reconstruction of tissue, etc.

In some patients one may observe a *fifth phase*, during which the regulatory mechanisms of hunger and satiety are apparently lost; as a result the patient gains weight and develops obesity.

To sum up, I feel it should be emphasised that surgical trauma, like any other comparable injury, affects the individual as a whole. It not only has an influence on the metabolic and hormonal balance, but also gives rise to neural and psychic changes.

References

1. HOFMEISTER, F.: Naunyn-Schmiedebergs Arch. exper. Path. (G.) 26, 355 (1890). — 2. THOMAS, C. B., and E. A. MURPLY: J. Chron. Dis. (U.S.A.) 8, 661 (1958). — 3. SCHONHEYDER, F., N. S. C. HEILSKOV, and K. OLESEN: Scand. J. Clin. Laborat. Invest. 6, 178 (1954). — 4. HOLDER, E., and K. SCHREIER: Langenbeck's Arch. klin. Chir. (G.) 294, 394 (1960). — SCHREIER, K., and H. KARCH: Langenbeck's Arch. klin. Chir. (G.) 280, 516 (1955). — 5. EVERSON, T. C., and M. C. FRITSCHEL: Surgery (U.S.A.) 31, 226 (1952) — 6. SCHREIER, K., and W. LESOINE: Strahlentherapie (G.) 102, 14 (1957). — 7. INGLE, D. J.: Pediatrics (U.S.A.) 17, 407 (1956). — 8. MASON, A. S., J. E. RICHARDSON, and C. E. KING: Lancet (G.B.) 1958/II, 649.

General Discussion

KINNEY: I would like to ask one particular question concerning circulating levels of alpha-amino nitrogen or of free amino acids. Dr. SCHREIER mentioned that, according to the work of EVERSON and FRITSCHEL, surgery had been followed by an acute drop in plasma alpha-amino nitrogen. I would like to ask him if he found any significant rises above the pre-operative level during early convalescence. As far as I remember, EVERSON and FRITSCHEL showed a return to about the same total level as they had observed pre-operatively. It looks as if nitrogen loss after injury could be a matter of more amino acids being supplied to the liver cytoplasm — no matter how they get there — which then prompts the increased urea production. If there is an increased concentration of circulating amino acids, this would be one fairly obvious explanation for the way in which extra amino acids get to the liver.

However, other people looking for such an increase have not been able to find it. MAN[1] and co-workers actually found a distinct decrease in

---

[1] MAN, E. B., P. G. BETTCHER, C. M. CAMERON, and J. P. PETERS: Clin. Sc. (U.S.A.) 8, 65 (1949).

circulating alpha-amino nitrogen in surgical patients, which prompted PETERS[1] to report this as a characteristic response to injury. LEVENSON[2], working on battle casualties, found that there was an approximately normal level of total plasma amino acids. With regard to the possibility that a wound hormone or some other unknown agent may be at work, LEVENSON, using column chromatography, was able to isolate a small peptide, or "conjugate", of varying length and of varying amino-acid composition in different patients, but it was clearly something that was not present in normal plasma. This seems to provide evidence in favour of the possibility that protein catabolism after injury may involve either incompleted normal or new pathways.

I would also like to ask if there is any evidence in late convalescence to suggest that when the body is in a strongly anabolic phase, with a rapid increase in lean tissue, fat is being utilised preferentially as a source of energy. We have wondered about this, because serial body-composition studies have tended to show that fat restoration in patients after a major loss of body tissue was nearly always delayed until the lean tissue had been partly or largely reconstituted. This has led to the division of convalescence into separate phases, putting the fat-gain phase after the lean-tissue restoration. This is of particular interest in view of the comments made here, and in the literature, that agents such as testosterone-propionate and growth hormone both tend to mobilise fatty acids and presumably therefore also tend somewhat to produce ketonuria. We wonder whether any material which promotes lean-tissue restoration has, either primarily or secondarily, a tendency to accelerate fat oxidation. Do others in the audience have any experience on this?

SCHREIER: If one plans to study amino acids in blood, it seems necessary to determine single amino acids and amino nitrogen as well. We all know that the blood amino nitrogen is very labile, and the values recorded therefore depend on the time at which you study the amino acids after a surgical trauma. If you carry out determinations immediately after the operation, you will get a decrease, as we found when we determined single amino acids using microbiological methods, among others. After some hours the amino acids, including especially some of the essential ones, tend to increase above the normal level. With reference to your remark about polypeptides, REHN[3] has reported roughly the same results as you stated. He found polypeptides in the blood which were very variable in their constitution. — I am afraid I cannot offer any explanation in answer to your other questions.

TREMOLIÈRES: I would suggest two working hypotheses to account for the phenomenon reported by Dr. KINNEY.

Firstly, with regard to the poor carbohydrate and high fat utilisation after surgery, it seems possible that we may be faced here with the same situation as we have observed in obese patients on a reduction diet or in very severe malnutrition or under certain conditions with cortisone therapy, i.e. a storage of carbohydrate in the fat tissues. Is this due to a reduced capacity to burn up carbohydrate or to an increased ability to store glycogen in some of the fat depots, as seen in hibernating animals?

Secondly, as for the relation between the extra nitrogen loss and the extra caloric expenditure, this is rather similar to the situation when you give

[1] PETERS, J. P.: Amer. J. Med. **5**, 100 (1948).

[2] LEVENSON, S. M., J. M. HOWARD, and H. ROSEN: Surg. Gyn. Obstetr. (U.S.A.) **101**, 35 (1955).

[3] REHN, J.: Langenbeck's Arch. klin. Chir. (G.) **290**, 466 (1959).

alcohol to an alcoholic; in this case we have seen that the phenomenon in question could be related to oxidation via a peroxidasic system. Xanthine-oxidase-catalase and amino-acid peroxidase both produce a loss of nitrogen and heat. So it might be interesting to find out whether or not a xanthine-oxidase-catalase system or amino-acid peroxidase appears in the plasma under these conditions[1].

WATERLOW: With regard to Dr. KINNEY's third question, in our children we have been interested in trying to find out whether the restitution of fat and of lean tissue go hand in hand. We haven't been able to make measurements of body composition other than of water and total solids, but we have tried to make an estimate of muscle and fat mass by crude methods, i.e. measurements of skin-fold thickness and limb circumference. The results do suggest that the babies first put on lean tissue and later put on fat. But this is only an indication.

QUERIDO: May I add a simple query to the general discussion? In situations involving a negative nitrogen balance, we are faced with the question of the strong protein-sparing action of carbohydrates. I would like to ask the biochemists what they think about this mechanism. Would you like to comment, Dr. YOUNG?

YOUNG: I suppose one should look for an effect of carbohydrate which is not shared by fat under ordinary conditions. Is that right?

QUERIDO: I would like to base it on the experience that a great reduction of nitrogen loss can be achieved in certain conditions with a small amount of carbohydrate calories. I don't know whether it has been tried with a small amount of fat.

YOUNG: I was thinking of the older experiments done during the last century. Carbohydrate makes available the precursors of amino acids — from the glycolytic pathway through pyruvate and from the pentose-phosphate cycle (or the so-called "shunt") — in a way that fat catabolism, through acetyl CoA, could not do. The keto acids, which could thus be formed from carbohydrate, could be converted into amino acids by transamination. Ribose could also be made available from carbohydrate for the synthesis of nucleotides, such as co-enzymes. All these processes might facilitate the retention of nitrogen, though this is only theoretical speculation.

BLOM: Isn't it pertinent to this question that all of us are eating carbohydrates every day, and perhaps our metabolism is therefore dependent on our taking carbohydrates. But in the absence of carbohydrates it is possible that adaptation occurs and that then perhaps you may find a protein-sparing effect of fat. There was an experiment done rather long ago by MACKAY[2] in which he reared half of a group of rats on a diet containing carbohydrate, while the other half got completely carbohydrate-free food; he found that the latter animals grew just as well as those receiving carbohydrate. In another experiment carried out by GEIGER[3], rats were given

[1] TREMOLIÈRES, J., and L. CARRÉ: Compt. rend. Acad. Sc. (Fr.) **251**, 2785 (1960). — Compt. rend. Soc. biol. (Fr.) **155**, 1022 (1961).
[2] MACKEY, E. M., R. H. BARNES, and H. O. CARNE: Amer. J. Physiol. **135**, 193 (1941).
[3] GEIGER, E., R. W. BARCROFT, and E. B. HAGERTY: J. Nutrit. (U.S.A.) **42**, 577 (1950).

the carbohydrate 2 hours after the protein meal; here, the rate of growth was quite different, because the rats were still utilising carbohydrates, albeit some hours later than the protein meal. In GEIGER's experiment the rats developed no adaptation to the lack of carbohydrate as he continued to supply it.

YOUNG: CUTHBERTSON and his colleagues [1] found that, when the carbohydrate and protein in the diet of the rat are given separately and at different times, the loss of nitrogen is greater than that observed if the protein and carbohydrate are given simultaneously. This is a good example of the protein-sparing action of carbohydrate — one that is not shared by fat. But no doubt animals or people can become well adapted to a low-carbohydrate diet.

WILSON: I should like to offer one other possible explanation for the protein-sparing action of carbohydrate. But before doing so, I would like to say that this subject was reviewed several years ago by MUNRO [2]. While fat does appear to spare protein, it does so only after a lag of several days, and clearly in the initial studies of starving animals, carbohydrate does have a much more marked effect than isocaloric amounts of fat. Several years ago, M. SIPERSTEIN and myself [3] studied the effect of glucose oxidation on protein synthesis in homogenates of rat liver. In this *in vitro* system, which may have no relationship to the protein-sparing effect of carbohydrate in the intact animal, one could enormously increase the synthesis of protein from non-amino-acid precursors, specifically from acetate-C^{14}, under circumstances in which glucose was oxidised. And since this was from acetate, we obtained evidence that what happened was that there was an increased conversion of alpha-keto-glutarate to glutamic acid via ammonia fixation, a reaction which requires either reduced TPNH or reduced DPNH, but which in the intact cell probably utilises only reduced TPNH. A tentative explanation offered at that time for the protein-sparing action of carbohydrate was that the glucose oxidation via the hexose-monophosphate shunt (and the subsequent generation of reduced TPNH) might conceivably accelerate protein synthesis in a unique way. Recently, LINGREL and WEBSTER [4] have reported that, in rat-liver microsomes, the addition of DPNH (and presumably also TPNH) accelerates protein synthesis, but not via this mechanism. And it is conceivable that the glucose oxidation via the hexose-monophosphate shunt might have a unique stimulatory action on protein synthesis by some mechanism other than the synthesis of amino acids.

KORNER: The possibility that control of metabolism can be exercised by varying amounts of oxidised and reduced NAD or NADP (as they ought to be called nowadays) is an attractive one which I would like to be able to accept, but I don't think it's a tenable position. There was a suggestion at one time that diabetes could be explained by lack of reduced NADP for synthesis of fat and nucleic acids, because of decreased oxidation of glucose

---

[1] CUTHBERTSON, D. P., A. McCUTCHEON, and H. N. MUNRO: Biochem. J. (G. B.) **34**, 1002 (1940).

[2] MUNRO, H. N.: Physiol. Rev. (U.S.A.) **31**, 449 (1951).

[3] WILSON, J. D., and M. D. SIPERSTEIN: J. Clin. Invest. (U.S.A.) **38**, 317 (1959).

[4] LINGREL, J. B., and G. WEBSTER: Biochem. Biophys. Res. Commun. (U.S.A.) **5**, 57 (1961).

by the hexose-monophosphate pathway, but I gather that this is not now accepted by many workers. I wonder if carbohydrate can in some way be sparing the substrates needed for protein synthesis. I feel that ultimately one has to explain the control of the rate of synthesis of a protein in terms of the availability of substrates for protein synthesis. Is it possible that, if fat is catabolised to acetyl CoA, this enters the Krebs cycle and is entirely metabolised to CO_2 and water by the two carbon-dioxide removing steps in the cycle, whereas if carbohydrate is catabolised to pyruvate, it can enter the cycle as malate or oxalo-acetate, and that these 4-carbon compounds are available for transamination to amino acids ?

QUERIDO: Apparently we have been joined by a guest, Dr. RITTENBERG, who would also like to say something.

RITTENBERG: It is very kind of you to let me make a few remarks on this last question. I, too, feel that what in the old days was referred to as the protein-sparing action of carbohydrate can today be explained by textbook reactions involving TPNH. Carbohydrate is the reducing agent which produces DPNH and TPNH from their oxidised forms. The empirical composition of proteins shows that they contain less oxygen per atom of carbon than glucose. Synthesis of protein from carbohydrate must involve reduction by either DPNH or TPNH.

Last year we initiated a study in *E. coli* of the quantitative utilisation of glucose in the oxidative pentose pathway. The oxidative pentose shunt is the principal source of TPNH. We found that in resting bacteria the pentose shunt is practically inoperative. As growth is initiated in the bacterial system, there is a great increase in the rate of utilisation of carbohydrate by the pentose shunt, and then, when growth ceases, the pentose shunt again goes down to practically zero. In this experiment you can observe the effect of TPNH on protein synthesis. Here, of course, as in all *in vivo* experiments you can't tell which comes first and which comes last. Either it is protein synthesis which turns on the pentose shunt or the pentose shunt which turns on the protein synthesis. But if you consider it from the simplest viewpoint, it would be most reasonable to suppose that the turning on of the pentose shunt is the thing that initiates protein synthesis.

KINNEY: I would like to ask Dr. RITTENBERG whether he feels that there is an analogy in mammalian cells. The reason why I am asking is that I believe from Dr. WILSON's comments that he feels the reduced TPN made available by the operation of the pentose shunt would be important for nitrogen fixation with alpha-keto-glutarate in higher animals. This is in contrast to the reactions for protein synthesis, which appear to involve transamination and the utilisation of amino acids in certain tissues which have arisen from protein break-down in others. Do you feel that the nitrogen incorporation by alpha-keto-glutarate, which apparently requires reduced TPN, is quantitatively important in mammalian cells ?

RITTENBERG: I can't answer this question, but on the general assumption that the Good Lord uses more or less the same reactions in every place, I would say "yes". I have no experimental evidence on this point.

WILSON: I would just like to say that in mammalian tissues the only reaction of which I am aware in which there is net amino-acid synthesis, in contrast to transamination in which there is no net amino-acid synthesis, is the fixation of ammonia to alpha-keto-glutarate. This is the primary mechanism by which non-essential amino acids are synthesised in mammalian cells. All other non-essential amino acids are formed by transamination. And so, if one looks upon it as a net sparing of nitrogen, it seems almost inevitable that this reaction must be crucial in some way.

KORNER: Could I ask Dr. RITTENBERG how he explains the TPNH control of protein synthesis in those tissues which appear not to have a hexose-monophosphate pathway? Muscle is supposed to have very little glucose metabolised by the hexose-monophosphate pathway, and very little TPNH is produced in this tissue — if one accepts the results in the literature.

RITTENBERG: In reply to this point, I would say that I believe most of the procedures currently used for measuring the pentose shunt are not very precise.

WILSON: I would just like to defend this hypothesis, which in my opinion may have no direct *in vivo* application. In order for this theory to be true, it only has to work in the one organ in the body in which non-essential amino-acid synthesis would be sufficient to provide amino-acid substrate for the rest of the tissues, so as to enable protein synthesis to occur. This is purely a theoretical possibility, but it is conceivable that most non-essential amino acids in a net sense are synthesised in the liver and are transported to the rest of the tissues, e. g. to the muscles. But even if this is not true, the older studies, which did not furnish evidence for the presence of the hexose-monophosphate pathway shunt in muscle, have in my opinion clearly been superseded. CHEFURKA[1] isolated each enzyme of the hexose-monophosphate pathway from some muscles, and using even the crudest of techniques — the differential rate of oxidation of the glucose 1 and glucose 6 of radioactive glucose — one can demonstrate the presence of a hexose-monophosphate shunt in pigeon muscle. The smaller the shunt, the more crucial this might be in a sense. But I would re-emphasise that I think this is a theoretical discussion.

DREYFUS: I must say that I did not quite understand the textbook discussion, because, after emphasis had been placed on TPNH generation, it was then said that the only reaction for protein synthesis is re-amination of keto-glutate, which uses DPNH.

WILSON: Purified glutamic dehydrogenase uses DPNH and TPNH equally. However, the quantity of reduced DPNH in the liver is so low that most physiologists interpret this as meaning that under *in vivo* circumstances it is actually TPNH which serves as the co-factor for this reaction.

---

[1] CHEFURKA, W.: Canad. J. Biochem. Physiol. **36**, 83 (1958).

The effects of anabolic agents in man

The evaluation of growth and maturity in children

By

J. M. TANNER

There are three questions that the clinician concerned with growth and development is commonly called on to answer.

1. The first arises when a child is referred to him for the first time. It is: "Is this child's size, shape, body composition, and physical maturity within normal limits for his age, sex, population and cultural subgroup ?".

2. The second arises when treatment with a hormone or a drug has been initiated. It is: "Has this treatment produced a significant effect on the rate of growth of the child in size, shape, body composition, or physical maturity ?".

3. The third question arises sometimes when perfectly healthy children wishing to enter certain occupations such as the ballet come to see him, and sometimes when treatment of a small or large child is in progress. It is: "What is the adult height of this child going to be, and will the treatment being given increase or decrease it ?".

None of these questions is particularly easy to answer at present — not even the first. As for the second, with which we shall be particularly concerned, one must begin by remarking that it is a long way from demonstrating a significantly increased retention of phosphorus and nitrogen over a period of perhaps 3 weeks to concluding that a child has grown significantly in height or even in muscle diameters over a period of perhaps a year.

We will deal first with measures of growth, such as height and weight, and later with measures of physical maturity, such as skeletal maturity (bone age), tooth eruption, and age at menarche. Lastly we shall discuss briefly the prediction of adult height.

Growth

"Distance" standards. Laboratories in several countries have produced standard percentile charts for the heights and weights

of their children. These give the distributions of height and weight
at each age, in terms usually of the 3rd, 10th, 25th, 50th, 75th, 90th,
and 97th percentiles. They are height-for-age and weight-for-age
standards and are called "distance" (or height-achieved) charts,
to distinguish them from velocity or rate-of-growth charts. They
are based on cross-sectional surveys of school children, eked out

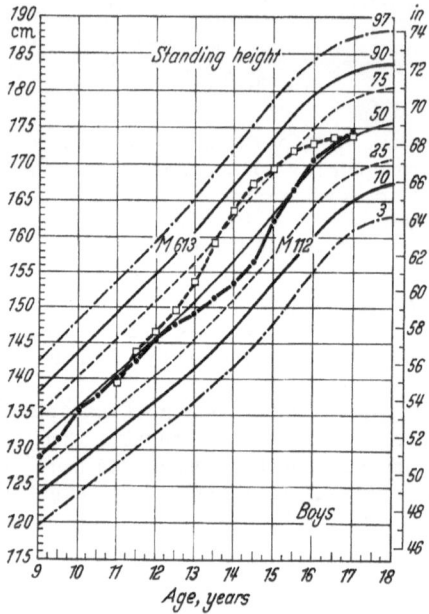

Fig. 1. Growth in height of two boys, one early- and the other late-maturing, plotted on the
TANNER-WHITEHOUSE standards for height in British children. (From Education and Physical
Growth. University Press, London, 1961)

with such information as can be obtained from well-baby clinics,
nursery schools, special growth studies, and studies of army recruits,
apprentices or students. A typical example is the TANNER-WHITE-
HOUSE standards for British children, one section of which is
illustrated in Fig. 1 (see TANNER, 1958; TANNER and WHITEHOUSE,
1959). In nearly all these standards, including the British one,
the ages from 6 to 14 inclusive are represented by much better
samples than earlier or later ages. The first year is often well
represented, but between 1 and 6 the samples are usually smaller
and often less representative of the population as a whole. The

British standards, for example, are based at these ages on two small longitudinal surveys comprising in all only some 350 children at each year of age. Since virtually the same 350 children were represented at each age, any non-representativeness of the sample, or even any random sampling error, affects the whole age range. The higher ages are even less well sampled in most surveys, since adolescents and young adults are very hard to survey after school-leaving age. The British standard here is based partly on children still at school, partly on army recruits, and partly on the known gains of a number of children followed longitudinally from 10 to 20.

Evidently 3rd and 97th percentiles based on such data must be subject to very considerable error of statistical estimation; yet it is these lines that are most used in clinical work. The error is less in the case of height than it is for weight, for one can take height to be normally distributed at a given age and hence calculate the percentiles from the standard deviations. This gives the positions of the lines more precisely than direct calculation of their values from the sample figures, which has to be done for a non-normal distribution (see TANNER, 1952). For weight, standard deviations are quite misleading, since the distributions at a given age are skewed, being more nearly log-normal than normal. Here the percentiles must be calculated directly from the data. Finally, all the lines in standards such as these must be smoothed simply by eye. It is perhaps surprising, but nevertheless true, that individual children plotted on the British charts do in fact follow the percentile lines closely until adolescence, and this gives one confidence in the general correctness of the curves.

There are a few other points that we should note. If the standardising group is cross-sectional, then from 5 to 6 years, for example, there are children covering the range of the whole year. The 50th percentile for height is correctly represented by the mean of their values and is plotted at 5.5 years. But the calculated standard deviation is too high, since it includes the variability of all children from 5 to 6, whereas, supposing our child were 5.5 exactly, we want to compare him against the variability of children aged 5.5 exactly. The percentiles must be corrected, therefore, to represent an "instantaneous" variation at each age (see TANNER, 1958; HEALY, 1962).

Much attention must be paid to technique of measurement. We routinely measure supine length up to and including age 3, and standing height from age 2 onwards, giving a slight overlap. For convenience, the British charts are in three sections, and the first, running from birth to 3, gives supine length, and the second (2 to 10) standing height. Supine length is in nearly all cases greater than standing height, with an average difference at ages 2 and 3 of 0.6 cm., provided that the child is stretched to his maximum stature when standing height is taken. If this is not done, the difference may be more, and is much more variable. The heels should be held down by one observer and gentle traction under the mastoid processes applied by another, with verbal encouragement to stretch up as much as possible. Only in this way can the time-of-day variation, due to postural change and tiredness, be minimised in children and adults.

Sitting height, a most important measurement, can also be taken reliably only from 2 or 3 years onwards, and up until then crown-rump length must be used. The average difference is 0.4 cm., but the variability of this difference is rather greater than the supine length-stature difference. Again, traction should be applied to get the maximum sitting height and the back placed in as straight a posture as possible. Special anthropometric instruments are available and are most advisable for measuring stature, sitting height, and other skeletal dimensions (WHITEHOUSE and TANNER, in press).

Lastly, in interpreting the child's position on a distance chart, account has to be taken of his family background, social group, and parents' size. There are still in most countries differences between the sizes of children in different social groups, the more favoured being larger at all ages; children with many siblings tend to be smaller than children with few (see TANNER, 1962, pp. 137—143). Allowance may also be made, though tentatively, for a child below the 3rd percentile if both his parents and perhaps some uncles and aunts are in the same position. This does not apply during the first year or two, when the parents' size correlates only slightly with the size of the child (TANNER, in press).

"Velocity" standards. On the first occasion one sees a child one is inevitably confined to the information one can get from a plot on a distance standard. But thereafter much more interesting information is obtained by measuring the child at intervals, usually of 6 months or a year, but perhaps of 3 months if a special treatment is being given. In most clinics, including at present the Growth Disorder Clinic at The Hospital For Sick Children, the child's progress is simply plotted as successive points on a distance chart. There are two disadvantages in this.

Firstly, the distance chart gives no statistical indication as to whether the velocity is abnormally large or small. If the child (up to 10) stays near a certain distance percentile for several years, one can presume that his velocity is about average. But one really wants to know much more than this; one wants to have percentiles for velocity just as we have for distance. Such charts, however, have to be founded on longitudinal data, for only longitudinal data give estimates of the variability (as opposed to the mean) of the velocity in a population (see TANNER, 1951). All one requires (at least up to age 10; see below) is a series of two-year longitudinal samples, a survey repeated on the same children after a year has elapsed. But nobody has done this, and all we have at present is percentiles up to age 4 founded on the relatively small longitudinal series being followed in London (MOORE, HINDLEY, and FALKNER, 1954), Zurich (HEIERLI, 1960), and other centres. Thus, astonishing though it may seem, if you treat a boy of age 6 and cause him to grow 10 cm. in a year, you cannot be sure how likely this is to have occurred by chance, even though you know that the average gain at this age is about 6 cm./year.

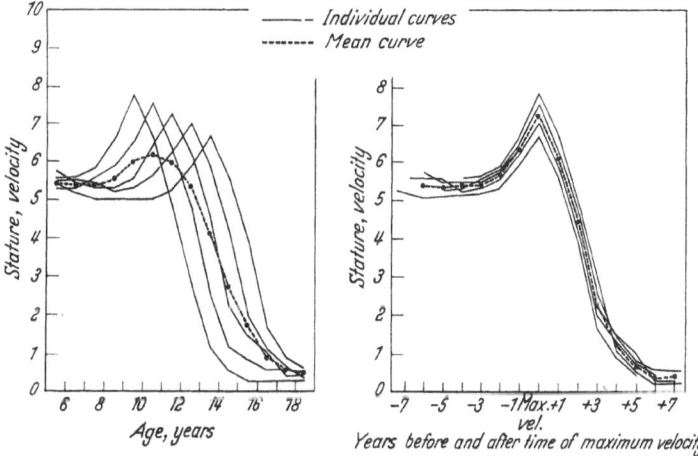

Fig. 2. Relation between individual and mean velocities during the adolescent spurt. Left, the height curves are plotted against chronological age; right, according to time of peak height velocity. (From Growth at Adolescence. Second edition. Blackwell's Sc. Publ., Oxford, 1962, after SHUTTLEWORTH)

The second disadvantage of plotting a child's position at successive ages on a distance chart is more profound. It applies particularly at adolescence and comes from the fact that different individuals begin their adolescent spurts of growth at widely different ages. The effect of this on "average" figures is illustrated in Fig. 2. The continuous lines show the velocity curves for five individuals from 6 to 18, each individual starting his spurt at a different time. The average of these curves, obtained simply by treating the values cross-sectionally and adding them up at ages 6, 7, 8, etc. and dividing by five, is shown by the heavy interrupted line. It is obvious that this line in no way characterises any individual; it smooths out the spurt, spreading it along the time axis.

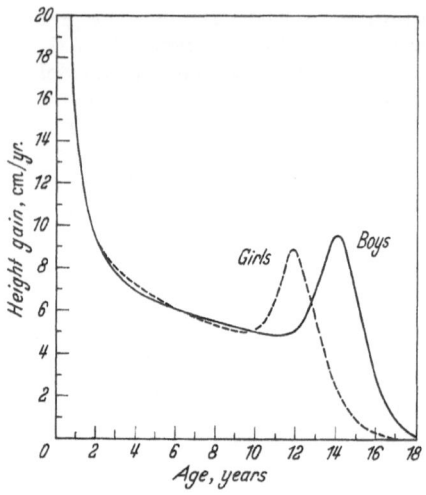

Fig. 3. "Fiftieth-percentile individual" height velocity curves for boys and girls, constructed from unpublished data of the Harpenden Growth Study and some published longitudinal material (see TANNER, 1962, p. 3; BAYLEY, 1956; and DEMING, 1957)

Averages computed from cross-sectional studies inevitably do this in the distance curves, and averages computed from two-year longitudinal studies would do it in the velocity curves. In consequence, a healthy child should *not* follow the percentiles in the published distance standards (including the British one) at adolescence (see Fig. 1 above). If he is an early-maturing boy he will rise through the percentiles and subsequently fall back to regain, on average, his position before puberty (M 613, Fig. 1). If he is a late-maturing boy he will fall through the percentiles first, and then regain rapidly his earlier status (M 112, Fig. 1). The only obvious way to overcome this difficulty is to plot skeletal age (see below) rather than chronological age as the time base along the X axis; this does, however, involve some degree of undesirable measurement error in the time axis.

Because of this objection to the usual standards, my colleague, Mr. WHITEHOUSE, and I have produced, with some trepidation, the velocity curves shown in Fig. 3. They purport to represent the curves of the 50th percentile *individual* boy and girl for height and weight. They are based at adolescence on the unpublished data of the Harpenden Growth Study, SHUTTLEWORTH's data (analysed by TANNER, 1962, p. 3), BAYLEY's (1956) graphs, and DEMING's (1957)

data. From six months to ten years they follow the results of
HEIERLI (1960) (up to 4) and the TANNER-WHITEHOUSE distance
standards, which are largely based on the longitudinal material
of the Child Study Centre group at the Institute of Child Health.

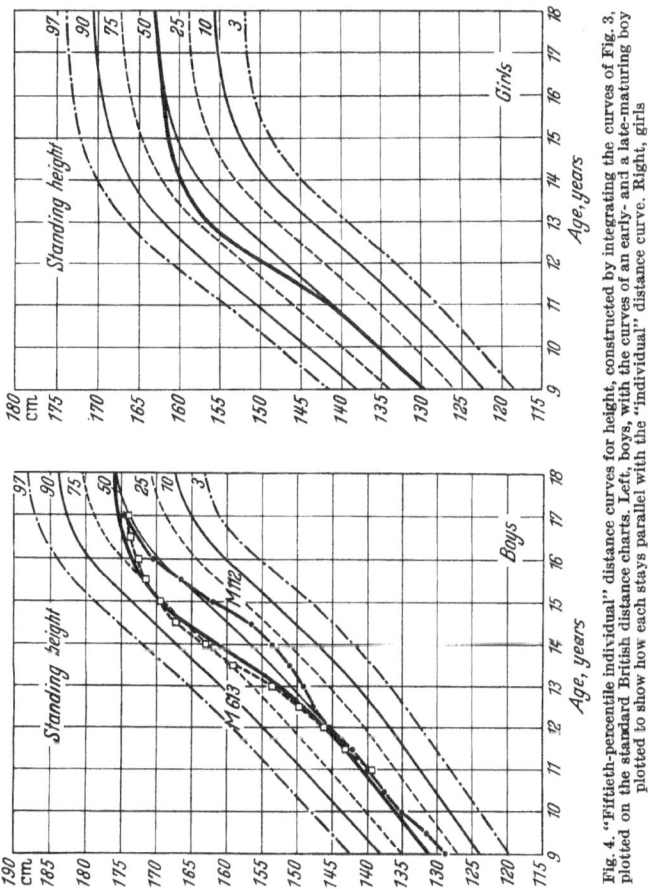

Fig. 4. "Fiftieth-percentile individual" distance curves for height, constructed by integrating the curves of Fig. 3, plotted on the standard British distance charts. Left, boys, with the curves of an early- and a late-maturing boy plotted to show how each stays parallel with the "individual" distance curve. Right, girls

They are probably a reasonable approximation to the truth, and
they probably represent the adolescent spurt quite truthfully; but
details such as the sex differences shown over the 3 to 9 year range
should not be taken too seriously until more data are available.

The velocity curves of individual children may be expected to follow generally parallel to these curves, even at adolescence. Some children reach a higher peak velocity at adolescence, of course, and others a lower one; one would like to have percentile lines as well as the mean. One systematic effect should be noticed, however; early maturers on average reach a higher peak velocity than late maturers, so that an individual curve which rises before the average curve is likely also to reach a slightly higher peak.

These "individual" velocity curves can be integrated to give distance curves which are free of the "out-of-phase" effect seen in Fig. 2. Such "individual" distance curves are shown in Fig. 4 over the age range 9 to 18, drawn on the British standard curves. It will be seen that the 50th-percentile child is expected to rise just to the 75th percentile at adolescence on the current standards. On the chart for boys the two early- and late-maturing boys of Fig. 1 are again plotted; now both of them follow nearly parallel to the new "individual" line, in contrast to their crossing of the old percentiles. If the further step of plotting on skeletal, rather than chronological age were taken, the curves of the two boys would nearly coincide with the 50th-percentile curve. In other words, the boys have each a fairly average adolescent growth spurt, though one is some two years ahead of the other in maturity.

In Figs 5 and 6 (from PRADER, TANNER, and VON HARNACK, in press) an example of the use of the charts is given, drawn from clinical practice. This concerns a hypothyroid boy, who had practically stopped growing when he was diagnosed and treated at age 12. In Fig. 5 the conventional, not the new, distance chart is used; a period of marked catch-up growth follows administration of thyroid.

In Fig. 6 the new velocity chart is used, and the whole process of height growth in this boy is made much more clear. A purely illustrational difficulty arises in plotting velocity, which does not occur in plotting distance. Whereas each distance plot represents a measurement taken at a particular point on the time axis, each velocity plot represents only the average velocity over the whole time passed between successive measuring sessions. Various suggestions have been made as to how best to represent this; it seems to me that plotting the average velocity at half-way between the two times of measurement, as in the figure, is the most logical method. However, it must not be assumed that each point represents the instantaneous velocity at that point of time: on the contrary, it will only do so if the velocity curve is linear over the period concerned. Thus, the first point in Fig. 6 represents the average velocity over the time interval indicated by the horizontal dashed line; undoubtedly at the beginning of this period the velocity was greater, and at the end less. Putting a dashed line of this

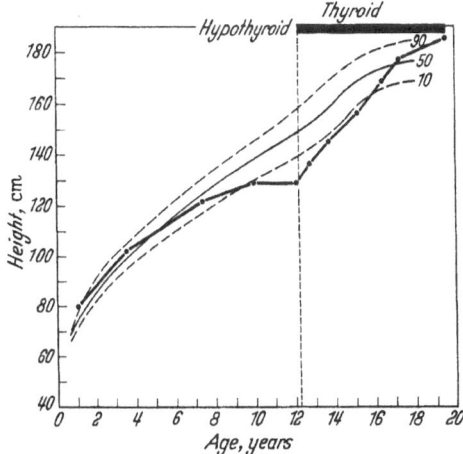

Fig. 5. Growth in height of a boy with hypothyroidism before and after treatment. Plotted on conventional "distance" standards. (From PRADER, TANNER, and VON HARNACK, in press)

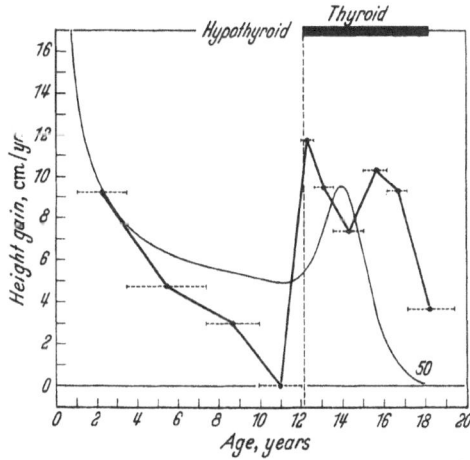

Fig. 6. Velocity of growth in height of a boy with hypothyroidism before and after treatment. Plotted on new "50th-percentile individual" velocity chart. (From PRADER, TANNER, and VON HARNACK, in press)

sort attached to each velocity point serves at least as a reminder that we are dealing only with the average velocity over the period indicated. Probably the only truly satisfactory way of dealing with this, if the data are good enough to allow it, is by fitting a curve mathematically to the distance plots, differentiating it, and plotting the derivative as the velocity (as is done in Figs 7 and 8 below). Separate curves would have to be fitted before and after treatment, and before and during adolescence. In Fig 6 the velocity is seen

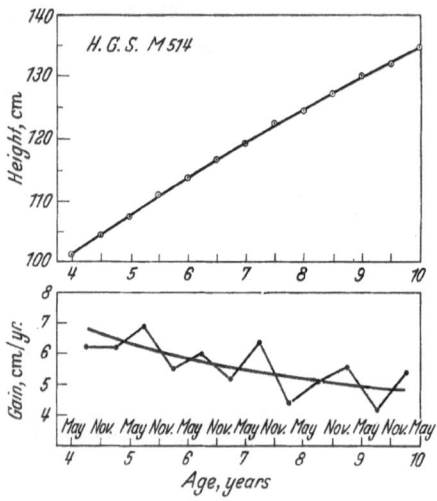

Fig. 7. Growth of boy in Harpenden Growth Study measured each six months by R. H. WHITEHOUSE. Above, "distance" plot of height achieved at each age; below, velocity plot of average velocities each six months. Solid lines: above, curve of form height $= a + bt + c \log t$ ($t =$ age) fitted to individual's distance points; below, the derivative of the fitted curve

to reach practically zero before treatment, then to increase very markedly (and still more than indicated, presumably, during the first few weeks after treatment). A clear adolescent spurt is seen, coming late, since maturation was much delayed, but having a quite normal appearance.

Seasonal variation and the design of trials. Growth in height, and in other skeletal measurements also, is a very regular process. The upper section of Fig. 7 shows the measurements of height of a boy in the Harpenden Growth Study taken every six months by a single observer, Mr. WHITEHOUSE. The circles represent the measurements, and the solid line is a simple mathematical curve of the form $h = a + bt + c \log t$ (where h is height and t age) fitted to them. None of the measurements deviates more than 4 mm. from this line, although the experimental error of measuring height may be 3 mm. even in experienced hands. The lower section of Fig. 7 shows the same data plotted as six-monthly velocities, with the solid curve derived from the curve in the upper section. One can at once see how even such an extremely regular series of measurements shows fluctuations, due chiefly to measuring error, when the velocity

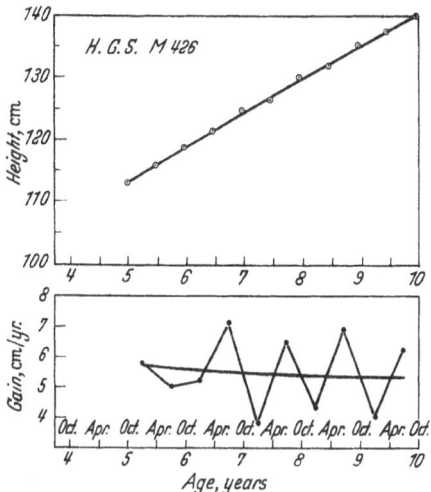

Fig. 8. Growth of boy in Harpenden Growth Study measured each six months by R. H. WHITEHOUSE. Above, "distance" plot of height achieved at each age; below, velocity plot of average velocities each six months. Solid lines: above, curve of form height $= a + bt + c$ $\log t$ ($t = $ age) fitted to individual's distance points; below, the derivative of the fitted curve. Note regular seasonal variation, Apr.-Oct. having greater velocity than Oct.-Apr. periods

is considered. However, if this was all we had to worry about we might well be able to detect increases in growth rate of, say, 4mm. in each three-monthly period (an average of about $1^1/_2$ cm./year). At the ages considered here, this represents an extra increment of about 25% over average.

However, another factor interferes to prevent us doing this over shorter periods than a full year, except in quite unusual cases. Fig. 8 is a similar representation to Fig. 7, but this time there are seen to be much larger fluctuations in velocity. Furthermore, these fluctuations are entirely regular, apart from the first, and represent, at least in the main, the effect that season of the year has

24*

on height growth in many, though not all, children. (Small and irregular six-monthly fluctuations could be caused by measuring error, a low value on one occasion causing a low velocity followed by a high one.) Deviations from the distance curve amount to 5 mm. above and below it, and the velocity of Apr.—Oct. growth averages no less than 3.2 cm./year above that in Oct.—Apr. Under these circumstances, unless we could depend on the seasonal fluctuation being absolutely regular, and unless we knew its magnitude through having studied the child for at least three years before treatment, we should be quite unable to distinguish a three-monthly extra increment of 4 mm. due to treatment from one due to a seasonal effect. Even an 8 mm. six-monthly extra increment could be perfectly well mimicked by a positive seasonal effect or obliterated by a negative one. For this reason we have found it necessary to alternate periods of a whole year for control and treatment in testing the effect of human growth hormone on dwarfed children (in at least some of whom the seasonal effect continues). If a drug is tested over a shorter period, a spurious effect may be described, or a real effect may be missed.

Shape and tissue composition. It is obviously desirable to form some idea of whether a treatment is disturbing the normal balance of the growth process, or whether, supposing it is causing growth at all, it is causing abnormal growth. The simplest way to determine this would seem to be to measure the effect on (a) the shape of the skeleton and (b) the growth of muscle and fat at various sites.

In most growth studies, sitting height is measured, giving leg length by subtraction from stature, and in addition bi-acromial and bi-iliac widths; sometimes the widths across the condyles of the humerus and femur are also taken. Growth in all these can either be assessed through standard distance or velocity charts for each measurement singly or else by reference to a regression of, for example, leg length on stature (for details see TANNER, 1952). Here again, velocities are better for both measurements than distances, but no such regression velocity standards have yet been published. Single-measurement distance standards suitable for American children have been given for sitting height, bi-acromial and bi-iliac diameters by BAYER and BAYLEY (1959). These authors also give distance standards for indices such as sitting height/stature, but there are statistical reasons for considering regression standards to be much more suitable. Endocrinologists have in the past frequently used span as a measure, but professional anthropologists have rightly spurned it. It is an inaccurate measurement at best, being a complex of a limb length and a body width together with an unknown postural contribution. It is better avoided. Arm length can usefully be measured, though except in rare instances gives an exactly similar result to leg length. Height from floor to pubes is sometimes used as a measure of leg length; this is probably less accurate than a well-taken sitting height, but is also quite serviceable.

From these measurements, considered singly or two at a time, one can obtain considerable insight into how growth in shape is proceeding. But a true consideration of shape would include all the measurements simultaneously in a mathematical formulation, such as MAHALANOBIS' generalised distance or D'ARCY THOMSON'S quantified transformed co-ordinates. No such formulation is available as yet, but clearly this is one of the next steps in the statistical study of children's growth.

Tissue composition may be studied either by chemical or by anthropometric means. Chemical studies are highly suitable for finding the proportions of water and fat in the body and following the growth in lean body mass. This is provided the precautions described by SIRI (for summary see TANNER, 1959) are taken, notably that of jointly determining water and fat. The biological errors involved in the assumptions which have to be made regarding lack of variation between individuals in proportions of bone and muscle, for instance, are very considerable and have also to be fully taken into account. Lastly, these methods yield information only about the total percentage of a substance in the body, and not about its distribution.

Both anthropometric and radiographic measurements may throw light upon changes in body composition during growth. Of all the anthropometric measures, body-weight, the most usually used, is the least useful. On a distance chart, either in child or adult, a high weight for a given height alone tells one very little: it may signify much fat or much muscle or even, to some extent, large cortices to the bones. In adults a velocity plot of weight admittedly may be useful, in that when an adult puts on weight it usually means he is getting fat. But in a child one can certainly not make this assumption: a much finer tissue-component analysis is required.

The easiest tissue component to measure, next to the skeleton, is subcutaneous fat. At various sites in the body a skinfold may be picked up from the underlying muscle and measured with constant-pressure spring calipers (the Harpenden Skinfold Caliper, see TANNER and WHITEHOUSE, 1955). Distance standards are now available for British children for two sites: over the triceps midway down the back of the left arm, and under the inferior angle of the left scapula (TANNER and WHITEHOUSE, 1962). The curves from these two sites differ, particularly at adolescence, and boys and girls differ also; the growth curves for subcutaneous fat are more complex than for stature.

Subcutaneous fat may also be measured, together with muscle width and widths of the medulla and cortex of the bones, by X-ray. The most convenient sites are the calf and the upper arm, but the components of the thigh may also be quite accurately measured in this way. Due protection must be given to the genitalia. The technique is described in TANNER (1962, appendix). No very adequate standards are yet available, but this also is clearly a rapidly advancing area of growth study and one of potentially much importance to clinicians treating children with androgens and other growth-promoting agents. Too little has so far been done by way of assessing directly the effects on muscle, bone, and fat during the course of prolonged treatment.

Maturation

Children differ greatly in the rate at which they travel along their course of growth to maturity. Some boys reach their final adult height before 15, others after 18, for example. Thus, at any chronological age children differ in their *developmental age* or maturity, i. e. in their degree of physical advancement.

There are various ways in which a child's maturity status or developmental age may be measured. The percentage of his adult height that he has reached is one measure, and a good one for use retrospectively in longitudinal research series, but of course useless for the clinic, since final height is not known. Rate of growth in height is not much of a guide to developmental age, nor is height itself. Early-maturing children, it is true, are on average taller than late-maturing ones at all ages. But if a child is tall for his age at say 6, this may be either because he is early-maturing or because he is a tall child of tall parents and is already demonstrating the fact. Paediatricians and endocrinologists still sometimes use "height age" and "weight age" as measures of development (height age being simply the age of the 50th-percentile child at the height shown by the child being examined). But these are fallacious measures, since they confound maturation and final size. A child who is tall for his age may either be advanced towards maturity (which is the interpretation made) or may simply be a tall child born of tall parents and of average advancement. Height age would be a permissible index only if all children ended up the same height.

Skeletal maturity. Skeletal maturity, or skeletal age, is free of this criticism and is the most commonly used indication of maturity. All children finish with the same degree of skeletal maturity (as it is measured), in other words, at the same skeletal age. Thus, no confounding occurs, and the measure is one of maturity only. It is in fact an essential and highly informative measure in all growth disorders.

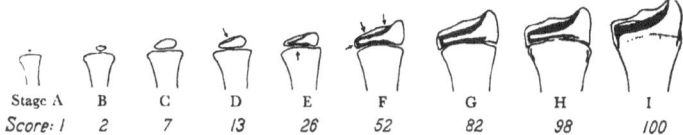

Stage A	B	C	D	E	F	G	H	I
Score: 1	2	7	13	26	52	82	98	100

Fig. 9. Stages of maturation of the lower end of the radius. The criteria for each stage are precisely described and each stage is given a score: the average of the scores of 20 bones of the hand and wrist gives directly the skeletal maturity of this area in percentage of adult maturity
(From TANNER, WHITEHOUSE, and HEALY, 1962)

Stage A
    (i) The centre is just visible as a single deposit of calcium, or more rarely as multiple deposits. The border is frequently ill-defined.

Stage B
    (i) The centre is distinct in appearance and oval in shape with a smooth continuous border.
    (The maximum diameter is less than half the width of the end of the radial shaft).

Stage C
    (i) The maximum diameter is half or more the width of the end of the radial shaft.
    (ii) The epiphysis has broadened chiefly at its radial side, so that this end is thicker and more rounded, the ulnar end more tapering.
    (iii) The centre third of the proximal surface is flat and slightly thickened, and the gap between it and the radial metaphysis has narrowed to about a millimetre.

Stage D
    (i) A thickened white line has appeared just inside the distal border of the epiphysis; this represents the edge of the palmar surface, and the newly appeared bone distal to it is the edge of the dorsal surface.

Stage E
    (i) The proximal border of the epiphysis is now differentiated into palmar and dorsal surfaces; the palmar surface is visible as a broad irregularly thickened white line at the proximal edge of the epiphysis.
    (ii) Both ends of the epiphysis, but particularly the ulnar one, have grown outwards and proximally since the last stage, so that the proximal border now conforms to the shape of the radial metaphysis along most of its extent.

Stage F
    (i) The dorsal surface now has distinct lunate and navicular articular edges joined at small hump. Radial to the navicular surface the styloid process carries the border in a distinct convexity distad.
    (ii) The medial border of the epiphysis has developed palmar and dorsal surfaces for articulation with the ulnar epiphysis; either palmar or dorsal surface may be the one which projects medially, depending on the position of the wrist.
    (iii) The proximal border of the epiphysis is now slightly concave.

Stage G
    (i) The epiphysis now caps the shaft on one (usually the medial) or both sides. (The styloid process is much further developed than in the last stage.)

Stage H
    (i) Fusion of the epiphysis and shaft has begun. A line is still visible across the shaft composed partly of black areas where the epiphyseal cartilage remains and partly of dense white areas where fusion is proceeding.

Stage I
    (i) Fusion of epiphysis and shaft is completed. Over the majority of its length the line of fusion has entirely disappeared, but some thickened remnant of it may still be visible.

All bones pass through a succession of stages in their development, and these can be visualised by radiography. To begin with,
the lower end of the radius, for example, has no visible (i. e.
calcified) epiphysis; then a small area of calcium appears; then
this area assumes a definite shape; the shape changes as the
calcified area enlarges; finally, the epiphysis fuses with the lower
end of the shaft. These stages are the same in all people, so far as
we know, and their sequence in each bone remains unaltered by
any except purely local disease. They are illustrated for the radius
in Fig. 9. The stage reached at any chronological age is a measure
of the child's skeletal maturity.

In the older systems for assessing skeletal maturity, the best
known of which are the Greulich-Pyle atlas of the hand and
wrist (second edition, 1959) and the Pyle-Hoerr (1955) atlas of the
knee, these stages are not given directly. Instead, a series of illustrations of the typical appearance of the hand or of the knee at each
year or six months of age are given, and the observer has to see
which of these illustrations most closely matches his child's X-ray.
The child is then said to have a skeletal age equal to the chronological age of the matching standard. The method has been very
useful, but is probably less accurate than methods rating the
bones separately; furthermore, it does not demonstrate the type of
growth curve shown by the skeletal maturing process, and is based
on a group of children selected from an above-average social group.

For these reasons my colleagues and I have recently introduced
a new system based on rating twenty bones or epiphyses of the
left hand and wrist (Tanner, Whitehouse, and Healy, 1962).
Scores have been provided for each stage of each bone (see Fig. 9)
so that the average of all 20 scores gives directly the percentage
maturity of this part of the skeleton. Separate scores may be
calculated for the long and the round bones, since some observers
believe (and others deny) that the carpals may be more influenced
than the long bones by certain pathological states. The scores have
been computed on the necessary assumption that in the normative
group each of the twenty bones is estimating a single numerical
quantity, namely, the "maturity" of the wrist; from this assumption a joint estimation procedure can be worked out giving the
scores.

In Fig. 10 the progress in skeletal maturity of a normal girl is
plotted; the standards, it will be noted, are plotted in logarithms,
which leads approximately to normal distributions at each age,
except at the top of the graph. The maturity scores computed in
this way show a marked adolescent spurt; indeed, their growth

curve resembles that of a linear physical measurement. Skeletal age may be directly taken from the chart, being the age at which the 50th-percentile child has the same score as the given child.

Below 1 year the assessment from the hand and wrist is inaccurate since too few centres are present. At this age the leg, including knee and ankle, is a better guide, though no useful standards for this have yet been produced. In general, of course,

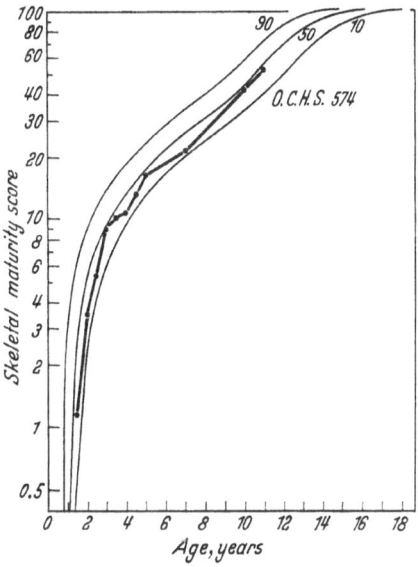

Fig. 10. Standard TANNER-WHITEHOUSE-HEALY skeletal-maturity charts, based on British children; the progression of a normal girl from the Oxford Child Health Survey is shown

other parts of the skeleton besides the hand may be used for skeletal ageing, but they yield closely correlated information in most cases, and standards are available only for the knee (PYLE and HOERR, 1955) and hip (ACHESON, 1957).

In Fig. 11 is shown an example from the clinic: a girl with CUSHING's syndrome, who had her adrenal tumour removed at the age of 4. Her skeletal maturity was already delayed at 2 years of age (skeletal age about $1^1/_4$ years) and markedly so at 4 (skeletal age $1^3/_4$ years). After operation a large catch-up occurred, and by 6 the child had reached average status. She went on to be an early maturer, being at the 90th percentile for skeletal maturity at $11^1/_2$ (skeletal age $12^1/_2$).

The factors which control skeletal maturation are not yet understood. Boys are from birth, or before, retarded relative to girls, and this seems to depend either directly or indirectly on genes on the Y chromosome, since XXY persons mature at the same

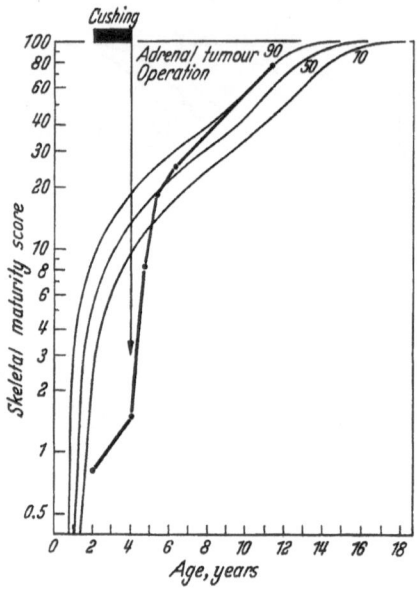

Fig. 11. Catch-up in skeletal maturity shown by a girl with CUSHING's syndrome after removal of the adrenal tumour. (From PRADER, TANNER, and VON HARNACK, in press)

rate as XY, and XO at the same rate as XX, at least until puberty (TANNER, PRADER, HABICH, and FERGUSON-SMITH, 1959). Androgens cause an increase in rate of skeletal maturing, and cortisone — above a certain dosage — a decrease. Hypothyroidism is associated with a decrease, but hyperthyroidism probably not always with an increase. Growth hormone seems to have little effect. In most circumstances, retardation of growth in size and growth in skeletal maturity occur together, and in the catch-up following recovery both usually increase together also. The details of this subject are, however, rather more complicated than there is space to describe here; they will be found in TANNER (1962, chapter IV).

Dental maturity. The appearance of the teeth may also be used as a basis for estimating physical maturity, either through a simple count of the number erupted, or by assessing the degree of maturity of the roots as seen in X-rays. Tooth maturity is correlated with skeletal maturity only slightly. Both are retarded in hypothyroidism, but in precocious puberty, either of cerebral origin or associated with adrenocortical hyperplasia, there is advancement of skeletal and sexual maturity without any corresponding effect upon the teeth (or the brain). In progeria, on the other hand, there is a retarded tooth age but a normal skeletal age (for full discussion see TANNER, 1952, pp. 84—85).

Secondary sex-character maturity. The appearance of secondary sex characters, and particularly of the menarche, may also be used as a measure of developmental age. In all studies of growth, pubic hair, breasts, and the male genitalia should be rated on a definite series of scales such as those detailed in TANNER (1962). The average age at menarche in England at present is 13.0 years with a standard deviation of about 1.0 years (and an approximately normal distribution). In nearly all cases secondary sex-character development runs closely parallel to skeletal maturity.

Prediction of adult height

Prediction of adult height may be made from the height of the child with fair accuracy from about 3 until puberty (see TANNER, HEALY, LOCKHART, MACKENZIE, and WHITEHOUSE, 1956). But at all ages the prediction is improved if the degree of skeletal maturity is also known, since skeletal maturity is a better guide to the percentage of his adult height that the child has reached. At adolescence, knowledge of the skeletal maturity is absolutely essential, since at age 14 in boys, say, one boy will be at 95% of his adult height and another at 85%.

BAYLEY and PINNEAU (1952) have given tables which relate GREULICH-PYLE skeletal age directly to percent of mature height, the latter being known since the data were from completed longitudinal studies. They add also a correction for whether a child is retarded or advanced relative to chronological age, since even though skeletal age correlates much better with percent of mature height than does chronological age, there is still a small residual independent effect of chronological age at given bone age, the older child being a little nearer adult height for his skeletal age. The BAYLEY-PINNEAU tables only classify children as over-a-year-retarded, over-a-year-advanced, or average. Thus, in these tables the adjustment in prediction for children with 2 years delay or more, such as are frequently seen in clinical practice, is probably too small. These tables are based on Californian children. BAYLEY (1962) has also given tables of percent mature height for given skeletal age in Boston children, but without the chronological-age correction.

For non-pathological children at least, the BAYLEY-PINNEAU predictions are sufficiently good to be useful in practice. We are able to advise the Royal Ballet School with some confidence on the probability that a girl aged 9 or 10 will end up between the established upper and lower limits in which they are interested. The error of prediction in 95% of cases from this age is about 3 cm. either way. We are at present reviewing our data so as to construct a new prediction based on the TANNER-WHITEHOUSE skeletal-maturity system and we hope that these will increase our accuracy.

Serial predictions of adult height have been used by clinicians testing androgens and other drugs. The idea is that, if a drug causes a relatively greater increase in height than in skeletal maturity, then this will be shown by an increase in predicted height, and the reverse will be shown by a decrease in predicted height. Though perfectly legitimate in theory, the method is not yet sufficiently accurate to support any statements about a drug affecting ultimate height based solely on predicted height. Such claims should only be based on the demonstrated final height reached. But serial predictions may indeed be useful in following the effect of a drug from year to year and in giving a warning if the skeletal maturity is being advanced more rapidly than the height.

Summary

The use of standard charts for assessing height and other measurements achieved at a given age is discussed. It is shown that no child should follow the percentiles on these charts at adolescence, since the individual growth spurt is greater than that shown by cross-sectionally-based data.

Velocity, or rate-of-growth, standards are preferable to height-achieved standards for clinical use, and a new chart is given showing the velocity in length or height growth of the 50th-percentile boy and girl from 6 months to maturity. The integral of this curve constitutes a true "individual" height-achieved curve in which a child does follow the percentiles at adolescence. Examples of this are given.

The magnitude of seasonal variations in height growth is discussed and reasons given why any drug suspected of causing increased or decreased growth rate must be tested over a whole year and contrasted with a control period of similar length.

Methods of assessing shape and tissue composition are described, and for following changes in these.

Methods for assessing skeletal maturity are described, including a new system based on scores given to stages of individual wrist bones. The usefulness and limitations of skeletal maturity in predicting adult height and in following children treated with androgens and other growth-producing agents are discussed.

Zusammenfassung

Die Benutzung von Standardkurven und -tabellen für die Beurteilung der Körpergröße und anderer in einem gegebenen Alter erreichter Masse wird besprochen. Es wird gezeigt, daß kein Kind während der Reifungsperiode den auf diesen Diagrammen angegebenen Percentillinien folgt, da der individuelle Wachstumsschub größer ist als derjenige, der sich aus den Werten ergibt, die auf Querschnittsangaben beruhen.

Für klinische Zwecke ist ein auf die Geschwindigkeit oder die Wachstumsrate bezogener Standard einem Standard vorzuziehen, der auf der erreichten Körpergröße basiert. Es wird eine neue Kurve vorgeschlagen, welche die Geschwindigkeit des Längen- oder Höhenwachstums auf der 50 Percentile (P_{50}) vom 6. Monat bis zur Reife angibt. Das Integral dieser Kurve stellt eine echte „individuelle" Kurve für die erreichte Körpergröße dar, nach welcher das Kind den Percentilen während der Reifungsperiode folgt. Beispiele dafür werden angegeben.

Das Ausmaß jahreszeitlicher Schwankungen im Längenwachstum wird besprochen und darauf hingewiesen, daß jede Substanz, von der vermutet wird, daß sie eine Zunahme oder Abnahme der Wachstumsrate hervorruft, während eines Jahres geprüft werden muß unter Vergleich mit einer gleich langen Kontrollperiode.

Methoden für die Beurteilung der Skelettreife werden beschrieben und ein neues Verfahren angegeben, das auf einer Gradeinteilung für die Reife einzelner Handgelenkknochen, z. B. der Epiphyse des Radius, beruht. Die Zweckmäßigkeit und die Grenzen, die Skelettreife als Maß für die Voraussage der Körpergröße heranzuziehen, werden besprochen, besonders im Hinblick auf die Behandlung von Kindern mit Androgenen und anderen wachstumsfördernden Stoffen.

Résumé

L'auteur discute l'emploi des courbes et tableaux standards pour déterminer la stature et les autres éléments correspondant à un âge donné. Il montre qu'à l'adolescence, aucun enfant ne suit les lignes de percentiles de ces diagrammes, l'accélération de la croissance étant pour chacun plus grande que l'accroissement calculé d'après des moyennes.

Pour l'usage clinique, un standard se rapportant à la vitesse ou au taux de croissance est préférable à un standard basé sur la taille. L'auteur présente un nouveau graphique indiquant la vitesse de la croissance en longueur ou en taille des filles et des garçons du percentile 50 (P_{50}), de 6 mois à la maturité. L'intégrale de cette courbe constitue une véritable «courbe individuelle» de la stature atteinte, et l'enfant en suit les percentiles pendant son adolescence. L'auteur donne des exemples.

Il discute de l'ampleur des variations saisonnières dans le développement statural et explique pourquoi toute substance dont on suppose qu'elle provoque une augmentation ou une diminution du taux de croissance doit être étudiée pendant un an en comparant les résultats avec les chiffres d'une période témoin de même durée.

Ce travail décrit les méthodes utilisées pour apprécier la maturité du squelette et indique un nouveau procédé fondé sur un étalonnage de la maturité de certains os du poignet, par exemple de l'épiphyse du radius. On discute de l'utilité et des limites de l'emploi de cette méthode pour prévoir la taille définitive, notamment au cours du traitement d'enfants par les androgènes et par d'autres substances qui stimulent la croissance.

Acknowledgements

I wish to thank most heartily my colleagues Mr. R. H. Whitehouse and Mr. John Harris for help with the computations and graphs made for this paper.

References

Acheson, R. M.: Clin. Orthop. (U.S.A.) 10, 19 (1957). — Bayer, L. M., and N. Bayley: Growth diagnosis. Univ. Press, Chicago, 1959. — Bayley, N.: J. Pediatr. (U.S.A.) 48, 187 (1956); — Mod. Probl. Pädiatr. (Switz.) 7, 234 (1962). — Bayley, N., and S. R. Pinneau: J. Pediatr. (U.S.A.) 40, 423 (1952). [Erratum corrected in J. Pediatr. (U.S.A.) 41, 371.] — Deming, J.: Human Biol. (U.S.A.) 29, 83 (1957). — Greulich, W. W., and S. I. Pyle: Radiographic atlas of skeletal development of the hand and wrist. Second edition. 256 pp. Stanford Univ. Press, Stanford, California, 1959. — Healy, M. J. R.: Amer. J. Physic. Anthrop. (in press) (1962). — Heierli, E.: Helvet. paediatr. acta. 15, 311 (1960). — Moore, T., C. B. Hindley, and F. Falkner: Brit. Med. J. 1954/I, 1132. — Prader, A., J. M. Tanner, and G. von Harnack: Catch-up growth following illness or starvation: an example of developmental canalization in man. (in press). — Pyle, S. I., and N. L. Hoerr: Radiographic atlas of skeletal development of the knee. 82 pp. Thomas, Springfield, Ill., 1955. — Tanner, J. M.: Human Biol. (U.S.A.) 23, 93 (1951); — Arch. Dis. Childh. (G.B.) 27, 10 (1952); — The evaluation of physical growth and development. In: Modern Trends in Paediatrics (second series). Ed. by A. Holzel and J. P. M. Tizard. Butterworth, London, 1958, pp. 325—344. — Proc. Nutr. Soc. (G.B.) 18, 148 (1959); — Education and Physical Growth. Univ. Press, London, 1961; — Growth at Adolescence. Second edition. pp. 325. Blackwell's Sc. Publ., Oxford, 1962. — Tanner, J. M., M. J. R. Healy, R. D. Lockhart, J. D. MacKenzie, and R. H. Whitehouse: Arch. Dis. Childh. (G.B.) 31, 372 (1956). — Tanner, J. M., A. Prader, H. Habich, and M. A. Ferguson-Smith: Lancet (G.B.) 1959/II, 141. — Tanner, J. M., and R. H. Whitehouse: Amer. J. Physic. Anthrop., N. S. 13, 743 (1955); — Lancet (G.B.) 1959/II, 1086; — Brit. Med. J. 1962/I, 446. — Tanner, J. M., R. H. Whitehouse, and M. J. R. Healy: A new system for estimating skeletal maturity from the hand and wrist, with standards derived from a study of 2,600 healthy British children. Part II. The scoring system. International Children's Centre, Paris. MS. (1962).

Use of anabolic agents in disorders of growth

By

A. Prader and R. Illig

Growth is an anabolic process. Anabolic agents can therefore be expected to accelerate growth. For practical purposes two types of anabolic agent are available: testosterone and its derivatives and pituitary growth hormone (GH).

Testosterone promotes growth and bone maturation and produces secondary male sex characteristics. Nature's experiments are the child with the adrenogenital syndrome and the boy with precocious puberty. In an attempt to prevent virilisation, it is the newer anabolic steroids, rather than testosterone, which are used today as growth-promoting agents in children. The use of these anabolic steroids in children represents in most instances a form of pharmacological treatment and not physiological therapy.

Pituitary GH promotes growth without accelerating bone maturation and without producing secondary sex characteristics. In this instance, Nature's experiment is the pituitary giant. Theoretically GH is the ideal agent with which to stimulate growth in all types of dwarfism. In practice, however, there are still some unsolved problems, the most important of which is the fact that only human GH (HGH) is effective in man. As long as HGH cannot be synthesised and has to be extracted from human pituitaries obtained at autopsy, its supply and the experience gained from its use are severely limited. So far it has mainly been given to hypopituitary dwarfs in a dosage which is probably within the physiological range.

As regards the effect of anabolic steroids and HGH on growth, I should like to emphasise that we, and many others, have had quite considerable experience with the use of anabolic steroids in short children, but little experience with the use of HGH in hypopituitary dwarfs and hardly any experience with the use of HGH in other dwarfs. Moreover, all our experience dates back only a few years. We have calculated, but have not yet really seen, the ultimate outcome of such treatment and the final adult height of children treated with these agents. Our knowledge about the use

of anabolic steroids and HGH to promote growth in children is
therefore still rudimentary.

Anabolic steroids

In the following discussion of the anabolic steroids, I shall
consider first their virilising action, and then their effect in accelerating growth and bone maturation, which involves the
important problem of the ultimate growth prognosis. A more
detailed discussion has already appeared elsewhere.

The *virilising effect* (increased frequency of erections, increase
in the size of the clitoris or the penis, pubic hair, changed voice,
etc.) is dosage and time dependent and can generally be prevented
if the dosage is kept low, i.e. at or below the following limits:
0.04 mg./kg./day for methandrostenolone, and 1 mg./kg./month
for nandrolone-phenylpropionate. These limits are only valid for
children, and it is our impression that hypothyroid children and
those suffering from brain damage are more sensitive than normal
children. Women tolerate higher doses without virilisation, and
in normal adult males the problem of virilisation does not arise.

Growth and *bone maturation* are nearly always accelerated, and
this effect also seems to be dosage and time dependent. Bone age
tends to be more accelerated than height age — a fact which lowers
the prediction of final adult height. The effect on bone age is easily
overlooked in short-term therapy, because it is frequently not seen
until some months after therapy has begun, and because it usually
continues for several months after treatment has been stopped.
This type of effect on growth and bone age has been known for
years, having been encountered in children undergoing treatment
with testosterone (SOBEL et al.) and in children with endogenous
overproduction of androgenic steroids (adrenogenital syndrome
and male precocity).

In the following figures we have used ordinary growth charts and
developmental charts. In the developmental charts chronological age is plotted
on the horizontal line and height age, bone age, etc. on the vertical line. The
normal development follows the diagonal line. This type of chart has certain
drawbacks, as Dr. TANNER has pointed out, but it does have the advantage
that both the course of growth and the course of bone maturation can be
plotted and compared on the same chart. Height and weight age have been
determined from the tables of STUART and STEVENSON and of HEIMENDINGER
and bone age from the atlas of GREULICH and PYLE. Expected adult height
has been calculated from age, bone age, and height with the help of the tables
of BAYLEY and PINNEAU.

The patient in Fig. 1, a hypopituitary dwarf, was treated with
testosterone at the age of 12 and has been kept under observation
without any further treatment up to the age of 21. The short period

of testosterone therapy exerted a profound influence on height and
bone age for many years — which shows that a long follow-up is

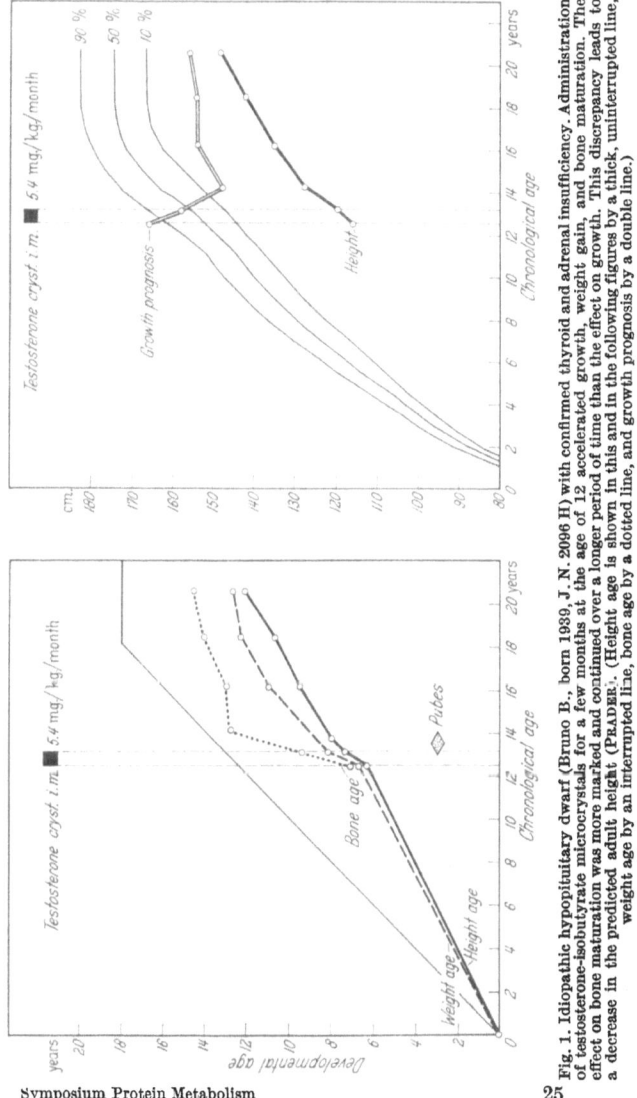

Fig. 1. Idiopathic hypopituitary dwarf (Bruno B., born 1939, J. N. 2096 H) with confirmed thyroid and adrenal insufficiency. Administration of testosterone-isobutyrate microcrystals for a few months at the age of 12 accelerated growth, weight gain, and bone maturation. The effect on bone maturation was more marked and continued over a longer period of time than the effect on growth. This discrepancy leads to a decrease in the predicted adult height (PRADER). (Height age is shown in this and in the following figures by a thick, uninterrupted line, weight age by an interrupted line, bone age by a dotted line, and growth prognosis by a double line.)

necessary for a full evaluation of the ultimate effect of such treatment. The immediate effects are a moderate acceleration of growth and a very marked acceleration of bone age. After the withdrawal of testosterone, the acceleration of height and bone age continues for some months, bone maturation being again faster than growth. The fact that the increase in bone age is much more marked than the increase in height age means a decrease in the calculated future

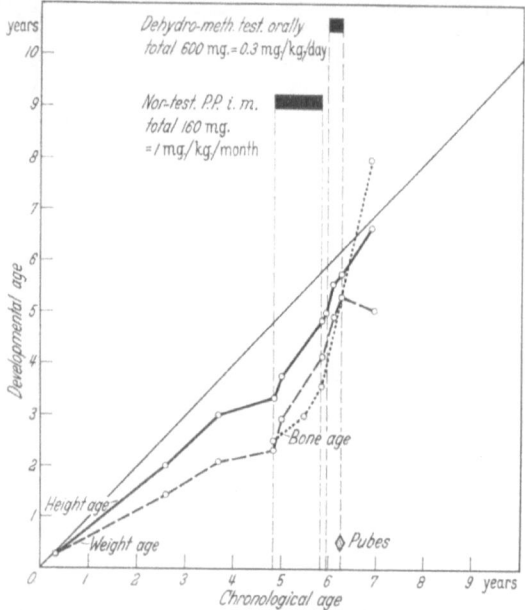

Fig. 2. Delayed development in a girl with multiple malformations (Anneliese St., born 1953, J. N. 7317/59). Treatment with nandrolone-phenylpropionate (nor-test. p.p.) and methandrostenolone (dehydro-meth. test.) for 1½ years improved the patient's growth, rate of weight gain, and bone maturation. Massive overdosage of methandrostenolone for 4 months was followed by a tremendous acceleration of bone age and by the transitory appearance of pubic hair

adult height. The conclusion is that the immediate growth improvement has only been achieved at the expense of a decreased adult stature. (It must be admitted that the calculation of adult height from the tables of Bayley and Pinneau may be unrealistic for hypopituitary dwarfs.)

In patients with a congenital endogenous overproduction of androgenic steroids (adrenogenital syndrome), growth is faster than normal, so that these children are excessively tall between 3 and 10 years of age. However, since bone age increases faster than height age, closure of the epiphyses occurs at the age of about 10 years, and growth therefore stops before the patient has attained normal adult size. Here, again, growth is accelerated by androgenic and anabolic steroids at the expense of a decreased adult stature.

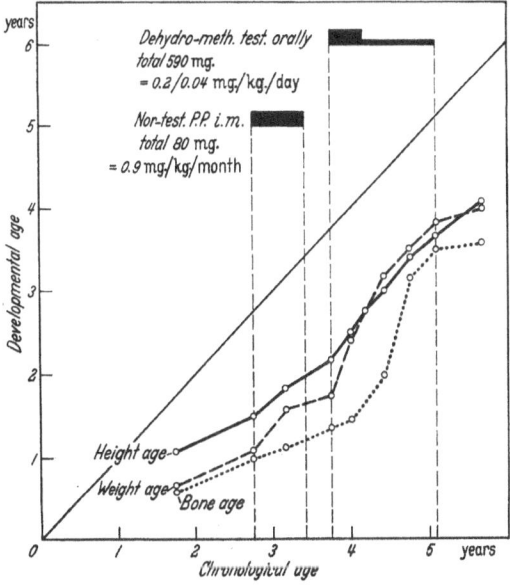

Fig. 3. Delayed development in a boy with a ventricular septal defect (Kurt G. born 1955, J. N. 8042 L). Treatment with nandrolone-phenylpropionate (nor-test. p.p.) and methandrostenolone (dehydro-meth. test.) for 3 years improved the rate of growth, weight gain, and bone maturation. Temporary overdosage of methandrostenolone explains the unusual speed of bone maturation over a short period

The degree of dissociation between the speed of growth and that of bone maturation, which determines *ultimate height*, also seems to be dosage dependent. With high virilising doses the increase in bone age is faster than the increase in height age, and the ultimate adult height is decreased (Fig. 2). With low non-virilising doses the increase in bone age is not faster (Fig. 3) or at least not much faster (Fig. 4) than the increase in height age, so that little or no effect is exerted on the predicted adult height. From our own experience we doubt whether it is ever possible to improve the expected adult height with anabolic steroids. However, it is not easy to make a definite statement on this point, because of the difficulties in evaluating bone age and in calculating future adult height in dwarfs, and because other steroids might have different properties.

Since high virilising doses decrease the expected adult height, they may be used in *excessively tall pubertal boys* in order to bring their expected adult height down to within the normal range. We

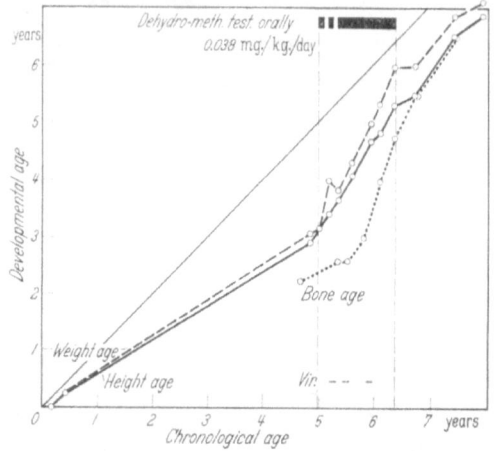

Fig. 4. Retarded development (constitutional) in a child of low birth-weight (Peter F., born 1954, J. N. 7106/59). Treatment with methandrostenolone in a dose of about 0.04 mg./kg./day for 16 months improved the height and bone age. In the beginning, the acceleration of bone age was slower than the acceleration of height age, whereas later it was faster. Some months after therapy had been stopped, the height and bone age increased again in parallel with each other

have done this successfully in a few selected patients. However, we are not prepared to recommend this therapy because of the accompanying testicular atrophy. Low non-virilising doses do not interfere with normal maturation of the gonads — a fact which we have frequently been able to confirm in pubertal patients. On the other hand, high doses suppress the production of pituitary gonadotropins and thus lead to gonadal atrophy. In our limited experience, however, such steroid-induced gonadal atrophy, even of several months' duration, is still completely reversible.

Since puberty normally appears at a specific stage of bone maturation (bone age of 11 in girls and 13 in boys), the acceleration of bone age induced by anabolic steroids leads to an earlier *onset of spontaneous puberty*. In other words, it is possible to hasten the appearance of spontaneous puberty by treating prepubertal children with anabolic steroids. This indirect effect should not be confused with the direct virilising effect caused by overdosage of the same steroids.

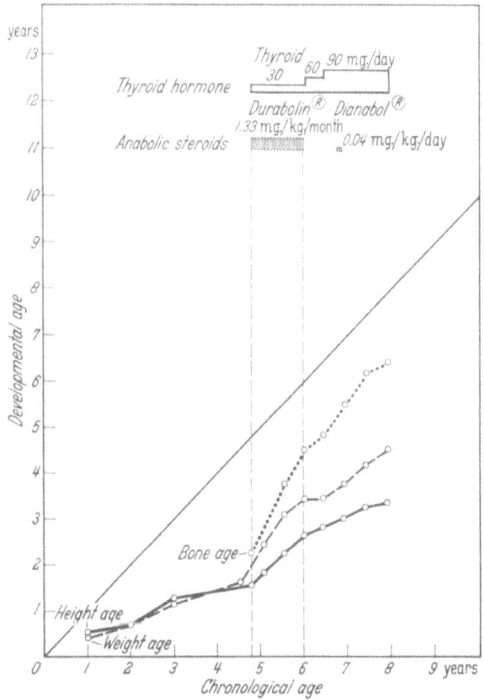

Fig. 5. Idiopathic hypopituitary dwarf (Anna S., born 1954, J. N. 5145/62), with confirmed adrenal insufficiency and questionable thyroid insufficiency. Combined treatment with thyroid and nandrolone-phenylpropionate for 1½ years improved the rate of growth and bone maturation. After treatment with anabolic steroid had been stopped, thyroid alone in increasing dosage had an effect similar to that of anabolic steroid, accelerating bone age more than height age

So far we have merely discussed growth and bone maturation under the influence of anabolic steroids alone. The situation is more complex in patients suffering from hypopituitary dwarfism with secondary thyroid and adrenal insufficiency, who receive other hormones at the same time as replacement therapy. In these patients high doses of *thyroid* frequently accelerate bone age more than height age (Fig. 5). In other words, in this situation, thyroid

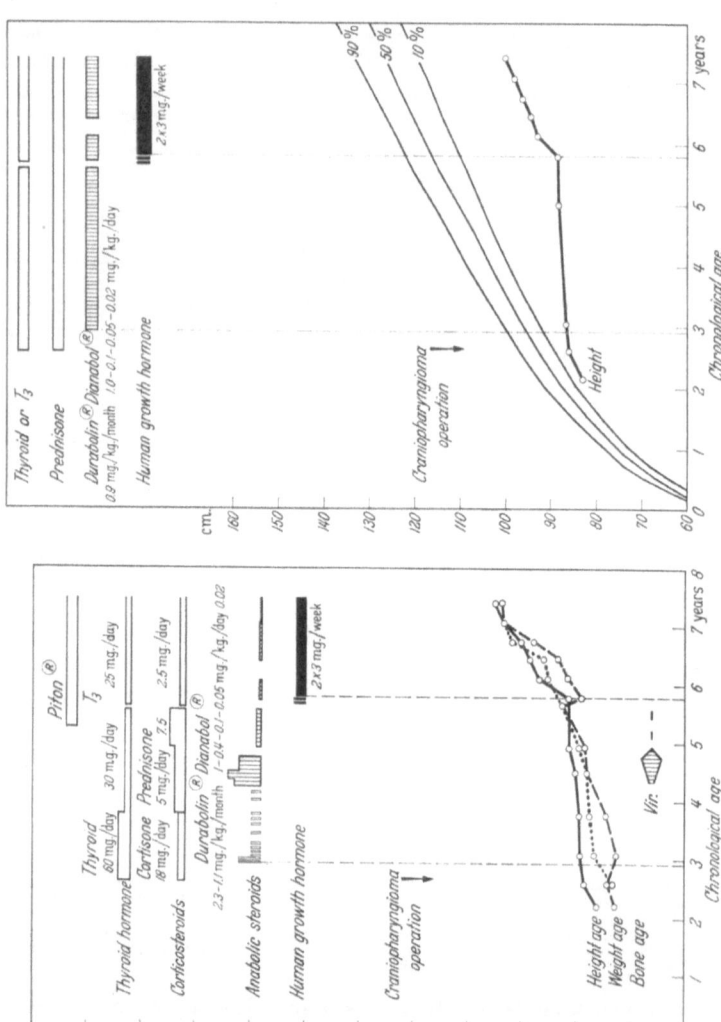

Fig. 6. Hypopituitary dwarf with craniopharyngioma (René St., born 1954, J. N. 1783/57; case No. 1 in Table 1). After surgical removal of the craniopharyngioma at $2^{1}/_{2}$ years of age, pituitary function deteriorated markedly, leading to a general anterior and posterior pituitary insufficiency. Three years of combined treatment with thyroid preparations, prednisone, and anabolic steroids up to the point of virilisation did not stimulate growth, which is in contrast to the response observed in patients with milder pituitary insufficiency. However, as soon as human growth hormone was added, the growth rate increased from 0.8 cm./year to 9.5 cm./year during the first months and to 6.5 cm./year for the total 22 months of treatment

may have an effect similar to that of anabolic steroids. On the other hand, by virtue of their anti-anabolic action, *glucocorticoids* may neutralise the effect of anabolic steroids in promoting growth and bone age.

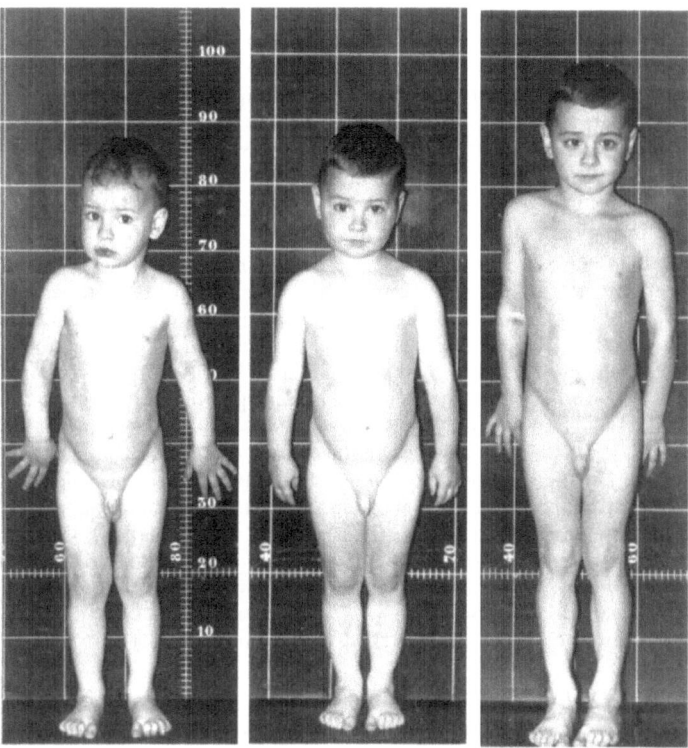

Fig. 7. The same patient as in Fig. 6 (René St.). *Centre:* at the age of 5 years and 8 months, i.e. at the beginning of treatment with growth hormone; *left:* at the age of 2 years and 7 months; *right:* 19 months after the beginning of treatment with growth hormone

The growth response to anabolic steroids depends not only on exogenous but also on *endogenous factors*. We have seen a number of patients who showed practically no growth acceleration in response to anabolic steroids. Among these unresponsive patients were healthy boys in their spontaneous pubertal growth-spurt, patients with gonadal dysgenesis, some dwarfs of low birth-weight, and some dwarfs with severe organic pituitary insufficiency (as opposed to idiopathic hypopituitary dwarfs, most of whom respond quite satisfactorily). We should like to suggest the following tentative explanations for these failures. In the normal pubertal growth-spurt there is a strong endogenous anabolic and androgenic influence at work, which cannot be increased by adding a small amount of exogenous steroid. In gonadal dysgenesis and in some dwarfs of low birth-weight there is possibly an inborn resistance of the bones to the growth-promoting effect but not to the bone-maturing effect. In severe pituitary insufficiency the lack of a growth response could be explained on the assumption that the presence of growth hormone, and perhaps of other hormones as well, is necessary to enable these steroids to exert their anabolic effect. Experimental work undertaken by the Evans group (Simpson et al.) and the results which Dr. Desaulles has shown us tend to support this assumption.

The patient referred to in Figs 6 and 7 is a hypopituitary dwarf who had a craniopharyngioma completely removed at the age of $2^1/_2$ years. As usual, the operation saved the patient from a neurological standpoint, but led to a deterioration in his pituitary function. During the following 3 years he grew less than 1 cm. per year in spite of full substitution therapy and in spite of anabolic steroids. At first sight this unresponsiveness to anabolic steroids might be attributed to the rather liberal glucocorticoid substitution therapy which the patient received, since it is known that the catabolic effect of glucocorticoids is more marked in hypopituitary patients (Blodget et al.; Lipset et al.). However, as the anabolic steroids were given in doses of up to 20 times the normal dosage (resulting, as was to be expected, in virilisation), we rather feel that the failure in this case was due to severe pituitary insufficiency and possibly to a severe lack of GH. The patient was later successfully treated with HGH.

Human growth hormone (HGH)

This brings us to the problem of HGH therapy. In 1958, Raben reported the first case of a dwarf who had been successfully treated with HGH, and some months ago he published a follow-up report on 11 dwarfs treated with HGH for periods of between 4 months and 4 years. Our own experience is less extensive and covers a shorter period. So far we have treated 6 patients for periods of between 4 months and 2 years. We have used Raben's preparation of HGH and have administered it to our patients in the following dosage: 1 daily intramuscular injection of 2 mg./m² during the first 5 days for metabolic investigations, and 2 injections of 5 mg./m² per week thereafter. The patient shown in Figs 6 and 7 has responded very encouragingly. His growth rate increased from less than 1 cm. per year to 9.5 cm. per year during the first few months and decreased again somewhat to an average of 6.5 cm. per year during the total 22 months of treatment. Bone age increased at the same rate as height age, possibly because of the simultaneous treatment with thyroid hormones and anabolic steroids.

We have treated in the same way 2 other patients with severe pituitary insufficiency and have obtained similar successes. In none of these patients were there any undesirable side effects. Among the favourable metabolic effects observed, the improvement in insulin hypersensitivity is most noteworthy.

In our own patients and in some other reported cases the growth rate, which at first was slightly above normal, decreased in the further course of treatment to slightly below normal. This is perhaps not surprising, because one sees the same phenomenon in hypothyroid patients treated with thyroid. We also wondered whether the dosage might perhaps be insufficient. Bearing in mind that the estimated daily turnover rate of HGH is 3 mg./m^2 (PARKER et al.), a dosage of 5 mg./m^2 twice weekly does seem rather low. In view of this and other considerations, we feel that our present method of treatment with HGH, though successful, probably constitutes only partial or, at best, full physiological replacement therapy for hypopituitary dwarfs and can therefore hardly be expected to stimulate growth in non-hypopituitary dwarfs who do not lack GH.

In order to test this hypothesis and to find out which patients profit most from such treatment, we have carefully examined the endocrine status of every patient, we have determined the serum GH concentration semiquantitatively using a modification of READ's haemagglutination inhibition method (SZÉKY et al.), and we have measured the nitrogen excretion in the urine under a constant diet before therapy and from the 2nd to 5th day of treatment with HGH; finally, we have tried to correlate all the results with the growth response during long-term therapy (Table 1). The overall result seems to confirm what we expected, i.e. that only hypopituitary patients, who lack GH, show a high rate of nitrogen retention and a good growth response under treatment with HGH. Non-hypopituitary dwarfs have variable GH concentrations, a smaller nitrogen retention, and a questionable growth response. The number of our patients is too small to permit of definite conclusions at the moment, but the results do suggest that the dosage used is within or below the range of physiological replacement therapy and that, therefore, only hypopituitary dwarfs can really profit from such treatment.

Table 1. *Effect of HGH* (Raben) *on nitrogen retention and growth in 11 dwarfs.* Dosage: 2 mg./m² daily for the first 5 days (to measure the effect on nitrogen retention); 5 mg./m² twice weekly after the first 5 days (to measure the effect on growth)

Probable cause of dwarfism	Age (years)	GH in serum	Increase in N retention		Rate of growth (cm./year)	
			mg./kg.	% of pre-treatment N excretion	Before therapy	With therapy +
Hypopit.						
1. Cranioph.	6	↓	132	41	$^1/_2$	$9^1/_2$
2. Idiopath.	8	↓	126	30	$3^1/_2$	—
3. Cranioph.	13	↓	80	27	$1^1/_2$	8
4. Idiopath.	13	↓	132	33	$2^1/_2$	$8^1/_2$
5. Idiopath.	7	↓	134	39	3	$7^1/_2$**
? Hypopit.						
6.	5	n	147*	46	$3^1/_2$	$3^1/_2$
Non-hypopit.						
7. Prenatal dwarfism	2	↓	100*	19	—	—
8. Prenatal dwarfism	5	(↓)	26	8	$5^1/_2$	—
9. Prenatal dwarfism	8	?	60	17	$4^1/_2$	3
10. Hypothyroid.	4	n	40	20	—	—
11. Glycogen disease	17	↑	49	12	$2^1/_2$	2

+ During the first 4—7 months.
* Higher HGH dosage (3 mg./m² daily).
** Total period of therapy only 5 days.

One of the most interesting patients in Table 1 is number 11, a dwarf with glycogen-storage disease of the classic type due to lack of glucose-6-phosphatase. Repeated GH assays of the blood yielded high values. Nitrogen retention with HGH was low, and the patient failed to grow during 4 months' treatment with HGH. However, he subsequently showed a gratifying acceleration of growth and bone age in response to anabolic steroids (Fig. 8).

The dwarfs with severe pituitary insufficiency — who do not respond to anabolic steroids but only to HGH — and this metabolic dwarf — who has enough GH and does not respond to HGH but only to anabolic steroids — demonstrate on a clinical level that the anabolic mechanisms, or the factors which control the anabolic mechanisms, of anabolic steroids and of GH are different.

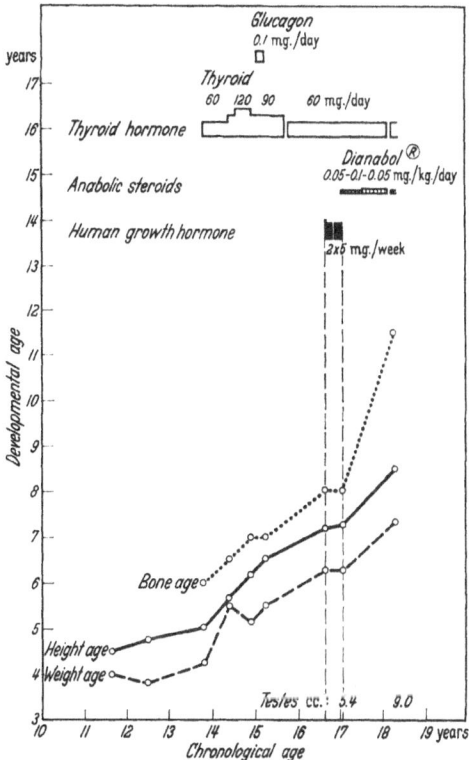

Fig. 8. Dwarf with glycogen-storage disease caused by a deficiency of glucose-6-phosphatase (Georg M., born 1944, J. N. 1051/60; case No. 11 in Table 1). Thyroid alone improved growth for some time. Growth hormone given for 4 months had no effect on growth, probably because growth hormone is not lacking in this patient. Methandrostenolone, however, immediately accelerated the rate of growth and bone maturation

Summary

Clinical experiences with anabolic steroids and human growth hormone (RABEN's HGH) are reported in children suffering from growth disorders due to various causes. The use of anabolic steroids is a form of pharmacological therapy, whereas that of HGH (in the usual dosage) constitutes a form of replacement therapy.

Administered in appropriate dosage, the anabolic steroids employed — methandrostenolone (0.04 mg./kg./day p.o.) and nandrolone-phenylpropionate

(1 mg./kg./month i.m.) — are non-virilising. Growth and bone maturation are nearly always accelerated, and this effect seems to be dosage and time dependent; bone age is often more accelerated than growth in height. The degree of dissociation between the rate of bone maturation and that of growth in height, which determines the ultimate height, also seems to be dosage dependent. It is doubtful whether it is ever possible to increase the expected adult height with anabolic steroids. High, virilising doses reduce the expected adult height and inhibit development of the gonads.

The growth-promoting effect of anabolic steroids is dependent not only upon the dosage and duration of treatment, but also upon endogenous factors. Little or no growth-promoting effect is observed in healthy boys during the pubertal growth-spurt, in girls with gonadal dysgenesis, in certain dwarfs of low birth-weight, or in pituitary dwarfs with severe pituitary insufficiency.

Human growth hormone (2 mg./m²/day or 2 × 5 mg./m²/week) promotes nitrogen retention and growth more strongly in pituitary dwarfism than in the other forms of dwarfism.

The fact that in certain cases growth effects can be achieved only with anabolic steroids, and in other cases only with HGH, would seem to indicate that a different mechanism of action is involved.

Zusammenfassung

Es wird über klinische Erfahrungen mit anabolen Steroiden und mit menschlichem Wachstumshormon (HGH von Raben) bei Kindern mit Wachstumsstörungen verschiedener Genese berichtet. Die Anwendung anaboler Steroide stellt eine pharmakologische Therapie und die Anwendung von HGH in der heute üblichen Dosierung eine physiologische Substitutions-Therapie dar.

In geeigneter Dosierung wirken die verwendeten Steroide Methandrostenolon (0,04 mg/kg/Tag p.o.) und Nandrolonphenylpropionat (1 mg/kg/Monat i.m.) nicht virilisierend. Wachstum und Knochenreifung werden fast immer beschleunigt in Abhängigkeit von Dosis und Zeit, wobei das Knochenalter oft rascher zunimmt als die Körpergröße. Der Grad der Dissoziierung zwischen der Geschwindigkeit der Knochenreifung und des Längenwachstums, der maßgebend für die endgültige Körpergröße ist, scheint ebenfalls dosisabhängig zu sein. Dabei ist zweifelhaft, ob es je gelingt, die zu erwartende definitive Körpergröße durch die Anwendung anaboler Steroide zu steigern. Hohe, virilisierende Dosen vermindern die zu erwartende Körpergröße und hemmen die Gonadenentwicklung.

Die wachstumsfördernde Wirkung anaboler Steroide ist nicht nur von Dosierung und Dauer der Behandlung abhängig, sondern auch von endogenen Faktoren. Sie fehlt ganz oder weitgehend bei gesunden Knaben im Wachstumsschub während der Pubertät, bei Mädchen mit Gonadendysgenesie, bei manchen Zwergen mit niedrigem Geburtsgewicht und bei hypophysären Zwergen mit schwerer Hypophyseninsuffizienz.

Menschliches Wachstumshormon (2 mg/m²/Tag oder 2 × 5 mg/m²/Woche) fördert bei hypophysärem Zwergwuchs die N-Retention und das Wachstum stärker als bei anderen Zwergwuchsformen

Die Tatsache, daß in bestimmten Fällen Wachstumseffekte nur mit anabolen Steroiden, bei anderen nur mit HGH erreicht werden, spricht für einen unterschiedlichen Wirkungsmechanismus dieser Stoffe.

Résumé

Les auteurs rapportent les résultats cliniques obtenus avec les stéroïdes anabolisants et l'hormone de croissance humaine (HGH de Raben) chez des enfants présentant des troubles de la croissance de diverse origine. Les stéroïdes anabolisants représentent une thérapeutique pharmacologique; l'hormone de croissance humaine, aux doses aujourd'hui usuelles, représente une thérapeutique physiologique substitutive.

A doses convenables, la méthandrosténolone (0.04 mg./kg. par jour per os) et le phénylpropionate de nandrolone (1 mg./kg. par mois i.m.), stéroïdes anabolisants, ne sont pas virilisants. La croissance et la maturation de l'os sont presque toujours accélérées, à un degré dépendant de la dose et de la durée du traitement. Mais l'âge osseux augmente souvent plus vite que l'âge statural. Le degré de dissociation entre la vitesse de maturation de l'os et la croissance en longueur — qui est déterminant pour la taille définitive du sujet — semble également dépendre de la dose. Il est douteux que l'on puisse jamais parvenir à augmenter la taille définitive d'un sujet avec les stéroïdes anabolisants. De hautes doses virilisantes réduisent la taille que le sujet aurait pu atteindre et inhibent le développement des gonades.

L'action stimulante des stéroïdes anabolisants sur la croissance dépend non seulement de la dose et de la durée du traitement, mais aussi de facteurs endogènes. Elle manque totalement, ou presque totalement, chez les garçons normaux au moment de la poussée de croissance de la puberté, chez les filles avec dysgénésie gonadique, chez quelques nains ayant présenté un poids de naissance insuffisant, et chez les nains hypophysaires présentant une insuffisance hypophysaire importante.

L'hormone de croissance humaine à des doses de 2 mg./m² par jour ou de 2 fois 5 mg./m² par semaine stimule chez les nains hypophysaires la rétention d'azote et la croissance de façon plus importante que dans les autres formes de nanisme.

Le fait que l'on exerce une action sur la croissance, seulement avec les stéroïdes anabolisants dans certains cas, seulement avec la HGH dans d'autres, est en faveur d'un mécanisme d'action différent de ces substances.

Acknowledgements

We gratefully acknowledge the assistance of Dr. J. SZÉKY (GH assay), of Dr. G. MÜRSET and Dr. A. FANCONI (management of some of the patients), and of Mr. F. HERMANN (chief laboratory technician).

References

BAYLEY, N., and S. R. PINNEAU: J. Pediatr. (U.S.A.) **40**, 423 (1952). — BLODGETT, F. M., L. BURGIN, and D. IEZZONI: N. England J. Med. **254**, 636 (1956). — GREULICH, W. W., and S. I. PYLE: Radiographic Atlas of Skeletal Development of the Hand and Wrist. 2nd Edition. Stanford University Press, 1959. — HEIMENDINGER, J.: Helvet. paediatr. acta **13**, 471 (1958); — Schweiz. med. Wschr. 88, 807 (1958). — LIPSETT, M. B., D. M. BERGENSTAL, and F. G. DHYSE: J. Clin. Endocr. (U.S.A.) **21**, 119 (1961). — PARKER, M. L., R. D. UTIGER, and W. H. DAUGHADAY: J. Clin. Invest. (U.S.A.) **41**, 262 (1962). — PRADER, A.: Acta endocr. (Den.) **39**, Suppl. 63, 78 (1961). — RABEN, M. S.: J. Clin. Endocr. (U.S.A.) **18**, 901 (1958); — N. England J. Med. **266**, 82 (1962). — SIMPSON, M. E., C. W. ASLING, and H. M. EVANS: Yale J. Biol. **23**, 1 (1950). — SOBEL, E. H., C. S. RAYMOND, K. V. QUINN, and N. B. TALBOT: J. Clin. Endocr. (U.S.A.) **16**, 241 (1956). — STUART, H. C., and S. S. STEVENSON: Physical growth and development. In: MITCHELL-NELSON's Textbook of Pediatrics. 5th Edition. W. B. Saunders Co., Philadelphia, 1950. — SZÉKY, J., L. HOLLÄNDER, and A. PRADER: Helvet. paediatr. acta **16**, 691 (1961).

Discussion

LARON: After Dr. TANNER's very interesting remarks one can certainly understand how difficult it is for us paediatricians or endocrinologists really to assess the effectiveness of treatment either with anabolic substances, such as testosterone derivatives, or with growth hormone, because our periods of treatment are relatively short. I would nevertheless like to make a few comments to supplement to some extent the data presented by Dr. PRADER. We have had experience of 65 children treated with fluoxymesterone, with

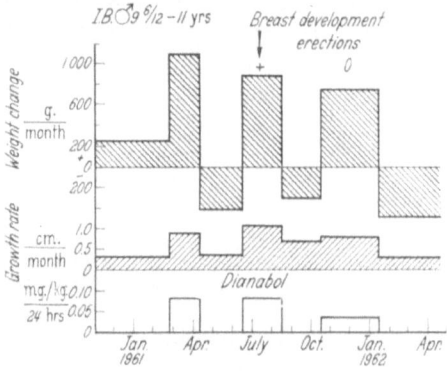

Fig. 1. Effect of various doses of methandrostenolone on body-weight and growth rate in a 9½-year-old boy

adequate pre-treatment and post-treatment control periods. Of these, one group of children was treated for 3 months, and 52 others were treated for 6 months. We found that the non-virilising dose of fluoxymesterone given by mouth for the 3-months' treatment period was 0.2 mg./kg. body-weight[1]; but when we treated the children for 6 months, the dose had to be reduced to about 0.1 mg./kg. body-weight. We have also treated 29 children with methandrostenolone (Dianabol). In Fig. 1 the growth rate — or growth velocity, as Dr. TANNER has referred to it — in cm. per month and the change in body-weight are given for a typical case. At various dosages we have a gain in body-weight, but after stopping the therapy a loss in body-weight ensues; this happened at 3 different dosage levels. We wonder whether this weight loss represents a loss of nitrogen or a loss of body-water, which we think is the more likely explanation. At all events, this suggests that we should pay close attention to the gain in body-weight during treatment with anabolic substances in patients with chronic diseases, as well as during treatment after operations, where the anabolic agent is administered only for short periods of

[1] LARON, Z.: Acta endocr. (Den.) **36**, 541 (1961).

time. I would also like to emphasise that this evidence certainly does not disprove the notion that nitrogen retention occurs.

Furthermore, we found that the growth response to smaller doses of methandrostenolone is not very much less than the growth response to larger doses; this suggests that, with smaller doses which produce fewer side effects, it is still possible to achieve adequate growth stimulation. Another finding which we have encountered was breast development in a large number of patients treated with methandrostenolone, to whom we admittedly gave bigger dosages than we are giving now. Signs of feminisation induced by methandrostenolone are summarised in Table 1. I should like to stress that

Table 1. *Urinary oestrogens during treatment with methandrostenolone (Dianabol)*

Patient	Sex	Age (years)	Dose (mg./kg.)	Urinary oestrogens (mcg./24 hrs)		
				before	during	after
Breast enlargement						
E. C.	F	10	0.08		3.3	
T. G.	F	$5^5/_{12}$	0.10		6.0	1.4
E. S.	F	$7^{11}/_{12}$	0.08		9.5	4.0
Z. L.	M	$8^2/_{12}$	0.09		4.2	1.3
M. F.	M	$9^9/_{12}$	0.08		4.7	
Y. B.	F	$9^5/_{12}$	0.08		8.0	2.3
Without breast enlargement						
O. N.	F	$3^{10}/_{12}$	0.08		6.0	0
C. K.	F	$6^8/_{12}$	0.05	1.5	3.0	
M. A.	M	$9^5/_{12}$	0.04		10.0	1.0

we had no such changes with fluoxymesterone. We have data on 6 children with breast development, a side effect which we have observed in a total of 14 children out of 29 treated. We were able to show that this was associated with increased oestrogen excretion in the urine. The oestrogen excretion was determined by the ITTRICH method (by courtesy of Dr. G. RUMNEY[1] and his associates) both during therapy as well as later, after administration of methandrostenolone had been discontinued. In patients in whom no breast development was noted, there was also a decrease in urinary oestrogen excretion after treatment. We have to admit, however, that the doses of methandrostenolone are high as compared with what we give today.

Another observation we made in two patients treated for a longer period with growth hormone was that, when the treatment was stopped after 6 months, complete inhibition of growth occurred, although growth was again resumed when treatment was reinstituted, whereas with anabolic substances derived from testosterone, growth continues after the drug has been withdrawn.

Concerning Dr. PRADER's question as to whether growth hormone produces antibodies when given over a prolonged period, we have tested this in our 3 patients, and we have found no antibodies even after one year of treatment.

[1] RUMNEY, G.: Endocrine Laboratories, Beilinson Medical Center, Petah-Tiqva, Israel.

GEMZELL: During the last years I have prepared human growth hormone and distributed it to several paediatricians in Sweden and in the other Scandinavian countries. This hormone has been tried on various patients, including hypopituitary dwarfs, as well as individuals with retarded growth due to other causes. I would like to ask Dr. TANNER and Dr. PRADER if they have observed any effect of human growth hormone on those patients who have not grown in response to the hormone. I have had several reports that there seems to be an increase in mental activity. This may of course be due merely to the very fact that something is being done for these patients.

TANNER: The mother of one of the patients that we have been treating spontaneously said very much the same thing. I tried quite hard to persuade her that this was purely a psychological effect, but without success. I wouldn't have laid any emphasis on this, if I hadn't heard Dr. GEMZELL saying the same. But this is only one case out of half a dozen.

PRADER: Dr. LARON mentioned gynaecomastia caused by methandrostenolone. I think this is not surprising if one gives it in high doses, because testosterone may also lead to gynaecomastia. — As regards Dr. GEMZELL's question, we have also seen an increase in activity and a better performance at school when we treated hypopituitary dwarfs with growth hormone. However, these children get so much attention and feel so happy to be growing that we did not know whether this "psychotonic" effect was directly due to the growth hormone or to a secondary psychological reaction.

RABEN: Dr. TANNER showed in one of his slides two 10-year-old boys of the same height. One went into an early puberty and one into a late puberty, but they both ended up at the same height. With a late adolescence and more years to grow one might have expected a taller final height. As this is apparently not so, it presumably means that the pubertal spurt is less marked in the person with the later adolescence. Is that usually the case?

TANNER: You will find in Fig. 1 (p. 362) that the later developer ended up very slightly taller than the early developer, and this represents the statistical facts correctly. Late developers on average do end up just a little taller than early developers, but the difference is very small, so that, by and large, you could say they ended up the same. Your conclusion from this is entirely correct. The peak velocity of growth in the early maturer is in fact systematically greater than in the late maturer (see Fig. 2, p. 365). It seems that the growth velocity potential diminishes as the child becomes older.

RABEN: I think this is very instructive. In hypopituitary patients, however, we do get growth with growth hormone, even at the age of 20 and 21, so that at least in that situation the growth potential is adequate. Dr. PRADER implied that there might be some danger in advancing the bone age of the hypopituitary patient with thyroid, but would it really matter, since the thyroid will only carry the bone age to perhaps the age of 12, in which case the epiphyses are still open? Would it reduce the potential growth if one were to advance the bone age from, say, 8 to 12 when spontaneous puberty will not occur?

PRADER: This question, Dr. RABEN, comes up again and again when one treats hypopituitary dwarfs. We do not really know how to calculate what height such a child will finally reach without treatment and what height it could reach under proper substitution therapy. The figures which we calculate from the tables of BAYLEY and PINNEAU[1] are theoretically interesting,

[1] BAYLEY, N., and S. R. PINNEAU: J. Pediatr. (U.S.A.) **40**, 423 (1952).

but perhaps unrealistic, because they are based on experience with normal children. It would be worth-while studying adult hypopituitary dwarfs, and x-rays taken when they were young children, in order to establish a basis on which to calculate how much an individual hypopituitary dwarf will grow if not treated. Perhaps Dr. TANNER can answer Dr. RABEN's question better.

TANNER: I agree exactly with what you said, Dr. PRADER.

QUERIDO: I have smuggled in somebody from my department this afternoon, who would like to make a remark.

VAN DER WERFF TEN BOSCH: There are just a few points I should like to make. First, I think there are two different types of pituitary dwarf: the so-called idiopathic type with panhypopituitarism, where we don't know what is wrong, and the other type, where you get pathological findings in the region of the pituitary itself; and I think there are marked differences between these two types as regards the effect of treatment. Now, we found that, out of 18 so-called idiopathic dwarfs in whom you only find dwarfism plus hypopituitarism, 15 had been born in an abnormal way: 11 were breech presentations and 4 others had broken arms or neurological signs involving the cranial nerves afterwards. I would contend that these patients present human examples of pituitary stalk section. It is known from the work of DANIEL and his colleagues[1] in London that after a trauma of the head one quite often gets disruption of the pituitary stalk. It is also known that after disruption of the pituitary stalk such a pituitary gland will display very little spontaneous activity, but may still respond to certain stimuli — for example, to propylthiouracil and to androgens in animal experiments. It is my contention that dwarfs with idiopathic pituitary dwarfism retain pituitary tissue which may respond to anabolic agents by secreting growth hormone. Dr. PRADER has touched on this point already with regard to the effect of androgens in hypophysectomised animals, and both Dr. PRADER and I have found that the same applies to children.

Dwarfism with proven pituitary disease occurs usually in girls, and in such cases you cannot promote growth with androgens. However, Dr. HAAK and I have made an observation of a girl of this type in whom we found that, although growth is not markedly stimulated, you do get the metabolic effects of the anabolic agent, so that we would like to think that there are two different effects: the anabolic effect and the effect on growth. The data in Fig. 2, I am afraid, are plotted in the conventional manner. I would agree with Dr. TANNER that it is not a very satisfactory method, because a pituitary dwarf will never be able to reach the normal standard for height, whereas the dwarf will always reach the normal standard for skeletal maturity, so that whatever you do there will inevitably be a dichotomy between bone age and height age. This is a girl who developed tuberculous meningitis at the age of 4, followed by diabetes insipidus at the age of 9; she came to see us when she was 13 years old. She was then a stunted girl with a bone age of 10; we have seen x-rays dating from 3 years previously, when the bone age was also 10. She had normal thyroid function, which explains the normal bone age at the age of 10 years, and she had apparently normal adrenal function. There were, however, good grounds for suspecting an abnormality of the pituitary, because in the region above the sella turcica there was an enormous amount of calcification at the base of the brain. We believe that this is an example of a

---

[1] DANIEL, P. M., M. M. L. PRICHARD, and C. S. TREIP: Lancet (G.B.) 1959/II, 927.

pituitary dwarf not of the idiopathic type; in fact, we have very little evidence to show that the girl is a hypopituitary case, except for the diabetes insipidus and perhaps the lack of growth; but she has normal thyroid and adrenal functions.

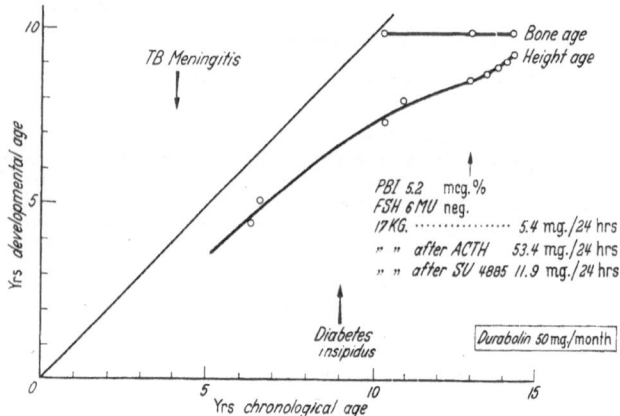

Fig. 2. Height and bone age of a girl with stunted growth. Diagnosis of pituitary dwarfism is based on functional (diabetes insipidus) and morphological (calcified base of brain) manifestations of hypothalamic damage following tuberculous meningitis. Normal thyroid and adrenal functions indicate that hypopituitarism is only partial

This pituitary dwarf has been under treatment since June 1961 with Durabolin, and we have followed up several parameters of growth. We have kept a record of the height and weight, we have taken skin-fold measurements, and we have made soft-tissue x-rays at various intervals; from these we have calculated the changes in fat, muscle, and so on (Table 2). In one year's time the girl grew 3 cm., which is quite normal for pituitary dwarfs of either the idiopathic or the other type. There were hardly any weight changes, but from

Table 2. *Effect of Durabolin (25 mg. twice monthly from June 1961) on a pituitary dwarf*

		April 1961	January 1962	May 1962
Height (cm.)		130.7	132.1	133.7
Weight (kg.)		31	32	32
% Fat	lower arm	29.5	22.9 (—22%)	20.6 (—30%)
	calf	19.9	16.1 (—19%)	14.3 (—28%)
	upper arm	20.9	16.7 (—20%)	13.2 (—37%)
% Muscle	lower arm	33.3	43.1 (+29%)	43.1 (+29%)
	calf	52.5	58.5 (+11%)	62.1 (+18%)
Urinary creatinine (mg./day)		507 (June 1961) 559*	624* (+23%)	

* For six days following injection of Durabolin.

the soft-tissue radiograms and from the skin-fold measurements we could calculate that during treatment since June 1961 fat decreased and muscle increased in the upper and lower arm and in the calf. The urinary creatinine increased, too, from 507 (this being an average for 6 days) to 624 mg./24 hrs. Skin-fold measurements over the triceps, the biceps, and the subscapular region revealed a decrease in these measurements, which, we think, reflects a fall in the fat content under the skin (Fig. 3).

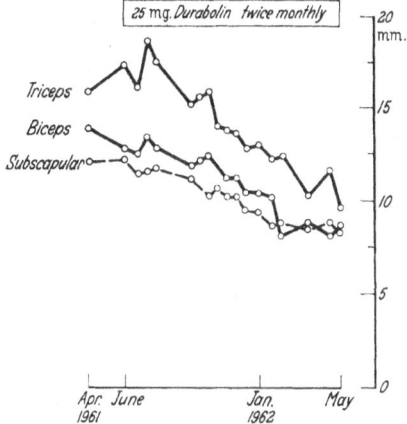

Fig. 3. Skin-fold measurements in pituitary dwarf (see Fig. 2) during treatment with Durabolin. The figures represent actual readings of the double thickness of skin plus subcutaneous fat

From this I think we can conclude that the androgen may exert two different effects: one on growth (not noted here) and the other on protein and fat metabolism. The first is possibly mediated by the pituitary secretion of growth hormone, whilst the latter may be independent of such secretion and may occur in the absence of pituitary tissue.

TANNER: The fall in fat and the increase in muscle under Durabolin which Dr. VAN DER WERFF TEN BOSCH showed is exactly the pattern of change that you find in the normal adolescent spurt in the male. But, of course, at that time growth is simultaneously being stimulated, so that in the normal situation there is a very close correlation between the growth-stimulating action and the changes in body composition. You never see a normal child with his growth spurt in height separated from the spurt in muscle or the decrease in fat; if one is early the other is early, and if one is late the other is late. By that I don't mean to imply that in an abnormal individual this could not happen.

PRADER: We have had exactly the same experience as Dr. VAN DER WERFF TEN BOSCH. There are two types of hypopituitary dwarf; idiopathic hypopituitary dwarfs (mainly boys) and organic hypopituitary dwarfs with a craniopharyngioma or some other intracranial lesion. We, too, feel that the most frequent cause of idiopathic hypopituitary dwarfism is a cerebral birth

26*

trauma. The growth response to anabolic steroids is usually good in the idiopathic type and usually bad in the organic type of hypopituitary dwarfism. —
Dr. VAN DER WERFF TEN BOSCH, in your patient with organic hypopituitary
dwarfism you observed no growth and no increase in bone age, but a clear
metabolic response. I am wondering whether this metabolic response is
normal or less than normal. There is another question I would like to ask you
in this connection: is the increase in bone age which is seen with anabolic
steroids an anabolic or a virilising effect? Some investigators feel that it is
connected with the androgenic activity. In our experience an organic
hypopituitary dwarf given anabolic steroids shows a normal virilising effect,
no growth effect, and a partial effect on bone age.

VAN DER WERFF TEN BOSCH: I don't think I can answer your first
question. We didn't find any signs of virilisation, with the exception of about
ten pubic hairs at the last examination — which is all we have noted from
the point of view of virilisation. Bone age actually increased by about
6 points in the TANNER and WHITEHOUSE system, but there was no difference
by the GREULICH and PYLE standards. I think that the age of 10 years is
rather unfortunate, since relatively few changes occur around this time, so I
don't think we should attach too much value to that anyway.

I don't know about the correlation between virilisation and bone age.
The difficulty is that bone age, of course, represents all sorts of things. One
of the problems was brought up earlier in connection with the thyroid and
bone age. Actually, I think that bone age as such is just a collection of
arbitrary units, and the collection is so chosen that during infancy, before the
age of 10, those units that are selected are responsive to thyroid hormone, so
that when you get a girl with GRAVES' disease you get an increased bone age.
If you get a girl of 15 with GRAVES' disease, you don't. It doesn't mean that
bone age is no longer affected; it just means that the indicator that was used
first is no longer used at a later age. I don't think that a constant bone age of
10 years means very much, and I shall be very interested to see what happens
during the next year in this patient. I wouldn't be surprised if she were to
shoot up at an increased speed later on.

PRADER: Was the nitrogen retention greater or similar to what you find
in control children?

VAN DER WERFF TEN BOSCH: No, not if you mean the amount of nitrogen
retained. I think Dr. HAAK would like to comment on this.

HAAK: Bearing in mind the difficulty of comparing the nitrogen-retaining effect of these compounds from one individual to another, we think that
in this case administration of the anabolic steroid resulted in nitrogen
retention which was not abnormal.

Definition, aetiology, and effects of osteopenia

By

G. C. H. BAUER

Introduction

By *definition* osteopenia is a clinical condition in which the amount of bone is less than normal; the expression "osteopenia" covers what ALBRIGHT and REIFENSTEIN termed "too little calcified bone".

While there is unanimous agreement as to the existence of this condition, great difficulties arise when one tries to adopt an objective standard for establishing the *diagnosis* of osteopenia in individual cases. Conventional radiological techniques, for example, do not seem to be sensitive to less than a 30—50% reduction in bone mass. When a rigorously standardised radiological technique is combined with microscopic studies of bone sample, the diagnosis "osteopenia" can probably be made with somewhat greater confidence (BARNETT and NORDIN, 1961).

In an occasional case, the *aetiology* of osteopenia may be shown to be an hormonal dysfunction like hyperthyroidism or hyperadrenocorticism from internal or external sources. In the large majority of cases, however, it is impossible at present to establish any causal relationship between hormonal dysfunction and osteopenia. ALBRIGHT and REIFENSTEIN believed that an imbalance between "anabolic" and "catabolic" steroids, brought about during the menopause, causes decreased bone formation and hence osteopenia in females about the age of 50. This theory has still not been confirmed by clinical or experimental evidence.

Probably most patients, even with marked osteopenia, rarely suffer any discomfort. Some, however, have *pain* and *fractures*. Pain is certainly due to pathological bone in rickets and osteomalacia, whereas this symptom is less clearly defined in osteopenia which is not due to vitamin D deficiency. At any rate it is certainly impossible to evaluate the degree or type of pain suffered by osteopenic individuals or even to differentiate such pain from pain caused by other factors operating in the elderly individual.

We decided, therefore, to differentiate a group of individuals suffering from osteopenia with the aid of one of the most severe symptoms of skeletal fragility, fracture of the neck of the femur. This decision started us on a series of investigations of fracture epidemiology (BAUER, 1960; ALFFRAM and BAUER, 1962). As I will explain in this paper, it is possible that we have thus been led away from the study of osteopenia!

Material and methods

Our investigation was carried out in Malmö, a highly industri-alised city in Sweden. It has a population of over 200,000, the composition of which in respect of age and sex is exactly known. For various reasons it is rare for the residents of Malmö to go anywhere but to the University Hospital Department of Ortho-paedics to seek help for a major fracture such as that of the neck of the femur or even a relatively minor one such as fracture of the forearm. Under these conditions we have been able to describe fairly accurately the incidence of these fractures, and in most cases is was deemed possible to evaluate roughly the degree of trauma which caused the fracture. In an attempt to evaluate the degree of displacement in the case of the forearm fractures, the incidence of reduction of these fractures was also recorded.

Results

Fracture of the neck of the femur

In agreement with previous studies dating back to the early 19th century we found few fractures of the neck of the femur before the age of 50. From then on the incidence of this fracture in females increased at an exponential rate: the incidence doubled with each 5-year increment in age throughout life. In men, also, the incidence rose sharply with advancing age, but it never even approximated to that in females of corresponding age (Table 1).

In age groups below 60, severe trauma and

Table 1. *Incidence of fracture of the neck of the femur in Malmö around 1955*

Age group	Annual fracture rate per 10,000		
	Females	Males	Females/males
<50	0.8	1.0	0.8
50—54	5.1	2.3	2.2
55—59	8.0	4.5	1.8
60—64	15.0	6.7	2.2
65—69	29.0	10.4	2.8
70—74	49.1	25.7	1.7
75—79	83.0	37.2	2.2
≦80	157.7	56.5	2.8

The data in this table are based on 1,332 consecutive cases studied over a period of 10 years by ALFFRAM (1962, to be published).

local affections of the hip area, such as cancer metastases or polio, were relatively common findings. With progressive age the degree of trauma involved in the fracture history seemed relatively minor. However, spontaneous fracture was uncommon even among very old people.

Fracture of the distal end of the forearm

The most striking features in the *incidence* (Fig. 1) of fractures of the forearm was the dramatic rise found in females after the

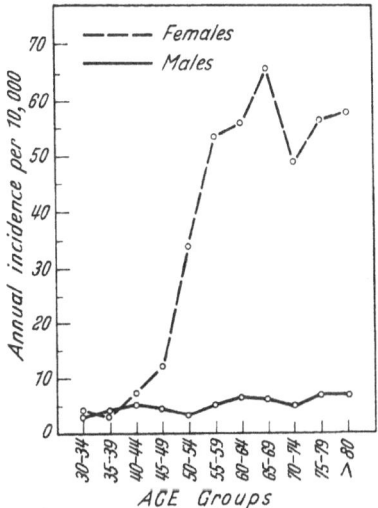

Fig. 1. Annual incidence of fractures of the distal end of the radius in the city of Malmö

age of 40, and the levelling off of this trend after about the age of 60. In contrast, the incidence in ageing males rose but little, so that, above the age of 60, females outnumbered males by a factor of more than seven.

The relative incidence of moderate *trauma* (Fig. 2) causing fracture of the distal end of the forearm, as indicated by the ratio of moderate to severe trauma, was found to rise in females after the age of 30, and a parallel, though somewhat later trend, was found in males.

In females the *displacement of the fragments* (Fig. 3) of the fracture of the distal end of the forearm was found to increase with age. In males there was rather a tendency towards less displacement in the higher age groups.

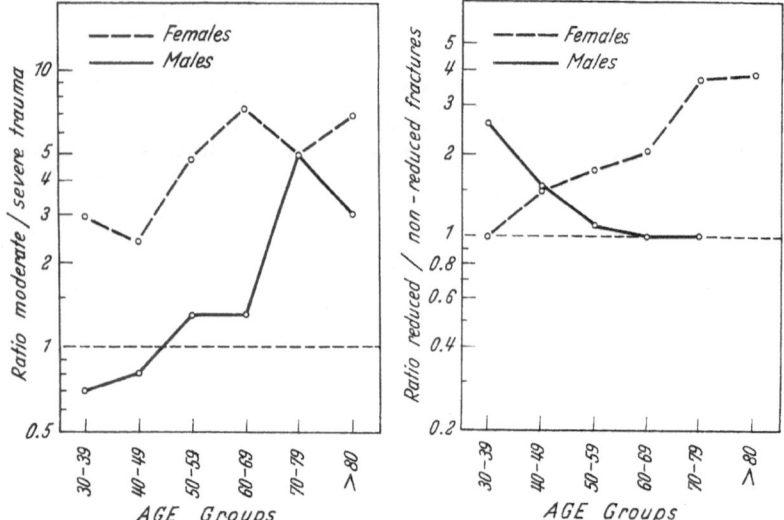

Fig. 2. Fracture of the distal end of the forearm. Ratio of moderate to severe trauma in males and females

Fig. 3. Fracture of the distal end of the forearm. Ratio of reduced to non-reduced fractures in males and females

Discussion

When compared with other criteria for age-changes in the human skeleton, it is obvious that the fracture incidence is by far the most revealing: results from grading roentgenological or histological signs of osteopenia or from bone-breaking tests do not show such a dramatic difference between old and young adults, or between females and males, as does the fracture incidence in a closed population. It is possible that this discrepancy may be due in part to difficulties inherent in the evaluation of x-ray films or to the choice of individuals tested for breaking strength or sampled for bone histology. I would like here, however, to point out another possible explanation. Perhaps osteopenia and skeletal fragility are less related than present dogma leads us to believe. This issue will quite likely be less confused if one realises that osteopenia is not a disease but a symptom of disease, caused by hormonal imbalance, protein deficiency, genetic disturbances, syndromes like rheumatoid arthritis, or — far more common — unknown factors.

In similar fashion, bone fragility may perhaps be caused by many different factors, among which, at present, only a genetic disturbance (osteogenesis imperfecta) is at all well known, whereas osteopenia is not necessarily obligatory or even common.

On this latter point I feel Fig. 1 is highly relevant as an argument against the belief that osteopenia and skeletal fragility need be associated: nobody, to my knowledge, has shown any data which even suggest that osteopenia is at all common in females between 40 and 50 years old.

As for the cause of skeletal fragility, I have little to offer but some negative evidence. ALFFRAM (1962) has found that the incidence of fractures of the forearm or of the neck of the femur seems unrelated to the marital status of the individuals studied or to their number of children. Also, it does not seem likely that food intake is a decisive factor.

The suggestion that osteopenia need not be an important feature in the aetiology of skeletal fragility leads one to the conclusion that the quality rather than the quantity of bone is the important factor in endogenous skeletal fragility. It seems not unlikely that a change in quality will have to be sought in the organic matrix rather than in the mineral constituents of the skeleton.

I believe it is important to work out in detail the fracture pattern of various populations, because such data may eventually lead to an understanding, and thus prevention, of an important disease. Also, any evaluation of drug or other effects on skeletal fragility will have to be based on accurate knowledge of fracture incidence in the population.

Summary

Osteopenia is a clinical condition in which the amount of bone present is less than normal; the term is applied to what ALBRIGHT and REIFENSTEIN described as "too little calcified bone". Conventional radiological techniques do not seem to be sensitive to less than a 30–50% reduction in bone mass, but a more reliable diagnosis can be achieved by means of bone biopsy. In most cases it is impossible to establish any causal relationship between hormonal dysfunction and osteopenia.

Epidemiological studies of fractures were undertaken, with particular reference to the incidence of fractures of the neck of the femur and forearm. Beyond the age of 50 years the incidence of fractures of the neck of the femur shows a sharp rise and is even higher in women than in men. With progressive age the degree of trauma involved in the fracture history seems relatively minor. In women over 40 years of age, the incidence of fracture of the forearm shows a dramatic rise, and by the age of 60 is some seven times higher than in men.

The incidence of fractures among comparable groups of the population is the most revealing criterion by which to assess age-changes in the human skeleton. It is suggested that, since osteopenia is not a disease but a symptom of disease caused by various factors (hormonal imbalance, protein deficiency, genetic disturbances, rheumatic disorders, etc.), it may not necessarily be responsible for increased skeletal fragility, nor need such fragility necessarily be associated with osteopenia.

Zusammenfassung

Als Osteopenie wird ein Zustandbild bezeichnet, bei dem weniger Knochen vorhanden ist als der Norm entspricht; es wird gleichgesetzt mit dem „zu wenig kalzifizierten Knochen" (ALBRIGHT u. REIFENSTEIN). Röntgenologisch läßt sich erst eine Verminderung der Knochenmenge um 30—50% erfassen, durch Biopsie kann die Diagnose sicherer gestaltet werden. In der Mehrzahl der Fälle ist es nicht möglich, ursächliche Beziehungen zwischen Osteopenie und hormonalen Störungen festzustellen.

Untersuchungen über die Epidemiologie von Frakturen unter besonderer Berücksichtigung der Häufigkeit von Schenkelhals- und distalen Unterarmfrakturen wurden vorgenommen. Jenseits des 50. Lebensjahres steigt die Zahl der Schenkelhalsfrakturen steil an, bei Frauen stärker als bei Männern, wobei im Alter Traumen seltener die auslösende Ursache darstellen. Bei Frauen über 40 Jahren nimmt die distale Vorderarmfraktur stark zu und ist im Alter von 60 Jahren etwa siebenmal häufiger als bei Männern.

Das Auftreten von Frakturen in einer vergleichbaren Bevölkerungsgruppe wird als das aufschlußreichste Kriterium für die Beurteilung von Altersveränderungen der Knochen angesehen. Es wird die Vermutung ausgesprochen, daß Osteopenie als ein Symptom, das durch verschiedene Veränderungen hervorgerufen sein kann (hormonale Störungen, Eiweißmangel, genetische Störungen, rheumatische Krankheiten), nicht unbedingt Ursache einer erhöhten Brüchigkeit der Knochen sein muß, und daß diese andererseits nicht notwendigerweise mit Osteopenie einhergeht.

Résumé

On appelle ostéopénie l'état clinique dans lequel la quantité d'os est inférieure à la normale; on applique aussi ce terme à "l'os trop peu calcifié" décrit par ALBRIGHT et REIFENSTEIN. L'examen radiologique ne révèle une diminution que si celle-ci est de 30—50% au moins; la biopsie permet un diagnostic plus sûr. Dans la plupart des cas, il est impossible de trouver un rapport de cause à effet entre une dysfonction hormonale et l'ostéopénie.

On a fait des recherches sur l'épidémiologie des fractures, en tenant plus particulièrement compte de la fréquence des fractures du col du fémur et des fractures distales de l'avant-bras. On a constaté que le nombre des fractures du col du fémur augmente considérablement après 50 ans, et dans une plus forte proportion chez les femmes que chez les hommes; des traumatismes sont plus rarement la cause de cet accident chez les vieillards. Le pourcentage des fractures distales de l'avant-bras augmente considérablement chez les femmes de plus de 40 ans; à 60 ans, il est environ 7 fois plus élevé que chez l'homme.

La proportion des fractures dans des groupes de population comparables est considérée comme le critère le plus concluant dont on dispose pour apprécier les altérations de l'os sous l'effet de l'âge. L'ostéopénie n'étant pas une maladie mais un symptôme de maladie dû à divers facteurs (déséquilibre hormonal, carence protéique, troubles génétiques, affections rhumatismales), on peut admettre qu'elle n'est pas forcément la cause d'une fragilité accrue de l'os, et que celle-ci à son tour n'est pas nécessairement en rapport avec une ostéopénie.

Acknowledgements

Financial support for this work was obtained from the Josiah Macy, Jr. Foundation, New York, N. Y., the Swedish Medical Research Council, and United States Public Health Service Grant D-1452.

Author's present address; Hospital for Special Surgery, 535 East 70 Street, New York 21, N. Y., U.S.A.

References

ALFFRAM, P.-A.: In preparation. 1962. — ALFFRAM, P.-A., and G. C. H. BAUER: J. Bone Surg. **44** — A, 105 (1962). — BARNETT, E., and B. E. C. NORDIN: Brit. J. Radiol. **34**, 683 (1961). — BAUER, G. C. H.: Clin. Orthop. (U.S.A.) **17**, 219 (1960).

Anabolic steroids in the treatment of osteopenia

By

J.-F. Dymling, B. Isaksson, and B. Sjögren

Albright and Reifenstein postulated that osteoporosis was caused by a humorally effected decrease in bone-matrix formation. This concept has never been proven, and has lately been challenged (Heaney and Whedon, Nordin, Dymling). The word "osteoporosis" is generally associated with this theory of aetiology, and it is for this reason that the term "osteopenia" has been introduced (Bauer, Carlsson, Lindquist). The meaning of osteopenia is "too little calcified bone" (Albright, Reifenstein).

The hormonal background of Albright's theory suggested that treatment with testosterone and comparable substances, such as the so-called anabolic steroids, might be effective. It is established that such treatment can induce a positive nitrogen balance in cases of osteopenia (Kochakian). In many instances retention of calcium or decreased loss of calcium has also been observed (Nowakowski). However, it is not self-evident that this effect should continue during long-term treatment. A decrease in the nitrogen-sparing effect of testosterone is common in metabolic studies in animals and man (Kochakian). A persistence of increased tissue anabolism has not been demonstrated during long-term treatment for osteopenia. Since roentgenological signs of healing have not been observed either, it is apparent that data on the metabolic effects of long-term treatment in cases of osteopenia are of particular interest.

Metabolic balance studies give a measure of the net difference between anabolism and catabolism. With the aid of bone-seeking isotopes the anabolism of calcium can be determined, and with a combination of the two techniques catabolism can be calculated. The isotope data permit estimation of accretion rate, exchangeable compartments, urinary clearance, and endogenous faecal clearance of calcium (Bauer, Wendeberg). Some of these parameters can be investigated in other ways. The accretion rate is the amount of calcium cleared from plasma and deposited in the skeleton per day. In normally mineralised bone this should be directly proportional to the bone-formation rate. Accretion rates have been studied in a number of cases with the aid of Ca^{47} and Sr^{85}. The excretory mechanisms of strontium differ from those of calcium, but accretion rates are comparable when tracer doses are used. This has been shown in simultaneous studies with the two isotopes.

Table 1. *Accretion rates of Ca in normal and osteopenic subjects*

	Number of observations	Accretion rates in litres of plasma cleared per day
Normal cases	24	4.6 ± 0.22
Idiopathic osteopenia.	32	3.4 ± 0.19
Rheumatic osteopenia	12	3.7 ± 0.35
Osteopenia post gastric resection.	12	5.1 ± 0.37

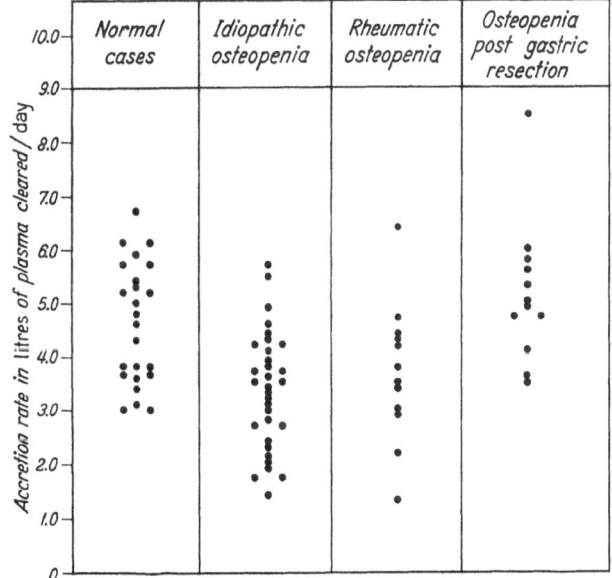

Fig. 1. Accretion rates in normal and osteopenic subjects

Accretion rates in normal and osteopenic subjects are shown in Fig. 1. The mean values are given in Table 1. The osteopenia was divided into three groups: osteopenia following gastric resection, osteopenia accompanying rheumatic disease, and idiopathic osteo-

Table 2. *Data on osteopenic subjects treated with Durabolin (25 mg. weekly) or Decadurabolin (25 mg. every third week)*

	Code No.	Age (years)	Sex	Weight (kg.)	Duration of treatment (months)	Isotope	Accretion rates (litre/plasma/day)		Exch. bone Ca (litre/plasma)		Urinary clearance (litre/plasma/day)		Faecal clearance (litre/plasma/day)	
							Untreated	Treated	Untreated	Treated	Untreated	Treated	Untreated	Treated
Durabolin	D — 61	49	M	81	6	Ca^{47}	2.3	1.9	25	20	1.3	2.3	1.6	2.0
	C — 149	65	M	70	6	Ca^{47}	6.0	4.5	31	24	0.6	6.0	2.1	2.4
	C — 133	67	M	70	11	Sr^{85}	2.7	4.3	23	23	2.5	2.6	2.4	1.5
	B — 44	68	M	85	7	Sr^{85}	3.1	2.8	19	22	5.9	7.3	3.1	2.3
	E — 7	19	F	46	6	Ca^{47}	3.9	5.8	21	18	2.2	4.3	1.2	2.5
	D — 96	52	F	55	6	Ca^{47}	1.7	3.1	17	15	1.1	1.0	1.6	2.0
	D — 73	54	F	87	6	Ca^{47}	5.5	3.4	22	22	2.9	1.2	0.4	0.6
	D — 43	58	F	65	22	Sr^{85}	3.4	2.5	23	26	5.2	6.5	2.7	3.4
	D — 25	59	F	95	14	Sr^{85}	5.7	3.1	25	26	7.7	8.6	2.2	2.3
	D — 40	59	F	52	7	Sr^{85}	3.8	3.5	26	24	5.3	5.3	1.9	2.6
	D — 31	65	F	76	15	Sr^{85}	3.9	3.8	17	19	4.0	2.9	1.1	1.3
Decadurabolin	D — 30	36	F	42	12	Sr^{85}	2.2	2.1	16	15	4.5	5.7	2.0	2.1
	D — 33	55	F	42	15	Sr^{85}	2.2	2.1	12	10	2.1	4.6	2.3	2.6
	D — 65	55	F	55	15	Sr^{85}	2.1	4.6	12	16	7.0	1.6	3.1	1.4

penia. Since the determinations were made with both Sr^{85} and Ca^{47} the values are given as litres of plasma cleared per day.

The accretion rates were about 30% lower than normal in idiopathic and rheumatic osteopenia, which is a significant difference. However, the loss of skeletal mass in the cases of osteopenia probably exceeds 30%. This is because we must rely on x-ray for the clinical diagnosis, since there is no way by which to estimate skeletal mass or total body calcium *in vivo*, and a 30% decrease in skeletal mass can hardly be diagnosed radiologically. Consequently the accretion rate per unit mass of bone was not found to be decreased in osteopenia. This does not accord with the earlier concept of the aetiology of this condition.

The next step in the investigation was to determine the effects of long-term treatment on skeletal anabolism. In 14 patients with osteopenia the accretion rates and exchangeable spaces were measured. Treatment was then begun with 19-nortestosterone-phenylpropionate (25 mg./week) or with 19-nortestosterone-decanoate (25 mg. every third week) and continued until a second study had been performed 6—22 months later[1].

Twelve patients improved subjectively during treatment. Cases C-133 and D-30 noted no change (Table 2). All the women developed some hirsutism, and in 3 of them the voice changed to a low and coarse pitch. In D-40 this change in voice was only slight, but in E-7 and D-96 it proved troublesome. These latter two have been under observation for 11 and 13 months, respectively, since the end of treatment, but no improvement has occurred. E-7 developed amenorrhoea during treatment.

The accretion rates showed no change and the exchangeable spaces remained unaltered (Figs 2 and 3 and Table 2). This would appear to indicate that long-term treatment with 19-nortestosterone-phenylpropionate or 19-nortestosterone-decanoate has no effect on the bone-formation rate. However, the patients did improve during treatment — a fact which might suggest another mechanism such as a decrease in bone resorption, i.e. an anti-catabolic effect of the steroids. As already pointed out, this can be investigated by the simultaneous isotope and stable balance technique.

At the metabolic unit of Sahlgrenska Sjukhuset in Göteborg a certain number of osteopenic patients were studied under treatment with anabolic steroids. Repeated metabolic balances and sometimes accretion-rate measurements were performed in the

---

[1] The steroids were generously supplied by N. V. Organon (Netherlands) through Pharmacia, Sweden.

same patient. Between the periods they spent in the hospital the patients were kept on the same regimen at home. The question

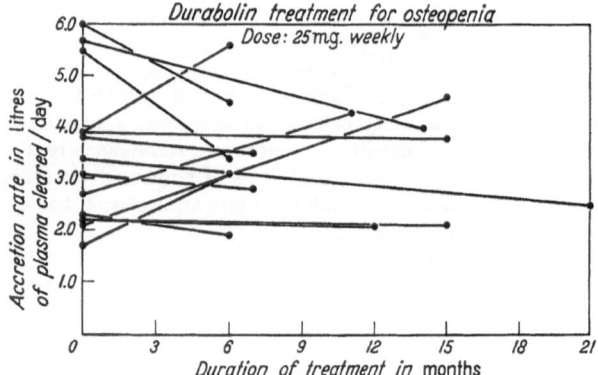

Fig. 2. The effect of long-term treatment with Durabolin on accretion rates in osteopenic subjects

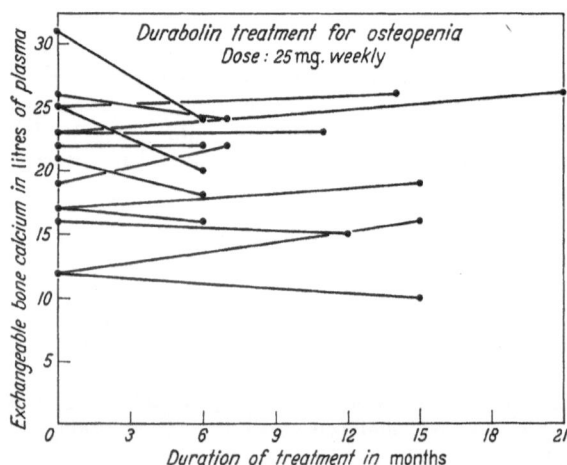

Fig. 3. The effect of long-term treatment with Durabolin on exchangeable bone calcium in osteopenic subjects

with which we were concerned was whether an initial effect on the nitrogen, calcium, and phosphorus balances was maintained throughout months and possibly years.

The first case to be considered (I.E.) developed osteopenia in the course of several years' treatment with cortisone. The patient was kept under observation for 16 months, during which time three separate metabolic periods, totalling 112 days, were studied.

Case I. E., female, aged 55; suffering from rheumatoid arthritis for the past 11 years; prednisolone therapy (up to 7.5 mg. daily) since June 1957. X-ray examination in October 1960 showed osteopenia with several vertebral compressions. There were few subjective symptoms. The patient was treated with anabolic steroids for several years before the first metabolic study was started in November 1960 (*Period A, 32 days*). She was then treated at home until November 1961, when the next study took place (*Period B, 32 days*). Thereafter she spent less than a month at home and was admitted for the third study in January 1962 (*Period C, 48 days*). During the entire study the patient remained up and about and was treated with prednisolone (7.5 mg. daily). 19-nortestosterone-phenylpropionate was given in doses of 50 mg. once a week during Period A and every second week between Period A and B as well as during Period B. The last injection was administered on the 23rd day of Period B, after which no further injections were given. Thus, Period C started 35 days after the last dose of anabolic steroid.

The protein and calcium intake was moderate during Period A. A supplement of 1.3 g. calcium daily, in the form of calcium lactate tablets, was introduced after Period A and maintained throughout the whole study (Table 3).

Table 3. *Case I. E.: Treatment and balance data (the balances are given as totals for each of the Periods A, B, and C, respectively)*

| Period | Days | Treatment | | | Balance | | | |
		Predniso-lone mg./day	Dura-bolin mg./day	Protein g./kg./day	Cal-cium mg./kg./day	Nitro-gen g.	Calcium g.	Phos-phorus g.	Potas-sium mEq.
A	32	2.5 × 3	50/ 7	1.7	20	+ 52	+ 2.8	+ 2.5	+ 204
At home	340	2.5 × 3	50/14	(1.5)	(50)				
B	32	2.5 × 3	50/14[1]	1.5	50	+ 46	— 5.6	— 3.1	+ 247
At home	24	2.5 × 3	0	(1.5)	(50)				
C	48	2.5 × 3	0	1.5	50	+ 3	— 5.3	—11.5	— 294

[1] on the 9th and 23rd day.

Case L. E., female, aged 49; the patient had been under treatment with triamcinolone for bronchial asthma since 1956. She developed severe osteopenia with back pain and had been treated with anabolic steroids since September 1960. A first 64-day metabolic study was undertaken in April 1961 (Table 4).

Case S. D., female, aged 67; bedridden for more than a year owing to senile osteopenia. She was treated with 19-nortestosterone-phenylpropionate during this period and was admitted to hospital for a metabolic study in March 1961 (68 days).

Table 4. *Cases L. E. and S. D.: Treatment and balance data (the balances are given as totals for each of the consecutive Periods A—1, A—2, andA—3)*

Period	Days	Treatment				Balance			
		Corticoid mg./day	Dura-bolin mg./day	Protein g./kg./day	Cal-cium mg./kg./day	Nitro-gen g.	Calcium g.	Phos-phorus g.	Potas-sium mEq.
L. E., 49 years ♀		Triam-cinolone							
A—1	24	4 × 2	0	1.0	20	— 24	+ 5.2	+ 0.4	
A—2	12	4 × 2	1/7	1.0	20	+ 6	+ 2.1	+ 0.7	
A—3	28	4 × 2	1/7	1.0	45	+ 17	+ 9.1	+ 4.9	
S. D., 67 years ♀									
A—1	36	0	1/7	1.5	16	+ 36	— 0.1	+ 1.3	+ 180
A—2	16	0	1/7	2.0	40	+ 43	+ 8.5	+ 6.0	+ 256
A—3	16	0	1/1	2.0	40	+ 61	+ 5.7	+ 4.8	+ 255

The patients were kept on a stable diet prepared in the ward kitchen and regularly analysed. Stools were collected in four-day periods and the urine in daily samples. The metabolic balance studies comprised nitrogen, calcium, phosphorus, potassium, and sometimes also fat. The chemical methods employed will be described elsewhere. Total exchangeable potassium was determined by a dilution technique using the isotope K^{42}.

The results obtained are listed in Tables 3—5 and Figs 4—5. Only the findings for Case I. E. will be discussed in detail:

A positive *nitrogen balance*, averaging 1.5 g. daily was found in Periods A and B. The patient showed a zero balance during Period C.

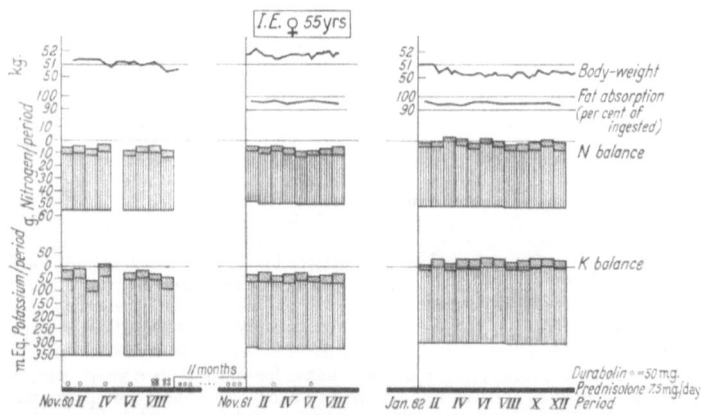

Fig. 4. Case I. E.: Nitrogen and potassium balances during the three different studies (Periods A, B, and C). Each column represents a 4-day period (Roman numerals) ▥: urine ▨: faeces

The *calcium* and *phosphorus balances* were positive in Period A (+ 88 and + 78 mg. daily, respectively), but negative in the two following Periods (B: — 175 mg. and — 100 mg.; C: — 110 mg. and — 240 mg. daily, respectively).

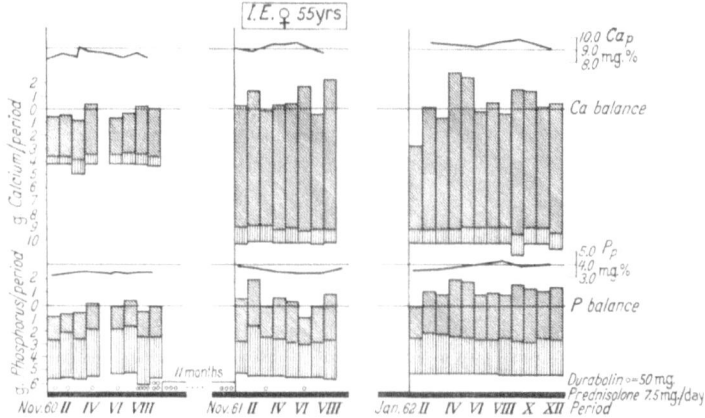

Fig. 5. Case I. E.: Calcium and phosphorus balances and plasma levels (Periods A, B, and C). Each column represents a 4-day period (Roman numerals) ▨ : urine ▩: faeces

The patient showed a positive *potassium balance* during Periods A and B (+ 10 and + 8 mEq. daily, respectively), whereas during Period C the balance was negative (— 6 mEq. daily).

When the *fat absorption index* was tested, it was found to be within normal limits.

The average values for *serum calcium* and *serum phosphorus* were a little higher in Period B than in Period A, and showed a further slight rise in Period C. The *serum alkaline phosphatase* levels were normal.

Total exchangeable potassium was not measured until at the end of Period B (1,635 mEq.). At the end of Period C the value recorded was 1,540 mEq., and fourteen days later 1,450 mEq.

The *accretion rate* was determined with Sr^{85}. It amounted to 7.4, 6.7, and 7.5 litres of plasma cleared per day in Periods A, B, and C, respectively. This corresponds to 630, 600, and 670 mg. calcium per day. From the average calcium balance in each Period the resorption of calcium from the skeleton was calculated at 540, 770, and 780 mg. daily (Table 5).

Table 5. *Case I. E.: Rate of calcium resorption from bone calculated from accretion rates and balance data*

Period	Accretion mg./day	Balance mg./day	Resorption mg./day
A	630	+ 90	540
B	600	— 170	770
C	670	— 110	780

The *body-weight* remained stable during and between Periods A and B. A weight reduction of 1.5 kg. was noted between Periods B and C. The daily caloric intake was kept at between 2,000—2,100 throughout these periods, except for the first half of Period C, when it amounted to 1,825 calories daily.

In case I. E. there was a marked difference in the retention of nitrogen during treatment with 19-nortestosterone-phenylpropionate (Periods A and B) as compared to Period C without such treatment. It should also be borne in mind that 19-nortestosterone-phenylpropionate had been given for more than a year prior to Period A. It appears therefore that the protein-sparing effect of 19-nortestosterone-phenylpropionate continued during prolonged administration of the steroid.

In Period A the calcium balance was only slightly positive. The duration of this positive balance is not known, but one year later — during Period B — the calcium balance was negative. In fact, 19-nortestosterone-phenylpropionate appeared to have no effect on calcium metabolism, as discontinuation of the steroid was not followed by any change in the parameters of calcium metabolism (Tables 3 and 5). It should be noted that calcium was given in abundance and that extra vitamin D was also supplied. Malabsorption was not present.

The potassium balance paralleled that of nitrogen in Periods A and B. In Period C there was a loss of potassium not accounted for by any loss of nitrogen and amounting to about 20% of the total exchangeable potassium. It should also be noted that there was an appreciable loss of phosphorus in Period C, which could not be explained by reference to the nitrogen and calcium balances.

The retention of nitrogen was of the same magnitude in both Periods A and B. It is impossible, however, to estimate the true gain of protein. The loss of nitrogen from the skin may amount to $1/2$ g. a day or more and cannot be neglected. The true nitrogen balance is thus less positive than is indicated in Table 3. The potassium balance must also be less positive than is apparent from the results. The unchanged weight during Periods A and B, as well as between Periods A and B, would also appear to indicate an unchanged cell mass during treatment with 19-nortestosterone-phenylpropionate. It seems that a reliable measure of tissue synthesis can scarcely be obtained by the use of balance techniques alone. Experience with repeated determinations of total exchangeable potassium as a parameter of cell mass suggests that such measurements may be of value in interpreting balance data during prolonged metabolic studies. In this study, however, exchangeable potassium was measured only from the end of Period B.

Calcium and phosphorus are also lost from the skin, and the calcium balance must be less positive in Period A and more negative in Periods B and C than is suggested by Table 5. It must nevertheless be concluded that the treatment given during Periods B and C was unable to prevent a continued loss of bone minerals. Nor do the difficulties involved in interpreting balance data invalidate the conclusion that 19-nortestosterone-phenylpropionate had a protein-sparing effect during the entire period of its administration. It seems impossible, however, to determine the degree of protein gain. Nor do we know how much protein would have been lost had no anabolic steroid been administered.

In conformity with results previously discussed, the accretion of calcium remained unchanged during Periods A, B, and C, irrespective of the administration of 19-nortestosterone-phenylpropionate.

The dissociation between the nitrogen and calcium balance found in Case I. E. was also observed in Case L. E. during treatment with 19-nortestosterone-phenylpropionate (Table 4). In the latter case the nitrogen balance became positive, although there was no improvement in the already positive calcium balance. An extra supply of calcium in Period A-2 increased the retention of calcium as well as of phosphorus. Similarly, in Case S. D. retention of nitrogen occurred during administration of 19-nortestosterone-phenylpropionate without retention of calcium, but an extra supply of calcium induced a positive calcium and phosphorus balance.

Long-term treatment for osteopenia with 19-nortestosterone-phenylpropionate and 19-nortestosterone-decanoate has been shown to have no effect on the rate of bone formation. Additional data suggest that bone resorption is also unaffected. Assuming that osteopenia is a reversible process at the stage when a clinical diagnosis can be made, our observations would not appear to favour long-term treatment with these anabolic steroids.

Summary

It is suggested that the term "osteopenia" (= "too little calcified bone") is preferable to that of "osteoporosis". By means of metabolic balance studies and the use of bone-seeking isotopes, the anabolism and catabolism of calcium was investigated. The accretion rate is defined as the amount of calcium cleared from plasma and deposited in the skeleton per day. Studies with Ca^{47} and Sr^{85} revealed that, in patients with idiopathic and rheumatic osteopenia, there was a decrease in the deposition of calcium in the bones in terms of clearance of plasma calcium but not per unit mass of bone.

In response to treatment with anabolic steroids (19-nortestosterone-phenylpropionate or 19-nortestosterone-decanoate) for 6—22 months, a subjective improvement was observed, but no change in the calcium accretion rate. When prednisone and 19-nortestosterone-phenylpropionate were administered together, the latter continued to exert a protein-sparing effect, but there was no increase in the accretion rate nor any decrease in bone resorption. These findings would appear to cast doubt on the value of the above-mentioned steroids in the treatment of osteopenia.

Zusammenfassung

Anstelle von Osteoporose wird die Bezeichnung Osteopenie vorgeschlagen und dadurch der „zu wenig kalzifizierte Knochen" charakterisiert. Durch Bilanzversuche und mit Hilfe knochenaffiner radioaktiver Isotopen werden Anabolismus und Katabolismus von Calcium untersucht. Die sich daraus ergebende Zuwachsrate von Calcium ist definiert durch die Menge von Calcium, die pro Tag aus dem Plasma verschwindet und im Knochen fixiert wird. Untersuchungen mit Ca^{47} und Sr^{85} ergaben verminderte Zuwachsraten von Calcium im Knochen bei idiopathischer und rheumatischer Osteopenie bezogen auf die Clearance von Plasma-Calcium, aber nicht bezogen auf die gesamte Knochenmenge.

Durch Anwendung anaboler Steroide (19-Nortestosteron-phenylpropionat oder -decanoat) während 6—22 Monaten ließ sich zwar eine subjektive Besserung der Beschwerden erzielen, aber keine Änderung der Calcium-Zuwachsrate des Knochens. Bei gleichzeitiger Anwendung von Prednison und von 19-Nortestosteron-phenylpropionat fand sich wohl für das anabole Steroid ein eiweißsparender Effekt, jedoch weder eine erhöhte Zuwachsrate für Calcium noch eine verminderte Knochenresorption. Auf Grund dieser Befunde wird die Zweckmäßigkeit bezweifelt, die genannten Steroide für die Behandlung der Osteopenie zu verwenden.

Résumé

Les auteurs proposent d'utiliser le terme d'ostéopénie plutôt que celui d'ostéoporose pour caractériser «l'os trop peu calcifié». Ils étudient l'anabolisme et le catabolisme du calcium par la méthode des bilans et l'emploi de radio-isotopes présentant une affinité pour l'os. Le taux de fixation du calcium est défini comme étant la quantité quotidienne de calcium disparaissant du plasma et déposée dans le squelette. Des recherches faites à l'aide du Ca^{47} et du Sr^{85} ont montré que chez les sujets atteints d'ostéopénie idiopathique et rhumatismale, le dépôt de calcium dans les os diminue si l'on s'en réfère à la disparition du calcium du plasma mais non pas si l'on s'en réfère à la quantité totale d'os.

L'administration de stéroïdes anabolisants (phénylpropionate ou décanoate de 19-nortestostérone) pendant 6 à 22 mois a permis d'obtenir une amélioration subjective, mais pas de modification du taux de fixation du calcium. L'administration simultanée de prednisone et de phénylpropionate de 19-nortestostérone a permis de réduire la perte de protéines, mais n'a donné ni accroissement du taux de fixation du calcium ni diminution de la résorption osseuse. A la suite de ces observations, l'auteur met en doute la valeur de ces stéroïdes dans le traitement de l'ostéopénie.

Authors' addresses:

J.-F. Dymling: Orthopaedic Research Laboratories, Allmänna Sjukhuset, Malmö. Supported by grants to G. C. H. Bauer from the Josiah Macy, Jr.

Foundation, New York, N. Y., the Swedish Medical Research Council, and United States Public Health Service Grant D-1452.

B. ISAKSSON and B. SJÖGREN: The Metabolic Unit, Medical Clinic II, Sahlgrenska Sjukhuset, Göteborg. Supported by grants from the Swedish Medical Research Council.

References

ALBRIGHT, F., and E. C. REIFENSTEIN: Parathyroid glands and metabolic bone disease. Williams and Wilkins, 1948. — BAUER, G. C. H.: Kinetics of calcium and strontium metabolism in man. Bone as a Tissue. McGraw Hill, 1960, p. 118. — BAUER, G. C. H., A. CARLSSON: and, B. LINDQUIST: Metabolism and homeostatic function of bone. Mineral Metabolism I. Academic Press Inc., 1961. — DYMLING, J.-F.: Accretion and excretory clearance rates and exchangeable spaces measured with Ca[47] and Sr[85] under normal and pathological conditions. Technical Review Series, 1962. International Atomic Energy Commission, Vienna (in press). — HEANEY, R. P., and G. D. WHEDON: J. Clin. Endocr. (U.S.A.) 18, 1246 (1958). — KOCHAKIAN, C. D.: Vitamins and Horm. (U.S.A.) 4, 256 (1946). — MITCHELL, H. H., and T. S. HAMILTON: J. Biol. Chem. (U.S.A.) 178, 345 (1949). — NORDIN, B. E. C.: Amer. J. Clin. Nutrit. 10, 384 (1962). — NOWAKOWSKI, H.: Acta endocr. (Den.) 63, Suppl., 37 (1962). — WENDEBERG, B.: Kinetics of Ca[47] and Sr[85] in man. Technical Review Series, 1962. International Atomic Energy Commission, Vienna (in press).

Discussion

QUERIDO: I remember that Dr. HOFFENBERG wanted to show two slides in relation to this study.

HOFFENBERG: We thought it would be of interest to study albumin pool sizes in osteoporotic as opposed to normal subjects and to see whether anabolic agents affected these pool sizes. Our study is very limited so far. We started with 8 patients: 2 dropped out for technical reasons, 1 experienced increased angina pectoris while taking the anabolic agent, and a fourth developed a marked increase in libido. In case anyone should ask why the last patient dropped out, I wish to point out that she was a 76-year-old widow!

We performed an initial study with I^{131}-albumin, administered methandrostenolone for a month, and then repeated the study. The dose used was 5 mg. twice daily, which was a rather large dose, admittedly, but we wished to use a dose that was likely to have an effect.

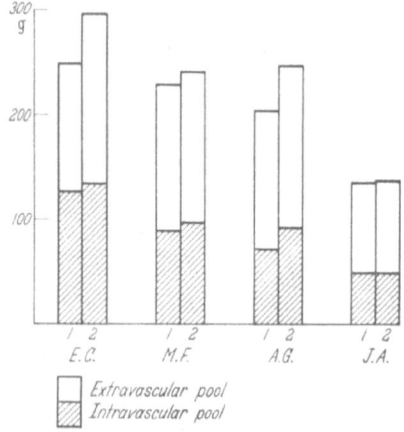

Fig. 1. Intra- and extravascular pool sizes before (1) and after (2) anabolic therapy

The results in the 4 patients studied are shown in Fig. 1. Patient E. C. was a male who had been castrated surgically for reasons which are obscure. Patients M. F. and A. G. have so-called "senile osteoporosis", and J. A. is a teen-age girl with severe rheumatoid arthritis. The block numbered 1 represents albumin pool size before therapy, and that numbered 2 follows a month of treatment. In 3 of the 4 patients there is an increase in total albumin pool, which is perhaps more marked in the extravascular compartment. Patient J. A. showed no change. Table 1 shows the detailed results, and it will be seen that weight changes were not constant. Plasma volume increased in all patients, whether expressed in absolute terms or in terms of body-weight. The plasma, total and extravascular pools have been demonstrated graphically in Fig. 1, and turnover rates did not show any consistent alteration at the end of a month of treatment.

Table 1. *Extra- and intravascular albumin pool sizes and turnover rate before (1) and after (2) 4 weeks' therapy with 10 mg. Dianabol daily*

		Weight	Plasma albumin conc.	Plasma volume		Plasma albumin pool		Total albumin pool		Extra-vascular pool		Turnover rate g./day
		kg.	g.%	ml.	ml./kg.	g.	g./kg.	g.	g./kg.	g.	g./kg.	g./day per kg.
E. C.	1	70.7	4.35	2896	41.0	126.0	1.78	249.0	3.51	123.0	1.73	8.4 0.12
	2	70.6	4.29	3133	44.4	134.4	1.91	296.4	4.21	162.0	2.30	7.7 0.11
M. F.	1	62.3	3.74	2373	38.1	88.8	1.43	228.3	3.67	139.5	2.24	7.4 0.12
	2	59.2	3.50	2755	46.5	96.4	1.63	240.6	4.07	144.2	2.44	9.6 0.16
A. G.	1	50.5	4.29	1659	32.9	71.2	1.41	204.7	4.05	133.5	2.64	7.1 0.14
	2	54.1	4.42	2065	38.2	91.3	1.69	246.3	4.56	155.0	2.87	7.4 0.14
J. A.	1	21.1	4.62	1058	50.1	48.7	2.31	134.8	6.39	86.1	4.08	3.2 0.15
	2	21.1	3.92	1257	59.6	49.3	2.34	136.7	6.48	87.4	4.14	3.4 0.16

MCCANCE: I would like to ask one question, or perhaps two. First of all, does anybody know what the effect of these anabolic steroids may be on the citric-acid metabolism of bone? I think this would be an interesting study, because the citric-acid metabolism of bone is affected by both Parathormone and vitamin D.

Secondly, is anything known about the calcium intake of the old ladies in Malmö who fracture the necks of their femurs so easily?

Lastly, I can add something to the intellectual pool about the effect of nutrition on the brittleness of bones. The bones of our animals, which have been maintained in a very undernourished state for a long time, are extremely brittle. They may break up as they are being cleaned, and I have tried to investigate this by measuring the breaking stress and other mechanical properties in the standard ways, but the results do not give me the information I want. I suspect myself that the brittleness is due to overcalcification of the cortex.

BAUER: The calcium intake is not known in these elderly women in Malmö. We do know, however, from our studies that skeletal fragility starts at the very early age of perhaps 30, and I believe that women in Malmö at this age, and for several decades afterwards, enjoy a diet which is rather richer in proteins and calcium than that of most women on earth. It seems that even in these elderly patients the turnover of the skeleton has certainly not stopped. DYMLING's findings are evidence for this, as is also the fact that, following fracture of the neck of the femur, the repair reaction as measured with radioisotopes starts very readily and doesn't seem to differ drastically from that observed in younger persons.

DYMLING: I have no direct answer to Dr. MCCANCE's question about the brittleness of bones. In all skeletal states in humans in which you see increased density on the x-ray plate, it is well known that these bones are brittle and fracture easily. Even so, at osteotomy these bones are very hard to saw through. I may suggest, however, that this it not entirely comparable with osteopenia, where the mineral content is normal.

As to Dr. HOFFENBERG's data, I would also like to show a case in this connection (Table 2). This is a patient treated with large doses of albumin intravenously over some period of time. She developed an abnormally high concentration of albumin in the plasma, but the accretion rate did not change.

Table 2. *Accretion rates and total body retention of Ca[47] before and after treatment with albumin*

Accretion rates per litre of plasma cleared daily	Total body retention in % of dose	
	5 days	10 days
3.1	66.9	58.0
2.5	81.2	73.5

Thus, even when the albumin concentration was raised markedly and when calcium was retained in the albumin fraction to an increased extent (on the 10th day the retention of Ca[47] was 58% and 74% respectively), the accretion rate did not change significantly.

Introduction to the General Discussion:
Some endocrine aspects of growth and bone

By

T. R. C. FRASER

I am sure it is the wish of all of you that my introduction to this discussion should be very brief. This afternoon we have been considering the effects of anabolic agents in man, a clinician's problem. We have been concerned on the one hand with growth and on the other hand with bones. Perhaps these are not so very different, since the growth experts have been measuring the skeletal height and skeletal age. I would like to go over some of the problems they have raised, considering growth first and the bones second, and at the same time to comment a little on some of our own experience.

Perhaps the aspects are really not so different as may seem at first sight; for Dr. BAUER's osteopenia may well be a problem of growth, if we are concerned with curing it rather than its cause. Obviously, if you have got thin bone, the clinician's problem is much more how to make bone grow than what was the cause of its osteopenia.

Now, for dwarfism from the clinician's point of view, and for anabolic agents, the standards of growth needed from the growth experts are standards which will help us to assess the efficacy or otherwise of various treatments. And listening to Dr. TANNER I still rather wished he would alter some of his standards. In the first place, I didn't like being asked to deal with derived data. When we measure a child, we measure its height; and it seems a nicer thing to set our assessments against standards in terms of what we measure rather than in terms of something inferred from it. This should still be possible, if he would only introduce a correction factor for puberty. This seems to me to be his real problem; puberty comes at an uncertain time and has a definite influence on the growth process. It would seem the simplest thing if the growth experts would devise two sets of tables, one giving height against chronological age for pre-pubertal subjects who have not yet had the obvious clinical signs of puberty, and the other giving height against years after the onset of puberty for people in whom these obvious clinical signs of puberty have appeared (perhaps the onset of periods or the change of the voice). If we looked at these two types of chart, rather than at complexly derived ones where we measure against skeletal age, I really think that original measurements (height in cms) would stand adequately against the norms.

Dr. TANNER mentioned that his accuracy of measurement was 4 mm.; if we were going to get an accuracy based on 2 measurements, I think that makes a total error of 6 mm. per measurement. This is not very suited to short time-intervals. He has already mentioned the snag about the seasonal effect.

Nevertheless, we are very grateful to him for looking into the correlations that exist between bone-age standards and height standards. To the ordinary person, it seems better to see the skeletal age charted in one graph and the height age in another graph against some sort of normal standard, rather than a compound chart of the two. Further, I would like to suggest that we should not regard bone age as a unitary concept. Surely, just as in the case of height, we must divide this into a pre-puberty and a post-puberty phase, but by a different procedure. From all we know about the effects of endocrines on growth, the closure of the epiphyses is a very different phenomenon from the maturation of the shape and the length of the bones. It should be very easy to take either the GREULICH and PYLE atlas or similar standards and to divide their criteria into those which are derived from epiphyseal closures, and which could give an epiphyseal-closure age, and the other ones which could give a bone-maturation age.

A patient with acromegaly is surely one of the best examples we have of the effects of a simple excess of growth hormone; in contrast with his state after the acromegaly has been corrected, we notice very striking changes in aspects of growth other than the height of the individual or his bones: there are obvious changes in the skin, for example. We have also used the height of a hypopituitary subject to assess the effects of various treatments. This underlines the importance that Dr. TANNER has ascribed to accuracy in measuring height, and also the need for a good period of observed growth before the treatments.

Now we should consider the bones for a moment. Obviously we all want to know not what happens to this somewhat theoretical accretion rate, or to the calcium balance, but whether our treatments cause the bones to get thicker and to break less easily. We have been hopeful about using a densitometry procedure for assessing standardised radiographs of the ulna immersed in water with the plate held at a distance (DOYLE, 1962). On the radiograph (Fig. 1) the density is compared against that of the aluminium step wedge. We have found, for example, when Cushing's syndrome in the child is

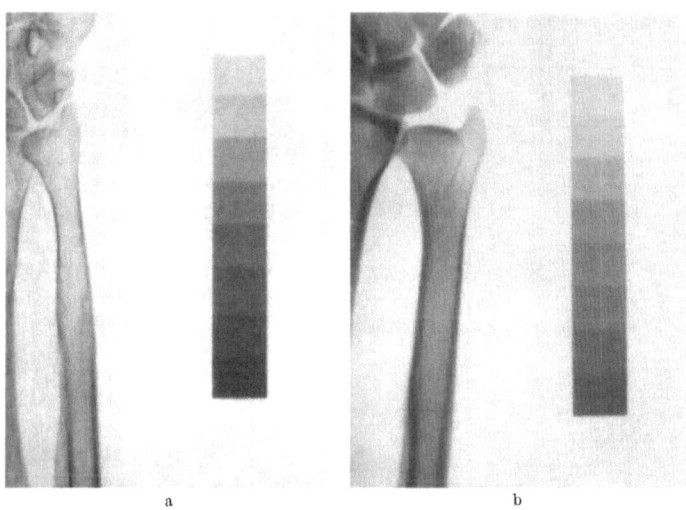

a   b

Fig. 1. X-rays for bone densitometry by the method of DOYLE (1961) — (a) an osteoporotic subject, (b) a normal subject. The films are taken with the arm held in water and the plate 12″ further behind. Subsequently, densitometric scanning along the long axis of the ulna can give charts such as those from another patient shown in Fig. 2

treated effectively, that not only is growth resumed but also the bone regains density (Fig. 2). It is a well-known fact that this doesn't occur in adults; possibly it is the restored ability to grow that enables these children to thicken their bones when their Cushing's syndrome is corrected. The more we study the problem of osteopenia in our hospital, the more we wonder whether the problem is not one of how to replace

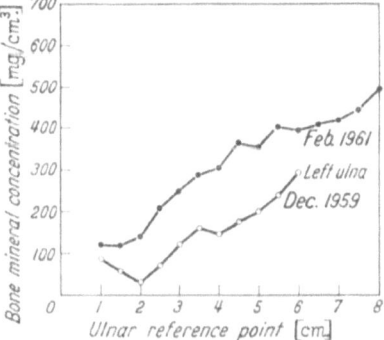

Fig. 2. Densitometric scans of the ulna x-rays of a young male subject aged 12 with Cushing's syndrome before and after its treatment by pituitary implant of Au-198 (May—July 1959). Note the initial thin or osteoporotic bone and the increased density after treatment

bone atrophy by regrowth, rather than how to find and correct
its causes.

Here it may be interesting to mention that, when we gave some
hypophysectomised rats growth hormone and followed their kidney-
tissue citrate, we found that
the previously high levels
fell along with the opposite
effect on the urinary excretion
of calcium (KARAM et al.,
1961). One of the well-known
effects of growth hormone is
to increase the urinary excretion of calcium; perhaps
it does this by lowering renal
citrate and so causing an increased reabsorption of calcium. On the other hand, this
is not the only effect that
growth hormone has on bones,
as indicated by these x-ray
pictures from an adult acromegalic, one showing the untreated state, and the other
after implantation in the pituitary had treated the acromegaly. It is of some interest
to notice the considerable
thinning of the cortex of the
bone which resulted. It does
seem therefore that here we
have, as with most things, a
mixed effect from the growth
hormone. It seems to increase
the amount of cortical bone, although if given in great excess it
may, of course, produce hypercalciuria and osteoporosis.

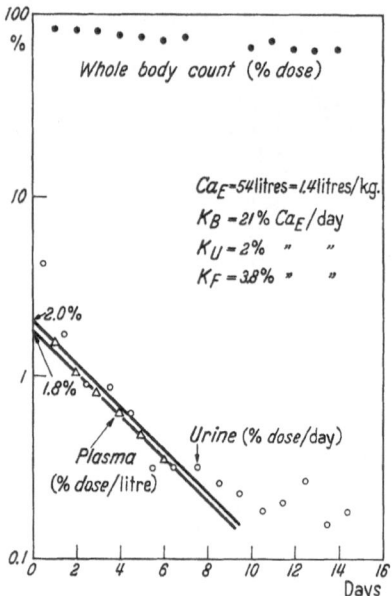

Fig. 3. Ca⁴⁷ measurements following an i.v.
dose given to a female subject aged 75 with
osteoporosis and thyrotoxicosis. Note how the
initial exponential fall in serum and urine concentrations becomes slower in the second week,
owing to a return of Ca⁴⁷ from deeper bone
(possibly due to resorption or implying a slow
exchange process)

May I also make a comment on the inferences we can draw from
measurements of the bone's accretion rate? We have very slender
evidence for inferring that these measure the rate at which bone is
formed. If we examine the Ca⁴⁷ measurements from a patient with
idiopathic osteopenia, we see the expected decline in the urinary
and plasma concentrations from which the accretion rate is
calculated. But if we take them from another patient, a thyro-
toxic osteoporotic, in whom the bone is turning over very much more
rapidly, there is not the same indication of a steady transfer rate

being measured (Fig. 3). I think we must regard this measurement of accretion rate as compounded of two things: an exchange into the deeper bone, and a true formation of new bone. At the moment I don't believe we can distinguish between the two in this test.

Finally, I would just like to come back again to emphasise the effect of age in this problem. Fig. 4 shows us some data from rats; we have been measuring their avidity for strontium by their 24-hour strontium space, which gives a measure equivalent to the accretion rate. There are measurements both from normal rats at various ages and also from rats which developed osteoporosis by being fed a calcium-deficient diet. You will see how strikingly both the normal avidity for calcium and the excessive avidity of calcium deficiency both decline with age, so that by the time they

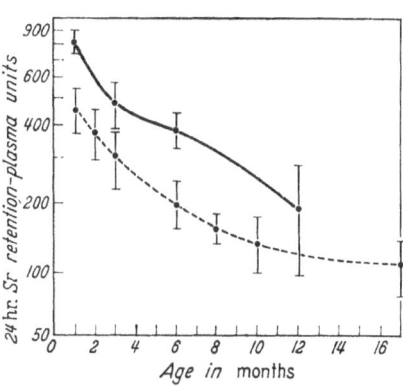

Fig. 4. Measurements of the 24-hour strontium space, which indicates bone avidity for calcium, from two groups of rats of various ages. Lower curve: normal rats; upper curve: rats fed a calcium-deficient diet. Note the greater avidity of the latter group and that both decline with age

come to old age you cannot distinguish one from the other. I think this is probably underlining a lesson as regards treatment for the osteopenic subject, who is so often elderly. How can we stimulate the growth of aged bone?

References

Doyle, F. E.: Brit. J. Radiol. **34**, 698 (1961). — Karam, J., M. T. Harrison, M. Hartog, and R. Fraser: Clin. Sc. (G.B.) **21**, 265 (1961).

General Discussion

TANNER: I'm afraid I feel impelled to disagree with nearly every particular of what Dr. Fraser said. I think his remarks are retrogressive and, if I may say so, quite unfortunate.

Firstly, he doesn't like what he calls "derived" data. Well, I don't think there is anything very "derived", or even particularly complicated, about the subtraction of one measurement from another. After all, many of you deal routinely in exponential functions and things like that. You saw the two slides of the hypothyroid boy (Figs 5 and 6, p. 369) — one on the distance chart and the other on the velocity chart. I can leave you to decide which gives you the more information.

Secondly, Dr. Fraser would like to have us give you one particular sign — yes or no — signifying puberty. For this, he chooses for the girl menarche, which of course occurs as the last thing when the rest of puberty is over; and for the boy, change of voice, which is not accurately measurable. Neither of these points would commend themselves to the people who study these things; indeed, change of voice is one of the worst signs of puberty to use in the boy. Worse than this, however, is the general approach, i.e. trying to divide the children, as he suggests, into two groups of pre- and of post-puberty; this is like dividing people into those who have got a high haemoglobin and those who have got a low haemoglobin. Nowadays we think it better practice to report the actual grammes of haemoglobin, because the haemoglobin concentration follows a continuous distribution. Of course, it is extremely simple — but would possibly be regarded as "derived" — to use skeletal age instead of chronological age on the growth standard, and that is what we do routinely. This does precisely what Dr. FRASER wants, that is, it allows for whereabouts in puberty the child is. Admittedly, you have got to use skeletal age instead of chronological age, but I don't think this is very complicated, either conceptually or in practice.

Thirdly, I also don't agree that closure of the epiphyses is in principle different from the other stages in skeletal maturation. The stages of skeletal growth in normal children progress in quite a smooth continuous manner and not by fits and starts. Under pathological conditions the child may halt at any one of the stages, not necessarily the last stage of fusion.

Lastly, I would like to add emphasis to what Dr. VAN DER WERFF TEN BOSCH said about the theoretical and practical difficulty of using so-called "height age" and "weight age". These measures really *are* derived, and they are derived in quite a fallacious way. I know they are much used by clinicians, but I am sure they shouldn't be. The reason for this is quite simple and straightforward. Advanced height age merely means that the child is tall for his age. Now if a child is tall for his age, it means one of two things: either that the child is actually advanced towards his final adult height — and this degree of advancement is what you are trying to measure — or merely that the child is a tall child of tall parents. There is a very high correlation, in fact nearly 0.8, between the height of a child at age 5, say, and his final height, and there is a correlation of about 0.6 between the mid-parental height and the child's height at this same age. Thus, in height age you are more influenced by the height of the parents than you are by the advancement of the child. This does not apply to the skeletal age, because everybody ends up with the same skeletal age as a consequence of the way it is assessed. But I do think that in height age and weight age we have the case where a derived measurement *has* been misleading. In other words, there are derivations and derivations!

BAUER: According to Dr. FRASER bone densitometry *in vivo* is not as difficult a problem as I suggested in my paper, whereas, on the other hand, the data on bone-formation rates presented here by DYMLING and me are based on "very slender evidence". Such a difference in opinion concerning basic parameters of metabolic bone disease must seem confusing to most members of this interdisciplinary Symposium. Perhaps reference to two recent publications[1, 2] will help to resolve this confusion more than would continued discussion here. It gives me a certain pleasure to point out, however, that Dr. FRASER's statement that bone is turning over more

---

[1] Radioisotopes and Bone: CIOMS Symposium, Princeton. Ed. by P. LACROIX and A. BUDY. Blackwell Scientific Publications, Oxford, 1962.

[2] Medical Uses of Ca[47]: A symposium organized and published by the International Atomic Energy Agency, Vienna, 1962.

rapidly in hyperthyroidism than in normal subjects is in fact based on tracer data; prior to the isotope era, ALBRIGHT, and therefore everybody, believed that the osteoporosis of hyperthyroidism developed at a lower than normal rate of bone formation.

We have been discussing for a number of years now why there is a break in the rate of loss of circulating tracer in hyperthyroidism and in PAGET's disease. It is quite obvious that resorption and/or exchange is responsible for this break. By analogy with what happens when labelled erythrocytes disintegrate after a life-span of, say, 128 days, this break can at present best be interpreted as due primarily to resorption of radioactive bone. When Dr. FRASER suggests that this break may illustrate why one should not be able to measure the rate of bone formation with the aid of radioactive tracers, he is using arguments which were *en vogue* some 10 years ago.

FRASER: Mr. Chairman, I very willingly condensed my remarks to 10 minutes but, if you really wish me to, I will be pleased to comment on these comments.

With regard to Dr. TANNER's remarks, I agree of course that you may have to derive data. What I would like to see used for growth recording is not "height age", but the actual height measured scored against the sort of normal chart he has shown, from which, however, puberty has been extracted by the sort of method I suggested. As to how to time puberty, this may be difficult and may possibly best be done by reference to bone age. But, however we do it, let us realise what we are doing. I would like to see two sets of growth charts: one for use pre-puberty and one post-puberty. Then we would not need any elaborations such as testing height against skeletal age.

Finally, about separating epiphyseal growth from other aspects of bone maturation — I dare say this is unimportant if you are studying normal children, but when you are assessing hypogonadal children, for example, it is different. We see "children" in their 30th year who haven't closed their epiphyses, but whose bone shape is fully matured. I find it difficult to see why these two aspects of bone maturation shouldn't be separated in terms of clinical standards. It may be, I entirely agree, that they would be very closely related in normal growth, but that isn't what the clinician is studying in most of his cases. He wants to dissect out different aspects of normal growth and to compare them against the normal in his cases, and that is why he would like you to do this dissecting out of puberty by a less mathematical method.

With regard to Dr. BAUER's comments on the bones, it is of course perfectly correct of him to deplore the fact that we have been calling osteopenia "osteoporosis" for such a long time. However, I believe that we should introduce a new term only for a good new reason, and not simply because the old term was bad. This usually needs some new evidence on the cause. Since ALBRIGHT made his comments in 1938, we have known this disorder as a bone atrophy, and that is what he called it.

Finally, I should just like to turn to the more fundamental point about measuring accretion rates. In the conventional diagram we picture an exchangeable calcium pool into which the injected calcium tracer very quickly mixes. Dr. BAUER's hypothesis is that we also have something else, which we might call deeper bone, all in one big, quite separate, pool. From the former into the latter we have new bone formation or accretion. All I am saying is that we also have a reverse exchange process going on all the time. This is evident from the slide I chose (Fig. 3, p. 430), because of the very big and rapid turnover. I doubt whether it is truly resorption which causes the kink on these curves. It is probably also going on in the earlier stages and is thus an ignored factor in the measurements of "accretion".

The effects of anabolic steroids on liver function

By

I. M. ARIAS

I would like to discuss briefly the effects of several anabolic steroids on liver function in man, studies regarding the mechanism of these effects, and the production of similar abnormalities in rats.

Shortly after the clinical introduction of 17α-ethyl-19-nortestosterone (norethandrolone), abnormal retention of injected Bromsulfophthalein (BSP) was observed in the serum of patients who received this steroid (*1, 2*). Subsequent studies revealed that BSP retention is related to steroid dosage, and hyperbilirubinemia is commonly observed when BSP retention exceeds approximately 40% (*2*). This is illustrated in Fig. 1, which combines unpublished

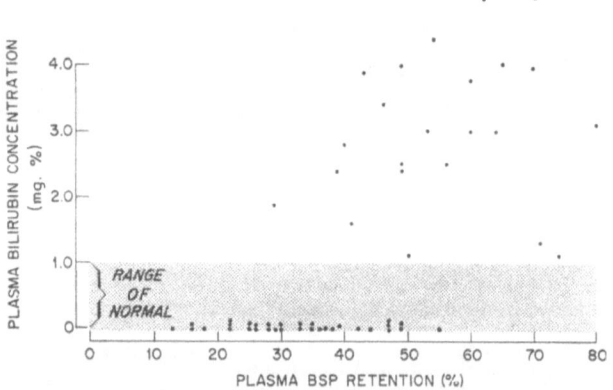

Fig. 1. Maximal plasma bilirubin concentrations and BSP retention in 55 patients who received 20 mg. norethandrolone per day for six weeks

observations of H. KAYDEN and the author in 55 patients who received 20 mg. norethandrolone per day for six weeks. The maximal plasma bilirubin concentration is plotted against the maximal BSP retention. Hyperbilirubinemia was observed in 21 of 29 subjects in whom plasma BSP retention exceeded 40%.

When 120 mg. of the steroid was administered daily to 8 additional patients, 5 patients manifested hyperbilirubinemia within three weeks (2). In these studies and similar investigations by others (3, 4), serum glutamic oxaloacetic transaminase (SGOT) and lactic dehydrogenase activities were occasionally slightly increased, whereas serum cephalin cholesterol and thymol flocculations, alkaline phosphatase activity, and the concentration of albumin, globulin, cholesterol and cholesterol esters were normal.

The administration of other synthetic anabolic steroids is associated with similar functional abnormalities (Table 1). Neither

Table 1. *The effect of anabolic steroids on liver function in man*

Drug	Dose (mg./day) and duration	Mean rise in BSP retention (%)	Mean rise in		Hyperbilirubinemia
			SGOT[1]	SLDH[2]	
Testosterone (5) . .	36—2 wks	1.6	0	0	Not reported
17α-Methyl-testosterone (6) . .	80—2 wks	12.0	18	—	Werner et al. (7)
Δ'-17α-Methyl-testosterone (4) . .	15—2 wks	8.2	—	3	Arias (8)
17α-Ethyl-19-nor-testosterone (4) . .	30—2 wks	17.6	—	14	Kayden (2)
17β-Hydroxy-α-methyl-2-oxa-5-andro-stane-3-one (2)	30—3 mths	20.0	11	—	Not reported

[1] serum glutamic oxaloacetic transaminase activity (U/ml.).
[2] serum lactic dehydrogenase activity (U/ml.).
( ) papers listed in References on page 445.

BSP retention nor hyperbilirubinemia are associated with testo-sterone administration (5). BSP retention occurs with clinical doses of 17α-methyl-testosterone (6), Δ'-17α-methyl-testosterone (methan-drostenolone) (4), and 17β-hydroxy-α-methyl-2-oxa-5-androstane-3-one (2). Whenever the effect of large doses of these steroids has been studied, hyperbilirubinemia has been frequently observed (2, 7, 8). These functional abnormalities disappear within three weeks after cessation of drug administration and appear to be unrelated to the subject's age or sex.

Light-microscopic examination of hematoxylin and eosin stained sections of liver obtained from subjects who developed hyper-bilirubinemia while receiving anabolic steroids reveals liver cells of normal appearance which contain variable amounts of bile pig-ment (9).

28*

Liver disease predisposes to the development of BSP retention
and hyperbilirubinemia. Administration of norethandrolone resulted in hyperbilirubinemia more commonly in patients with
antecedent BSP retention than in those without antecedent BSP
retention (2). If anabolic steroids are administered to icteric
patients, hyperbilirubinemia increases. Fig. 2 illustrates the effect
of norethandrolone on the serum bilirubin concentration in a

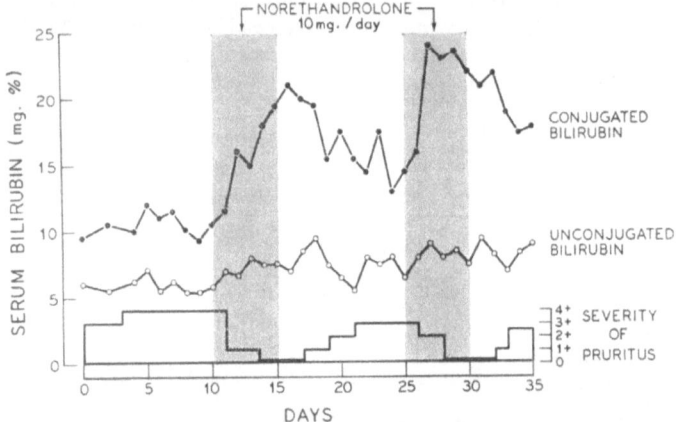

Fig. 2. The effect of norethandrolone on the serum bilirubin concentration in a patient with
post-necrotic cirrhosis and generalized pruritus

patient with post-necrotic cirrhosis and generalized pruritus. On
two occasions, pruritus was relieved by steroid administration;
however, the concentration of conjugated bilirubin in the serum
increased significantly.

Thus, the administration of large doses of several synthetic anabolic steroids is regularly associated with BSP retention and
frequently with hyperbilirubinemia without liver-cell necrosis. These
functional abnormalities are reversible, related to steroid dosage,
and occur more readily in patients with liver disease than in normal
subjects.

To elucidate the mechanism responsible for these functional
effects, bilirubin and BSP metabolism have been studied in
patients receiving anabolic steroids (10, 11).

The hepatic metabolism of bilirubin has been reviewed elsewhere (12). Unconjugated bilirubin in plasma is bound to albumin
and enters the liver cell by unknown mechanisms. Within the cell,

bilirubin is primarily conjugated with glucuronic acid and, to a lesser extent, with ethereal sulfate and other as yet unidentified polar groups. The transfer of glucuronic acid from uridine diphosphate glucuronic acid (UDPGA) to bilirubin is catalyzed by glucuronyl transferase, a microsomal enzyme. Bilirubin must be conjugated to be excreted into the bile (*13, 14*). The ability of the liver cell to excrete bilirubin conjugates probably limits the overall transport of bilirubin from blood to bile (*14*).

HSIA, DOWBEN, and SHAW (*15*) demonstrated that various anabolic steroids inhibit glucuronyl transferase activity *in vitro* and suggested that reduced formation of bilirubin glucuronide may account for hyperbilirubinemia in patients receiving these steroids. We have studied glucuronide formation *in vivo* and *in vitro* in four patients who manifested hyperbilirubinemia during administration of either 17α-methyl-testosterone or 17α-ethyl-19-nortestosterone (Table 2). Glucuronide formation was studied *in vivo* by the menthol tolerance test. Following ingestion of one gram of menthol,

Table 2. *Glucuronide formation in patients with drug-induced hyperbilirubinemia*

Drug	Dose (mg./day) and duration	Serum total bilirubin	Menthol tolerance test	Hepatic glucuronyl transferase activity
17α-Methyl-testosterone .	40—2 wks	15.0	51	220
17α-Ethyl-19-nor-testosterone	60—3 wks	7.5	42	185
17α-Ethyl-19-nor-testosterone	20—8 wks	24.1	38	236
17α-Ethyl-19-nor-testosterone	100—6 wks	6.0	46	195
	Normal:	0.1—0.9 mg. %	49.6±7.1 %	220±41 units

normal subjects excrete 49.6 ± 7.1 (S.D.) % of ingested menthol in the urine as menthol glucuronide (*16*). The menthol tolerance test was normal in the four jaundiced patients. Hepatic glucuronyl transferase activity was assayed in specimens of liver obtained by aspiration biopsy using 4-methyl-umbelliferone as a glucuronide receptor (*16*). No difference in glucuronyl transferase activity was observed between the jaundiced patients and normal subjects. These studies fail to demonstrate an impairment in glucuronide formation in patients who manifested jaundice during steroid administration.

To study this problem further, the serum bile pigments were identified chromatographically using the technique of SCHMID (17). In each case conjugated bilirubin accounted for more than 70% of the total serum bile pigments (Table 3). More than 70% of the conjugated bilirubin in serum was hydrolyzed by alkaline or beta glucuronidase. These studies demonstrate that bilirubin in the patients' serum is primarily conjugated with glucuronic acid, and no abnormality in glucuronide formation was observed.

Table 3. *Serum bile pigments in patients with drug-induced hyperbilirubinemia*

Drug	Serum bilirubin		% Conjugated bilirubin hydrolyzed by:	
	Total (mg. %)	Conjugated (%)	Alkaline	B-Glucuron-idase
17α-Methyl-testosterone . .	15.0	73	80	82
17α-Ethyl-19-nortestosterone	7.5	80	70	75
17α-Ethyl-19-nortestosterone	24.1	75	85	80
17α-Ethyl-19-nortestosterone	6.0	83	80	90

Fig. 3 is a schematic representation of the postulated pathogenesis of jaundice in these patients. Conjugation is unimpaired; however, the capacity of the liver cell to transport conjugated bilirubin into the bile is reduced, and conjugated bilirubin enters the plasma.

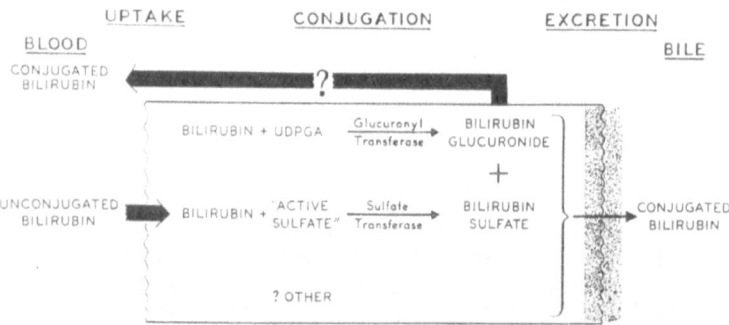

Fig. 3. Schematic representation of the postulated pathogenesis of jaundice in patients receiving C-17 alkylated anabolic steroids

The hepatic metabolism of BSP, which is schematically represented in Fig. 4, involves uptake and storage of BSP, intracellular formation of a glutathione conjugate, and subsequent

excretion of free and conjugated BSP into the bile. The storage of
BSP in parenchymal cells is directly proportional to the plasma
concentration of BSP, whereas transfer of BSP from liver cells to
bile occurs by a rate-limiting transfer mechanism (18). Conjugation
of BSP with glutathione is enzymatically catalyzed (19) and is an
important determinant of the maximal rate at which BSP is
transferred from the blood to bile (20). A method for the simultaneous estimation of hepatic BSP storage and BSP Tm has been
described by WHEELER et al. (18). Values for the relative hepatic

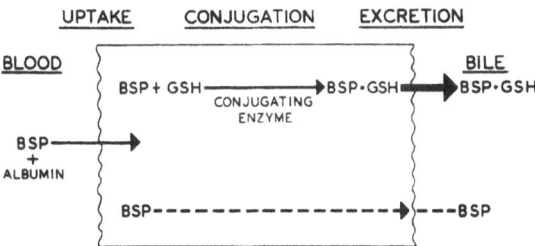

Fig. 4. Schematic representation of the hepatic metabolism of BSP

storage of BSP (expressed as mg. BSP stored/mg. % BSP in the
plasma) and BSP Tm (expressed as mg. BSP excreted/min.) are
derived from the concentration of BSP in plasma serially obtained
during separate intravenous infusions at three different rates. Both
BSP storage and Tm are reduced in patients with various acquired
liver diseases (18). This technique was used to study the effect of
norethandrolone and methandrostenolone on BSP metabolism in
twelve patients (11). The results are presented in Fig. 5. Hepatic
BSP Tm was reduced virtually to zero in the patients who received
the anabolic steroids; however, BSP storage remained normal.
One week after discontinuation of steroid administration, BSP
Tm returned to normal in the two patients in whom this was studied.
The only clinical disorder in which a similar pattern of BSP metabolism has been observed is chronic familial jaundice (DUBIN-
JOHNSON syndrome), in which an inherited abnormality in hepatic-
cell excretory transport has been postulated (18, 21).

    The reduction observed in hepatic BSP Tm in patients re-
ceiving anabolic steroids could result from a deficiency in hepatic
glutathione content, reduced activity of the conjugating enzyme,
and/or an abnormality in hepatic-cell excretory function. To
clarify this problem, the ratio of the concentration of plasma con-

jugated BSP: total BSP was estimated at the end of each hour
of the BSP infusion study in two patients before and during administration of norethandrolone (Fig. 6). Prior to steroid administration, the ratio of conjugated BSP: total BSP increased
when Tm was exceeded (i.e. in infusion periods 1 and 3, when the

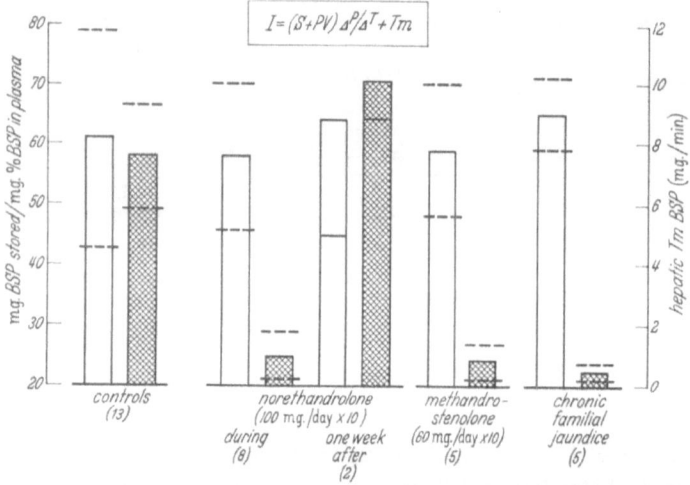

Fig. 5. Estimates of the relative hepatic storage (S) and hepatic transport maximum (Tm)
for BSP in normal subjects, in patients receiving norethandrolone or methandrostenolone,
and in patients with chronic familial jaundice (Dubin-Johnson syndrome)

infusion rates were 15 and 10 mg. BSP/min., respectively). Following steroid administration, the ratio increased in each infusion
period. When the infusion rate was zero in period 2, almost 90%
of the retained BSP was conjugated. Because BSP was retained in
plasma primarily as a conjugate after steroid administration, it is
unlikely that either hepatic glutathione content or activity of the
conjugating enzyme is limiting. In rats, neither hepatic glutathione
content nor conjugating enzyme activity is decreased by administration of comparable doses of steroid (22). These observations
suggest that the major effect of anabolic steroids on BSP metabolism is to reduce the maximum rate at which conjugated BSP
is transported from the liver cell into the bile.

Because of an interest in the biochemical mechanism responsible for the functional effects observed after administration of
anabolic steroids, attempts were made to produce similar abnormalities in rats (10). Normal Wistar and homozygous Gunn

rats were used. Homozygous Gunn rats have chronic unconjugated hyperbilirubinemia due to an inherited deficiency in glucuronyl transferase (*13*). The bile duct and an external jugular vein were cannulated. Unconjugated and conjugated bilirubin were separately infused intravenously at 2.5 and at 3.0 mg./min., respectively. Bile was collected at five-minute intervals and the maximal biliary excretion of bilirubin was estimated. In normal rats infused with

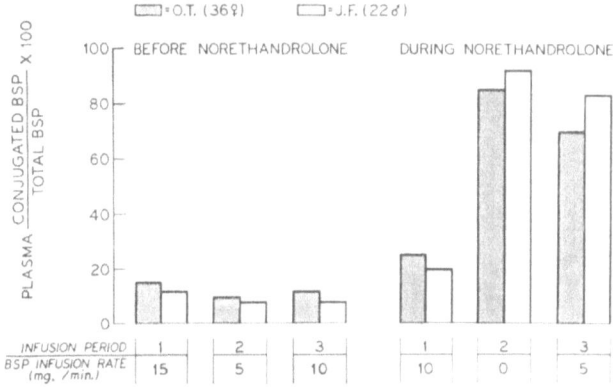

Fig. 6. Ratio of the concentration of plasma conjugated BSP: total BSP at the end of each hour of a BSP infusion study in two patients before and during administration of norethandrolone

either conjugated or unconjugated bilirubin, and in Gunn rats infused with conjugated bilirubin, the maximal biliary excretion of bilirubin was approximately 70 gamma of bilirubin excreted/100 gram of rat/min. (*14*).

Normal Wistar and Gunn rats were treated with various doses of anabolic steroids and the effect on the excretory maximum for bilirubin was studied. Fig. 7 presents the mean bilirubin excretory maximum $\pm$ one standard deviation in untreated rats and when testosterone, 17α-methyl-testosterone, norethandrolone, methandrostenolone, and 17β-hydroxy-α-methyl-2-oxa-5-andro-stane-3-one (SC-11585) were administered in the doses indicated. The 17α-substituted steroids were associated with significant reduction in the excretory maximum for bilirubin, whereas no reduction occurred after testosterone administration. Because the excretory maximum for bilirubin was similarly reduced when unconjugated bilirubin was given to normal rats and when conjugated bilirubin was administered to Gunn rats, it may be concluded that the steroids do not affect bilirubin conjugation. Norethandrolone was administered in increasing doses to Gunn rats, conjugated bilirubin

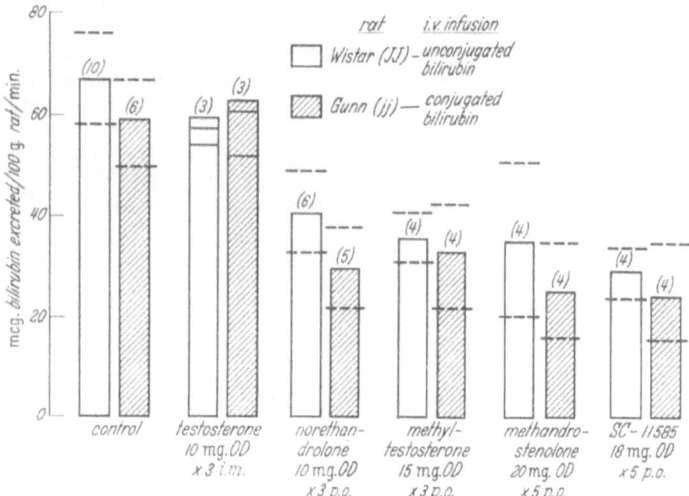

Fig. 7. Maximal bilirubin excretion in untreated rats and after administration of various anabolic steroids

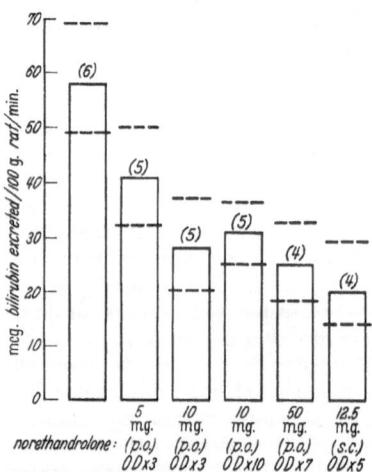

Fig. 8. Effect of norethandrolone on the excretory maximum for bilirubin following administration of conjugated bilirubin to Gunn (jj) rats

was infused, and the excretory maximum for bilirubin was estimated (Fig. 8). A dose-related reduction in the excretory maximum was observed.

In other studies, the biliary excretions of injected BSP, conjugated BSP, and phylloerythrin were similarly reduced after steroid administration (8). These observations suggest that anabolic steroids interfere with the excretion of various substances from the liver cell into the bile.

In an attempt to correlate these functional abnormalities with morphologic changes, liver biopsies were obtained from normal rats treated with norethandrolone and examined by light and electron microscopy (23). As previously mentioned, hematoxylin and eosin stained sections of liver were histologically normal except for occasional bile pigment within parenchymal cells. The following cytochemical abnormalities were consistently observed. When normal rat liver is examined for "ATP-ase" activity, the bile canaliculi are demonstrated. After steroid administration, the canaliculi are fragmented and dilated and the sinusoids are stained. When normal rat liver is examined for acid phosphatase activity, the pericanalicular lysosomes are demonstrated. After steroid administration, the lysosomes appear more abundant and are distributed through the parenchymal cells. When normal rat liver is stained for alkaline phosphatase activity, segments of bile canaliculi show enzyme activity. After steroid administration, canaliculi are fragmented and dilated and the sinusoids show enzyme activity. Electron-microscopic examination reveals occasional dilatation of bile canaliculi, blunting or absence of canalicular microvilli, and dispersion of lysosomes through the cell. Other intracellular organelles appear normal. The morphologic changes predominantly involve bile canaliculi and lysosomes. These structures have been considered to be important in bile secretion (12). Similar electron-microscopic changes have been observed by other investigators in liver obtained from rats and humans treated with norethandrolone (24).

Summary

(1) Large doses of synthetic anabolic steroids regularly produce BSP retention, and frequently conjugated hyperbilirubinemia, without morphologic changes in hematoxylin and eosin stained sections of liver.

(2) Most of the serum bilirubin in patients with jaundice due to steroid administration is bilirubin glucuronide. *In vivo* and *in vitro* studies of glucuronide formation yield normal findings in these patients.

(3) In patients receiving norethandrolone or methandrostenolone, hepatic BSP Tm is virtually zero, BSP storage is normal, and BSP is retained in the plasma primarily as a conjugate.

(4) Administration of synthetic anabolic steroids to normal and Gunn rats is associated with a dosage-related reduction in the bilirubin excretory maximum and reduced biliary excretion of injected BSP, conjugated BSP, and phylloerythrin.

(5) In humans and rats treated with norethandrolone, histochemical and electron-microscopic examinations revealed changes in bile canaliculi and lysosomes.

(6) These physiologic and morphologic studies suggest that large doses of synthetic anabolic steroids interfere with the poorly understood mechanisms by which bilirubin conjugates and other metabolites are transported from the liver cell into the bile. Study of the biochemical changes associated

with the functional abnormalities may provide important information regarding the mechanism and regulation of normal hepatic-cellular excretory transport.

Zusammenfassung

1. Große Dosen von synthetischen anabolen Steroiden bewirken regelmäßig Retention von Bromsulphalein (BSP) und häufig erhöhte Plasmakonzentration von gebundenem Bilirubin. In mit Haematoxylin-Eosin gefärbten Leberschnitten sind dagegen keine morphologischen Veränderungen nachweisbar.

2. Bei Patienten mit Ikterus infolge einer Steroidbehandlung liegt Bilirubin im Serum zum größten Teil als Glucuronid vor. Bei diesen Patienten fanden sich weder in vitro noch in vivo Störungen der Glucuronidbildung.

3. Bei Patienten, die Noräthandrolon oder Δ1-17α-Methyltestosteron (Methandrostenolon) erhielten, ist der Transport von BSP aus den Leberzellen in die Galle fast vollständig aufgehoben, während Aufnahme und Speicherung von BSP in den Leberzellen normal sind. BSP liegt im Plasma hauptsächlich als Konjugat mit Glutathion und als Glucuronid vor.

4. Behandlung von normalen Wistar-Ratten und von Gunn-Ratten mit synthetischen anabolen Steroiden führt zu einer dosisabhängigen Verminderung der maximalen Bilirubinausscheidung und einer reduzierten Gallenausscheidung von injiziertem BSP, konjugiertem BSP und von Phylloerythrin.

5. Behandlung mit Noräthandrolon ruft beim Menschen und bei der Ratte histochemisch und elektronenmikroskopisch erfaßbare Veränderungen in den Gallenkanälchen und den Lysosomen hervor.

6. Diese Untersuchungen sprechen dafür, daß hohe Dosen von synthetischen anabolen Steroiden die noch weitgehend unbekannten Vorgänge beeinträchtigen, durch die Bilirubinkonjugate und andere Metaboliten von der Leberzelle in die Galle transportiert werden. Das Studium der den funktionellen Störungen zugrundeliegenden biochemischen Veränderungen kann wichtige Auskünfte über den Mechanismus und die Regulation der normalen hepatozellulären Ausscheidungsvorgänge geben.

Résumé

1. De fortes doses de stéroïdes anabolisants de synthèse déterminent régulièrement une rétention de bromsulfonephtaléine (BSP) et souvent une augmentation dans le sang de la bilirubine glycuro-conjuguée. Dans les coupes du foie colorées à l'hématoxyline-éosine en revanche, on ne peut mettre en évidence d'altérations morphologiques.

2. Chez les sujets atteints d'ictère après un traitement stéroïdien, la bilirubine sérique se présente en majorité sous forme glycuro-conjuguée. On n'a pas constaté de troubles de la glycuro-conjugaison chez ces malades, tant in vivo qu'in vitro.

3. Chez les sujets qui ont reçu de la noréthandrolone ou de la Δ 1-méthyl-17α-testostérone (méthandrosténolone), le transport de la BSP des cellules hépatiques dans la bile est tombé presque à zéro, tandis que la réception et l'accumulation de la BSP dans les cellules du foie étaient normales. Dans le plasma, la BSP se trouve surtout conjuguée avec le glutathion et sous forme de glycuronide.

4. L'administration de stéroïdes anabolisants de synthèse à des rats normaux et à des rats de Gunn diminue l'excrétion maximale de bilirubine proportionnellement à la dose, et réduit l'élimination biliaire de la BSP injectée, de la BSP conjuguée et de la phyllo-érythrine.

5. Chez l'homme et chez le rat, la noréthandrolone provoque des altérations des canalicules biliaires et des lysosomes décelables à l'examen histochimique et au microscope électronique.

6. Ces recherches indiquent que de fortes doses de stéroïdes anabolisants de synthèse peuvent influencer les processus encore mal connus par lesquels les produits de conjugaison de la bilirubine et d'autres métabolites sont transportés de la cellule hépatique dans la bile. L'étude des modifications biochimiques en rapport avec les troubles fonctionnels peut fournir des renseignements importants sur le mécanisme et sur la régulation des processus d'excrétion hépato-cellulaires normaux.

Acknowledgements

The studies from the author's laboratory were performed in association with Drs. E. ESSNER, S. GOLDFISCHER, M. KIRSCHNER, L. KRESCH, A. NOVIKOFF, and J. SCHERB and Mrs. W. FURMAN and were supported by research grants from the National Institute of Arthritis and Metabolic Diseases of the National Institutes of Health, United States Public Health Service (A-2019, H-3838, M-2562 and A-2966) and the G. D. Searle Co. of Chicago, Illinois.

The author is a senior research fellow of the New York Heart Association and Heart Fund, Inc.

References

1. KORY, R. G., M. H. BRADLEY, R. N. WATSON, R. CALLAHAN, and B. J. PETERS: Amer. J. Med. **26**, 243 (1959). — 2. KAYDEN, H.: personal communication. — 3. DOWBEN, R. M.: J. Clin. Endocr. (U.S.A.) **18**, 1308 (1958). — 4. MARQUARDT, G. H., C. I. FISHER, P. LEVY, and R. M. DOWBEN: J. Amer. Med. Ass. **175**, 115 (1961). — 5. HAYES, T.: personal communication. — 6. CARBONE, J. V., G. M. GRODSKY, and V. J. HJELTE: J. Clin. Invest. (U.S.A.) **38**, 1989 (1959). — 7. WERNER, C. S., F. M. HANGER Jr., and R. KRITZLER: Amer. J. Med. **8**, 325 (1950). — 8. ARIAS, I. M.: unpublished data. — 9. SCHAFFNER, F., H. POPPER, and E. CHESROW: Amer. J. Med. **26**, 249 (1960). — 10. ARIAS, I. M., S. GOLDFISCHER, E. ESSNER, and A. NOVIKOFF: J. Clin. Invest. (U.S.A.) **40**, 1023 (1961). — 11. SCHERB, J., M. KIRSCHNER, and I. M. ARIAS: unpublished observations. — 12. ARIAS, I. M.: In: Progress in Liver Disease. Ed. by POPPER, H., and F. SCHAFFNER. Grune and Stratton, New York, N. Y., 1961. — 13. SCHMID, R., J. AXELROD, L. HAMMAKER, and R. J. SWARM: J. Clin. Invest. (U.S.A.) **37**, 1123 (1958). — 14. ARIAS, I. M., L. JOHNSON, and S. WOLFSON: Amer. J. Physiol. **200**, 1091 (1961). — 15. HSIA, D. Y., R. M. DOWBEN, and R. SHAW: Nature (G.B.) **187**, 693 (1960). — 16. ARIAS, I. M.: unpublished observations. — 17. SCHMID, R.: J. Biol. Chem. (U.S.A.) **229**, 881 (1957). — 18. WHEELER, H., J. MELTZER, and S. BRADLEY: J. Clin. Invest. (U.S.A.) **39**, 1131 (1960). — 19. COMBES, B., and G. S. STAKELUM: J. Clin. Invest. (U.S.A.) **40**, 981 (1961). — 20. COMBES, B.: Gastroenterology (U.S.A.) **42**, 471 (1962). — 21. ARIAS, I. M.: Amer. J. Med. **31**, 510 (1961). — 22. SCHERB, J., and I. M. ARIAS: unpublished data. — 23. GOLDFISCHER, S., I. M. ARIAS, E. ESSNER, and A. B. NOVIKOFF: J. Exper. Med. (U.S.A.) **115**, 467 (1962). — 24. SCHAFFNER, F., H. POPPER, and V. PEREZ: J. Laborat. Clin. Med. (U.S.A.) **56**, 623 (1960).

Discussion

OVERBEEK: I wonder if it is quite correct to refer to the effects of "anabolic steroids" as such, or whether one ought not to specify that it is anabolic steroids with a 17α-methyl or an ethyl group with which one is dealing. For this reason I should like to ask whether you observed the same effect with some of the nandrolone esters.

ARIAS: Thank you for correcting me, Dr. OVERBEEK. I should mention that our interest in these studies has been to find compounds which interfere physiologically with hepatic cellular excretory function. We have not studied the effect of other structural relationships in the steroid molecule. Except for testosterone, the only hormones we studied were C-17 substitutions.

GROSS: Do you know of any steroids with a 17α-alkyl substitution which have no anabolic activity but which have the same effect on liver function ? My second question is: how does the impairment of liver function observed with 17-alpha substituted derivatives compare with that observed with other substances which are known to produce BSP retention and occasionally jaundice, such as chlorpromazine or other phenothiazine derivatives ? Thirdly, in the Gunn rats there was no close dose-response relationship with respect to the excretion of conjugated and non-conjugated BSP. I wonder if in other experiments you were able to show a closer dose-response relationship ? And my last question: how do the 17-alpha substituted steroids which you investigated interfere with the excretion of other glucuronides, and in particular how do they interfere with steroid metabolism, as we know that the steroids are also eliminated for a major part in the form of glucuronides ?

ARIAS: We have not studied other steroids in terms of these effects. The relationship to drugs like chlorpromazine is an important one which requires some comment. The occurrence of BSP retention or hyperbilirubinaemia in patients receiving chlorpromazine and related drugs is rare, whereas the effects of C-17 substituted anabolic steroids can almost be predicted. If normal individuals receive 17α-ethyl-19-nortestosterone in a daily dose of 100 mg. for 10 days, BSP Tm is regularly reduced to zero. We have not performed similar studies in patients receiving chlorpromazine.

The fact that there is not a log dose-response curve is not too surprising, because we are using a relatively crude technique for measuring excretion.

We have not studied the effect of these steroids on the metabolism of other steroids either in man or in animals. In mammals bilirubin must be conjugated to be excreted in the bile, and about 80% of the bilirubin found in freshly obtained bile is present as an ester glucuronide. This is not the case for many steroids. The bile is the preferential route of excretion for bile pigment, but not for most steroids. I would therefore speculate that the functional effect which we have observed may not be related to altered steroid metabolism or excretion.

MANDEMA: I think that the studies of Dr. ARIAS show very convincingly that some of the anabolic steroids exert a profound effect on BSP metabolism. I would like to ask him whether he thinks that his studies entirely rule out the possibility of some decrease in the uptake of BSP by the liver in these

patients. The reason why I'm asking this is that I am still not fully convinced that a normal relative storage capacity and a strongly reduced Tm,
as measured by the method of WHEELER and co-workers[1], completely rules
out the possibility of a slightly diminished uptake of BSP by the liver. Using
this method, the liver has to be more or less saturated with BSP before one
can take measurements, and a BSP-saturated liver may not be entirely
suitable for studying the first phase of BSP metabolism, namely the uptake
of BSP from the blood into the liver cells. In the one patient we have studied
so far, using only a single rapid injection, we found after 45 minutes a very
marked retention of BSP, of which 68% was in the unconjugated form and
only 32% was conjugated.

A second point I should like to bring up is that in DUBIN-SPRINZ disease
the patients may be jaundiced, although BSP retention after 45 minutes may
be only slight. The patients treated with anabolic steroids develop jaundice only when BSP retention is very marked. This points to a somewhat
different mechanism in these two types of abnormal BSP metabolism. If one
assumes that in DUBIN-SPRINZ disease there is a block in the secretion of
BSP from the liver cell into the bile and no interference with the uptake of
BSP by the liver, then one might speculate that in patients treated with
anabolic steroids there must also be some diminished uptake of BSP to
explain the discrepancy between BSP and bilirubin retention in these two
conditions. I would like to ask you whether you know of any studies on
patients who were given anabolic steroids with only a single injection of BSP
and, if so, what the results were; and do you think that a diminution in the
uptake of BSP — possibly only a slight diminution — may be partly responsible for the alteration in BSP metabolism after giving anabolic steroids ?

ARIAS: I completely agree with you, Dr. MANDEMA. It is possible that
the steroids may affect the uptake of BSP from the plasma into the liver
cells. Dr. LEEVY and his associates[2] demonstrated a delay in the early phase
of the disappearance of BSP from the plasma following a single injection in
patients who had received 17α-ethyl-19-nortestosterone. The fact that BSP
is retained in the plasma primarily as a conjugate suggests that the major
functional abnormality is at the other pole of the liver cell. What is lacking,
of course, are means of studying the transfer of any substance from the
plasma into the liver cells.

DREYFUS: I would like to ask Dr. ARIAS whether the metabolism of
these steroids is known. Could it be possible that they are transported by the
bile and that the mechanism of transfer would compete also with that of
conjugated bilirubin ? Are they excreted in the urine, like androgenic steroids,
or are they excreted in any other way, e.g. in the bile ?

ARIAS: If I understand you correctly, you are asking whether the steroid
is metabolised and excreted in a way which could possibly compete with
bilirubin. I am sorry that I do not have any information regarding this
question. Perhaps some of our colleagues can provide the answer.

OVERBEEK: There are some experiments — not my own — which seem
to show what is exactly the difference with these 17α-ethyl and methyl
compounds. They are excreted by the bile and appear in the faeces, whereas
the nandrolone esters, for instance, and others, such as testosterone, follow
the other pathway and appear in the urine. So this may indeed explain why

---

[1] WHEELER, H. O., et al.: J. Clin. Invest. (U.S.A.) **39**, 1131 (1960).

[2] LEEVY, C., et al.: J. Laborat. Clin. Med. (U.S.A.) **56**, 193 (1961).

the 17α-methyl and ethyl compounds have these particular effects, whereas other steroids probably do not.

SUBRAMANIAM: I would like to mention two observations on what happens when these hormones are used in cases of nutritional cirrhosis of the liver. When we use 17α-ethyl-19-nortestosterone the patients become worse. They suffer from loss of appetite and increased oedema, whereas when testosterone is given in a daily dose of 100 mg. intramuscularly for 10—12 days, they show clinical improvement, i.e. their appetite increases and the oedema subsides.

VERMEULEN: May I revert to the question of Dr. GROSS concerning the interference of anabolic steroids with the conjugation of adrenocortical hormones ? We studied the influence of 17α-methyl-19-nortestosterone on the metabolism of cortisol, and we did in fact observe an increase in the excretion of unconjugated cortisol metabolites after the administration of 17α-methyl-19-nortestosterone. However, I doubt whether this is due to an impaired conjugation as such, or whether it is related to an alteration in metabolism secondary to increased protein binding of the cortisol in the plasma.

Then I would like to ask Dr. ARIAS if he also studied the influence of anabolic steroids on the enzyme levels in the blood, i.e. on the transaminases (SGOT and SGPT), alkaline phosphatase, leucine-amino-peptidase (LAP), etc. With DEMEULENAERE[1] we did some experiments on the influence of Dianabol on serum enzymes in rabbits (cf. Table 1). As can be seen in the

Table 1. *Influence of Dianabol on serum enzyme levels in rabbits*

Dose of Dianabol administered	Duration	SGOT	SGPT	Alk. phosph.	LAP
		Control values			
		16—32 U.	13—33 U.	1—5 U.	160—240 U.
0.1 mg./kg. . .	2 days	14	22	4.5	182
	4 days	27	37	3.6	220
	8 days	83*	84*	—	285
0.2 mg./kg. . .	4 days	34	39	2.6	165
	8 days	94*	90*	2.7	315*
0.4 mg./kg. . .	2 days	88	75	4.8	212
	8 days	137*	135*	3.6	370*

* Statistically significant difference from normal values.

table, administration of Dianabol for 8 days, in a dose of 0.1 mg./kg. daily, significantly increases the serum transaminase levels, whereas a dosage of 0.4 mg./kg. significantly increases the serum transaminase levels after only 2 days. It is interesting to note that the alkaline phosphatases remain unaffected, whereas leucine-amino-peptidase levels are significantly increased.

The increased transaminase levels are generally interpreted as a sign of cellular damage.

ARIAS: We have some experience in humans. Our findings are similar to yours. I would add a note of caution regarding the physiological significance of increased enzymatic activity in the blood after steroid administration.

---

[1] DEMEULENAERE, L., J. LAGAE, and A. VERMEULEN: unpublished data.

Because patients with hepatic-cell necrosis show increased activity of various enzymes in the blood, it is widely believed that the changes in serum enzyme activity are always representative of parenchymal liver-cell function. I do not believe that this is always the case, and the physiological significance of enzymatic activity in the blood is not well understood. Perhaps you may have some suggestions.

LEATHEM: I would just like to ask a question as to duration of administration. I think your rats were injected for a period of 5 days, weren't they? What would happen on chronic treatment? Secondly, have you examined patients who have received therapy repeatedly? In other words, how do they react if they are re-treated?

ARIAS: We have not studied the effect of long-term administration in rats. We have a group of rats which have been receiving 17α-ethyl-19-nortestosterone for one year, and we intend to study them in the near future. We have studied two patients who received a second course of 17α-ethyl-19-nortestosterone 2 months after a reduced BSP Tm had been observed during the first course of steroid administration. A similar functional abnormality was again demonstrated during the second course of steroid treatment.

Anabolic steroids in the treatment of renal failure

By

J. Hess Thaysen

Rational therapy in kidney diseases associated with renal failure is subject to great limitations. As long as this remains the case, the physician has to concentrate his efforts on mere symptomatic management of the complicating uraemia. The cause of the uraemic intoxication is not known, but is in all probability somehow connected with the retention of waste products from protein metabolism. It is therefore a cardinal principle in the treatment of uraemia to reduce the rate of protein break-down. This can be achieved by dietary means and, to a limited extent, also by the use of hormones exerting an anabolic effect.

It is the purpose of the present communication to report on the effect of anabolic 19-nor-steroids in the symptomatic treatment of acute and chronic renal failure. Owing mainly to the pronounced oliguria which accompanies most types of acute renal failure, dietary problems are in general more complicated in this condition than in chronic renal insufficiency. For this reason the two types of renal failure will be discussed separately.

Material and methods

Acute renal failure

Patients. Between May 1955 and May 1962, 275 patients with acute uraemia were admitted to the dialysis unit in Medical Department A of Rigshospitalet; 15 had "pre-renal uraemia" chiefly due to sodium and water depletion, 28 had "post-renal uraemia" due to ureteral obstruction, and 232 were suffering from acute uraemia due to disease of the renal parenchyma (Table 1).

Among the 232 cases of renal disease the patients with acute exacerbations in chronic renal failure were excluded from the present study (apart from one case), since they presented problems which differ from those of acute oliguric renal failure. Patients with vascular and glomerular damage were also excluded (except for three cases), since a considerable number had received prednisone or other steroids with a documented catabolic action. We are thus left with a series of 173 patients comprising 169 cases of tubular

nephropathy, 2 cases of total bilateral renal cortical necrosis,
1 case of total anuria due to accidental operative lesion of the
entire renal arterial blood supply, and 1 case of acute renal failure
following intravenous pyelography in a patient with multiple
myeloma.

Table 1. *Diagnosis in 232 cases of acute uraemia due to renal disease*

1. *Acute exacerbation of chronic renal disease*

Chronic pyelonephritis	20	
Chronic glomerulonephritis	4	
Myeloma kidney	3	
Amyloidosis	2	
Oxalosis	1	
		30

2. *Acute glomerular or vascular lesions*

Acute glomerulonephritis	22	
Panarteritis, L.E.D., Schönlein-Henoch	5	
Lesion of renal vessels	4	
Bilateral cortical necrosis (post-partum)	2	
		33

3. *Acute tubular nephropathy*

Post-traumatic (mostly traffic accidents)		15	
Following "medical diseases"		21	
Following surgery or "surgical diseases"		98	
Post-partum (or abortem)		20	
Intoxications			
nephrotoxins	7		
incompatible blood	2		
narcotic poisons	6		
		15	
			169
			232

The 169 cases of tubular nephropathy are listed in Table 1 in
decreasing order of severity with respect to the primary diseases
which had precipitated renal failure. The post-traumatic cases were
mostly severe traffic accidents with multiple fractures and con-
tusions. The medical cases comprise extensive coronary thrombosis with profound shock, septicaemia, tetanus, severe acute
gastro-enteritis, and other serious illnesses. The surgical cases
represent a mixed group ranging from severe "surgical diseases"
and extensive and complicated operations to relatively mild
"surgical diseases" and uncomplicated operations. In the post-
partum (abortem) and intoxication groups, the condition which
had precipitated renal failure was in most cases already under control by the time the patient was transferred to the dialysis unit.

Definitions. Retention of urea-N in the body fluids does, of course, cease as soon as equilibrium is established between urea-N production and urea-N excretion. The former depends on the intensity of catabolism, whereas the latter depends primarily upon renal function, but also upon the concentration of urea-N in the serum. At average rates of urea-N production, and provided serum urea-N is kept below 200 mg./100 ml. (if necessary by haemodialysis), a spontaneous drop in serum urea-N from the 200 mg./100 ml. level usually ensues when creatinine clearance exceeds 5 ml./min. Somewhat arbitrarily this clearance level is therefore used in the present paper to distinguish between the *"oliguric phase"* (retention phase), characterised by small urine volumes, slow improvement of renal function, and a gradual increase in the serum urea-N concentration, and the *"diuretic phase"* (excretory phase), characterised by large urine volumes (usually around 2—3 litres), more rapid improvement in renal function, and a gradual decrease in the serum urea-N concentration.

Principles of treatment. In the *oliguric phase (retention phase)* fluids were restricted according to conventional principles so as to permit maintenance of normal hydration. Sodium was omitted (unless sodium loss occurred) and hyperpotassaemia was combated by rectal resin lavage or, preferably, by oral resins. Acidosis was normally not corrected. All patients were mobilised as much as possible. If they were unable to walk, they were made to perform exercises in bed.

Supplementary haemodialysis was carried out with different types of dialyser, all so equipped that ultrafiltration of fluid was possible, when indicated. ALWALL's machine (urea clearance 100—120 ml./min.), which was first used, was later replaced by KOLFF's disposable twin-coil kidney (urea clearance 150—180 ml./min. with two units), and during the last year by BRUN's modification of the LEONARD SKEGGS apparatus (urea clearance 350—500 ml./min.). With the use of more and more efficient dialysers we have followed the general tendency to dialyse on increasingly liberal indications (limit decreased from about 200 to about 130—150 mg. urea-N per 100 ml. serum). For this reason the frequency of dialysis has not been measurably affected by the change in technique *per se*.

In the *excretory (diuretic) phase* urinary and extrarenal losses of water and electrolytes were quantitatively replaced and the patients were offered an *ad lib.* diet with as much protein as they could take.

With respect to the administration of calories, protein, and anabolic steroids during the oliguric phase, different principles have been followed over the years:

1. *From mid-1955 to mid-1956* the patients received 50% glucose via an indwelling caval catheter in quantities restricted by the requirements for maintenance of normal hydration. An average of 600 ml. of 50% glucose = 300 g. = 1,100 Cal. was given per day.

2. *From mid-1956 to mid-1957* the patients received restricted fluids orally or intravenously with the maximum quantity of glucose which they could ingest or which the peripheral veins could tolerate. An average of 100 g. glucose = 400 Cal./24 hours was given.

3. *From mid-1957 to mid-1961* the patients received about 400 glucose Cal./day supplemented by daily intramuscular injection of 50 mg. nandrolone (Durabolin).

4. *Since mid-1961* we have been more liberal in giving the patients calories and protein during oliguria, but have continued to use nandrolone in the same dose as before.

Chronic renal failure

Patients. 16 patients with severe chronic renal failure were treated with varying calorie and protein intakes before and during administration of anabolic steroids and were subjected to balance studies. The anabolic steroid used was either 17α-ethyl-19-nortestosterone (norethandrolone, Nilevar) given orally or nandrolone given intramuscularly.

2 of the 16 patients had chronic glomerulonephritis, 1 had subacute renal failure resulting from staphylococcal septicaemia, and 13 had the type of interstitial fibrosis with contracted kidneys generally diagnosed as "chronic pyelonephritis", although signs and symptoms of infection were frequently slight or even absent; 11 of these 13 patients were middle-aged or elderly women.

24-hour endogenous creatinine clearance was below 10 ml./min. in all patients. Marked uraemia and varying degrees of anaemia were present. Except in the cases of glomerulonephritis, hypertension was slight or absent.

Principles of treatment. Many of the patients with chronic pyelonephritis had renal sodium loss leading to extracellular dehydration and aggravation of their uraemia. Considerable improvement was often observed in response to correction of the sodium and water deficits. Metabolic studies were therefore never carried out until the sodium and water losses had been replaced and until a complete sodium balance and normal serum sodium had been maintained for long enough to ensure that the patient was in a steady state. Acidosis was corrected with sodium bicarbonate, but an additional sodium supplement (as NaCl) was frequently necessary to maintain the sodium balance.

Metabolic technique

The patients were examined in the metabolic ward, which works in close contact with the dialysis unit. All diets were served from the diet kitchen by the dietician or her assistants. To ensure quantitative fractionation of the urine in 24-hours periods, all patients had an indwelling catheter in the bladder throughout the entire period of the balance studies.

Analyses of diet and excreta for nitrogen, and determinations of non-protein nitrogen (NPN) in the serum, were carried out in the laboratory of the metabolic ward by the method of KJEL-DAHL. Determinations of urea in serum, urine, and other excreta were performed in the Central Clinical Laboratory of Rigshospitalet using the micro-diffusion technique of CONWAY. The caloric intake was calculated from standard tables. The nitrogen balance was calculated from the measured intake minus the measured output in urine, stools, and other excreta and was corrected for accumulation (or decumulation) of urea-N in the body fluids. (Owing to differences in the volumes of distribution of the various non-protein nitrogen products, calculations on the basis of total NPN were not considered feasible.)

24-hour accumulation (or decumulation) of urea-N was calculated as the product of Δ-serum urea-N (in grammes per litre) and total body water (in litres). In dialysed patients total body water was calculated from the total quantity of urea-N removed (in grammes) divided by the reduction in serum urea-N (in grammes per litre). In patients who were not dialysed, total body water was estimated as being 60% of body-weight in men and 55% in women.

Complete nitrogen balance studies were carried out in all patients with chronic renal failure and in all the patients with acute oliguric renal failure who received dietary protein.

Complete nitrogen balances were, however, only undertaken in 11 of the 38 patients with acute oliguric renal failure who received no protein during the oliguric phase (those of Tables 3, 4, 5, and 6). In the remaining 27 patients the rate of catabolism was estimated exclusively on the basis of urea-N production per 24 hours. Urea-N production was calculated as the sum of the urea-N excreted in the urine (usually a very small quantity) and the urea-N accumulated in the body fluids. In patients receiving no protein, the difference between a complete nitrogen balance and a calculation of urea-N production lies only in the fact that extrarenal nitrogen losses and urinary excretion of nitrogen other than urea-N remain undetermined by the latter method. This difference proved to be small in the 11 patients in whom complete nitrogen balance studies were carried out in addition to calculation of urea-N production, and it represented a constant error which remained unaffected by variations in the glucose supply or by the administration of anabolic steroids.

Results

Magnitude of urea-N production in oliguric patients receiving about 400 glucose Cal. per day

Urea-N production could be calculated in 143 of 173 patients with acute oliguric renal failure. In the remaining 30 patients calculation was impossible either because of inadequate serum and urine analyses or because a more liberal diet and/or anabolic steroid was administered almost from the onset of renal failure, leaving too short an observation period on the restricted glucose intake.

The results are illustrated in Table 2 and Fig. 1, where the patients are grouped according to the severity of the primary illness which had precipitated renal failure. Urea-N production was calculated in g./24 hours as well as in mg./litre TBW/24 hours (in order to rule out possible variations due to differences in body-size and sex distribution between groups).

It is apparent from Table 2 that there is a considerable variation in urea-N production within groups; on the whole, however, there is a correlation between urea-N production and the severity of the primary illness (Fig. 1).

Two conclusions of importance as regards the metabolic evaluation of diets and drugs in patients with acute renal failure can be drawn from these findings:

1. Results obtained in healthy, non-uraemic subjects can hardly be directly applied to patients with acute renal failure, who are frequently "hypercatabolic" and thus in a different metabolic situation.

2. Owing to the considerable variation in urea-N production from patient to patient it is preferable that each patient should

Table 2. *Urea-N production in various groups of patients with acute oliguric renal failure treated with 400 glucose Cal./day* (All cases were due to tubular nephropathy, except for those indicated with asterisks in footnote below).

Group	Total	Calculated	Urea-N production in g./24 hrs			Urea-N production in mg./litre TBW/24 hrs		
			Mean	S.D.	Range	Mean	S.D.	Range
Post-traumatic . . .	15	14	15.4	8.8	4.1—37.9	364	167	128—743
"Medical"	22*	18	12.9	7.3	4.8—35.0	348	133	137—761
"Surgical" (I). . .	52	41	11.1	3.2	4.9—19.5	292	57	164—414
"Surgical" (II) . .	47**	43	9.7	4.6	3.4—26.3	245	98	118—625
Post-partum or abortem . . .	22***	15	7.4	2.6	4.9—14.0	219	86	72—259
Intoxication . . .	15	12	7.1	3.7	1.2—13.9	196	69	108—309
	173	**143**	**10.6**	**5.5**	**1.2—37.9**	**275**	**110**	**72—761**

* Including 1 case of acute renal failure following i. v. pyelography in a patient with multiple myeloma.

** Including 1 case of total anuria following accidental operative lesion of entire renal blood supply.

*** Including 2 cases of extensive bilateral renal cortical necrosis post-partum.

"Surgical" (I) indicates extensive, complicated, or repetitive surgery.

"Surgical" (II) indicates cases in which surgery was more or less uncomplicated.

serve as his own "control" during *periods* of non-treatment and treatment. Very large series of patients would probably be necessary if one were to compare non-treated and treated *patients*.

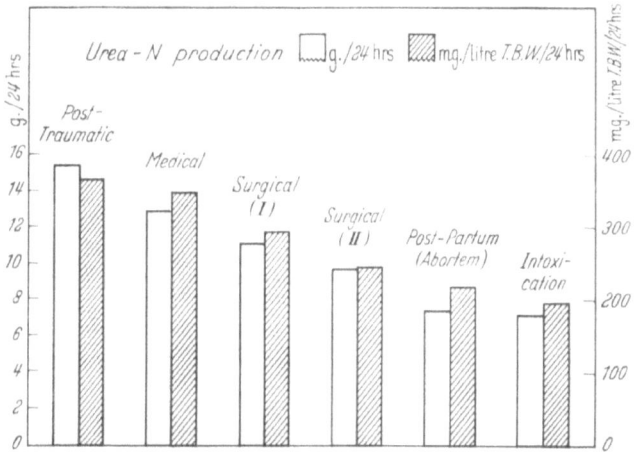

Fig. 1. Urea-N production in various groups of patients with acute oliguric renal failure (for details see Table 2)

Rate of urea-N production and the nitrogen balance in oliguric patients receiving about 400 glucose Cal. per day

Since each patient has to serve as his own control in a metabolic evaluation of the effect of diet and anabolic hormone, it is essential to study the spontaneous variations in urea-N production which may occur with time on an unvaried regimen.

For a period of one year all patients in our department received about 400 glucose Cal. per day (orally or intravenously) throughout the entire oliguric phase.

Among these patients, cases were selected on the basis of the following criteria:

1. Oliguric phase of more than 10 days' duration, in order to ensure an adequate observation period.

2. Absence of complicating severe infection or gastro-intestinal haemorrhage in the course of oliguria, since these uraemic complications may give rise to sudden and uncontrollable fluctuations in urea-N production.

10 patients fulfilled these criteria. The oliguric phase was divided into two periods of equal length in each patient. The mean

Table 3. *Urea-N production during oliguria in patients treated with 400 glucose Cal./day*

No.	R.H. Dial. Unit No.	Sex	Age	Primary disease	24-hour creat. clearance <5 ml. per min. (days)	Mean increase in serum urea-N (mg./100 ml./24 h.)				Mean urea-N production (g./24 h.)				No. of dialyses	* Alive † Dead (day)
						Whole period	First half period	Second half period	Change per cent	Whole period	First half period	Second half period	Change per cent		
1	21	m	32	Urolithiasis, pyelotomy	15	16.6	16	17	+6	7.4	6.9	7.6	+10	0	*
2	27	f	35	Haemorrhage post-partum	24	33.4	36	32	−11	9.3	9.2	9.3	+1	3	*
3	28	f	51	Strictura choledochi, hepaticojejunostomy	10	20.0	25	14	−44	5.2	5.2	5.2	0	0	*
4	30	f	30	Haemorrhage post-partum	17	22.4	24	21	−12	8.3	8.3	8.2	−1	2	*
5	31	m	36	Myeloma kidney, i.v. pyelography	10	20.9	20	21	+5	9.8	9.6	9.9	+3	0	†(10)
6	32	f	40	Cholelithiasis, cholecystectomy	12	20.8	19	21	+11	5.8	5.7	5.9	+3	1	*
7	38	m	61	Traffic accident, multiple fractures, cerebral haemorrhage	10	40.5	38	43	+13	17.1	15.9	18.0	+13	1	†(10)
8	39	f	46	Intestinal obstruction, necrosis intestini tenuis, resectio instestini	13	41.4	40	43	+7	13.1	13.1	13.0	−1	3	†(13)
9	44	m	64	Strictura choledochi, choledochoduodenostomy, necrosis pancreatis	14	34.2	35	33	−6	14.7	15.4	13.8	−10	2	†(14)
10	57	f	43	Eclampsia, shock	25	24.4	25	24	−4	12.0	12.1	12.0	−1	4	*
Average			44		15	27.5	28	27	−4	10.3	10.1	10.3	+2	1.6	

daily rise in serum urea-N and the mean daily urea-N production in these two periods were compared (Table 3).

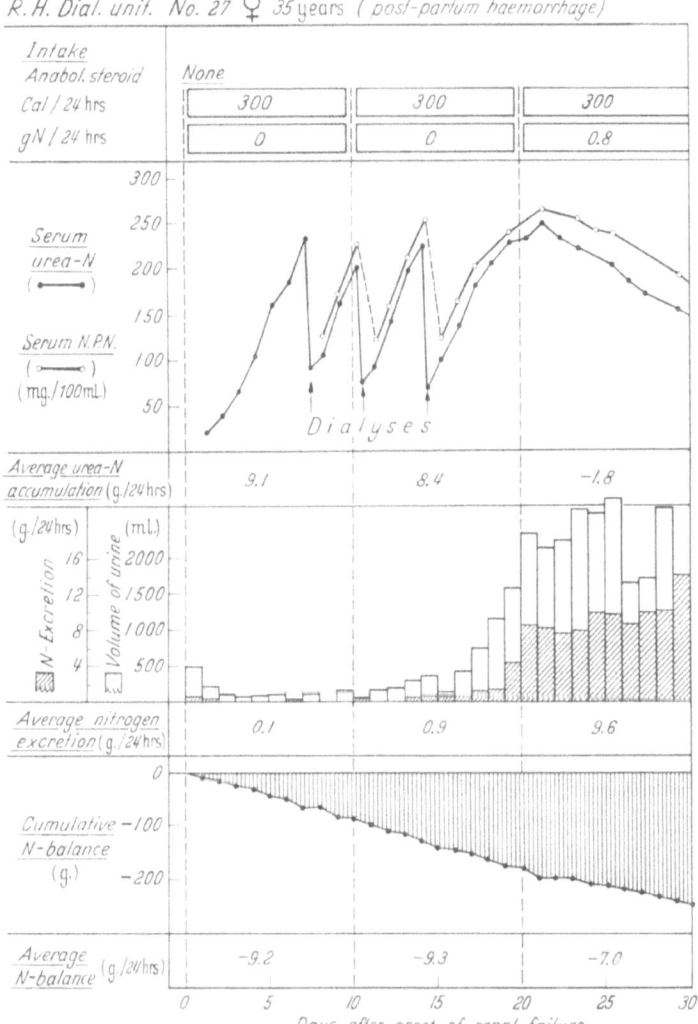

Fig. 2. Nitrogen balance in a 35-year-old woman with acute renal failure treated with 300 glucose Cal. per 24 hrs throughout oliguria

It appears from Table 3 that the rate of urea-N production varies very little in the course of an oliguric phase of from 10 to 25 days' duration, provided the therapeutic regimen is kept unchanged and provided severe complications are absent.

Complete nitrogen balance studies were carried out in 3 of the 10 patients and confirmed the results obtained by calculation of urea-N production. An example is shown in Fig. 2.

Effect of intracaval glucose administration on urea-N production and nitrogen balance in oliguric patients

For about one year we resorted to intracaval infusion of 50% glucose solution in quantities restricted by the need to maintain normal hydration during oliguria. The average quantity administered in the oliguric phase was about 600 ml. 50% glucose = 300 g. glucose = 1,140 Cal. per day.

6 patients treated with intracaval glucose infusion fulfilled the criteria adopted for the selection of the "control material" (see Table 3): 5 of these patients showed an average reduction in urea-N production of 31% (15—48) when the period of intracaval glucose infusion was compared with the preceding period in which the patients received only about 400 glucose Cal. per day; in 1 patient, who was "hypercatabolic" as compared with the others, no effect could be observed (Table 4).

Thus an apparently significant reduction in catabolism can be obtained by raising the quantity of glucose calories from about 400 to about 1,100 per day. Despite the beneficial effect of intracaval glucose infusion, we have abandoned this form of therapy, because of the disadvantages and risks which the method involves.

Complete nitrogen balances were carried out in 2 of the 6 patients and confirmed the results obtained by calculation of urea-N production.

Effect of an anabolic steroid on urea-N production and nitrogen balance in oliguric patients receiving about 400 glucose Cal. per day

For about 4 years we employed the anabolic steroid nandrolone (Durabolin) in a dose of 50 mg. i.m. daily in patients receiving about 400 glucose Cal. per day orally or intravenously.

20 patients treated with nandrolone fulfilled the criteria used for the selection of the "control material" (see Table 3).

In 10 of these patients an average reduction in urea-N production of 57% (24—72) was observed when the period on nandrolone + 400 glucose Cal./day was compared with the preceding "control

Table 4. *Urea-N production during oliguria when glucose administration was increased from about 400 Cal./day to about 1,100 Cal./day* (intracaval glucose)

No.	R.H. Dial. Unit No.	Sex	Age	Primary disease	24-hour creat. clearance below 5 ml./min. (days)	Mean increase in serum urea-N (mg./100 ml./24 h.)			Mean urea-N production (g./24 h.)			No. of dialyses	* Alive † Dead (day)
						400 glucose Cal.	1,100 glucose Cal.	Change per cent	400 glucose Cal.	1,100 glucose Cal.	Change per cent		
11	13	f	24	Eclampsia, shock	10	18	13	—27	6.8	5.8	—15	0	*
12	14	m	55	Coronary thrombosis	12	27	16	—41	11.3	8.9	—21	1	† (12)
13	18	f	61	Gastro-enteritis	22	33	22	—33	11.8	7.3	—38	1	† (22)
14	20	m	49	Infection	15	19	11	—42	8.2	4.3	—48	2	† (15)
15	26	f	34	Total bilateral renal cortical necrosis	51	15	10	—33	6.3	4.1	—35	4	† (51)
Average			45		22	22	14	—36	8.9	6.1	—31	1.6	
16	25	m	58	Cholangitis, abscessus hepatis et pancreatis	15	28	24	—14	12.8	11.8	— 8	0	† (18)

period" on 400 glucose Cal./day (Table 5). The average urea-N production of 4.4 g./day during nandrolone therapy was in fact the lowest average observed in any group among our whole series of patients with acute oliguric renal failure (cf. Table 2 and Fig. 1).

In the remaining 10 patients nandrolone was apparently ineffective (Table 6).

The patients in whom nandrolone was ineffective were "hypercatabolic" as compared with the group in which an effect was obtained. The average urea-N production of the non-responsive patients during the pre-treatment period was 13.7 g./24 hours, whereas it was only 10.2 g./24 hours during the pre-treatment period in the case of those patients in whom a favourable response to nandrolone was obtained. These average values do, however, cover considerable variations. Certain of the patients in the responsive group (Nos 17, 23, 24, and 25) were as "hypercatabolic" as the average patient in the non-responsive group (this is also true when urea-N production is calculated in mg./litre TBW/24 hours to rule out possible variations in urea-N production due to differences in body-size and sex).

Complete nitrogen balance studies were carried out in 6 of the 20 patients and confirmed results obtained by calculation of urea-N production. An example is shown in Fig. 3.

During the entire 4-year period nandrolone treatment was given to a total of 56 patients with an oliguric phase of more than 10 days' duration. Only in 20 patients (those shown in Tables 5 and 6) could a definite answer regarding the effect of nandrolone be given. In the remaining 36 patients, complications likely to reduce the reliability of the evaluation (infection, haemorrhage) were present to varying degrees.

An attempt was, however, made to estimate the effect of nandrolone in all 56 patients by comparing urea-N production before and after administration of the anabolic steroid. The results are shown in Table 7, where the patients are grouped according to the severity of the primary disease which had precipitated renal failure. Two facts are apparent from Table 7.

First, a favourable response to nandrolone appears to occur with decreasing frequency as the primary disease becomes more and more severe and the patient more and more "hypercatabolic" (compare Table 7 with Table 2 and Fig. 1).

Second, a favourable response to nandrolone is less frequent in patients with complications (6 responses out of 36) than in patients free from complications (10 responses out of 20).

Table 5. Urea-N production during oliguria before and after administration of nandrolone (responsive cases)

No.	R.H. Dial. Unit No.	Sex	Age	Primary disease	Creat. clear. below 5 ml./min. (days)	Mean increase in serum urea-N (mg./100 ml./24 h.)			Urea-N production (g./24 h.)			No. of dialyses	*Alive †Dead (day)
						Before nandrol.	After nandrol.	Change per cent	Before nandrol.	After nandrol.	Change per cent		
17	71	m	22	Traffic accident	24	35.5	13.4	—62	13.8	5.6	—59	4	*
18	90	f	62	C. cap. pancreatis, gastro-entero-anastomosis et cholecysto-enterostomosis facta	18	31.5	13.5	—57	10.6	4.2	—60	1	†(18)
19	103	m	68	Prostatectomy	22	13.8	4.7	—66	5.6	1.9	—66	0	*
20	104	m	28	CCl$_4$ poisoning	15	14.8	7.3	—51	5.3	4.2	—25	1	*
21	114	f	62	Biliary colic	21	18.3	4.8	—74	7.4	2.9	—61	1	*
22	146	m	41	Laminectomy	20	18.7	12.9	—31	10.2	7.7	—24	1	*
23	150	m	29	HgO poisoning	13	26.9	9.4	—65	12.2	4.5	—63	1	*
24	151	m	45	Ren arcuatus, accidental operative lesion of entire renal blood supply	27	28.8	8.9	—69	12.2	3.4	—72	1	†(27)
25	180	f	23	Anuria post-abortem	23	43.9	14.9	—66	14.0	5.1	—63	2	*
26	205	f	71	Cancer vesicae, cystectomia partialis	15	27.5	6.5	—76	10.5	3.7	—65	1	*
Average			45		20	25.9	9.6	—63	10.2	4.4	—57	1.3	

Table 6. *Urea-N production during oliguria before and after administration of nandrolone (non-responsive cases)*

No.	R.H. Dial. Unit No.	Sex	Age	Primary disease	Creat. clear. below 5 ml/min. (days)	Mean increase in serum urea-N (mg./100 ml./24 h.)			Urea-N production (g./24 h.)			No. of dialyses	* Alive † Dead (day)
						Before nandrol.	After nandrol.	Change per cent	Before nandrol.	After nandrol.	Change per cent		
27	76	m	57	Appendicitis, peritonitis	16	33.4	28.6	−14	14.0	11.4	−18	2	†(16)
28	80	m	79	Obstructive jaundice, cholangitis	12	28.0	30.8	+10	12.4	13.7	+10	2	†(12)
29	81	m	41	Gastric resection, peritoneal haemorrhage	13	33.9	37.6	+11	16.8	19.5	+16	2	†(28)
30	99	m	45	Cholecystectomy	15	37.8	34.0	−10	16.6	15.0	−9	3	†(15)
31	109	f	49	Cellulitis, sepsis tetanus	11	32.7	26.5	−19	12.0	10.0	−16	2	†(11)
32	139	f	62	Cholecystitis perf., subphrenic absc.	10	35.3	37.0	+5	12.4	13.8	+11	1	*
33	145	m	56	Traffic accident, multiple fractures	10	32.9	38.1	+16	13.2	14.4	+9	2	†(10)
34	153	m	32	Traffic accident, multiple fractures, ruptured spleen and colon, contusion left kidney	17	41.0	39.0	−5	16.0	15.6	−2	3	*
35	173	m	48	Complicated herniotomy, incompatible blood	19	28.0	23.2	−17	12.3	11.5	−7	3	*
36	202	m	64	Gastrectomy	16	26.8	30.9	+15	11.1	12.3	+11	4	*
Average			53		14	32.9	32.6	−1	13.7	13.7	0	2.4	

In those patients who showed a favourable response to nandrolone the effect of the anabolic steroid on urea-N production was

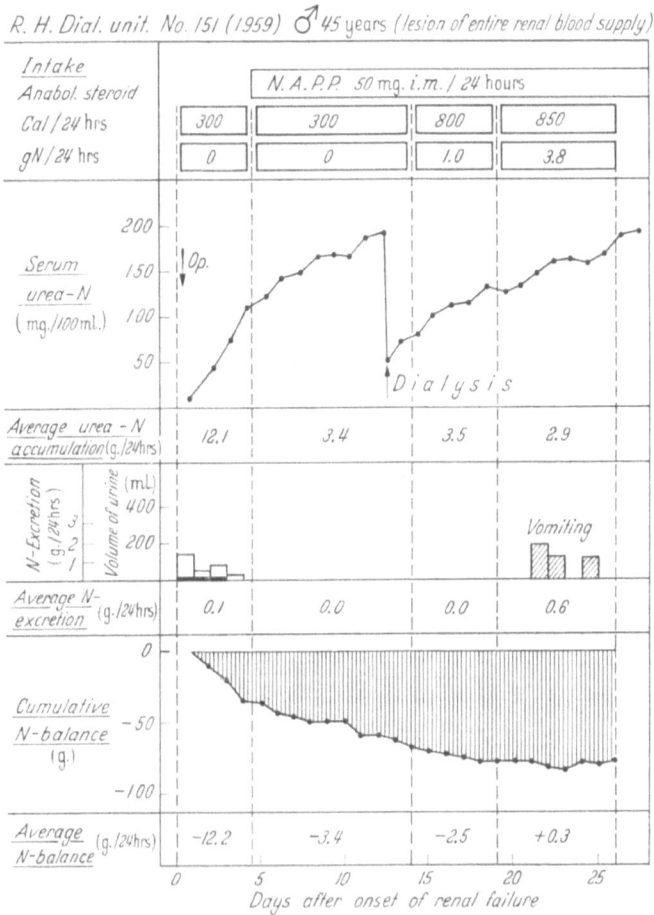

Fig. 3. Nitrogen balance in a 45-year-old man with acute renal failure treated with nandrolone-phenylpropionate (N.A.P.P.) and varying calorie and protein intakes

very prompt and generally set in within 1 or 2 days (see example in Fig. 3).

Table 7. *Effect of nandrolone in various groups of patients with oliguria of more than 10 days' duration (same groups as shown in Table 2)*

Group	Total No. of cases	Creat. clear. below 5 ml./min. for more than 10 days		Effect of nandrolone				Per cent effect
				Uncomplicated cases		Complicated cases		
		Total	Treated with nandrolone	Effect (cf. Table 5)	No effect (cf. Table 6)	Effect	No effect	
Post-traumatic	15	10	7	1	2	0	4	9
"Medical"	22	10	3	0	1	0	2	0
"Surgical" (I) . . .	52	29	23	1	4	3	15	17
"Surgical" (II) . . .	47	17	16	5	3	1	7	38
Post-partum or abortem	22	8	3	1	0	1	1	66
Intoxication . . .	15	5	4	2	0	1	1	75
	173	**79**	**56**	**10**	**10**	**6**	**30**	**29**

Effect of an anabolic steroid on the nitrogen balance in oliguric patients receiving protein-containing diets

Patients with prolonged acute renal insufficiency frequently experience marked weight losses, which may continue far into the diuretic phase despite a rapid improvement in renal function and "biochemical uraemia" (see example in Fig. 2). In our patients with oliguria of more than 10 days' duration, maximal weight loss averaged 9.4% of the normal body-weight, the extremes being 4.5% and 35%. The resulting cachexia may give rise to such serious complications that a more liberal administration of calories and proteins during oliguria may appear justified even at the risk that this might result in an increased need for haemodialysis.

Figs 3 and 4 show the effect of more liberal feeding on nitrogen balance and urea-N accumulation during oliguria. Patient No. 106/1958 (Fig. 4) was fed by stomach tube, and patient No. 151/1959 (Fig. 3) by mouth. In both patients the amount of protein fed during the protein balance periods was approximately equal to the "endogenous protein breakdown" in the isocaloric nonprotein balance periods.

It appears from Figs 3 and 4 that an increase in protein intake by a quantity which

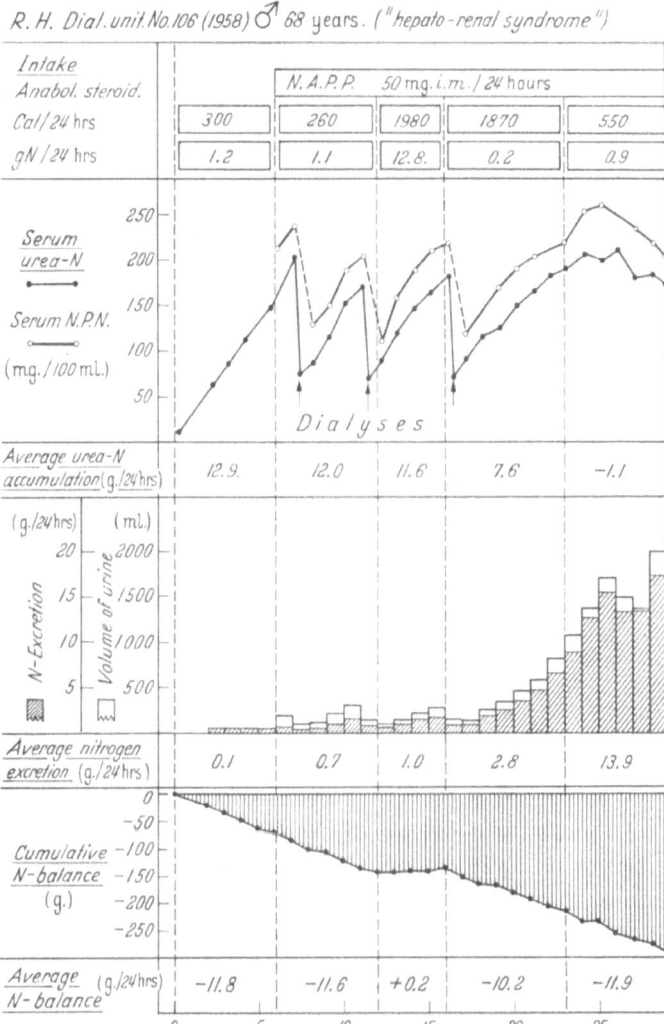

Fig. 4. Nitrogen balance in a 68-year-old man with acute renal failure treated with nandrolone-phenylpropionate (N.A.P.P.) and varying calorie and protein intakes

is about equal to endogenous protein break-down during low-protein feeding can result in nitrogen equilibrium without measurable increases in the rate of urea-N accumulation. This indicates that the protein administered is almost completely "anabolised" (or that it prevents "catabolism" of an almost equivalent amount of body protein).

Whether this phenomenon is due to an increased spontaneous tendency to anabolise exogenous protein in semi-starved and protein-depleted patients, or whether it represents the effect of the simultaneously administered nandrolone, cannot be stated, since controlled balance studies with and without nandrolone were not carried out. The results obtained in a semi-starved patient with chronic renal failure (Fig. 5) do, however, suggest that the administration of anabolic steroid may be the more important of the two possibilities.

Encouraged by these findings, we have since been feeding our patients on a more liberal protein diet, even during the oliguric phase of acute renal failure. Owing to the fact that optimal hydration has to be maintained during oliguria, that the appetite of the patients is frequently reduced, and that the feeding of glucose, fat, and protein by stomach tube may give rise to diarrhoea, it has generally proved impossible to supply a normal diet during oliguria. The average daily quantities which have been given in our cases amounted to 800 Cal. and 4.2 g. of protein-N, supplemented by 50 mg. nandrolone per 24 hours. The calories and the protein were given either by mouth or as glucose and amino acids i. v. The impression gained from this therapy is that neither urea-N production nor potassium accumulation are measurably increased, despite the administration of these limited quantities of protein. The "biochemical" and clinical response to this form of therapy, however, will have to be evaluated in larger series before any final conclusions can be drawn.

Effect of an anabolic steroid on nitrogen balance and urea-N accumulation in patients with chronic renal failure

Fig. 5 illustrates the effect of 17-ethyl-19-nortestosterone (nor-ethandrolone, Nilevar) in a patient with markedly impaired renal function due to chronic pyelonephritis.

The patient, who was severely underweight and protein-depleted when the balance study was started, developed a positive nitrogen balance when a high-protein and high-calorie diet was given. It is apparent from Fig. 5, however, that she anabolised considerably more of the ingested protein when norethandrolone

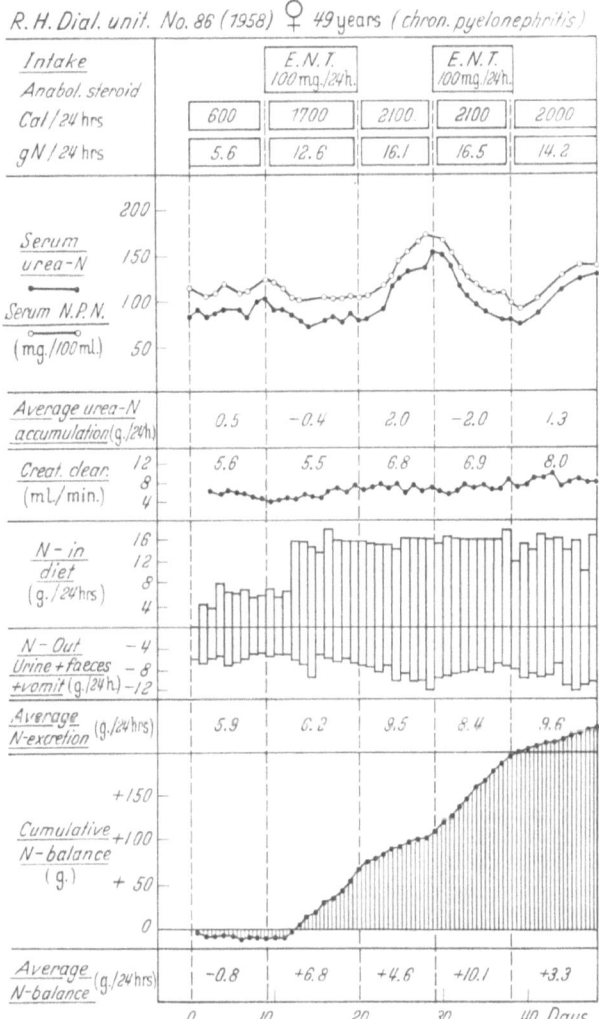

Fig. 5. Nitrogen balance in a 49-year-old woman with chronic renal failure treated with nor-ethandrolone (E.N.T.) and varying calorie and protein intakes

was administered than during control periods without norethandrolone. Over the entire period of high-protein and high-calorie feeding, i.e. from the 12th to 46th day, the patient retained 220 grammes of nitrogen $= 1,375$ grammes of protein $= 6,900$ grammes of lean tissue, but the weight gain recorded within the same 34 days was only 3,000 grammes. We have observed a roughly similar discrepancy between the expected weight gain due to N-retention and the actual weight gain in all 5 patients in whom norethandrolone or nandrolone elicited an effect.

Only 5 of our 16 patients with chronic renal failure responded to treatment with norethandrolone (50—120 mg. daily) or nandrolone (50 mg. daily). The other 11 patients did not respond; in most of them either infection was present or the renal disease or uraemia were complicated in other ways.

Discussion

Acute renal failure

Modern techniques of haemodialysis undoubtedly represent the greatest advance hitherto made in the management of patients with acute oliguric renal failure. Despite this fact, however, all measures capable of retarding the development of uraemia and minimising protein depletion and weight loss are still of great importance. Three main approaches have been adopted:

1. Feeding of non-protein calories

a) *Glucose-oil mixtures, administered orally or via stomach tube*, have been advocated by Borst (*1*) and by Bull, Joekes, and Lowe (*5*). These diets can reduce catabolism, but since they have to be given in concentrated form in order to maintain normal hydration during oliguria, they frequently provoke diarrhoea and vomiting.

b) *Intracaval infusion of 50% glucose solution*. If optimal hydration is to be maintained, only about 500—600 ml. of this solution, yielding about 1,000 non-protein calories, can be administered per day during the oliguric phase. It has been questioned whether this quantity of glucose calories has an effect on catabolism over and above that achieved by the administration of 400 glucose Cal./day. However, the experiences of Merrill (*13*), the metabolic studies of Gjørup (*9*) and of Blagg, Parsons, and Young (*3*), and the results of the present investigation indicate that this is indeed the case. In the present work, an increase in glucose Cal. from 400 to 1,000 daily was followed by a reduction of about 30% in urea-N production. But it must be emphasised that this effect was seen only in patients with a low pre-treatment urea-N production, whereas the "hypercatabolic" patient failed to respond. Despite the demonstrated effectiveness of intracaval glucose administration, the advantages of the method would appear to be outweighed by its disadvantages:

1. An indwelling caval catheter represents a route for the introduction of bacteria, and the uraemic subject is notoriously susceptible to septicaemia.

2. Thrombosis of the caval vein has been observed and may lead to serious complications, e.g. embolism.

3. Where such fluid as is permissible is administered exclusively by the parenteral route, this imposes a grave psychological strain on the conscious oliguric patient.

c) *Administration of as many glucose calories as can be given by mouth or via the peripheral veins.* Owing to the disadvantages and risks of the above-mentioned methods, many haemodialysis units have restricted the administration of non-protein calories during oliguria to as much glucose as the patient can take orally or as he can be given via the peripheral veins. Given the restricted fluid volume, the average is about 400 glucose Cal./day. Even this limited amount of glucose does have a definite, albeit not very big, protein-sparing effect [see, for example, BLAGG, PARSONS, and YOUNG (3)].

In the present consecutive series of patients with acute oliguric renal failure observed during treatment with approx. 400 glucose Cal./day, we found an average urea-N production of 10.6 grammes daily (equalling 66.3 grammes of protein or about 330 grammes of lean tissue) and an average daily rise in serum urea-N of 275 mg per litre (equalling 590 mg. of urea per litre). More effective means of reducing catabolism are thus badly needed, particularly in the "hypercatabolic" patient, in whom lean-tissue loss is considerable and the rate of progression of uraemia so rapid that haemodialysis becomes necessary every 3—4 days.

2. Feeding of non-protein calories supplemented by administration of an anabolic steroid

The use of anabolic steroids in the symptomatic treatment of renal failure was first advocated by FREEDMAN and SPENCER (7) and by MERRILL (14), who used testosterone in relatively low dosage and with questionable effect. The new and less androgenic 19-nor-steroids have been employed by GJØRUP and THAYSEN

(*10*, *11*) and by McCracken and Parsons (*15*) in comparatively high doses and with good effect in selected cases. So far as I am aware, growth hormone has never been used in this connection.

When patients receiving about 400 glucose Cal./day are treated with norethandrolone (Nilevar) in a dosage of 30 mg. daily [Blagg and Parsons (*2*)] or nandrolone (Durabolin) in a dosage of 50 mg. daily [Gjørup and Thaysen (*12*)], a decrease of about 50 to 70% in urea-N production may be observed. A response, albeit somewhat less pronounced, may even be obtained in patients who receive no calories at all [Blagg, Parsons, and Young (*3*)]. Whether an increase in glucose calories over and above 400 Cal./day would further enhance the effect of an anabolic steroid is not known.

The effect of the anabolic steroid on urea-N production is relatively prompt and sets in within 1 or 2 days after administration of the drug. The slower progression of biochemical uraemia is paralleled by a delay in the rate at which clinical symptoms of uraemia develop, and the need for supplementary haemodialysis is reduced.

The anabolic steroids, however, are effective only in a limited number of patients. Gjørup and Thaysen (*12*) showed that nandrolone was active only in patients with relatively uncomplicated acute renal failure and with a comparatively low pre-treatment rate of catabolism. "Hypercatabolic" patients apparently did not respond to treatment. In the patients discussed in the present study, it has been shown that only about $1/_4$ of all patients in a consecutive series respond to therapy, and that the nature of the primary disease and the presence of uraemic complications have a decisive influence in determining whether or not 19-nor-steroid is effective (Table 7). Complications in the primary disease or in the uraemia do definitely affect the rate of catabolism (Table 2), but it is apparent that factors other than the pre-treatment catabolic rate also influence the response to anabolic steroids. This highly variable response from patient to patient may explain the controversial evidence in the literature regarding the effectiveness of anabolic steroids in acute renal failure (*6*, *8*, *16*). Several reports are based on one or, at the most, only a few cases.

The problem as to why anabolic steroids are only effective in a limited number of patients has not been clarified by the present study. Patients with complications in their primary disease and in their uraemia may be in a "situation of stress" leading to increased adrenal secretion of catabolically active steroids. It has been shown that administration of a steroid endowed with catabolic activity (e.g. prednisone) reduces the anabolic effect of an anabolic steroid

[BRØCHNER-MORTENSEN, GJØRUP, and THAYSEN (4)]. Extensive trauma or infection may give rise to a considerable break-down of protein in devitalised tissue — a type of "catabolism" which possibly cannot be prevented by treatment with an anabolic hormone. All such considerations are, however, still speculative.

In conclusion, of the results of the present study it must be stated that the use of anabolic steroids in patients receiving about 400 glucose Cal./day represents only a limited advance in the symptomatic treatment of acute renal failure. The anabolic steroids are active only in some of the patients, i.e. in those with slowly progressing uraemia who are least in need of therapy. In "hypercatabolic" patients, who are in dire need of an anabolic effect because the rapidly progressing uraemia necessitates frequent dialyses, the anabolic steroids are almost always ineffective.

3. Feeding of more liberal protein diets, supplemented by administration of an anabolic steroid

Before the introduction of haemodialysis, uraemic intoxication constituted the principal risk in acute renal failure. The weight loss induced by a protein-free, low-calorie diet was of minor importance as compared with the risk of progressive uraemia. Moreover, the weight loss would never reach excessive degrees, since few patients survived for more than 8—10 days after complete renal shut-down. The introduction of highly effective and comparatively safe techniques of haemodialysis wrought a decisive change. The development of uraemic intoxication can now be prevented and the patient kept alive during several weeks of severe renal failure. In prolonged oliguria of 25—30 days' duration, the chief problem is no longer uraemia, but the pronounced weight loss, which may continue for some time into the diuretic phase. In our patients, the maximal weight loss averaged 9.4% of the original body-weight, but losses amounting to 25—35% were sometimes observed. The resulting cachexia may well contribute to the serious complications (in particular septicaemia) which are by no means uncommon during the diuretic phase when biochemical uraemia is rapidly improving. More liberal feeding (including protein) during the oliguric phase would appear to be justified in an attempt to counteract the weight loss, even if this means that the uraemia progresses more rapidly and that dialysis has to be performed at more frequent intervals.

An increased protein intake does tend to aggravate biochemical uraemia in patients with chronic renal failure (cf. Fig. 5), and, as might be expected, the same also applies to cases of acute renal insufficiency [Blagg, Parsons, and Young (3)]. When, however, an anabolic steroid is administered in addition to the protein diet, the rate of progression of the uraemia is not measurably enhanced as a result of the proteins supplied. This is shown by the results of the present study (Figs 3 and 4) and by the recent observations of Blagg, Parsons, and Young (3).

Further experience will be needed, however, before it can be ascertained whether the anabolic steroids always show this "anabolic" effect. It is also too early to make any statements about the clinical benefit to be derived from combining "free-feeding therapy" during oliguria with an anabolic steroid. The fact that the external nitrogen balance is improved by the anabolic steroid does not necessarily mean that the nitrogen retained is utilised for the repair of those tissues and protein pools which are particularly sensitive to starvation and protein depletion.

Chronic renal failure

A positive nitrogen balance may temporarily be achieved by the use of an anabolic steroid, but if the drug is administered over a prolonged period in unaltered dosage a new steady state is apparently arrived at. In view of this, it appears reasonable that the anabolic steroids should have been tried first and foremost in acute reversible renal failure, whereas they would appear to be of less value in chronic renal failure, in which the likelihood of partial or complete recovery of renal function is small.

In patients with *relatively well-preserved renal function* (GFR above 10—15 ml./min.), a moderate restriction of dietary protein usually suffices to prevent severe clinical symptoms of uraemia and does not interfere with the nitrogen balance or with maintenance of a stable normal weight. In this situation, anabolic steroids have little to add to the therapeutic regimen, and we do not consider them indicated.

In order to prevent manifest uraemic intoxication in patients with *very severe chronic renal failure* (GFR below 10 ml./min.), the dietary protein intake must frequently be reduced so drastically that a negative nitrogen balance and weight loss will follow. The often marked decrease in the appetite of the patient accentuates the adverse effects of protein restriction. In this situation, an anabolic steroid can apparently enable the patient to ingest more protein, resulting in a positive nitrogen balance and weight gain without accentuation of the uraemia (Fig. 5). The appetite and general well-being of the patient also appear to improve. The increased calorie intake promotes weight gain and improves the nitrogen balance, but it should be emphasised that we have observed the protein-sparing action of anabolic steroids in isocaloric balance periods. The anabolic effect of the anabolic steroid is, however, only transient, so that the ultimate prognosis will depend almost entirely on the extent to which it is possible to achieve a simultaneous improvement of the underlying renal disease. Further, it must be stressed that only about $1/3$ of our patients with severe chronic renal failure responded to treatment with an anabolic steroid.

As might have been expected, it can be concluded from our results that anabolic steroids are only of limited value in the management of chronic renal failure.

Summary

The effect of anabolic steroids (19-nortestosterone-phenylpropionate = nandrolone-phenylpropionate = Durabolin and 17 α-ethyl-19-nortestosterone = norethandrolone = Nilevar) in the symptomatic management of uraemia due to acute and chronic renal failure was studied by metabolic techniques.

In *acute oliguric renal failure*, the rapid progression of uraemia, necessitating repeated haemodialyses, and the pronounced weight loss, induced by conventional protein-free, low-caloric diets, represent major problems.

In patients receiving the conventional diet of 400 glucose Cal. per day, nandrolone (50 mg. per day) reduced urea-N production by about 50—70%, whereas an increase in the glucose supply to 1,000 Cal. per day (intracaval infusion of 50% glucose) decreased urea-N production by only about 30%. The slower progression of "biochemical" uraemia occurring in response to nandrolone, was paralleled by a slower progression of the clinical uraemic symptoms, and the need for supplementary haemodialysis was reduced. Nandrolone was, however, only effective in 29% of the patients, i.e. chiefly in those who had a low rate of catabolism before treatment. Patients who were "hypercatabolic" — owing to their primary disease or to some complication in their uraemia — did not apparently respond at all. Thus, the use of an anabolic steroid constitutes a measure of only limited practical importance with which to supplement the conventional conservative management of patients suffering from acute renal failure.

In order to prevent the progressive cachexia induced by a diet of 400 glucose Cal., protein-containing diets were administered even during actual oliguria. When nandrolone (50 mg. per day) was given simultaneously, the rate of progression of the uraemia was not apparently increased by the proteins fed, indicating that most of the dietary protein was being anabolised. The consistency of this effect, and the clinical value of combining an anabolic steroid with "free-feeding therapy" in the oliguric patient, will, however, have to be evaluated in larger groups of patients.

In 5 of 16 patients with very severe *chronic renal failure*, nandrolone (50 mg. per day) or norethandrolone (50—100 mg. per day) enabled the patient to take more protein, resulting in a positive nitrogen balance and weight gain without an increase in "biochemical" uraemia. But the improvement in the nitrogen balance was only transient and was followed by a new steady state in response to further treatment with the anabolic agent in the same dosage. In view of this, there is no apparent reason for using anabolic steroids in the long-term treatment of patients with chronic renal failure. These drugs, however, may be tried in the underweight and severely uraemic patient in order to produce a temporary improvement, which can be exploited by means of other therapy directed towards the underlying renal disease.

Zusammenfassung

Die Wirkung anaboler Steroide (19-nortestosteron-phenylpropionat = Nandrolonphenylpropionat = Durabolin, 19-nor-17α-aethyltestosteron = Norethandrolon = Nilevar) wurde im Rahmen der symptomatischen Behandlung der Uraemie infolge akuter oder chronischer Niereninsuffizienz durch Messung einzelner Stoffwechselgrößen untersucht.

Bei der *akuten Niereninsuffizienz mit Oligurie* stellen das rasche Fortschreiten der Uraemie, das wiederholte Haemodialysen erforderlich macht, und der starke Gewichtsverlust infolge der üblichen proteinfreien Diät mit niedrigem Kaloriengehalt die Hauptprobleme für die Behandlung dar.

Bei Patienten, die die übliche Diät von 400 Glukosekalorien pro Tag erhalten, vermindert Nandrolon (50 mg täglich) die Produktion von Harnstoff-N um ungefähr 50—70%, während eine Steigerung der Glukosezufuhr auf 1000 Kal. pro Tag (intracavale Infusion von 50% Glukoselösung) nur eine Verminderung der Bildung von Harnstoff-N um etwa 30% herbeiführt. Das durch Nandrolon bedingte langsamere Fortschreiten der „biochemischen" Uraemie geht parallel mit einer verzögerten Entwicklung der klinischen Uraemiesymptome, wodurch die Notwendigkeit zu zusätzlicher Haemodialyse vermindert wird. Nandrolon war jedoch nur bei 29% der Patienten wirksam, und zwar vorwiegend bei denjenigen, die vor der Behandlung einen niedrigen Eiweißabbau hatten. Patienten, die infolge ihrer primären Erkrankung oder auf Grund anderer Komplikationen in einem „hyperkatabolen" Zustand waren, sprachen auf die Behandlung nicht an. Die Anwendung anaboler Steroide stellt somit bei Patienten mit akuter Niereninsuffizienz neben den üblichen konservativen Methoden nur eine zusätzliche Maßnahme von begrenzter praktischer Bedeutung dar.

Um die fortschreitende Kachexie zu verhüten, die sich als Folge der 400-Glukosekalorien-Diät einstellt, wurden proteinhaltige Diäten verabreicht, und zwar auch bei Bestehen einer Oligurie. Gleichzeitige Gabe von Nandrolon (50 mg/Tag) beschleunigte das Fortschreiten der Uraemie infolge vermehrt zugeführten Eiweißes nicht, ein Hinweis darauf, daß das meiste Nahrungseiweiß angebaut wurde. Die Konstanz dieser Wirkung und der klinische Wert der Kombination eines anabolen Steroids mit einer „freien Ernährungstherapie" bei Oligurie bedürfen jedoch noch der Bestätigung an einer größeren Zahl von Patienten.

Bei 5 von 16 Patienten mit sehr schwerer *chronischer Niereninsuffizienz* konnte bei täglicher Gabe von Nandrolon (50 mg) oder Norethandrolon (50—100 mg) eine eiweißreichere Diät gestattet werden, wobei eine positive Stickstoffbilanz und Gewichtszunahme ohne Anstieg der „biochemischen" Uraemie resultierten. Die Verbesserung der Stickstoffbilanz war jedoch nur vorübergehend, da sich trotz Weiterbehandlung mit gleichen Dosen der anabolen Steroide ein neues Gleichgewicht einstellte. Es liegt daher kein Grund vor, anabole Steroide in der Langzeitbehandlung von Patienten mit chronischer Niereninsuffizienz zu geben. Ein Versuch mit diesen Substanzen ist jedoch beim untergewichtigen und schwer uraemischen Patienten

indiziert, um eine zeitweilige Besserung zu erzielen, die für andere thera-
peutische Maßnahmen gegen das zugrundeliegende Nierenleiden ausgenutzt
werden kann.

Résumé

Les effets des stéroïdes anabolisants (phénylpropionate de 19-nortesto-
stérone = phénylpropionate de nandrolone = Duraboline et 17α-éthyl-19-
nortestostérone = noréthandrolone = Nilévar) dans le traitement sympto-
matique des hyperazotémies par insuffisance rénale aiguë ou chronique
ont été étudiés au moyen de techniques métaboliques.

Dans l'insuffisance rénale aiguë anurique, la rapide élévation de l'urée
sanguine, nécessitant des hémodialyses répétées, et l'importante perte de
poids provoquée par le régime classique hypocalorique sans protéines,
constituent des problèmes majeurs.

Le nandrolone (50 mg par jour) a pu réduire, chez les malades recevant
le régime habituel glucosé à 400 Cal. par jour, la production d'azote uréique
dans une proportion de 50 à 70%, alors qu'une augmentation de l'apport
de glucose, jusqu'à 1000 Cal. par jour (perfusion cave de glucose à 50%),
diminue la production d'azote uréique seulement de 30%. La progression plus
lente du taux de l'urée sanguine qu'a réalisée le nandrolon es'est accompagnée
parallèlement d'une évolution plus lente des signes cliniques d'urémie et
d'une moindre nécessité d'hémodialyses supplémentaires. Le nandrolone s'est
cependant montré actif seulement dans 29% des cas, surtout chez ceux qui
présentaient avant le traitement un catabolisme peu intense. Les malades
en "hypercatabolisme" — en raison de leur maladie causale ou par complication de leur urémie — n'ont absolument pas réagi au traitement. L'utilisation des stéroïdes anabolisants ne représente donc qu'un adjuvant au
traitement classique des insuffisances rénales aiguës, d'intérêt pratique
limité.

Afin d'empêcher la cachexie progressive qu'entraîne l'alimentation
glucosée à 400 Cal., on a administré des régimes contenant des protéines,
même en pleine oligurie. Quand le nandrolone (50 mg par jour) a été administré simultanément, l'élévation du taux d'urée n'a apparemment
pas été accélérée par l'adjonction de protéines, ce qui indique que la plus
grande partie des protéines alimentaires a été métabolisée. La constance
de cet effet ainsi que l'intérêt clinique de l'association d'un stéroïde anabolisant et d'un «régime libre» chez des malades anuriques, restent cependant
à être appréciés sur un plus grand nombre de cas.

Chez 5 malades sur 16 présentant une insuffisance rénale chronique
très sévère, le nandrolone (50 mg par jour) et le noréthandrolone (50 à
100 mg par jour) ont permis aux malades de prendre plus de protéines,
d'où un bilan azoté positif et un gain de poids, sans élévation du taux
de l'urée sanguine. L'amélioration du bilan azoté s'est montrée cependant
transitoire et a été suivie d'un nouvel état d'équilibre, sous la même dose
du stéroïde anabolisant. Dans ces conditions, il n'apparaît aucune raison
d'administrer des stéroïdes anabolisants en cure prolongée chez des malades en insuffisance rénale chronique. Ces produits peuvent cependant
être essayés chez les malades présentant une hyperazotémie importante et
un poids insuffisant, afin d'obtenir une amélioration temporaire pouvant
être mise à profit pour d'autres thérapeutiques dirigées directement contre
la maladie rénale causale.

References

1. Borst, J. G. G.: Lancet (G.B.) **254**, 824 (1948). — 2. Blagg, C. R., and F. M. Parsons: Lancet (G.B.) **1960**/II, 577. — 3. Blagg, C. R., F. M. Parsons, and G. A. Young: Lancet (G.B.) **1962**/I, 608. — 4. Brøchner-Mortensen, K., S. Gjørup, and J. Hess Thaysen: Acta med. Scand. **165**, 197 (1959). — 5. Bull, G. M., A. M. Joekes, and K. G. Lowe: Lancet (G.B.) **257**, 229 (1949). — 6. Eaton, J. C.: Proc. Roy. Soc. Med. **52**, 511 (1959). — 7. Freedman, P., and A. G. Spencer: Clin. Sc. (G.B.) **16**, 11 (1957). — 8. Gerber, W. von, and P. Cottier: Helvet. med. acta **27**, 539 (1960). — 9. Gjørup, S.: Acta med. Scand. **161**, 223 (1958). — 10. Gjørup, S., and J. Hess Thaysen: Uskr. Laeger (Den.) **120**, 1499 (1958). — 11. Gjørup, S., and J. Hess Thaysen: Lancet (G.B.) **1958**/II, 886. — 12. Gjørup, S., and J. Hess Thaysen: Acta med. Scand. **167**, 227 (1960). — 13. Merrill, J. P.: Fifth Conf. Renal Funct., Josiah Macy, New York (1953). — 14. Merrill, J. P.: The Treatment of Renal Failure. Grune and Stratton, New York (1955). — 15. McCracken, B. H., and F. M. Parsons: Lancet (G.B.) **1958**/II, 885. — 16. Szold, E.: Lancet (G.B.) **1959**/I, 368.

Discussion

ASCHKENASY: I should like to make two remarks in connection with Dr. THAYSEN's paper. The first concerns the lack of parallelism between the action of androgens on the nitrogen balance and on body-weight. This well-known discrepancy is perhaps due partly to the retention of nitrogen in non-protein form, but it may possibly also be explained by the fact that the protein-anabolic action of androgens interferes with the fat-catabolic action of the same hormones. It is probable that both humans as well as laboratory animals receiving androgens lose fat while they are gaining nitrogen.

My second remark is related to the effect of 17α-ethyl-19-nortestosterone on the kidney. I should like to ask Dr. THAYSEN whether he has had an opportunity to compare the nephrotrophic effect of this steroid with that of other androgens. In rats the 17α-ethyl-19-nortestosterone has rather a mild nephrotrophic effect. I did not observe any statistically significant changes in the weight of the kidneys in castrated male rats treated with this steroid over a period of one month. Nor was 17α-methyl-testosterone any more effective in this respect (Table 1).

Table 1. *Weights of kidneys and urinary bladder (mean values in mg. $\pm$ S.E.M. per 100 g. of body-weight) in castrated male control rats and in castrated rats receiving oral treatment with either 17α-methyl-testosterone or 17α-ethyl-19-nortestosterone for 4 weeks (5 mg. daily). The dry weights are given in brackets.* [From ASCHKENASY, A.: Thérapie (Fr.) 14, 332 (1959)].

Organ	Controls (7 rats)	17α-Methyl-testosterone (9 rats)	17α-Ethyl-19-nortestosterone (8 rats)
Kidneys	756 $\pm$ 17	779 $\pm$ 28	779 $\pm$ 21
	(184 $\pm$ 3)	(193 $\pm$ 6)	(193 $\pm$ 3)
Urinary bladder .	39 $\pm$ 3	38 $\pm$ 7	38 $\pm$ 2

It was only after 4 months' administration of large doses of 17α-ethyl-19-nortestosterone that the kidneys became significantly hypertrophied in female as well as in male (non-castrated) rats (Table 2).

Table 2. *Weights of kidneys and urinary bladder (mean values in mg. $\pm$ S.E.M. per 100 g. of body-weight) in control (non-castrated) rats and in rats treated with 17α-ethyl-19-nortestosterone (ENT) for 4 months (25 mg. or 100 mg. per 100 g. of dry diet). The dry weights are given in brackets.* [From ASCHKENASY, A.: Thérapie (Fr.) 14, 332 (1959)].

Group	Kidneys		Urinary bladder	
	Males	Females	Males	Females
Controls	979 $\pm$ 61	1006 $\pm$ 27	28 $\pm$ 2	38 $\pm$ 3
(4 M + 9 F)	(245 $\pm$ 11)	(243 $\pm$ 6)		
25 mg. ENT	1160 $\pm$ 34	1167 $\pm$ 33	39 $\pm$ 3	34 $\pm$ 1
(10 M + 10 F)	(290 $\pm$ 8)	(299 $\pm$ 11)		
100 mg. ENT	1296 $\pm$ 33	1334 $\pm$ 42	36 $\pm$ 4	50 $\pm$ 7
(10 M + 10 F)	(318 $\pm$ 7)	(125 $\pm$ 8)		

In the light of these experimental data it is perhaps not very surprising that you did not obtain notable therapeutic results with this anabolic androgen.

THAYSEN: It was very interesting for me to hear Dr. ASCHKENASY's explanation for the discrepancy between the measured weight gain and the expected weight gain assuming that all retained nitrogen was stored in protein form as lean tissue. I myself have no explanation to offer for this phenomenon. The average daily nitrogen gain was about 5 g., and I do not think that we can have missed that much merely because of a technical error.

With respect to your second comment, I want to emphasise that the reduction in urea accumulation, which was noted in a limited number of patients treated with 19-nor-steroids, was due entirely to a decrease in urea production ("anabolic" effect) and not to an increase in urea excretion ("renotrophic" effect). This applies both to 17α-ethyl-19-nortestosterone as well as to nandrolone-phenylpropionate. We have not tried any other nor-steroids besides these. Testosterone has, however, been used in patients with acute renal failure, as mentioned in my paper. Whereas it may be of limited value as an anabolic agent in reducing urea production, I have never seen any comment to the effect that it has a beneficial influence on renal functional recovery. I do not believe there is any evidence showing that androgens, be it testosterone or the 19-nor-steroids, have an effect on the repair of diseased renal tissue.

MCCANCE: May I congratulate the last two speakers on their communications before I pass on to the question of positive nitrogen balances and their relation to gain in tissue mass ? One of the distressing things about nitrogen balances is that, if they continue to be positive over a prolonged time, the result can seldom be equated with gains in tissue mass. It is one of the great problems in growing children and animals to explain the apparently positive balances, because the products retained cannot be found in the body by analysis or accounted for by rate of growth. If the intakes are increased, the retentions are increased, even in long-term studies; this has been discussed by several workers, and it is thought that the apparently positive balances are possibly due to very small losses. There is therefore a real difficulty about the interpretation of positive balances, for they may not give a true evaluation of the metabolic action of an anabolic steroid.

LABHART: Dr. THAYSEN, in what percentage of cases is the rise in the urea concentration, and in what percentage the hyperpotassaemia, the indication for dialysis in your experience ? Is it possible to inhibit the rise in potassium by administering anabolic steroids which inhibit tissue catabolism and hence enable one to avoid or postpone dialysis ?

THAYSEN: As regards the indication for dialysis, we go by the urea concentration in the serum. To start with, we used to dialyse whenever serum urea exceeded 4 g./litre, but in latter years we have decreased the limit to about 3 g./litre. If one dialyses on the strength of these rather liberal, so-called "prophylactic" indications, then hyperpotassaemia does not normally become much of a problem. If hyperpotassaemia does develop despite "prophylactic" haemodialysis, it can usually be corrected quite easily with the use of resins. Actually, we have only been forced to dialyse twice (out of 250 dialyses) on a "potassium indication". Both patients were severely traumatised with extensive tissue necrosis and large haematomas. The release of potassium from this devitalised tissue was so extensive and rapid that orally administered resins, intravenous glucose, and other measures proved ineffective and compelled us to dialyse before the serum urea had reached the critical limit of 3 g./litre, merely in order to combat the hyper-potassaemia.

You also asked whether the rate at which hyperpotassaemia develops is slowed down in parallel with the reduction in the rate of urea accumulation in those patients who show a favourable response to 19-nor-steroids. This is a very important question, since it bears on the mode of action of the 19-nor-steroids as discussed by Dr. ASCHKENASY. If the delay in the rate of urea accumulation were due to an "anti-catabolic" or "anabolic" action of the nor-steroid, one would expect the rate at which hyperpotassaemia develops to be slowed down as well. If the retardation in the rate of urea accumulation is not due to a "lean-tissue sparing" action on the part of the nor-steroid, but is due merely to the fact that the nitrogen from catabolised lean tissue is stored away in some unidentified nitrogen pool, then hyperpotassaemia might well continue to develop at an unaltered rate despite the retarded urea accumulation. It is not very easy, however, to give a clear reply to this question. The correlation between the rise in serum urea and the rise in serum potassium is not a simple and straightforward one in patients with acute renal failure. Extracellular potassium accumulation depends not only upon the quantity of potassium that is set free by catabolism of lean tissue (and here we are not even certain that the K/N ratio is always a constant of about 2.5—3.0), but also upon factors affecting the distribution of potassium between the extra-cellular and intracellular compartments (e. g. the degree of acidosis, the blood glucose level, the serum sodium, etc.). In our management of these patients we have in most cases altered a number of the factors which influence potassium distribution independently of the administration of 19-nor-steoirds. Direct removal of potassium from the body by means of resins has also been resorted to in some patients. Thus, I am afraid I must admit that the measures which we adopt as clinicians in order to counteract the danger of hyperpotassaemia make it impossible for me to give a clear-cut and well-documented reply to your question.

WATERLOW: Dr. THAYSEN, have you any experience with the use of intravenous albumin in these cases to combat the depletion ? Dr. WYNN of St. Mary's Hospital, London, has told me that, when treating some cases of acute renal failure of this kind, intravenous albumin did not increase the accumulation of urea to the same extent as an equal amount of nitrogen given by mouth would have done. If this is so, it suggests that the accumulation may in part be due to a failure of synthesis rather than to an increased rate of break-down.

THAYSEN: We have given albumin only to combat hypo-albuminaemia, which is frequent in acute renal failure, but we have not used albumin as a source of protein.

QUERIDO: Do you think that in those patients reacting favourably there was a difference between males and females ?

THAYSEN: We have obtained favourable responses to 19-nor-steroid treatment in acute renal failure both in men and in women. In a consecutive material, however, I should not be surprised if one were to find a preponderance of women among the good responders. You will have noted from Tables 2 and 7 that the best results with 19-nor-steroid are obtained in the two groups with a low pre-treatment rate of catabolism: post-partum cases and post-intoxication cases. The first group, of course, only comprises women; the second has a roughly equal number of men and women. Thus, women are liable to dominate numerically in the type of acute renal failure in which a favourable response to 19-nor-steroids can be expected.

Action of anabolic steroids on red-cell production

By

R. W. RUNDLES

During the past decade or so, considerable progress has been made in understanding some of the factors that influence erythropoiesis. "Specific" substances are, of course, required for the manufacture of red blood cells, such as iron, pyridoxine, and vitamin B_{12}, which are either incorporated directly into the substance of the erythrocyte or function as co-enzymes in the biosynthesis of essential cell constituents. Among the factors which control or regulate erythropoiesis are tissue hypoxia, a plasma factor known as erythropoietin, and the adrenal cortical and androgenic steroids. Each of these is now recognized as having a fairly well-defined and characteristic effect on hematopoietic tissues (Table 1).

Table 1. *Semi-quantitative comparison of the effect of androgens with that of other agents on the formed elements of blood and plasma proteins*

	Red cells	Hemoglobin	Platelets	Neutrophils	Eosinophils	Lymphocytes	Plasma cells	Plasma Proteins	
								albumin	globulin
Testosterone, Fluoxymesterone . .	↑↑↑	↑↑	0	0	0	0	0	0	0
"Anabolic" steroids. .	?	?					0	0	0
Erythropoietin. . . .	↑↑	↑	?	?					
ACTH, cortisone, hydrocortisone, prednisone, prednisolone, etc..	↑↑	↑↑	↑↑	↑↑	↓↓	↓↓	↓	0 to ↑	0 to ↓

↑: Increased number or concentration
↓: Decreased number or concentration
0: No definite effect

In reference to the androgens, there has been clinical and experimental evidence for many years that they influence erythropoiesis. Adult males normally have a hemoglobin concentration of 15 to 17 g. %, i.e. at least 2 g. % higher than in females, and the volume of packed red cells in men is 45 to 56%, as compared with 38 to 45% in women. There are comparable differences in total red-cell mass. Before puberty, and late in life after sexual involution has occurred, the hemoglobin concentration and packed cell volume tend to be the same in both sexes. In men subjected to orchiectomy the hemoglobin concentration and packed red-cell volume falls. The number of leukocytes and platelets, however, is about the same in both sexes at all stages of life.

Renewed interest in the erythropoietic effects of the androgenic hormones was stimulated by the report of KENNEDY and associates, about 5 years ago, that women with metastatic breast carcinoma given pharmacologic doses of androgens develop in a matter of 2—4 months a conspicuous increase in hemoglobin concentration, erythrocyte count, hematocrit, and red-cell mass, and a pronounced erythroid hyperplasia in the bone marrow. The number and types of leukocytes in the circulating blood and the number of platelets were not definitely affected.

While anemia is ordinarily not a major problem in women with breast carcinoma, the erythropoietic stimulation by androgens was so pronounced that KENNEDY and his associates extended their investigations to study the hematologic effects of androgens in patients with various types of refractory anemia. A useful degree of therapeutic response was demonstrated in several patients with impaired bone-marrow function.

More extensive and detailed investigations of the effect of androgens in a variety of hematologic diseases have been reported recently by GARDNER and PRINGLE and by SHAHIDI and DIAMOND. Improvement in the hematologic status of patients with variants of myelogenous leukemia, myeloid metaplasia, multiple myeloma, paroxysmal nocturnal hemoglobinuria, and hypoplastic and aplastic anemia has been observed. In the latter entities, the best results have been in children with impaired bone-marrow function resulting from the use of chloramphenicol. In these patients, androgen therapy was necessary for a long period of time, during which some degree of spontaneous recovery might have occurred, and prednisone was given concurrently.

The administration of maximally tolerated pharmacologic doses of androgens for periods of 2 to 3 months or longer seems to afford a selective erythropoietic stimulus for anemic patients generally — provided that there is some residual marrow function — without necessarily affecting the basic disease process. Erythroid hyperplasia develops in the bone marrow, and large red cells (somewhat undersaturated with hemoglobin) appear in the circulating blood. The leukocytes and platelets are affected to a minimal extent or not at all, and the total concentration of plasma protein and the electrophoretic composition are not notably affected by androgen therapy. Quantitative comparison of the hematologic effects of different androgens in various heterogeneous disease states has been difficult and inconclusive.

The possible relationship of androgenic stimulation to other factors that influence erythropoiesis was illustrated by an interesting patient studied by GARDNER and NATHAN for a period of more than 4 years. This 62-year-old retired executive developed an anemia in February, 1957, which proved to be refractory to the administration of iron, vitamin B_{12}, folic acid, prednisone, etc. Symptomatic improvement occurred after fluoxymesterone and testosterone-enanthate had been given for several weeks, but periodic blood transfusions were still necessary. Pyridoxine was then added to his regimen, in view of the persistence of hypochromic red cells. With this dual therapy his erythroid status was considerably improved, but became apparently normal only when crude liver extract was given by mouth in addition. Relapse occurred when any one of the three therapeutic agents was omitted.

The oxygen content of the arterial blood has long been recognized as another important factor in the regulation of erythropoietic activity. Anoxemia, which may be produced by any of a variety of ways, leads to erythrocytosis. Conversely, an increase in the oxygen tension in the inspired air has been shown to depress erythropoiesis in sickle-cell anemia, pernicious anemia, and in the normal subject. The effect of anoxemia on the bone marrow is not a direct one, however, but involves some intermediate mechanism.

The existence of an erythropoietic stimulating factor in the blood plasma has become well established in recent years. This substance is an alpha globulin of low molecular weight. The kidney seems to be directly or indirectly involved in its production. In bioassays of erythropoietin there is some possibility that alpha globulin fractions may be contaminated with thyroxin, vitamin B_{12}, or corticosteroid-binding globulin. High levels of erythropoietin have been found in the plasma of patients with different types of hypoplastic anemia, thalassemia major, leukemia, and polycythemia vera, and in the cord blood of the newborn. The quantity present in circulating blood has been variable, however, and a normal physiologic role in the regulation of erythropoiesis seems to be unlikely. Some investigators interpret their experiments to indicate that erythropoietin, when released from the tissues following acute blood loss, hemolysis, etc., provides a "panic" mechanism for stimulating red-cell regeneration. The effect of erythropoietin is that of promoting the differentiation of stem cells in the bone marrow into pronormoblasts. When large amounts of erythropoietin are given experimentally, macrocytes are produced, and these cells appear to have a comparatively short life-span (STOHLMAN).

It is now well established that ACTH and the adrenal cortical hormones, hydrocortisone, prednisone, prednisolone, etc., all have a similar pattern of hematologic effect. The proliferation of some cell types is stimulated, while that of others is suppressed. In CUSHING's syndrome, in both the iatrogenic and spontaneously occurring varieties, characteristic hematologic deviations include neutrophilic leukocytosis, which may be quite pronounced, thrombocytosis, increased red-cell mass, eosinopenia, and lymphopenia. There is a tendency toward reduction in the number of plasma cells in the bone marrow and hypogammaglobulinemia.

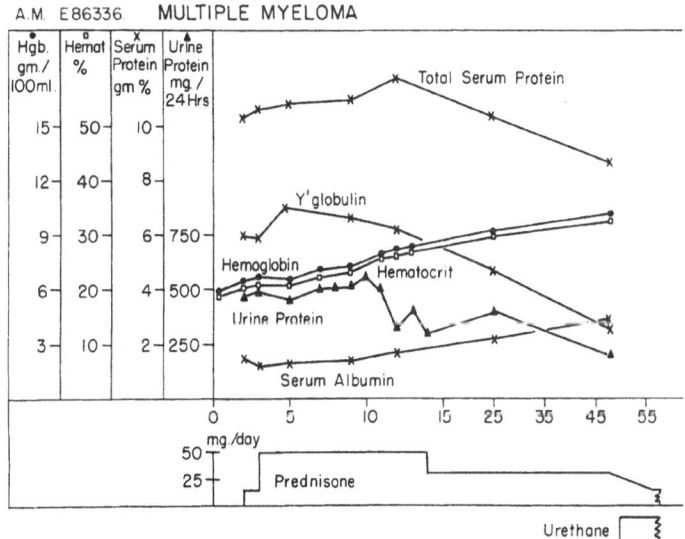

Fig. 1. Effect of prednisone therapy in a patient with multiple myeloma

The adrenal cortical steroids are extraordinarily useful in the treatment of the lymphocytic proliferative diseases, malignant lymphomas and chronic lymphocytic leukemia, and the malignant plasma cell growth in multiple myeloma. The typical hematologic effects of prednisone are illustrated in a patient with plasmocytic myeloma treated with this agent alone for a period of 6 weeks (Fig. 1). The subject was a 59-year-old colored laborer whose major complaints related to severe anemia. During the early period of therapy his hemoglobin and serum albumin concentration increased by about 50% and later became normal. The amount of abnormal globulin in his serum fell by about 50%, as did the quantity of BENCE JONES protein excreted in the urine (Fig. 1).

The therapeutic effect of androgens and of other anabolic steroids in multiple myeloma, which has been studied to a rather limited extent by ourselves and others, appears to be restricted to improved erythroid status and some alleviation of the symptoms of skeletal demineralization. There has generally been no effect on the abnormal serum or urinary proteins or on the growth of abnormal plasma cells.

Summary

Pharmacologic doses of androgens, given to the human subject for a period of 2 to 3 months or longer, provide an erythropoietic stimulus of some therapeutic usefulness. Erythroid hyperplasia develops in the bone marrow, and macrocytes somewhat undersaturated with hemoglobin appear in the circulating blood. In one patient, additional responses to pyridoxine and to a crude liver extract have been reported.

The hematologic effect of androgens relates almost exclusively to the red-cell elements. This resembles the effect of erythropoietin experimentally and contrasts with the wider spectrum of hematologic change produced by ACTH and the adrenal cortical hormones.

Zusammenfassung

Beim Menschen können hohe Dosen von Androgenen, wenn sie während 2—3 Monaten oder länger gegeben werden, eine therapeutisch brauchbare Anregung der Erythropoese hervorrufen. Im Knochenmark entwickelt sich eine erythroide Hyperplasie, und im strömenden Blut treten Makrocyten auf, mit einem etwas unter der Norm liegenden Haemoglobingehalt. Bei einem Patienten ließ sich zusätzlich zur Androgen-Behandlung ein Effekt von Pyridoxin und von rohem Leberextrakt nachweisen.

Die Androgene beeinflussen ausschließlich die roten Zellelemente. Dieser Effekt ist in gewissem Sinne dem experimentell mit Erythropoietin beobachteten vergleichbar und steht im Gegensatz zu dem breiteren Spektrum haematologischer Veränderungen, das ACTH oder Nebennierenrindenhormone hervorrufen.

Résumé

Chez l'homme, de fortes doses d'androgènes administrées pendant 2 à 3 mois ou plus stimulent l'érythropoïèse, effet utilisable en thérapeutique. On constate dans la moëlle osseuse une hyperplasie érythroïde et dans le sang circulant des macrocytes dont la teneur en hémoglobine est légèrement inférieure à la normale. Chez un malade, on a pu mettre en évidence, en plus de l'effet androgénique, un effet de la pyridoxine et de l'extrait de foie cru.

Les androgènes agissent presque exclusivement sur les éléments cellulaires rouges. Cet effet ressemble à celui qu'on obtient dans les expériences avec l'érythropoïétine; il est contraire au large spectre de modifications hématologiques produit par l'ACTH et par les hormones corticosurrénaliennes.

References

BRENNAN, M. J., and W. L. SIMPSON: Biological Activities of Steroids in Relation to Cancer. Academic Press, 1960, p. 477. — ERSLEV, A. J.: N. Y. Acad. Sc. Trans. **24**, 131 (1961). — GARDNER, F. H., and J. C. PRINGLE Jr.: A.M.A. Arch. Int. Med. **107**, 846 (1961). — GARDNER, F. H., and J. C. PRINGLE Jr.: N. England J. Med. **264**, 103 (1961). — GARDNER, F. H., and D. G. NATHAN: Amer. J. Med. Sc. **81**, 477 (1962). — HATHAWAY, W. E., and J. H. GITHENS: Amer. J. Dis. Child. **102**, 389 (1961). — HUGULEY, C. M., Jr., A. J. ERSLEV, and D. E. BERGSAGEL: J. Amer. Med. Ass. **177**, 23 (1961). — ISRAELS, M. C. G., and J. F. WILKINSON: Lancet (G.B.) **1961**/I, 63. — KENNEDY, B. J., and A. S. GILBERTSON: N. England J. Med. **256**, 719 (1957). — SHAHIDI, N. T., and L. K. DIAMOND: N. England J. Med. **264**, 953 (1961). — STOHLMAN, F., Jr.: N. Y. Acad. Sc. Trans. **24**, 312 (1962). — WINKERT, J., A. S. GORDON, and E. WINKERT: N. Y. Acad. Sc. Trans. **24**, 135 (1961).

Discussion

GEMZELL: Several years ago SJÖSTRAND and I[1] were interested in the presence of a possible erythropoietic factor in the human pituitary. We obtained six different fractions from the human pituitary, injected them into hypophysectomised rats, and measured the total amount of haemoglobin by a CO-method designed by SJÖSTRAND. This method enables one to make repeated determinations on the same animal without any disturbances in the haematopoiesis. We found that following hypophysectomy the total amount of haemoglobin decreased by about 30%. When the fractions from the human pituitary were injected, some increase was found only in those fractions that contained growth hormone. The increase, however, was only in proportion to the increase in body-weight. ACTH from pig pituitaries caused a very questionable increase in the total amount of haemoglobin. Testosterone and oestrogens were also injected into gonadectomised male and female rats. Testosterone produced a marked increase in the total amount of haemoglobin, whereas oestrogen produced a decrease.

RUNDLES: Your observations are very interesting and apparently agree with the observations made so far in the human subject.

GROSS: I wonder if there was any confirmation of the work of EVANS[2], published a few years ago, on the erythropoietic factor in the pituitary which he claimed to have isolated. Could this factor be related to what Dr. GEMZELL has been describing?

GEMZELL: Well, I am quite familiar with this work, which was done in HERBERT EVANS's laboratory. It was mainly VAN DYCK who did the investigation. I think there is no indication at present that there are any erythropoietic factors in the pituitary.

ASCHKENASY: I should like to ask your permission to show three slides referring to the haematological effects of testosterone-propionate and 17α-ethyl-19-nortestosterone in the rat.

Fig. 1 indicates the degree to which testosterone-propionate (TP) stimulated recovery from nutritional anaemia in male rats fed a 7% casein diet after 2 months of protein deprivation[3].

Fig. 2 shows that in the same rats TP enhances not only the restoration of erythrocyte counts (in comparison with the counts in control rats injected with olive oil), but also the restoration of neutrophils, whereas the same hormone has an inhibitory effect on the restoration of the blood eosinophil counts, which are strongly reduced after protein starvation.

17α-ethyl-19-nortestosterone (ENT) is also endowed with a certain degree of erythropoietic potency. I did not study the effects of this steroid on

---

[1] GEMZELL, C. A., and T. SJÖSTRAND: Acta endocr. (Den.) **16**, 6 (1954).

[2] VAN DYKE, D. C., M. E. SIMPSON, A. N. CONTOPOULOS, and H. M. EVANS: Blood (U.S.A.) **12**, 539 (1957).

[3] ASCHKENASY, A., and F. DRAY: Sang (Fr.) **25**, 461 (1954).

recovery from protein-starvation anaemia, but in non-undernourished rats I was able to demonstrate some erythropoietic action, but only in female and not in male rats and only after very protracted administration of large doses of ENT (25 or 100 mg. per 100 g. of dry diet).

As Table 1 shows, after 4 months of such treatment, the increase in the haematocrit values is no more pronounced in treated male rats than in controls. On the other hand, in intact females there is a certain difference between

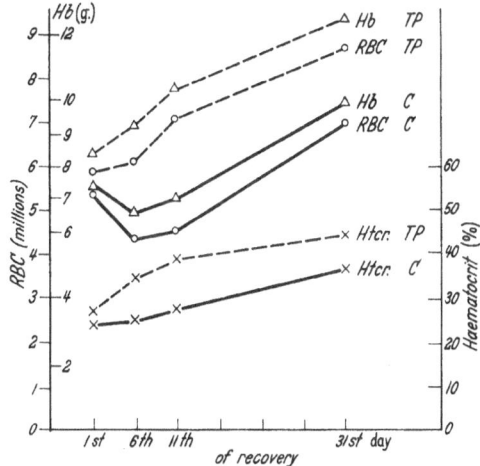

Fig. 1. Effect of injections of testosterone-propionate (TP) (2.5 mg. twice a week) on recovery from anaemia (RBC, haemoglobin, and haematocrit) in male rats receiving a diet poor in protein (7% casein) after 60 days of protein starvation. C — Controls injected with olive oil. From ASCHKENASY-LELU, P., and A. ASCHKENASY: World Rev. Nutrit. Dietet. 1, 33 '(1959)

the controls — which do not exhibit any changes in their mean haematocrit values from the beginning till the end of the experimental period — and the hormone-treated rats, which show an increase in these values.

A similar sex difference in the effects of ENT has been observed in the same rats with regard to the body-weight increase, which was highly stimulated by this androgen in females but not in males[1].

Finally, I should like to mention certain new experimental data. After 7 days of protein deprivation, rats show a striking inhibition of erythropoiesis as demonstrated by a sharp decrease in the blood reticulocyte counts and in the incorporation of radioactive iron into the red cells. If one injects 5 mg. of a long-acting androgen (testosterone-oenanthate or ENT-propionate) on the first day of the protein-free diet, then the production or erythrocytes decreases after one week to a significantly lesser extent than in non-injected rats. Nevertheless, this beneficial effect of both androgens is relatively small in comparison with the striking protection provided by injections of erythropoietin under the same experimental conditions.

---

[1] ASCHKENASY, A.: Thérapie (Fr.) **14**, 332 (1959).

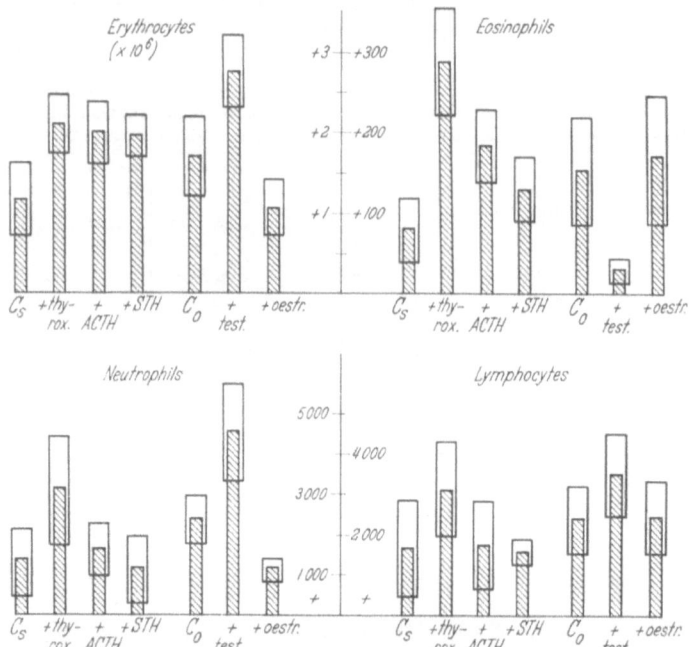

Fig. 2. Effect of several hormones (thyroxin, ACTH, STH, testosterone-propionate, and oestradiol-benzoate) on blood regeneration after prolonged protein starvation. Restoration is effected with a low-protein diet (7%) over a period of 30 days. The hatched columns represent the mean quantitative increases per mml. between the 1st and 31st day of restoration. The blank columns correspond to twice the standard errors of the means. C_S = controls injected with physiological saline; C_O = controls injected with olive oil. From ASCHKENASY, A.: Amer. J. Clin. Nutrit. 5, 14 (1957)

Table 1. *Mean values ($\pm$ SEM) of haematocrits (p. 100) on the first and the last day of oral administration of 17α-ethyl-19-nortestosterone (ENT) for 4 months to male and female non-castrated rats (25 mg. or 100 mg. per 100 g. of dry diet). Number of rats shown in brackets*

Group	Males			Females		
	1st day	Last day	Difference	1st day	Last day	Difference
Non-treated controls (4 M and 9 F)	32.1 ± 1.2	38.5 ± 0.9	+ 6.4	38.5 ± 0.8	39 ± 0.4	+ 0.5
25 mg. ENT (10 M and 10 F)	32.9 ± 1.1	36.9 ± 0.6	+ 4.0	36.4 ± 0.8	40.9 ± 0.8	+ 4.5
100 mg. ENT (10 M and 10 F)	34.8 ± 0.8	40.1 ± 0.8	+ 5.1	36.5 ± 1.1	41.1 ± 0.5	+ 4.6

GROSS: May I ask Dr. ASCHKENASY about the significance of his data in the last figure? There was the same difference in the non-treated males as in the treated females. Is the difference in the haematocrit values observed in treated females significant?

ASCHKENASY: In Table 1, the comparison should be made not between the non-treated males and the treated females, but in each sex between treated and non-treated animals. One can then see that in males ENT does not amplify the physiological increase in the haematocrit values, whereas this steroid does seem to have a notable effect in females.

With regard to the sex differences in the non-treated animals, it is well known that the erythrocyte and haemoglobin values increase during growth to a greater extent in males than in females. Thus, one may conclude that protracted administration of large doses of ENT leads to a "haematological masculinisation" of the female rats, just as it does with respect to the increase in body-weight.

KASSENAAR: I should like to ask Dr. RUNDLES about his first patient who reacted so well when testosterone was given together with pyridoxine. Have you any idea about the life-span of the red cells before and during treatment? I was amazed by the fact that, after stopping the testosterone, a much rapid decline in the red-cell count occurred than could have been expected had the mean life-span been around 120 days.

RUNDLES: Dr. GARDNER's [1] studies showed a moderate reduction in red-cell survival before treatment and normal survival during treatment. He also had a mild reticulocytosis of 2—5%, so there was some acceleration of haemolysis.

MANDEMA: Dr. NIEWEG [2] in our department recently used anabolic steroids, but no corticosteroids, to treat a patient with congenital hypoplastic anaemia associated with other congenital anomalies (the so-called Fanconi syndrome), and we were very much impressed by the results. Now recently Dr. JONXIS [3] in Groningen found that in the Fanconi syndrome there is a relatively large amount of foetal haemoglobin. Is there any information as to whether the effect of anabolic steroids in this sort of patient primarily affects the production of foetal haemoglobin or of adult haemoglobin, or if the two go together? The second question I wanted to ask you is: is there any information available about the interrelationship of anabolic steroids and erythropoietin?

RUNDLES: I don't believe the foetal haemoglobin has been determined serially in individuals who have been given androgens. There are individuals who have a high percentage of foetal haemoglobin that persists throughout adult life. It might be more practical to study them than other types of patient, who, even when they are quite anaemic, may have only 5—10% foetal haemoglobin. I don't know of any comprehensive study of the level of erythropoietin in the plasma as related to the erythropoietic effect of testosterone. This is a gap in our knowledge.

LEATHEM: Just a question concerning hypophysectomy being followed by microcytic hypochromic anaemia. If I remember rightly, one can correct

---

[1] GARDNER, F. H., and D. G. NATHAN: Amer. J. Med. Sc. **243**, 456 (1962).
[2] NIEWEG, H. O.: personal communication.
[3] JONXIS, J. H.: personal communication.

the anaemia by a combination of testosterone and thyroxin, but the improvement is better in the presence of an increased protein intake. I wonder if more recent investigations support the nutritional enhancement of hormonal action.

RUNDLES: I don't have any specific information about this possibility. Most of the individuals treated with androgens, etc. eat an ordinarily adequate diet. They are not conspicuously undernourished or malnourished, but precise details of their protein intake are not available.

PRADER: I should like to comment on this from a clinical point of view. One disease which demonstrates the androgenic effect on haemoglobin is the congenital adrenogenital syndrome. In children with this syndrome, the haemoglobin is higher than in normal children[1]. If I understood you correctly, Dr. RUNDLES, you said that testosterone and fluoxymesterone are both effective in treating aplastic anaemia, but that you would doubt whether other anabolic steroids have the same effect. We have seen one patient with FANCONI's aplastic anaemia, in whom methandrostenolone was just as effective and less virilising than testosterone. My other question is about the effect of testosterone on aplastic anaemia. Is that an androgenic or an anabolic effect? Has anybody tried using growth hormone to treat a patient with aplastic anaemia? I think that such an experiment would give the answer to my question.

RUNDLES: These are very pertinent but unanswerable questions at this time. The high haemoglobin levels that you notice are probably comparable to the so-called polycythaemia of CUSHING's syndrome. This is not really a polycythaemia comparable to polycythaemia vera, and it does not involve the white-cell elements or platelets. I have no real information as to whether the androgenic and anabolic effects on red-cell production do go in parallel. We might infer from the findings with reference to breast carcinoma that the androgenic, anabolic, and erythropoietic effects of steroids are not inseparable.

ARIAS: Several reports have indicated an increase in the serum alpha-2-globulins following administration of testosterone, oestrogens, or synthetic anabolic steroids in man. This increase represents in part an increased synthesis of ceruloplasmin[2]. Was an increase in alpha-2-globulins observed in your experience following treatment with testosterone?

RUNDLES: In general, within the limits of the accuracy of our measurements using paper or zone electrophoresis, there is no real change in the amount of the alpha globulins or other proteins. Almost all sick people have variable and unpredictable changes in the alpha globulins that are difficult to interpret.

DESAULLES: My question is only a question aside, not directly relevant to what Dr. RUNDLES has said. I am wondering if one can establish any connection between muscle mass and the possible haematopoietic action of testosterone. I do not think that it is possible in this context to make a basic differentiation between primates and other animals. In all growing animals, the differences are only of degree and are not fundamental. We must be aware that in male rats the muscle mass is greater than in females, but these animals do not show the rapid growth and weight increment seen in male primates at the time of puberty. In rats, the growth spurt seen at puberty

---

[1] PRADER, A.: Helvet. paediatr. acta **8**, 386 (1953).

[2] STERNLIEB, I., et al.: J. Clin. Invest. (U.S.A.) **40**, 1834 (1961).

overlaps somatic growth, which is still active, so that there is an almost continuous shift from one period into the other.

But another point disturbs me much more: in certain groups of higher vertebrates, for instance in birds of prey (eagles and falcons), the females are bigger and more powerful than the males. This also applies particularly to whales, especially fin whales (Mysticetes, Balaenopteridae) as well as certain genera of toothed whales (Odontocetes)[*, **].

We have no indication as to the existence of a growth spurt during puberty in these animals. All we know is that their growth is very rapid and that they attain sexual maturity remarkably precociously in relation to their size. I should be very interested to know what pattern of pituitary and sexual hormones leads to this increased muscle mass and haemoglobin production.

---

[*] According to SLIJPER, E. J. (Whales. Hutchinson, London, 1962, p. 85): "While the two sexes in Cetaceans do not differ in colour, they often differ in length. It seems that in Mysticetes, the average female is 3—6 feet longer than the male, while the female Odontocete is shorter than her mate. As a rule, however, the difference is smaller than in Mysticetes, and in porpoises, for instance, there seems to be no difference at all. In other species, e.g. in Sperm Whales, Bottlenose Whales, Killer Whales, and False Killer Whales, the difference between males and females is about 19, 13, 8, and 3 feet respectively. Ziphiids of the genera Berardius and Ziphius form the exception, for the males are shorter.

[**] During the years 1946—1949, male Blue Whales *(Balaenoptera musculus)* measuring over 26 metres in length have averaged only 0.3% of the total catch in the Antarctic, whereas females over 26 metres have averaged 16%, the largest measuring 30.5 metres and being a female (according to the International Whaling Statistics 19., 20., 21., 22. Det Norske Hvalrads Statistiske Publikationer, Oslo, 1948—1949).

Concluding Discussion

QUERIDO: As Chairman, I should first like to claim the privilege of putting a question to the audience, or rather to Dr. KORNER and Dr. WILSON. We began by talking about the mode of action by which growth hormone and androgens promote protein synthesis in the cell. Afterwards we discussed the action of these agents on overall metabolism in the organism. Dr. RABEN supported the views of the Cambridge school that the liberation of energy from the fatty acids might be an important factor in the effect of growth hormone on the body. The first question I want to ask Dr. KORNER is how he reconciles these two observations. My second question is on testosterone. With this compound a fat-mobilising action has also been observed. Also the word "anticatabolic" has been used. May I therefore ask Dr. WILSON how he would like to reconcile these different actions of testosterone, i.e. promotion of protein synthesis, anti-catabolic action, and fat-mobilising action?

KORNER: I'm sorry, but I need to think about that question.

RABEN: While Dr. KORNER is thinking, I wonder if I might add another question? Dr. KORNER, when you gave your paper I had the impression that you had shown that hypophysectomised rats treated with growth hormone had ribosomes in their livers which now were capable of synthesising more protein, presumably as a result of growth-hormone treatment; and I thought you meant that the ribosomes must therefore be a limiting feature in protein synthesis and that the increase in their ability to synthesise protein was the way growth hormone might work. However, you also said that you wouldn't be surprised if a liver from an animal treated with a catabolic agent were capable of synthesising more protein simply as a result of the break-down products from the other tissues coming to the liver; this seemed to suggest that you didn't think the ribosomes were limiting in the first place and that the extra provision of substrate would allow more protein to be synthesised.

KORNER: I think that the ribosomes are the limiting factor in protein synthesis, and I feel that the proportion of ribosomes in the cell that are able to incorporate is under the control of a variety of factors. One of these factors is growth hormone, but it need not act directly. It could, for instance, act through the mediation of insulin, i.e. stimulating fat break-down so that glucose oxidation by muscle is inhibited, thus causing an increase in the blood glucose level and, consequently, in the amount of insulin secreted by the beta cells. At the same time, growth hormone controls the hypoglycaemic effects of insulin. Certainly insulin is in some way able to change the ribosomes to a form capable of synthesising more protein.

Another possibility is that the amino-acid level can, in some way, also determine the rate of protein synthesis. I wonder if the ribosomes are constantly being broken down and re-activated from some precursor which looks like a ribosome from the point of view of size and ribonucleoprotein content, but not from the point of view of being able to synthesise protein. The rate at which ribosomes are activated could be under hormonal control, whereas the stability could be under the control of the amino-acid level in the tissue. MUNRO and his group[1] in Glasgow have shown, for instance, that the rate of synthesis of RNA is under the control of the animal's caloric intake, whereas the stability of the RNA is under the control of the amino-acid level in the tissue and the blood. Something similar could be happening with the ribosomes, so that when the liver of the rat is flooded with amino acids as a result of treatment with corticosteroids, more liver ribosomes are stabilised in an incorporating form. When one gives growth hormone to a rat by a series of processes, one eventually arrives at a situation where the number of ribosomes being made into an active form from an inactive precursor is increased. I hope that answers your very searching question.

LEATHEM: Cortisone administration to the hypophysectomised rat does *not* permit the transfer of amino acids to the liver. The liver does *not* take up more protein than in the normal or in the adrenalectomised animal. Thus, in the absence of growth hormone, the flooding of the body with the excess amino acids cannot enhance protein in liver. I know this doesn't answer the question, of course, but maybe it adds to the complexity of the problem.

KORNER: Does it not depend on the dose of cortisone? Would your observation be entirely true no matter how long the rats had been hypophysectomised and no matter what dose of corticosteroid you used?

LEATHEM: I can't answer that very extensively, except to say that we have tried it at periods of 10 days and at 60 days. We have used doses of 0.5 mg. and 1 mg. daily, each in periods of 20 days.

KORNER: I observed increased incorporation of amino acids into the protein of liver microsomes of hypophysectomised rats after treatment with cortisol, but the dose and other conditions affected the response obtained.

QUERIDO: Dr. WILSON, would you be ready to answer the other question I asked?

WILSON: I would take a very cautious and conservative view with regard to the physiological interpretation of *in vitro* studies on the prior injection of hormones and the subsequent effects on protein synthesis, because in the intact animal the effects on growth may be due to anti-catabolic actions either on the turnover of proteins or on protein synthesis. For this reason I have tried to be very cautious about any sort of physiological interpretation of the ultimate anabolic or androgenic action of testosterone in the studies which I have performed. The question as to whether or not the effect on protein synthesis is a primary hormonal effect or, alternatively, whether it is the remote consequence of a whole series of other chemical reactions, is one that cannot be answered and must remain unsettled. It is obvious that effects which are first demonstrable even after two hours may be the passive consequence of an involved chain of reactions working at several different levels, and this may not be just a secondary effect; it may be an effect of the nth order. However, it is also possible on *a priori* grounds that these are primary effects or very close to primary effects, because the lack of an effect of the direct addition of a hormone to an homogenate, or to slices, may mean only that we do not have the hormone in the proper form in which it would be active *in vivo* (such as bound to a protein). So I think this is still unsettled. The experiments I have performed were not actually designed in an attempt to answer the much more important and crucial question as to the ultimate

---

[1] CLARK, C. M., D. J. NAISMITH, and H. N. MUNRO: Biochim. biophysica acta (U.S.A.) **23**, 587 (1957).

mechanism of action of steroid hormones. They were done in an effort to elucidate one limited aspect of the action of steroid hormones, and that question was simply: what step in protein synthesis is rate-limiting in these organs in immature animals ? I think the maximum interpretation one can place on such experiments is that in these tissues the peptide-bonding step does appear to be rate-limiting for protein synthesis; but I want to emphasise again very explicitly that whether this is a primary effect, which I doubt, or whether it is secondary to a whole series of other metabolic effects is still entirely unsettled.

DREYFUS: I think that the questions put by our Chairman are difficult to answer in terms of the cell-free system, as with cell-free systems we are still in a biochemical era and not yet in a physiological one. To make these cell-free systems work, we give them pre-digested food and we furnish them with ready-made amino acids and with energy suppliers such as ATP; but we are quite unable at present to determine the physiological importance of the substrate which serves as food in this tissue. We don't know whether it's glucose or fatty acids, which seem to be quite important in the case of growth hormone. You are all aware of the long-term studies of ZIERLER and his associates[1], which appear to prove that muscle tissue lives mainly on fatty acids and not on glucose. I am afraid that it would be premature at this time to attempt an explanation in terms of ribosomes, although the very valuable studies which have been described at this Symposium may well contribute towards an understanding of this question.

MCCANCE: I wish to make a few remarks of a general nature, perhaps rather inappropriately, but I think this is the time to make them. My interest in this field is as a physiologist, not a pathologist, and more particularly as one interested in infant physiology. We have heard in this conference a description of fascinating new work, some of it of a very fundamental nature and nearly all of it concerned with points of detail . . . great detail . . . microsomal detail! I admire very much the people who can manipulate these things! We have, however, heard practically nothing about the function of anabolic agents in normal physiology. In particular, we have heard nothing about the control of their release by the hypothalamus, which I am sure is involved, and the effects of chronological age, which is certainly vitally important. The teeth, for example, develop closely in accordance with chronological age, even in profound abnormalities of growth; and whether there is any growth hormone involved in their development seems to be nobody's business.

I regard the somatotrophic hormone as a metabolic hormone, and I ask myself why this makes it a growth hormone. Can anybody answer this ? It is well known that puncture of the hypothalamus does not lead to obesity in the young animal. KENNEDY[2] showed this. Hypothalamic control of the food intake is only required when the rate of growth falls off enough to make it necessary to reduce the calorie intake below peak levels. Somatotrophic hormone is not required for growth in the very young, but it is normally present in the pituitary. Is it secreted ? I don't know; it may be. I suggest that it only makes its presence felt as a growth hormone when the velocity of growth falls below a critical level, and that its physiological function then is to maintain growth at a desirable rate when the somatic cells by themselves would fail to do so.

---

[1] ZIERLER, K. L., and R. ANDRES: J. Clin. Invest. (U.S.A.) **35**, 991 (1956).
[2] KENNEDY, G. C.: J. Endocr. (G.B.) **16**, 9 (1956).

I suggest further that the anabolic steroids come in to supplement the effects of the somatotrophic hormone at the appropriate age, to direct growth along certain channels, and then to stop it by making sure that the epiphyses fuse. These anabolic steroids are of course produced in response to pituitary gonadotropic hormones, and my final suggestion is that the whole process is co-ordinated through the hypothalamus, about which we have heard absolutely nothing at this conference. I have a feeling that at some future conference it will be in the centre of the picture. It's extraordinary how much has been learned about the function of the hypothalamus in recent years by puncture experiments and electrical stimulation, and I feel that perhaps this is a line for the future.

QUERIDO: Anyone who wants to extend on Dr. McCANCE's remarks?

TREMOLIÈRES: There is another point that has been left out of consideration, i.e. the effect of steroid hormone on appetite. The integrative processes which regulate appetite in the hypothalamus seem to have a basic bearing on the effect of these hormones. When they are clinically active, it is when they increase the appetite. This was shown by Dr. THAYSEN in renal failure, and the same applies to cases in which they prove active in undernutrition.

Physiologically, the best way to get the desired response is to increase the food intake. It is not possible to induce nitrogen retention of physiological interest without increasing the food intake.

Increasing the food intake without improving the appetite has generally led to vomiting, diarrhoea, and malabsorption. It therefore would appear to be most important to study the many integrated factors which condition the appetite.

OVERBEEK: I would like to add something to what Dr. McCANCE has said. I have the impression that during this conference it has not always been realised that, although growth hormone and anabolic steroids both have either anabolic or anti-catabolic activity, they may act differently. For instance, they have very different properties inasmuch as the effect of growth hormone on fat metabolism proves to be much more pronounced than that of the anabolic steroids if you look at the free fatty acids appearing in the blood after the administration of either of these substances. On the other hand, the various anabolic steroids themselves usually display a number of different activities. Thus, we are not only faced with the problem already discussed this morning, i.e. which effects are to be ascribed to the androgenic activity and which to the anabolic activity, but we also have to consider pituitary inhibition, the anti-oestrogenic and the progestational activity, which may all differ in degree in the different steroids. For this reason, I feel one cannot possibly assume that growth hormone on the one hand and all the anabolic steroids on the other have the same kind of action; it may be entirely different.

LARON: I have the feeling that we should not go away from this conference with the idea that a great deal of what has been said is purely negative and that we must acknowledge ourselves as more or less ignorant about the fundamental pathways of anabolic substances and growth hormone. Such a defeatist attitude would leave us poor clinicians with the question: how are we going to deal with these substances? Are we going to refrain from using them and to wait until their pathways of action have been clarified? Or are we still bound to carry on using them?

If I may, I would like to come back again to a clinical subject discussed yesterday, namely, the use of anabolic substances and the problems of growth. Bearing in mind the fact that growth disorders are among the problems most commonly encountered in a paediatric endocrine clinic, and while fully appreciating the limitations to the usefulness of anabolic substances as explained so well by Dr. PRADER in his paper, I would nevertheless say that we are still left with certain situations in which these substances can be used. I am thinking, for example, of situations like delayed adolescence, where anabolic agents may be considered for use as substitution therapy. By slightly advancing their adolescence and promoting their growth, great help can be given to these children, who very often develop an inferiority complex. If there is nothing else available, anabolic drugs might also be used on a limited scale in children with retarded growth and retarded bone age. However, adequate checks on side effects and repeated estimations of bone age would seem to be essential when using these drugs.

QUERIDO: If I may, I should like to explain the theory which, in the light of all I have so far heard at this meeting, I have formed concerning the so-called anabolic agents. I think that no evidence has been presented to show that the androgenic action and the anabolic action are fundamentally different at the cellular level. It seems to me that these compounds are only different because their affinity for the tissues, or the ability of the tissues to metabolise them, is different.

My second impression is that, with regard to the action of so-called anabolic steroids, we have to distinguish between three groups of individuals, i.e. males, females, and pre-puberal children. Anything you can do with an androgen you will also be able to do with the so-called anabolic steroids, with the restriction that, if they have a changed affinity for certain tissues, they will not stimulate all the tissues in the way that androgens would otherwise do. In normal males, androgens have little or no metabolic action, and therefore anabolic steroids also have little or no action. In females, physiological effects that can be produced with androgens will also be seen with these substances, but in a restricted area because of their structure. In children, the result will be analogous to puberty and may also involve a restricted area of tissues, because the molecules are changed. On the other hand, however, in this Symposium we have had examples of tissues with pathological nitrogen metabolism in which anabolic steroids produce no effect, because it would not be consonant with their physiological action for them to do so. This would be my working hypothesis on the basis of what I have heard at this conference, but maybe you can persuade me to think otherwise.

PRADER: I should like to re-emphasise what the Chairman has just said. I think it is extremely important to distinguish between the action of anabolic steroids in children, in adult females, and in adult males. The effect of these steroids depends largely on the endocrine situation, which of course is different in children, in women, and in men. As I have mentioned, children are very sensitive, women less sensitive, and adult males not sensitive at all to the virilising effect. From yesterday's discussion it seems that the bones of children do respond to anabolic steroids, whereas those of normal adults do not. Yesterday, the Chairman stated in the form of a question that a normal adult male does not show any anabolic response to anabolic steroids. This is certainly different in children. As regards the treatment of acute renal failure, Dr. THAYSEN has just presented his rather disappointing results in adults. I should like to ask him whether children respond better than adults in this respect. I ask this question, because in our country internists have stated that these steroids are of no value in acute renal failure[1], whereas paediatricians have reported a very good effect in children with the same condition[2]. All these examples demonstrate that the effect of these steroids is dependent on age and sex.

THAYSEN: I believe that the 19-nor-steroids may have an effect in both men and women with acute renal failure, at least when administered in the very high doses which we have employed. However, the effect is inconstant and appears to depend somehow on the nutritional and perhaps also on the hormonal status of the patient. With respect to children, I must admit that I do not know. For some odd reason "tubular necrosis" is very rare in children before puberty. In our experience, the most common cause of acute renal failure in that age group is acute glomerulonephritis. We have treated our patients suffering from acute glomerulonephritis with prednisone in an attempt — possibly vain — to improve the renal lesion. Prednisone has a documented catabolic effect, so I do not want to draw any conclusions as to the effect of anabolics in a prednisone-treated group of patients.

DESAULLES: From the point of view of the experimental endocrinologist it seems to me that this meeting has confirmed how important it is to evaluate the endocrine balance very thoroughly when treating either animals or patients with any of the compounds which we have been discussing. It is my impression that during growth — as well as in adults, whether female or male — the total endocrine balance of the organism has a really vital bearing on the outcome of the effect of the substance we are studying. And if we do not understand the endocrine balance sufficiently well, it is very difficult indeed to know exactly what we are doing.

[1] GERBER, W., and P. COTTIER: Helvet. med. acta **27**, 539 (1960).

[2] GAUTIER, E., and O. TÖNZ: Helvet. med. acta **27**, 535 (1960).

Summing up

By

F. G. Young

Dr. Querido, Members of the Symposium, at this point I feel it would be appropriate to start by quoting the words of the English poet A. E. Houseman, who began a lecture, which he had been invited to give, by saying: "It is first my duty to acknowledge the honour done me by those who invited me to give this address and to thank them for this token of their good will. My second duty is to say that I condemn their judgement and deplore their choice." Dr. Querido, I am not sure whose judgement I should now condemn, but it is quite clear which choice I deplore. But I think I can console myself by the fact that it would be presumptuous of anybody to attempt to summarise the proceedings of a conference of this sort, which has included 25 or 30 papers and much discussion over some days. Certainly it would be impossible for me to do this. I could of course have given an additional paper, pointing out that it is all due to the action of insulin on the incorporation of amino acids, but I decided that it was too late a stage to do this. There seems, however, to be some advantage in pulling out a telescope and looking at our proceedings through it from the wrong end, so as to be able to pick out some points that emerge particularly clearly, or at least which interest me, both negatively or positively.

To begin with, I cannot do better than take some of the points in the Chairman's admirable introductory talk, and consider how far some of the questions raised in it have been answered. You pointed out, Dr. Querido, that you hoped we should discuss biological mechanisms concerned in protein metabolism at the cellular level and in the organism in health, in disease, and during repair. Although we have discussed some matters from the cellular level, such discussions are not always easily and directly related to experience of the intact organism or in the clinic; indeed, as has been emphasised more than once, they are often rather theoretical at the present moment. But as an excuse for this I should point out that it is only within the last 20 years or so that the cellular

mechanisms involved in biosynthesis have become susceptible to
investigation. This was made possible by the realisation that a
source of energy must be coupled with a biosynthetic system, and
such a system is usually not the reverse of that operative in degradation, as was assumed, perhaps naturally, at first. So any lack
of complete integration of the picture is perhaps not surprising at
the present time. Then there is also, Mr. Chairman, the point that
you emphasised to the effect that the enzymes which are necessary
for protein synthesis are themselves proteins, so that protein
metabolism must in a sense necessarily be a little self-centred.
Protein biosynthesis in a way hoists itself up by its own boot-
straps. Examination of the cellular mechanisms involved in
protein biosynthesis largely depends today on methods which
involve the use of homogenates and other preparations from
rat tissues. Under such conditions, the reactions with different
tissues are quite similar, and this may indicate either a limitation
of our methods or a fundamental similarity in the pattern of
biosynthetic mechanisms in different tissues.

The structures and organisms that we have considered during
these four days show how widely ideas about protein biosynthesis
must be applicable. I have noted a few of them: the comb of the
cock, the levator ani of the rat, and the shoulder muscles of the
gorilla have jostled in our thoughts with patients suffering from
trauma of different types, old ladies who break their femurs much
more often in Malmö than they are alleged to do in the U.S.A.,
ballet dancers who grow in the spring, and, recently, whales,
whose lack of teeth may be important. We must consider many
different organisms before coming to any agreed view about protein
biosynthesis and its control, and not only must we discuss different
organisms, but we must also consider different tissues in each organism. There are quite a few that we have not discussed, and only
a short while ago Dr. McCANCE pointed out that teeth have received
scant attention so far. Variations in sex, age, endocrine balance,
health, and disease of all varieties are relevant, and our programme
has been well designed to cover much that is of significance in these
matters. We are all aware of the importance of species differences,
and Dr. TANNER has frequently stressed this, and the significance
of steroids, in the spurt in muscle growth at puberty in the human,
which is not obviously or easily paralleled in animal investigations.

In discussing the nature of the protein in the human body, you
pointed out, Mr. Chairman, that collagen constitutes about one-
third of the body protein in the human being. But we have not
heard very much about collagen, though we now realise that col-

lagen, or what we have called it in the past, is probably a heterogeneous material, some fractions of which are highly active metabolically. Your question as to which proteins are particularly
affected in disease cannot, I think, be answered simply at present;
nor do we know why a positive nitrogen balance can be obtained at
very different levels of nitrogen intake. These are all questions that
you raised in your introductory address, the answers to which are
not very clear even now.

Why protein synthesis sometimes fails to catch up with protein
degradation, and what the controlling mechanism is which limits
protein synthesis under such conditions, is certainly still not clear
to me from our deliberations. Where does the nitrogen go which
is apparently retained when there is evidence of the retention
of nitrogen without a corresponding formation of lean tissue ?
Dr. McCANCE today took an iconoclastic attitude with regard
to nitrogen balance studies. The possible influence of systematic
errors is a point that one must obviously bear in mind in this respect.
Just 30 years ago, ADAMS and POULTON[1] suggested that there was
an error in the classic method of calculating the heat production of
the animal body from the non-protein respiratory quotient. Their
conclusions suggested that above a non-protein RQ of about 0.80,
the higher the RQ, the greater the discrepancy between the
results of direct and indirect calorimetry, the calculated value for
heat production being lower than that measured. If these views
were correct, some systematic error would have increased as the
RQ rose. Is this similar to the problem that faces us now with
respect to studies on nitrogen balance ?

I personally am glad that we did not spend too much time in
trying to define what we have been talking about. "Growth" is a
most difficult thing to define usefully; I have never been able to
decide, in my own experiments, whether a rat is growing when its
bones elongate, but the total energy content of its body falls, or,
alternatively, when its body energy steadily rises but its bones do
not increase in size. Frankly, I do not think that it is worth discussing these points. But what stops growth normally ? This is a point
that we have almost ignored, even though it might have come in
for consideration as something which follows the pubertal spurt in
growth which we have discussed.

With respect to the so-called anabolic steroids, and indeed other
active substances, I think that their specificity of action is likely to

---

[1] ADAMS, T. W., and E. P. POULTON: J. Physiol. (G.B.) 77, 1 P—2 P
(1933).

be determined rather by reference to the barriers that must be crossed before the site of action is reached than by reference to the site of action itself. That is to say, the substances which reach the sites of action involved in protein biosynthesis, and can then exert a stimulant effect, stimulate not simply a part of the biosynthetic process in question, but the whole of it. Dr. DESAULLES made this point clear, and I agree with him. In this connection, I wonder whether a so-called anabolic steroid with a side-chain of a peptide nature, say at C-17, has been produced, since such a substance might fail to reach the sites of action in sexual tissues. The peptide might prevent the steroid from gaining access to sexual tissues, in contrast to what now tends to occur with such steroids.

Where and what are the sites which respond to hormonal stimulation of muscle growth ? Surely they are not in adipose tissue alone and concerned only with the mobilisation of free fatty acids, although such a process is probably important in providing the energy needed for this growth. The mobilisation of free fatty acids from adipose tissue and other sources is rather a non-specific reaction, at least *in vitro.* I still believe that insulin is concerned in growth, although the fact that in some circumstances growth hormone mimics the action of insulin on glucose utilisation is of complicating importance. I have been attracted by the fact that insulin exerts an action *in vitro* in stimulating amino-acid incorporation, a process that can be examined under controlled conditions. Is it possible that metabolites of growth hormone are formed, some of which act on fat metabolism and others on amino-acid utilisation, with or without the action of insulin ? Ribosome activity and formation and the size of the amino-acid pool are all relevant.

Methods are obviously fundamentally important to our discussions. I agree with Dr. WATERLOW's comment that, if you have something which gives self-consistent measurements, then it is at least worth investigating, though one must be careful about measuring something that has a built-in self-consistency. If I may be frivolous for one moment, I should like to describe a situation in a railway train in England in which a man was tearing up pieces of paper and throwing the pieces out of the open window. After a time, somebody said: "You really shouldn't do that, since it makes the railway untidy." "I have to do this, you see, to keep the elephants away," was the unexpected reply. The complainer then said: "But there aren't any elephants around." "I agree," replied the man, "that shows how efficient my method is." I am not implying that any of the methods we have been discussing are quite in

that category, but I think we ought always to bear in mind the limitations of the methods that we are using. In the investigation of
complex matters, we have to use the tools that are available, but we
should not blind ourselves to the danger of adopting under all
conditions tools that were designed for one specific task. It is
necessary to iodinate serum albumin in order to investigate the pool
size, rate of disappearance, and so forth, but when one considers the
important role that the secondary structures of proteins may play in
their metabolic reactions, we should be cautious.

Dr. TUCHMANN-DUPLESSIS touched on the question of growth
hormone in embryonic life and youth, and what growth hormone
is doing in adult people, assuming that it is secreted in the adult.
In the very young rat, as Dr. McCANCE stressed again just now,
hypophysectomy does not immediately and completely stop
growth. The rats of our stock grow to about 100 g. if they are
hypophysectomised below this weight. I think there is no reason
to suppose that hormones are controlling metabolism at a very
fundamental level. They are more like the police, which control a
metabolic crowd, though the movement of the crowd goes on,
whether or not the police are present. The police are needed to provide for an orderly flow, particularly under conditions where some
deviation from the normal is likely to take place. With reference
to growth in embryonic tissues, I recall the experiments of Dr.
CAVALLERO, in which the administration of glucagon to chick
embryos, and also to dwarf mice, stimulated growth. We have
heard little about glucagon − perhaps because it is not normally
regarded as important for growth.

Growth hormone, I think, must have functions other than those
that we normally associate with its name. I have long thought that
it may well be of importance in the maintenance of satisfactory
metabolic conditions during fasting[1].

We have heard remarkably little about the thyroid. I think that
40 years ago the growth of a cretin given thyroid would have been
one of the things we should have spent much time on, but now
we are more interested in other matters. But it is of interest that
catabolism is often a concomitant or a permissive factor for
anabolism. I was interested to see the effect of cortisone in some
of Dr. LEATHEM's experiments in this connection.

Like Dr. McCANCE and Dr. TREMOLIÈRES I am a little surprised
that the hypothalamus and the control of appetite are matters
which we have almost entirely neglected.

[1] YOUNG, F. G.: Biochem. J. (G.B.) **39**, 515 (1945); KETTERER, B., P. J.
RANDLE, and F. G. YOUNG: Erg. Physiol. (G.) **49**, 127 (1957).

Mr. Chairman, I hope that what I have said will not draw any of your searching questions. It seems to me that the more relevant the questions, the more unanswerable they are, particularly those that you have put to us. But we must accept the fact that a great barrier to our learning the truth is sometimes what we believe we understand already. And one of the important actions of this Symposium has been its catabolic effect on apparent fact! This may be an essential part of the biosynthesis of knowledge to which we have devoted the last four days. We are indeed grateful to you, Mr. Chairman, for the penetrating questions you have put to us from time to time.

List of authors

Subject index

REPRINT FROM

PROTEIN METABOLISM

AN INTERNATIONAL SYMPOSIUM

CHAIRMAN

A. QUERIDO · LEYDEN

EDITED BY

F. GROSS · BASLE

SPRINGER-VERLAG / BERLIN · GÖTTINGEN · HEIDELBERG 1962
PRINTED IN GERMANY

NOT IN CIRCULATION

PROTEIN METABOLISM
SOME PROBLEMS BROUGHT INTO FOCUS
OPENING REMARKS

BY

A. QUERIDO

WITH 1 FIGURE

Diuresis and Diuretics
Diurese und Diuretica

An International Symposium

Herrenchiemsee, 17.—20. Juni 1959. Veranstaltet mit Unterstützung der CIBA. Leitung HERBERT SCHWIEGK, München. Herausgegeben von EBERHARD BUCHBORN, München, KLAUS DIETRICH BOCK, Basel.

With 88 figures. XII, 382 pages (92 pages in English) 8°. 1959
Cloth DM 25,50

Essential Hypertension

An International Symposium

Berne, June 7th — 10th, 1960. Sponsored by CIBA. Chairman F. C. REUBI, Berne. Edited by K. D. BOCK, Basle, P. T. COTTIER, Berne.

With 81 figures. VIII, 392 pages 8°. 1960.*   Cloth DM 33,80

Shock, Pathogenesis and Therapy

An International Symposium

Stockholm, June, 27th — 30th 1961. Sponsored by CIBA. Chairman U. S. VON EULER, Stockholm. Edited by K. D. BOCK, Basle.

In English. With 120 figures. VIII, 387 pages 8°. 1962.*
Cloth DM 37,50

In the USA and Canada this book is distributed by Academic Press Inc., Publishers, New York

* Available also in a German edition

SPRINGER-VERLAG · BERLIN · GÖTTINGEN · HEIDELBERG

REPRINT FROM
PROTEIN METABOLISM
AN INTERNATIONAL SYMPOSIUM
CHAIRMAN
A. QUERIDO · LEYDEN
EDITED BY
F. GROSS · BASLE

SPRINGER-VERLAG / BERLIN · GÖTTINGEN · HEIDELBERG 1962
PRINTED IN GERMANY
NOT IN CIRCULATION

THE EFFECT OF GROWTH HORMONE
ON PROTEIN SYNTHESIS

BY

A. KORNER

WITH 6 FIGURES

REPRINT FROM

PROTEIN METABOLISM

AN INTERNATIONAL SYMPOSIUM

CHAIRMAN

A. QUERIDO · LEYDEN

EDITED BY

F. GROSS · BASLE

SPRINGER-VERLAG / BERLIN · GÖTTINGEN · HEIDELBERG 1962
PRINTED IN GERMANY

NOT IN CIRCULATION

REGULATION OF PROTEIN SYNTHESIS
BY ANDROGENS AND ESTROGENS

BY

J. D. WILSON

WITH 7 FIGURES

REPRINT FROM

PROTEIN METABOLISM

AN INTERNATIONAL SYMPOSIUM

CHAIRMAN

A. QUERIDO · LEYDEN

EDITED BY

F. GROSS · BASLE

SPRINGER-VERLAG / BERLIN · GÖTTINGEN · HEIDELBERG 1962
PRINTED IN GERMANY

NOT IN CIRCULATION

HISTOLOGICAL ASPECTS OF THE ACTION
OF ANDROGENS AND OESTROGENS

BY

J. A. SZIRMAI

WITH 11 FIGURES

REPRINT FROM

PROTEIN METABOLISM
AN INTERNATIONAL SYMPOSIUM

CHAIRMAN
A. QUERIDO · LEYDEN

EDITED BY
F. GROSS · BASLE

SPRINGER-VERLAG / BERLIN · GÖTTINGEN · HEIDELBERG 1962
PRINTED IN GERMANY
NOT IN CIRCULATION

INTRODUCTION TO THE GENERAL DISCUSSION:
THE IMPORTANCE OF EXPERIMENTAL CONDITIONS
FOR THE DEMONSTRATION OF HORMONE EFFECTS
IN VITRO AND IN VIVO

BY
W. DIRSCHERL

WITH 5 FIGURES

REPRINT FROM

PROTEIN METABOLISM

AN INTERNATIONAL SYMPOSIUM

CHAIRMAN

A. QUERIDO · LEYDEN

EDITED BY

F. GROSS · BASLE

SPRINGER-VERLAG / BERLIN · GÖTTINGEN · HEIDELBERG 1962
PRINTED IN GERMANY

NOT IN CIRCULATION

PROTEIN MALNUTRITION AND REPLENISHMENT
WITH PROTEIN IN MAN AND ANIMALS

BY

J. C. WATERLOW

REPRINT FROM

PROTEIN METABOLISM

AN INTERNATIONAL SYMPOSIUM

CHAIRMAN

A. QUERIDO · LEYDEN

EDITED BY

F. GROSS · BASLE

SPRINGER-VERLAG / BERLIN · GÖTTINGEN · HEIDELBERG 1962
PRINTED IN GERMANY

NOT IN CIRCULATION

THE BEARING OF THE PLANE OF NUTRITION
ON GROWTH AND ENDOCRINE DEVELOPMENT

BY

R. A. MCCANCE AND E. M. WIDDOWSON

WITH 1 FIGURE

REPRINT FROM

PROTEIN METABOLISM

AN INTERNATIONAL SYMPOSIUM

CHAIRMAN

A. QUERIDO · LEYDEN

EDITED BY

F. GROSS · BASLE

SPRINGER-VERLAG / BERLIN · GÖTTINGEN · HEIDELBERG 1962
PRINTED IN GERMANY

NOT IN CIRCULATION

METABOLIC EFFECTS OF GROWTH HORMONE IN MAN

BY

M. S. RABEN, P. R. MINTON, M. L. MITCHELL, AND H. JUAREZ-PENALVA

WITH 7 FIGURES

REPRINT FROM

PROTEIN METABOLISM

AN INTERNATIONAL SYMPOSIUM

CHAIRMAN

A. QUERIDO · LEYDEN

EDITED BY

F. GROSS · BASLE

SPRINGER-VERLAG / BERLIN · GÖTTINGEN · HEIDELBERG 1962
PRINTED IN GERMANY

NOT IN CIRCULATION

NITROGEN BALANCE STUDIES:
THE RELATION OF PROTEIN METABOLISM
TO EXCESSIVE SHIFTS OF SODIUM AND POTASSIUM

BY

P. S. BLOM, J. DE GRAEFF, A. A. H. KASSENAAR,
AND H. A. SONNEVELDT

WITH 9 FIGURES

REPRINT FROM

PROTEIN METABOLISM

AN INTERNATIONAL SYMPOSIUM

CHAIRMAN

A. QUERIDO · LEYDEN

EDITED BY

F. GROSS · BASLE

SPRINGER-VERLAG / BERLIN · GÖTTINGEN · HEIDELBERG 1962
PRINTED IN GERMANY

NOT IN CIRCULATION

STUDIES OF PROTEIN METABOLISM
IN HUMANS WITH THE
AID OF N15-LABELLED GLYCINE

BY

A. HAAK, A. A. H. KASSENAAR, AND A. QUERIDO

WITH 4 FIGURES